Hess's Paint Film Defects

Their Causes and Cure

Hess's Paint Film Defects

Their Causes and Cure

THIRD EDITION

edited and revised by
H. R. HAMBURG *and* W. M. MORGANS

with contributions by
T. A. BANFIELD *and* S. T. HARRIS
and
Chapter 8, Hazards to the paint user *by*
H. BRUNNER

London
Chapman and Hall

First German edition 1938
First English edition 1951,
published by Chapman and Hall Ltd
11 New Fetter Lane, London EC4P 4EE
Reprinted twice
Second edition 1965
Third edition, revised by H.R. Hamburg and W.M. Morgans, 1979

Typeset by C. Josée Utteridge-Faivre
and printed in Great Britain by
Richard Clay (The Chaucer Press), Ltd,
Bungay, Suffolk

ISBN 0 412 13880 8

Distributed in the U.S.A. by
Halsted Press, a Division of
John Wiley & Sons, Inc., New York

Library of Congress Cataloging in Publication Data

Hess, Manfred, ed.
Hess's paint film defects, their causes and cure.
Translation of Häufige Anstrichmängel und Anstrichschäden.
'A Halsted Press book.'
Bibliography: p.
Includes index.
1. Lacquer and lacquering. 2. Painting, Industrial.
3. Industrial toxicology. I. Hamburg, Herman R.
II. Morgans, Wilfred Morley, 1907- III. Title.
IV. Title: Paint film defects, their causes and cure.
TP939.H413 1979 667.6 78-26994
ISBN 0-470-99018-X

Contents

Black and white and colour figures appear between pages 242 and 243

Foreword by M. Hess

My recollections of paints encompass the period from the time they were mainly based on drying oils and run gums (almost a sort of 'witches' brew') to the present, when the main ingredients are often highly sophisticated polymers. During this time demands on paint films have, in conjunction with their much more varied uses and methods of application, been increased steadily, as have the numbers of potential defects. In many cases, however modern the surface coating, the underlying causes of paint film defects remain the same.

I noticed as a young paint chemist that complaints by users of my own products were often stereotyped, and I decided after investigating the cause and remedy of such failures to put my observations to paper for the benefit of these users. In so doing I became deeply involved in the subject, the field widened more and more, and in the end this led to the first edition of *Häufige Anstrichmängel und Anstrichschäden, ihre Ursache und Verhütung.* This and the following improved editions in English, German, French and Spanish were welcomed by those interested in the subject.

In this era of specialization there are not many all-rounders who are qualified to survey such a big and varied field. I have been particularly lucky that my old friends and colleagues, Dr Herman Hamburg and Dr Wilfred Morgans, took over the arduous task of editing and revising my original brainchild in its third English edition. I am therefore confident that this will again prove to be a valuable aid to many.

Manfred Hess
Northwood, 1978

Preface to the third edition

Since publication of the first English edition this book has become the standard reference work on paint film defects throughout the world. The very considerable advances in coatings technology since the second English edition was published in 1965 have necessitated a revision of the book, a task which from the outset was recognized as formidable. The very wide field to be covered required specialist knowledge as well as wide experience, and we were fortunate in being able to enlist the services of a group of contributors who were well qualified for the task.

Due to his advancing age Mr Manfred Hess, the originator of this work, felt unable to take an active part in the preparation of the new edition. He entrusted not only a large part of the necessarily extensive revision of the text, but also the editorial work, the planning and compilation of the index to us jointly.

A variety of causes has prevented the main contributors to the second edition, Mr W.A. Edwards and Mr T.W. Wilkinson, from revising their sections. Nevertheless, much of what they and others have contributed to previous editions has enabled us to build on valuable foundations. Much new material has been added; the illustrations section has been expanded and enhanced by the addition of several colour plates.

Mr S.T. Harris revised the sections concerned with industrial finishes and in particular powder coatings, and Dr T.A. Banfield contributed the sections on marine paints and compositions.

In the last decade increasing importance has been attached, and rightly so, to the health and safety aspects of all industrial processes, and to the safeguarding of the users of industrial products. A great deal of legislation on this subject has been enacted in many countries and much guidance is needed. We therefore thought it desirable to devote a large amount of space to a chapter entitled 'Hazards to the paint user' and succeeded in persuading Dr Henry Brunner to undertake this task.

Our thanks are due to a number of friends and companies for helpful suggestions and illustrations; in particular we would mention the following:

Mr J.S. Abbott, British Steel Corporation
Mr J. Boxall, Building Research Establishment, Princes Risborough Laboratory
Mr T.R. Bullett, Paint Research Association
Dr G. Geissler, Hoechst AG
Mr G.V.G. Hill, BIE Anti-Corrosion Ltd.
Mr M.J. Mann, British Standards Institution
Mr W.R. Moon, Hoechst (U.K.) Ltd.
Mr J.L. Prosser, Paint Research Association
Mr A.F. Sherwood, Paint Research Association
Mr P.A. Sinclair, Laporte Industries Ltd.
Mr W.R. Springle, Paint Research Association
Mr J.T. Tooke-Kirby, Hoechst (U.K.) Ltd.
Mr A. Wells, BTP Tioxide Ltd.
Mr P. Whiteley, Building Research Establishment, Building Research Station

Some changes in design have been made to bring the book into line with current forms of presentation, though the arrangement of the material is essentially unchanged. The decimal notation has been used for sub-divisions of chapters, and the index entries are now by page rather than by section references.

It is hoped that these changes will facilitate the locating of specific items and that the book will prove a useful guide to preventing as well as remedying paint film defects.

H.R. Hamburg
W.M. Morgans

Editors and co-authors

How to use this book

A single fault is in many cases due to a number of causes. In some cases several defects occur simultaneously, so that the solution of the trouble is rendered more difficult. The headings of the individual sections indicate the various kinds of faults. Each section sets out the possible causes of these faults and suggests methods of preventing and eliminating them.

There are two ways of finding one's way through the labyrinth of possibilities:
(a) In most cases it is advisable:

(i) To consult the subject index at the end of the book for the fault in question.
(ii) To study the section or sections concerned thoroughly, omitting nothing, so as not to miss possible alternative causes.

(b) If, on the other hand, it is desired to determine what faults may be possible, the Table of Contents at the front of the book should be consulted and used as follows:

(i) Look for the appropriate chapters.
(ii) Study the relevant sections.
(iii) Look up all the sub-sections of interest.

The Author Index as well as the Subject Index are located at the end of the book.

The term 'coating' is used only in relation to paint, lacquer, enamel and varnish-like products.

Definitions, mainly relating to defects, have been incorporated in the text. They are mostly based on the excellent *Glossary of Paint Terms* published by the British Standards Institution as British Standard No. 2015:1965. This should be referred to, if further information on technical terms is needed.

Most paint materials are potentially dangerous, and paint users should ensure that they apply these materials in a manner that is safe for all concerned. While every effort has been made to ensure the accuracy of the information given in this book, the Editors, Contributors and Publishers in no way accept liability for loss or injury resulting from the use made by any person of such information.

1

Faults which develop during storage

1.1 Cloudiness and sedimentation in varnishes and boiled oils

Causes

(a) Excessively cold storage (particularly at night) can lead to cloudiness or precipitation. In many instances, slight cloudiness is of no practical importance and often disappears when the material is removed to a warm room. Quicker results are obtained if the tin is partly immersed in a pail of hot water. Heating over a flame of any kind is highly dangerous since the volatile solvents are highly inflammable. If serious sedimentation has occurred this may not disappear on warming; filtration is necessary and this can be done satisfactorily only by the manufacturer.

A storage temperature of about 20°C (70°F) is required for oil-based or oil-modified varnishes or paints and they should be kept away from radiators and stoves. The containers should not be left to stand out of doors after work. Store rooms, especially those of light construction, frequently cool off considerably overnight, particularly during the cold season. A varnish placed directly on the cold floor is bound to undergo chilling to a low temperature which is retained at least during the morning. The containers should therefore be placed on shelves or boards.

(b) Varnishes received in a cloudy condition may have been chilled in transit (in which case the cloudiness should disappear on warming) or the cloudiness (and possibly sedimentation) may be due to the use of unsuitable thinners. In this case the varnish will not remain clear after warming and the manufacturer should be consulted. (It may be clear when warm but the cloudiness will return on cooling.) Such a varnish is not likely to give a bright durable film.

(c) Addition of excessive quantities of driers containing lead, particularly to long oil varnishes and alkyds, or addition of excessive quantities of driers generally associated with very cold storage may cause cloudiness. Slight cloudiness in varnishes (and in boiled oils containing lead) is generally of no importance and can be ignored. More serious cloudiness can be removed in many cases either by warming in hot water or by addition of a small quantity of calcium naphthenate solution. The presence of this material is an effective preventative of the separation of lead soaps, the main cause of cloudiness developing at room temperature. In some cases calcium naphthenate will aid the dissolution of mild sedimentation, but where this is serious, filtation may be necessary. Serious sedimentation can result in pinholes in the film, the pinholes surrounding the separated particles of lead soaps.

(d) The presence of moisture in a varnish can also lead to cloudiness, but in this instance it is not usually very severe. The commonly-used thinner, white spirit, can contain a small quantity of water and when it is added to a varnish the water may separate and form a cloudy emulsion. Traces of water are held very tenaciously and are difficult to remove by heating. They can often be removed by pouring the varnish through a funnel containing a dehydrating agent such as activated alumina. In paints the presence of small amounts of moisture is not obvious but it can lead to defects such as poor flow and deficient gloss.

(e) Addition of unsuitable thinners can produce cloudiness in the early stages followed by more serious defects. For example, addition of white spirit to some medium or medium/short oil length alkyd varnishes can result in an initial cloudiness; further addition of the solvent can result in precipitation of the resin. Varnishes of this type have limited tolerance for white spirit and aliphatic hydrocarbon solvents generally, and require a proportion of stronger aromatic types (e.g. Shellsol [1], Solvesso [2] or Aromasol [3]) in the thinners used.

(f) The stability of varnishes depends on the mutual solubility of the non-volatile binder and the volatile solvent or thinner. If too much non-solvent was present initially, cloudiness or possibly thickening and separation may develop slowly. In some mild cases it may be corrected by addition of aromatic solvent, but in severe cases reformulation is necessary. Generally, the more complex the varnish, the more sensitive it is to maltreatment.

Notes

[1] Shell Chemicals Ltd.
[2] Esso Petroleum Company.
[3] ICI Ltd.

1.2 Thickening, possibly followed by separation, in oleoresinous and synthetic varnishes

Possible causes

(a) Excessively cold storage can cause separation in some varnishes.

(b) Long exposure of the material to air either in badly closed or in open containers: loss of solvent takes place and this allows access of oxygen to the varnish, resulting in slow oxidation and consequent thickening. A surface skin usually forms, but the material may also thicken.

(c) The use of unsuitable solvents accompanied by low temperature can lead to thickening and separation of resin. Solvent power decreases with fall in temperature and the use of too much non-solvent can accentuate the effect.

(d) Slight incompatibility between two or more resins can lead to progressive increase in thickness, although the solution initially may be clear. Ultimately the system may gel. This can happen with some alkyd/vinyl and alkyd/chlorinated rubber systems. The cure may lie in a change in the ratios or possibly in the type of one of the components.

(e) Some resin solutions (e.g. alkyds) tend to thicken irreversibly on long storage, probably as a result of progressive polymerization; the end product may be a gel or a two-phase system in which the resin has become insoluble in the solvents.

If a varnish has gelled or separated, little can be done to restore the original condition; however, a varnish which has merely thickened will often respond to small additions of a strong solvent such as xylol or a high aromatic petroleum solvent.

1.3 Changes of consistency

Thickening of pigmented systems can be due to physical or chemical changes or to a combination of the two. The chemical changes result in 'feeding', which is described below. Many of the physical changes are reversible, and a well-known example of such a process is *thixotropy*. A thixotropic paint sets up to a gel on standing; this gel is broken down on application of shear in the form of stirring or brushing but recovers as soon as the shear is removed. Thixotropic paints permit the application of thick coats without the risk of running or sagging. They are also known as 'Non-drip' or 'Jelly' paints.

Feeding is the progressive increase in viscosity caused by chemical reaction between basic pigments and acidic constituents of the medium. Zinc oxide, some grades of zinc phosphate and calcium plumbate can react in this way. The degree of feeding can vary from a slight increase in thickness (often induced if a paint is too thin) to the ultimately useless 'livered' or 'gelled' condition. (The terms 'livering' and 'gelling' are frequently applied to the same phenomenon. The latter term is the more widely used). Undesirable feeding can sometimes be checked, if taken in the early stages, by addition of small quantities of benzoic, salicylic or citric acid, which act as peptizing agents.

(a) Alkyds of high acid value can feed, not only with zinc oxide and calcium plumbate, but also with some grades of titanium dioxide, where a basic oxide is present as part of the surface treatment. The feeding with titanium dioxide is less severe than with the other two pigments.

(b) Some metallic pigments, notably aluminium, bronzes and zinc can induce thickening in media of relatively low acidity. This is sometimes accompanied by gassing, which can lead to bursting of containers. The presence of moisture can also cause gassing and irreversible gelling in paints containing metal powders. A small quantity of quicklime is sometimes incorporated to react with and remove the moisture, but the amount used must not be sufficient to cause any saponification if an ester-type medium has been used. Moisture can also be absorbed by introducing molecular sieves with which there is no risk of chemical reaction [1]. It is common practice, where there is any risk of thickening or gassing, to supply the metal and medium in separate containers. These are mixed immediately before use.

(c) Paints containing carbon black as the principal or sole pigment tend to show a progressive increase in viscosity on storage. This is thought to be due to adsorption by the black of the lower molecular weight fractions of the polymer. Since these act as solvents for the high molecular weight fractions the solubility of the latter decreases, resulting in an increase in viscosity. With high concentrations of carbon black an irreversible gel can result. The fault can be mitigated by the addition of certain metal compounds, e.g. 'Nuact', [2], at the dispersion stage. The additive is adsorbed by the carbon black particles, thereby reducing the tendency to adsorb other materials, including drier metals.

(d) Phthalocyanine blues, in high concentrations, are prone to gelation particularly with bodied media such as alkyds. The cause appears to be similar to that affecting carbon black (above). This type of feeding is accentuated in highly polymerized media.

(e) At higher ambient temperatures e.g. in the tropics, gelation can occur with and without basic pigments, particularly with alkyds.

(f) Silicone type aluminium paints have been known to gel on storage if the solder used in the cans contains a high percentage of lead.

(g) Irreversible gelling of chlorinated rubber paints can be attributed to a reaction of traces of hydrogen chloride split off from the chlorinated rubber, with iron, zinc or aluminium. The metals originate either from galvanized or black iron containers, or are present as metallic powder pigments.

(h) 'Two-pack' lacquers and paints have a restricted life once the components have been mixed, e.g. cold-setting urea-formaldehyde resin lacquers, amine-cured epoxy resins, air-drying isocyanate coatings, polyester resins and etch primers based on polyvinyl butyral. Only the quantities required for rapid use should be prepared. Two-feed spray guns may also be used in some instances (particularly with polyesters), so that the components mix only after leaving the gun.

(i) In emulsion paints, thickening can be caused by flocculation of the pigment. Addition of a small quantity of an anionic or a non-ionic surfactant will often restore the consistency.

(j) Emulsion paints prepared with hard water (and sometimes water containing ferric ions) will thicken if there is insufficient sequestering agent (e.g. Calgon [3]) and particularly if the protective colloid concentration is low. A further addition of Calgon will often help, provided the thickening has not reached the stage of polymer coagulation.

(k) Containers for emulsion paints must be effectively lacquered internally to prevent any contact between the contents and steel. Pinholes or 'holidays' can occur at points where lugs for handles are attached to the cans. A highly localized concentration of ferrous ions can cause precipitation of emulsion at these points.

(l) A thinning of emulsion paints is usually the result of bacterial degradation of the water-soluble cellulose thickener. The sodium salt of carboxymethyl cellulose is more susceptible than the cellulose ethers, but in all cases the trouble can be avoided by the use of an adequate concentration of preservative.

Notes

[1] Doerr, Y.J. (1969), *Am. Paint J.*, **56** No. 42, **26**. Molecular sieves are marketed under the trade name 'Zeolith' (Bayer).
[2] Durham Raw Materials Ltd.
[3] Albright & Wilson Ltd.

1.4 Separation, sedimentation, precipitation, clouding of liquid driers

(a) Defects under this heading principally arise from the causes described in Section 1.1. Pure lead driers in varnishes, drier solutions and boiled oils have the greatest tendency in this direction. Of the other most commonly used driers, manganese behaves very well and cobalt still better, while intermediate positions are taken by the combinations of the various metals.

When selecting driers for the purposes in question, attention has to be paid to the fact that lead and manganese accelerate the drying of the films from the inside, i.e. from the bottom upward, while cobalt exerts the reverse effect. Lead and manganese driers possess a much stronger drying action in combination with one another. Unnecessary addition, which can only be harmful, should not be made.

b) The acid constituent of a drier (e.g. the naphthenate or linoleate radical) is much less important than the metal content, although at different temperature and with variation of media, the type and amount of the various drier metals will vary widely for optimum drying properties.

(c) One of the principal demands made upon liquid driers is that apart from satisfactory drying properties and minimum discoloration of the paints and varnishes containing them, the mixtures must remain clear and free from precipitate or sediment.

In the case of unusually high concentrations, clouding or separation cannot be entirely prevented. Correctly manufactured liquid driers made up in white spirit as a rule keep bright provided their containers are well stoppered.

(d) If the containers are badly stoppered, crusty skins begin to form within a short time, which grow thicker and may drop to the bottom of the container, to be replaced by fresh surface skins.

(e) Calcium is not regarded as a drier metal but exerts a stabilizing effect, particularly on lead driers. If, for example, a proportion of calcium naphthenate is added to a lead/cobalt mixture or to varnishes or oils containing lead driers, the tendency to cloudiness or precipitation is considerably reduced.

(f) As a result of the trend toward leadless paints, lead driers are being replaced by less toxic metals e.g. zirconium. Naphthenates are also being replaced by the paler and less odorous octoates and versatates.

1.5 Changes of consistency of bitumen and pitch varnishes [1]

Incorrect after-thinning may lead to gelling. Gilsonite-type asphalts dissolved wholly in white spirit will almost certainly feed up. On the other hand, it has been observed that pitch varnishes, e.g. those based on stearin-pitch, may lose viscosity on storage even within weeks in airtight containers, but may regain it on continued long storage.

Care is necessary when thinning bituminous varnishes. Naphtha is preferable to white spirit, as this will reduce the viscosity of the material more rapidly. Some varnishes, based on coal tar or water-gas pitches can only be thinned with special grades of naphtha, and these are preferably obtained from the manufacturer. Gelled varnishes may be warmed before thinning, which facilitates the incorporation of the thinner. It may also be possible to prevent gelling in Gilsonite-based coatings by including a small proportion of an alcoholic solvent in the formulation.

Note

[1] H. Abraham in *Asphalts and Allied Substances*, 5th ed., p. 467, stresses that products which are hard and possess at the same time tough and rubber-like properties are, when used alone, apt to gelatinize after they have been dissolved in the volatile solvent. Such paints which, freshly made, may appear liquid will after some time change over to a jelly-like mass particularly with hard rubbery fatty acid pitches, wurtzilite asphalt and gilsonite.

1.6 The fall of viscosity during storage of cellulose lacquers

(a) Stored for a considerable time in closed vessels, an appreciable lowering of viscosity is usually to be noticed in nitrocellulose lacquers without loss of solvent by evaporation or any fundamental changes in the original qualities having taken place. This proceeds quite rapidly at first, within ten days, for instance at 50°C, after which it remains constant. The phenomenon is displayed by all collodion cottons and particularly those of high viscosity (possibly caused by incomplete solvation).

The only means to counter it is to select a cellulose ester which has a higher viscosity if it is essential that the viscosity of the lacquer must be high after storage.

(b) Some substances, principally alkaline materials such as pyridine, but also basic salts in pigments, and acids, are capable of degrading the nitrocellulose. This would change important film properties and lower the tensile strength [1].

Note

[1] Jordan, O., *The Technology of Solvents.*

1.7 Skinning of varnishes, paints, fillers and putties

(a) This is usually caused by the addition of large amounts of driers to oil-base varnishes, which may be necessary for other reasons; also when wood oil and similar oils are present, which is often the case in synthetic resin varnishes. The development of anti-skinning agents such as the oximes has greatly reduced the danger.

(b) This fault is promoted by insufficient closing or re-closing of the containers, so that contact with the air becomes possible [1].

Skinning may be the first step towards gelation. Such skins appear in their first stages nearly as thin as if they were mono-molecular; they become, however, thicker and thicker until the whole mass may gel. Skin formation may be particularly troublesome in partly-filled containers of thixotropic paints based on thixotropic alkyds. The skins rapidly become extremely thick.

(c) Most alkyd-based paints are prone to skin unless a suitable 'anti-skinning' agent is incorporated. These are commonly oximes, e.g. methyl ethyl ketoxime, and function through complex formation with the cobalt in the drier. The cobalt is thereby rendered ineffective in the liquid paint, but when a film is applied and exposed the complex breaks down, releasing the volatile oxime, and the cobalt then fulfils its normal role as a surface drier.

Anti-skinning agents, in the appropriate quantities recommended by the manufacturer, can be stirred into liquid paints which are giving trouble by rapid skinning. Addition of a small quantity of butanol or high boiling solvent can also be effective but additions must be small to avoid over-thinning the paint.

(d) Storage under very warm conditions can accentuate skinning.

(e) The surface of oil fillers and oil putty in containers should be covered with a grease paper, a polythene sheet, or a thin layer of linseed oil. Cellulose and spirit lacquers do not form insoluble skins unless used in a combination with media containing drying oils, etc.

Note

[1] OCCA, (1976), *Introduction to Paint Technology*, 4th edn., p. 159.

1.8 Settling, changes of viscosity, colour and finish, of solvent-based paints

Settling, etc., causes changes of consistency and reduction of covering power in the case of pigmented paints, and also a relatively higher oil content of the total in the supernatant medium (important in the case of analysis of some specification paints). This may affect not only the decorative properties but also the value as a protective coating. The wetting properties of the pigments are an important factor not only in the grinding process but also on storage.

The viscosity of paints can fall on storage due to wetting out of certain pigments. This may be responsible for hard settlement where low oil absorption pigments, especially of high specific gravity, wet out after initial grinding. This may account for the fact sometimes observed, that certain paints mature on storage.

The stability of pigment suspensions and settling out are determined by a number of factors:

(i) Particle size and shape of suspended pigment (disperse phase).
(ii) Viscosity of the liquid phase (dispersion medium).
(iii) Difference in density between pigments and liquid phase.
(iv) Interaction between pigments and liquid phase.
(v) Interaction between different pigments in the system.
(vi) Effect of additives.
(vii) Storage conditions.

Gradual wetting out of pigments can take place during storage. Incorrect ballmilling techniques in the manufacture can lead to excessive wetting out. In both cases hard settlement is likely to occur. Heavily pigmented paints, such as flat finishes and undercoatings, are prone to hard settlement.

The wetting out of pigments during storage can lead to a fall in viscosity and settlement is accelerated. Flocculation (aggregation) of pigments has the reverse effect. A balance between well dispersed and flocculated pigment is therefore desirable. The character of the liquid phase, especially its wetting properties (surface tension), has an influence on settling tendencies. Pigments will behave differently in different vehicles [1]. Mixtures of paints with various pigments, each having the same consistency, may exhibit higher or lower consistencies after mixing.

Thixotropic ('non-drip') paints possess a structure which virtually inhibits pigment settlement. Addition of a proportion of thixotropic alkyd to a 'settling' paint will confer a degree of structure and this will reduce the tendency of the pigment to settle. T.C. Patton [2] has considered in some detail the suspension of pigments in paint vehicles and explored the effects of pigment flocculation and thixotropic structure on the process of settlement. He also described a simple gauge for measuring settlement. Another instrument has been described by Veiel and Papenroth [3] and by Bayer Farbenfabriken [4].

Specific dispersing agents have been recommended to reduce proneness to settlement. As an example, citric acid, partially esterified with a polyethyleneglycol, may be mentioned. Such additives can be added either during manufacture or to the finished product [5].

(a) Insufficient stirring or shaking during storage: the degree of sedimentation increases with the duration of the storage. The pigments finally form a more or less solid sediment on the bottom of the container which cannot always be stirred up again. The supernatant material then contains more oil or varnish and is thus glossier, and in the case of oil-type media, richer in oil, than the average paint material of the same general composition, while the lower strata of the paint have a lower oil content as well as being flatter, so that numerous complications may arise unless all the pigments are stirred into the fluid mass again. This fault is particularly undesirable in the case of matt enamels and varnishes, especially those required for optical purposes and other fine work, where the finishes must be matt or even free from any reflection. Enamels containing media on an alkyd basis tend particularly to become more glossy on storage.

(b) If sedimentation takes place in paints containing pigments which are difficult to wet or heavy, for instance, lead pigments, titanium whites, chrome pigments, reds, ultramarines, many kinds of ochre, impure earth pigments, some natural red oxides of iron, graphite, micaceous iron ore, mica, etc., or mixtures of these pigments, the colour tints of the paint materials may also be changed, as the lighter pigments [6] tend to remain floating in the vehicle while the heavier ones are hard to stir back into the vehicle so as to form a homogeneous mass. Badly ground pigments and those which have a low oil absorption behave similarly.

In view of the difficulties frequently encountered in stirring the sediment back into the vehicle, it is always preferable to prevent appreciable sedimentation from the beginning. Round or square containers, which must, of course, be well closed, should be stored lying on their sides and be rolled or turned at certain suitable

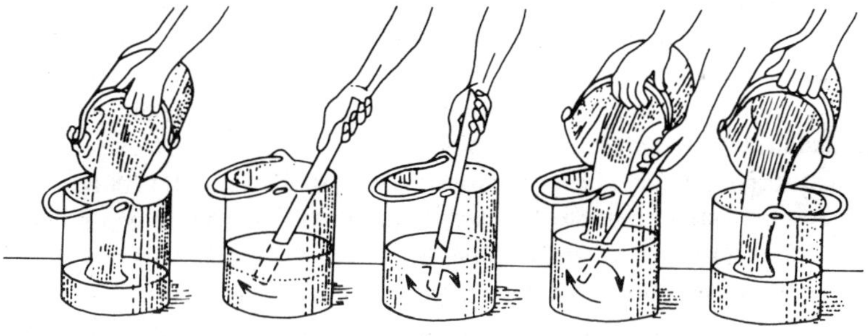

(1) Pour ¼ of the contents into another container.	**(2) Then loosen the pigment at the bottom with the stirring rod.**	**(3) Stir thoroughly by revolving and beating movements.**	**(4) Gradually add the first ¼ to the main body, stirring constantly.**	**(5) Pour from one can into the other a number of times.**

Fig. 1.1 Thorough mixing of paints by repeated pouring from one vessel into another and intense stirring.

intervals. Tins or cans should be placed bottom-up. The paint must also be stirred prior to pouring into the pots and during application. It is clear, of course, that stirring the paint into a homogeneous mass must be effected correctly (see Fig. 1.1).

The process described above, known to older technicians as 'boxing', has fallen to some extent into disuse due to the availability of vibratory shakers ('Red Devil') [11] and mechanical stirrers which greatly assist in the rapid and efficient redispersion of pigments. Storage tanks for bulk quantities of paint should be fitted with slow speed mechanical stirrers.

Even if no noticeable deposit is felt at the bottom of a can when inserting a stirring rod, it is essential to stir thoroughly, because the unavoidable gravitational sedimentation of pigments leads to a higher pigment content in the lower part of the container. Manual stirring is effective in this case, and an upward and circular motion of the stirrer is advised.

The presence of extenders such as asbestine, china clay etc. which confer structure on paints, if incorporated during the grinding process, and/or of anti-settling agents, is beneficial. The most frequently used products of this kind are aluminium stearate, hydrogenated castor oil and modified bentonite clays, known as 'Bentones'. In the case of red lead, which is much inclined to heavy settlement in anti-corrosive primers, it is important to ensure that the pigment is ground in acid refined linseed oil (with an acid value of approximately 8) and that a small quantity of asbestine is present. The addition of half to one part of stearic acid to 100 parts of non-setting red lead is claimed to result in exceptionally good keeping qualities [7].

(c) Some sedimentation may also occur where lakes of organic dyes on natural or precipitated barium sulphate, alumina, etc., are used.

Blanc fixe, which does not settle as readily as natural barium sulphate (barytes), may be present in amounts up to 95 per cent by weight of the total pigment.

(d) Excessively warm store-rooms increase the fluidity of the vehicle and thus promote the settling tendencies of the pigments. It is also possible that undesirable reactions between certain pigments (heavy metal pigments such as zinc white, white lead, etc. and vehicles (such as some synthetic resin varnishes) are promoted. The protective coating of the pigment particles as in the case of non-setting red lead paints (of lead soaps or lead addition compounds) may also be dissolved by the thinners such as white spirit [7].

The most suitable temperature for storing paints is about 20°C (70°F). Whilst in countries with a cold or moderate climate trouble often arises in winter by too low storing temperatures, the sufficiently cool storage of paints is of special importance in countries with tropical climates, and care has also to be taken during transit in hot ships' holds. Red lead paints ought to contain a minimum of thinners if destined for hot countries.

(e) When a paint storage vessel is earthed hard pigment deposits may be formed, soft ones when it is insulated. This is due to electrostatic causes [8].

(f) Sedimentation in the dipping tank in the case of highly pigmented paints, such as undercoats: these paints must be maintained in motion either by hand or by mechanical stirring devices or some effective circulating apparatus. This is particularly important in the case of conveyor production lines.

(g) Thinning of varnishes or enamels with unsuitable materials (e.g. short oil alkyd varnishes with white spirit) may result in throwing the medium solids partly or wholly out of solution. A sediment of this type cannot usually be stirred up again.

The precipitated body may form a tough deposit together with the pigment. In some cases the addition of a good solvent combined with stirring may help to recondition, but see below (h).

(h) Excessive thinning: apart from the extreme reduction of the specific gravity and viscosity of the paint medium so that the pigments cannot remain in suspension, colloidal factors may have to be taken into consideration. The property of a medium in so far as it serves as a protective colloid can thus be partly or totally destroyed. The Scientific Adviser to the Greater London Council has repeatedly pointed out in his annual reports that excessive thinning by operators is the most common cause of faulty paint on site.

(i) Addition of moist pigment to the vehicle on grinding oil-base paints or enamels: these pigments tend to separate quickly and solidly, due to bad wetting.

Even small amounts of water in red lead paints catalyse the reaction between lead and linseed oil and cause the paint to set or harden in half the normal time [7].

There are, however, cases where the medium consists of certain synthetic materials, where a limited moisture content of the pigments is advantageous for the grinding process.

All dry colours must be stored in dry condition and especially protected. e.g. against ground moisture until they are used.

(j) Paints which contain ferricyanide blues (Prussian, Paris, etc.) may become paler during storage in closed containers–probably due to the reduction of the blue by the oil (because of lack of atmospheric oxygen) [9], particularly in presence of zinc oxide. The risk increases if the medium dries fast, such as an alkyd resin containing conjugated double bonds. As soon as sufficient air gets into contact with the wet film, the original colour returns. This defect can only be really serious in very pale shades of, say, blue, green or purple, as only a limited quantity of blue can be reduced by the medium. The packing material may also exert a certain influence [9].

Since the reduction of the pigment is promoted by insufficiently washed-out salts and by pigments such as titanium dioxide, zinc oxide and lithopone, attention must be paid to the purity of the Prussian, Paris, etc., blue employed. If the shade required is to be made with pigments of the type mentioned, it may be advisable to use another class of blue.

(k) Some red azo cum white pigment systems in long oil alkyd media have shown fading on storage if asbestos type fillers were present. This defect was more pronounced with white pigments of low tinting strength and if storage had been above room temperature [10].

(l) Severe colour change can occur with paints containing certain phthalocyanine blues in the presence of aromatic solvents. This is due to a change of crystalline form of the blue causing loss of colour on storage or a change to a weaker and redder shade.

Do not use aromatic solvents with phthalocyanine colours without a previous test. Non-crystallizing types of these pigments are available.

Notes

[1] Houston Society of Paint Technology (1970), *J. Paint Tech.* **42**, 55.
[2] Patton, T.C. (1966), *J. Paint Tech.* **38**, 387.
[3] Veiel, U. and Papenroth, W. (1968), *Fatipec Congress Book,* Sect. 3, p. 19.
[4] Bayer Farbenfabriken, Technical Leaflet, (1968).
[5] Benckiser G.m.b.H., German Patent No. 1,228,736.
[6] High grade lithopones and zinc whites do not tend to settle.
[7] Chater, T.W.J. (1941), *JOCCA* **24**, 146.
[8] Becker, E.A. (1955), *Farben Zeitung* **1**, 30.
[9] Reduction from ferric to ferrous state results in the colour change from blue to white. Re-oxidation can be effected by addition of peroxides or by exposure to atmospheric oxygen.
[10] I.C.I. Techn. Inf., Dyehouse No. 273.
[11] Wentworth Instruments Ltd., Slough, Berks.

1.9 Storage problems of water-based paints

Water-based coatings may deteriorate on storage due to viscosity changes of some of the ingredients. These changes may be due to chemical or physico-chemical effects or to attack by micro-organisms, bacteria, moulds, etc. [1]. The following other defects may occur:

Rusting of the container on prolonged storage (or rapidly, due to chemical attack).

Coagulation of the disperse phase due to storage at low temperatures causing freezing of the aqueous phase, or due to the flushing or instability of dispersed pigments, particularly organic colours.

Coagulation due to addition of solvents.

Gassing, sometimes accompanied by the production of a foul odour, due to putrefaction and decomposition of colloidal thickeners or sulphur-containing fungicides.

Discoloration by can corrosion or mould growth.

(a) Viscosity reduction, loss of structure especially at elevated temperatures, of water paints and distempers: paints of these types based on proteins such as casein, glue, gelatine, etc., may lose gel structure or become fluid when originally in semi-prepared paste consistency.

When loss of structure is accompanied by foul odour, this is attributable to bacterial attack of the protein binder or other thickening agents (see below).

Avoid storage at elevated temperatures for prolonged periods, e.g. close to steampipes, radiators etc. Add a preservative such as phenol or cresylic acid to the product during manufacture.

(b) Emulsion paints may lose viscosity, resulting in poor application properties and low hiding power. The most usual cause is the reduction of viscosity of the thickening agent. This is attributed to microbiological degradation due to enzymes produced by micro-organisms. Sodiumcarboxymethyl cellulose is particularly prone to attack by enzymes, but other cellulosic thickeners can also be affected [2]. Good housekeeping and cleanliness during manufacture are important factors in preventing contamination of emulsion paints with micro-organisms.

As a precautionary measure, preservatives are usually added by the manufacturer. Chlorinated phenols, parachlor-metacresol and orthophenyl-phenate have been recommended. Phenyl-mercury compounds are effective, but their mammalian toxicity is high. More recently modified barium metaborate has been claimed to be a good inhibitor of enzymes which degrade cellulosic thickeners [3]. An effective 'in-can' preservative gaining widespread acceptance is based on 1, 2-benzisothiazolin-3-one [4].

(c) Viscosity increases, etc: water paints and distempers can thicken and harden due to drying out of the material in the container when the lid has been left off, or when the closed container has not been airtight.

Emulsion paints may give viscosity increases due to thickening agents or to pigment flocculation. Thus water-soluble alginates thicken at certain pH changes, and also in presence of polyvalent ions [5].

The use of sequestering agents such as alkali metal polyphosphates and pyrophosphates may control thickening due to polyvalent ions. On the other hand, polyphosphates change over a period of time into ortho-phosphates which are not only nutrients for micro-organisms but can also promote pigment flocculation which leads to an increase in viscosity.

(d) Can corrosion. All water-based paints may ultimately cause rusting of unprotected mild steel or in plate containers and become discoloured.

This will depend upon:

(i) The surface protection of the metal.
(ii) The use of rust inhibitors in the formulation.
(iii) The pH of the paint, and any tendency to an acid drift on storage.

Many water-based paints have an alkaline pH, and above pH 7·5 corrosion of containers is not excessive. Good-quality tin plate or suitably coated mild steel drums are usually satisfactory at this level of pH. Most polyvinyl acetate homopolymer and copolymer emulsions are supplied in the pH range of 4·5–6·0, and the pH of paints based on them is normally somewhat higher due to the buffering action of pigments, extenders, dispersing agents, etc. Severe rusting occurs after the paint's pH has drifted on storage to become more acidic, i.e. to below 6·0. It is recommended that corrosion inhibitors, such as sodium nitrite and sodium benzoate and/or vapour phase inhibitors for the ullage area, be included in the paint formulation but sodium nitrite is only effective if the pH is kept at 7 or higher. Care is needed in the selection of corrosion inhibitors and the quantities employed, if other defects such as gassing are to be avoided. Inclusion of a small proportion of calcium carbonate as an extender in the formulation assists in neutralizing acidic hydrolysis products of vinylacetate.

Although good quality tinplate containers will resist corrosion it is a sensible precaution to use, for water paints, cans lined with a stoved phenolic or epoxy varnish. Steel drums must, of course, be so lined if they are to contain any water based product. In recent years plastic drums, moulded from polyethylene or polypropylene have begun to enjoy some popularity. This type of container eliminates all corrosion problems.

(e) Coagulation, instability, rubberiness: instability of emulsion-based paints including oil-bound distempers may be due to an emulsion breakdown caused either by freezing or pigment/binder coagulation or addition of solvents.

The freezing of emulsion-based paints of all kinds (oil-bound distempers, water paints, emulsion paints) causes the coagulation of the disperse phase unless the paint has been specifically formulated to be freeze-thaw stable. 'Soft' distempers, i.e. oil-free distempers such as paste ceiling whites, etc., are usually freeze-thaw stable due to the absence of a dispersed oil phase. After thawing out, the breakdown may take the form of a sticky oily mass in the case of oil-bound distempers and water paints, or a rubbery, plastic mass in the case of polyvinyl acetate or similar polymer and copolymer emulsion paints.

Some coloured pigments may be responsible for causing coagulation of the disperse phase of emulsion-based paints, especially synthetic polymer emulsions such as polyvinyl acetate. Organic pigments, yellow, reds and also the phthalocyanine blues and greens may give trouble. The breakdown is similar in appearance to that given by freezing, i.e. a rubbery mass is formed, and the paint is unusable.

All water-based paints should be stored under frostproof conditions unless stated to be freeze-thaw stable by the manufacturer. The mechanism of damage by freezing is well described by A.C. Fletcher, S.H.A. Hirsch and J.E.O. Mayne [2]. Certain co-polymer emulsions of vinylacetate with ethylene are very resistant to freezing and will also give satisfactory film formation at 0°C or below.

Hydrophobic pigments such as arylamide yellows (Hansa G, 10G, etc.), toluidine (Helio) reds, phthalocyanine blues, greens, etc., should always be stabilized for emulsion paints with suitable surface active agents in order to make the pigment surface more hydrophilic. These organic pigments and some fine particle size synthetic inorganic pigments tend to 'flush' into the disperse phase causing coagulation of the resin/plasticizer/pigment mass. C. Bondy [6] states that the tendency of these pigments to flush into the resin phase increases with the higher plasticizer concentration.

(f) Gassing of emulsion paints due to the breakdown of certain ingredients may cause can explosions and distended containers due to internal pressure, accompanied in many cases by the foul odour of putrefied material. As gassing and/or putrefaction is more rapid at elevated temperatures, paints which give satisfactory storage in temperate climates may give trouble in tropical conditions.

Bacterial breakdown may be minimized by careful selection of preservatives.

(g) Shade variations on storage may be caused by pigments of unsuitable stability or by can corrosion causing staining by iron oxide. Some organic red pigments used in oil-bound distempers may become more blue toned in the presence of iron and proteins. Ultramarine blue is attacked under acid conditions, causing loss of colour and the foul smell of hydrogen sulphide.

Casein and glue have been responsible for a white emulsion paint turning pink on storage [2]. Some apparent loss of tinting pigment, resulting in a lighter colour due to flocculation of the coloured pigment or to deflocculation of the white pigment present may also occur. Naphthol colours are said to be sensitive to the combined action of acetic acid and sodium nitrite, both of which may be present in a polyvinyl acetate emulsion paint, e.g. Naphthol Green B under these conditions can change to a muddy brown colour [7].

Careful selection of pigments of good chemical stability is very important due to pH changes on storage, and also the presence of numerous reactive products in emulsion paint formulations. Naphthol pigments will give satisfactory stability when sodium nitrite is replaced by another corrosion inhibitor, e.g. sodium benzoate.

(h) Water paints and oil-bound distempers may decompose due to the action of moulds or bacteria. The proteins used as binders and emulsifying agents are nutrient materials for these micro-organisms, and their breakdown can result in the production of foul odour and premature emulsion breakdown.

The use of preservatives is necessary, some fungicides and bactericides being more suitable than others for specific micro-organisms. The amount used is also important. The problem is comparable to the gassing and putrefaction in emulsion paints, although excessive gassing is rarely experienced in water paints and distempers, and break-down is usually apparent by odour only. A change of preservative, or the use of an increased concentration of the preservative combinations is often satisfactory.

Notes

[1] Carter, G. (1973), *JOCCA* **56**, 302.
[2] Fletcher, A.C., Hirsch, S.H.A. and Mayne, J.E.O. (1954), *JOCCA* **37**, 300.
[3] Ross, R.T. (1971), *Am. Paint J.*, **55**, No. 37, 23.
[4] Romanelli, F.H. (1972), *Am. Paint J.*, **56**, No. 47, 56.
[5] A drop in pH increases the viscosity of alginate solutions. Alginates also tend to gel in the presence of polyvalent ions which may occur as pigment impurities.
[6] Bondy, C. (1955), *JOCCA* **38**, 503.
[7] Fletcher, A.C. (1953), *JOCCA* **36**, 56.

1.10 Darkening and other changes of varnishes, paints and thinners

(a) Nearly all paints, varnishes and lacquers keep well in tinplated containers with few exceptions (such as spirit varnishes).

(b) Some solvents or resins and paint vehicles with high acid values react with the iron of drums and tanks and even to some degree with galvanized iron or zinc, resulting in discoloration, darkening or milkiness. In the case of solvent naphtha, brown discoloration may be observed within one week of contact with iron whilst it takes several months to show a yellow discoloration in contact with zinc coated (galvanized) casks.

(c) Certain spirit varnishes, notably those based on the highly acidic shellac, can darken and thicken in tinplate containers unless they contain a small quantity (0.25–0.5%) of ethylanilinephosphate.

(d) Genuine turpentine becomes red in iron containers.

(e) Cellulose lacquers are especially susceptible to iron, because they often contain ester solvents which tend to saponify, giving free acids.

(f) Oil-free alkyd media are known to have completely corroded tinplate during storage, and turned red.

(g) White deposits can form if nitrocellulose or other lacquers containing ethyl lactate are kept in tin containers for lengthy periods.

All the changes described can be minimized by arranging storage away from heat (which can also lead to build-up of pressure inside the container) and by avoiding contamination with water. In some cases the addition of 0.1% citric acid to sequester iron is helpful. In all cases lined containers are recommended to combat effectively the development of these faults.

Where necessary the maximum shelf life of the product should be indicated on the label.

A cardinal rule for reducing the occurrence of defects during the storage of all coatings and related products is to use the older stocks first.

1.11 Seediness, bittiness, pigment flocculation

'Seediness' or 'bittiness' in a paint is a condition in which numbers of apparently solid particles or 'bits' are distributed in the liquid. The 'bits' may be pigment flocculates, particles of skin or adventitious matter. (Sometimes small air bubbles, which can result in pinholes, are mistakenly taken for 'bits' in the wet film. These must be carefully distinguished).

(a) Pigment flocculation (Figs. 1.2 and 1.3) may be responsible for seediness or bittiness in gloss paint films. Bittiness due to dust, etc., can also be accentuated by pigment flotation. This defect can develop after a paint has been manufactured, apparently satisfactorily, to good dispersion standards; it may occur after hours or days or on prolonged storage. Certain pigments of very high oil absorption are particularly prone to this defect, notably high grade carbon blacks, Prussian blues and some organic pigment dyestuffs and toners. This is shown by the fact that black and high chroma paints are more likely to exhibit bittiness due to pigment flocculation than are whites and pastel shades. The defect appears to be the result of faulty pigment/vehicle relationships. It has been shown by W. Jettmar [1] that certain pigments adsorb the lower molecular weight fractions of the medium and the adsorbed layer is held very tenaciously. He states 'This adsorption results in the formation of relatively immobile flocculates which increase the viscosity of the system and lead to the formation of semi-solid zones enriched in pigment'. Although this statement refers to flocculation *per se,* the denser types of flocculates can produce seediness.

(b) Seediness has resulted from the presence of a basic pigment in an alkyd finish [2]. A soya alkyd when pigmented with titanium dioxide and a small amount of colloidal zinc oxide became 'seedy' on storage. The 'seeds' contained titanium dioxide, zinc stearate and zinc palmitate. Warming the paint gave only a temporary improvement.

(c) 'Shock seeding' is a form of seediness which can occur in the dilution or 'letting down' of highly pigmented ball mill bases, and is often described as 'pigment shock' [3]. It occurs when medium is added too rapidly to a ball mill base which contains a relatively low ratio of binder to pigment. It is also influenced by the nature of the solvent present, aliphatic solvents being less likely to cause seeding than aromatic. When a mill base has seeded it should be reground; often the addition of a non-ionic or cationic surfactant facilitates the de-flocculation. Addition of 'let down' medium should be made gradually, the mill being run after each addition.

(d) Seeding may be due to poor dispersion of colloidal anti-settling and matting agents, e.g. hydrogenated castor oil, colloidal silicas and clays. The presence of moisture greatly accentuates this defect. Re-milling the paint will help, possibly with addition of wetting agent or polar solvent. If water may have been present in the pigments or extenders, addition of small quantities of alcohols or, preferably glycol ethers, is often beneficial with or without additional milling. Hydrogenated castor oil requires to be converted to a gelled condition before it functions as an effective anti-settling agent. It will not form a gel unless sufficient heat is developed in the machine.

(e) Seediness can result from overthinning, the use of unsuitable thinners or excessively cold storage conditions, as described in earlier sections.

(f) Bittiness in powder coatings can be caused by deterioration of the product on prolonged storage, i.e. generally in excess of one year at 20°C, or correspondingly shorter periods with rapid cure systems or when the powder is stored at elevated temperatures (Fig. 1.4).

The presence of fine particles can cause agglomerates which may produce bittiness. Oversprayed powder should be mixed with fresh powder and re-sieved before use, to minimize this fault.

Notes

[1] Jettmar, W. (1969), *J. Paint Tech.,* **41**, 559.
[2] O'Neill, L.A. and Brett, R.A. (1969), *JOCCA*., **52**, 1054.
[3] Daniel, F.K. (1956), *Off. Dig.,* **10**, 837.

2
Difficulties arising during application

2.1 Faulty handling of containers

(a) Wooden casks or steel drums should be stored in sheds or in the shade. Where this is impossible they should, on hot days, be sprayed with water from time to time.

(b) Unloading of drums should proceed with care, avoiding unnecessary jolts.

(c) Bungs in metal drums should be inserted and removed by using a suitable key; otherwise proper closing and sufficient tightening of the bungs will be impossible and in paint drums skinning of the contents could take place.

(d) Paint should not be allowed to stand in open containers or paint kettles for any length of time to prevent loss of volatile solvents, drying and skinning.

2.2 Damage to paint rollers

Fast application to large areas is possible with the use of rollers, which have come into widespread use during the last 20 years [1]. The technique requires less skill than brush application. A wide variety of roller coverings is available. The most common roller width is 9 in, but 12 in, 14 in and 18 in sizes are on the market. Roller coverings are supplied with different pile heights to accommodate different degrees of surface roughness.

The principal coverings are:

Lambswool	solvent resistant, for exterior woodwork, good paint carrying capacity
Mohair	solvent resistant, for enamel and gloss paints on smooth surfaces
Polyurethane foam	for high gloss paints on smooth surfaces; not resistant to ketonic solvents

100% nylon fibre and other synthetic fibres mixed with nylon	for general use and emulsion paints

(a) Allowing rollers to remain uncleaned after use may cause damage. It is imperative to clean rollers immediately after use with the appropriate solvent. In the case of emulsion paint, hot water and soap should be used. Mechanical removal of hardened paint from the surface can damage the pile, resulting in uneven application.

(b) Swelling of rollers, particularly those made of rubber and gelatine; this can lead to uneven film thickness and is due to solvent attack on the roller covering. Select correct cover for roller. Similar considerations apply to rollers on roller coating machines.

Note

[1] Speight, N. (1969), *Decorating Contractor,* **68**, 809–812.

2.3 Damage to brushes

(a) Permitting brushes to dry up after use: directly after using and cleaning with solvents, brushes should be freely suspended in a mixture of equal parts of refined linseed oil and white spirit. If kept in oil which is too thick, dried lumps of oil may collect in the brushes and they become 'lousy'.

Stiff brushes can often be softened:

(i) With a suitable solvent emulsion cleaner.
(ii) With a neutral organic paint remover which should preferably be free from wax.

Immediately after softening by one of the above agents, the brushes should be washed with water or with water and soap or detergent. Traces of wax should be thoroughly removed by solvent before washing with water. Residual water should be well shaken off.

Brushes should not be kept in water as this may damage the materials used for holding the base of the bristles and cause a subsequent loss of bristles.

(b) Permitting brushes to harden after use with water paints: it is necessary to wash all such brushes with warm water immediately after use.

(c) Using the same brushes for various incompatible paints: it should be made a practice to use separate brushes for every type of paint material, or to clean them carefully between every change.

(d) Early loosening of bristles of oil-paint brushes due to washing out in water. Brushes to be used in oil paint should not be washed out in water when new until after they have been thoroughly worked in an oil paint. Normal cleaning procedure in solvent can then be carried out, followed by washing in soap and water if necessary. This practice tends to prevent the defect.

(e) In the case of new brushes the hairs sometimes lose minute particles which contaminate the varnish films, forming sandy, gritty surfaces. Loss of bristles on account of old age is no defect.

Comprehensive recommendations regarding the care of brushes have been made by the British Brush Manufacturers Research Association (90 Cowcross Street, London EC1).

2.4 Spraying techniques and spraygun problems

The properties of a sprayed film, especially if air pressure is used, depend on the following factors. Some of them also apply to other spraying methods:

(i) Size of droplets directly at the nozzle (0·005 to 0·01 mm).
(ii) Speeds of the droplets between nozzle and sprayed surface (1/200th to 1/100th second); impingement force (33 to 130 ft/s = 10 to 40 m/s).
(iii) Angle of impingement (30 to 45 degrees).
(iv) Uniformity of distribution of the droplets at the points of impingement depending on the number of headings given.
(v) Spraying distance and air pressure can influence the drying time to a certain extent.
(vi) Working speed and experience of the operator.
(vii) Ratio of available compressed air to coating (0·5 l of paint may need 500 l of air).
(viii) Temperatures of the coating, air (for the spraying) and of the surrounding atmosphere (see also below).
(ix) Shape, size and temperature of objects to be sprayed.
(x) Number of coats.
(xi) Nature of base or undercoating.
(xii) Character of film with regard to its gloss.
(xiii) Viscosity of the coating (see below).
(xiv) Properties of the solvents which have been added in the thinner.
(xv) Size of the spray fan.
(xvi) Diameter of the fluid tip nozzle of the spray pistol.
(xvii) Intensity of the draught in the spraying booth, which depends on the strength of the exhaust and the cleanliness kept in it. This item is of special importance if the spraying goods are large or if the work is being carried out on a conveyor.
(xviii) The compressed air used must be carefully filtered and must not contain any oil or moisture.
(xix) Paint must always be strained to ensure freedom from coarse particles of any description. Otherwise frequent blockages of the spray nozzle will occur.

(a) Non-continuous, dotted film due to excessive pressure in relation to the fluid nozzle or to the very fast drying of some types of coatings, particularly those drying by evaporation if they contain a high percentage of very volatile diluents.

Too high spraying pressure or too close spraying can produce blowholes in some polyester lacquers.

It is necessary either to diminish the pressure or to insert a larger nozzle. Smooth flow of rapid-drying varnishes is promoted by low spraying pressure, but see (c) below.

For polyester lacquers a good average spraying distance is 15 in (approx. 38 cm), except for small areas.

(b) Determine whether the horse-power and the receiver for the compressed air are large enough for the volume of air required in order to neutralize any differences of compression.

(c) Excessively thick coating materials.

If the paint, etc., material is too thick for spraying, it is either thinned with suitable thinners or, if possible, heated to, e.g. 80°C (175°F), to reduce its consistency to that necessary for spraying. The latter applies if the paint material contains a sufficient percentage of high boiling constituents and suitable hot spray apparatus is available. Coatings with high solids content are particularly suited for this technique, and thicker films can be deposited in one application. 'Hot spray' lacquers and paints should be specially designed and for the achievement of optimum results, selection of the correct solvent blend is very important. Good products show little tendency to blushing (in the case of lacquers), in general do not give rise to "orange peel', and are economical. In the application, care has to be taken to avoid bubbling. Apart from the flow rate through the nozzle which is most important, other factors such as draughts, slow stroking and high room temperatures can all contribute to bubbling.

Recently a new air-assisted spraying system, known as 'Airmix' has been described [1]. This is claimed to reduce drastically the spray mist usually associated with normal air-assisted spray guns. It is also claimed that the overall air consumption is reduced and that only minimal exhaustion from the spray booth is needed.

Airless spraying [2] is based on the atomization of coatings at a pressure of 100 to 200 atmospheres (up to 3000 lb/in^2). The material is brought to this pressure by a hydraulic pump, usually driven by compressed air. Less air, approximately 500 l per 5 l of paint, is required than in air-assisted spraying, but a high pressure supply is necessary. More recently an airless spray unit has been described in which the high pressure is achieved by means of an electrically driven piston pump [3]. Lower operating costs are claimed for this system.

The advantages of airless spraying are speedy application of thick films with a minimum of overspray and good adhesion to the substrate, particularly concrete. For good film appearance, air-assisted spray application is preferred. Highly viscous mastics, elastomeric sealants and high solids coatings can be applied by this method and, if necessary, electric heaters can be inserted in the pressure line between pump and spraygun to reduce viscosity at the nozzle and speed up drying.

Careful selection of correct solvents [4], pressure and matching nozzle sizes (often by trial and error) is essential for satisfactory results and products must be free from pigment and filler aggregates to minimize time-wasting nozzle blockages. Abrasive fillers can cause heavy nozzle wear. Great care is necessary when spraying high build coatings. At the PRA laboratories a tendency to the formation of so-called 'vacuoles' has been observed (see Fig. 2.1) which appear to consist of occluded air and impair the film integrity [5].

Spraying with superheated (dry) steam at 105–110°C also allows the application of heavier smooth coats with less overspray and bounce-off, since the steam contracts on leaving the nozzle of the gun. This results in reduced paint consumption (15–25% is claimed). Nozzle pressure is given as 50 to 70 lb/in^2 (approx. 3.5–4.5 atm). Spraying with petroleum solvents vaporized at 70 to 75°C in place of air is also practised. This process requires lower pressures.

(d) Excessively cold paint material.

(e) Water-thinnable stoving coatings which are based on high polymer resins frequently offer greater application difficulties than solvent-thinned materials if sprayed. The main complaint is cratering, but bubbling, particularly after stoving, is quite common.

It is important to adjust the viscosity correctly. It usually needs to be higher than conventional enamels. If it is too low, cratering and cissing may occur; if too high, bubbling with partially cracked bubbles may blemish the film. More than ever must traces of grease be avoided on the substrates. The flash-off time may have to be extended.

Spraying table

The following table (Table 2.1) gives recommended fluid nozzle diameters and air pressures. The figures offered are average values and refer to a distance between aperture of nozzle and the sprayed surface of 8 to 10 in (20 to 25 cm). If spraying is carried out at too short a distance, runners and similar faults are bound to occur, while too great a spraying distance leads to 'dry spraying'. Small and medium-size objects are preferably placed about 30 to 35 in (70 to 85 cm) above ground, ensuring most comfortable spraying at an angle of 30 to 45 degrees.

Large areas, such as bridge structures, wagons, coaches, automobile bodies, etc., are sprayed with flat or wide jets. The use of pressure-feed containers will be economical. Round jets are used for small areas.

Touching up the blemishes, etc., in films is effected by spraying with round jets, using a small nozzle of, e.g. only 0·5 mm diameter.

Attention is called to the fact that in the U.S.A., for example, air pressures of 7 atm (100 lb/in^2) have been recommended for paints containing solvents of sufficiently high boiling points (C.F. Scribner).

Table 2.1 Spraying

Type of coating (approximate instructions)	Diameter of fluid nozzle (mm)*	Pressure †	
		Atmospheres ‡	lb/in²
Filler	2–3	3	45
Coloured enamels on oil and synthetic resin basis, depending on consistency	1–3	1–3	15–45
(a) Fluid	1–1·5 §	1–1·5	15–25
(b) Thick	2–3	2·5–3	35–45
Finest black oil or synthetic enamels			
(a) Matt	1–1·5	2–3	30–45
(b) Glossy	1–1·5	1–1·5	15–25
Cellulose enamels			
(a) Large areas and bodies	2–3	2·5–3	35–45
(b) Small areas	1–1·5	2–2·5	30–35
Non-washable, i.e. soft distempers or synthetic oil-bound water paints, emulsion paints, lime wash	1·5	2·75–3	40–45
Air-drying and stoving asphalt varnishes			
(a) Very viscous, preferably to be sprayed hot	2–3	3	45
(b) Normal consistency, to be sprayed cold	2–3	3–4	45–60
Lacquers for aircraft	1·5 §	3·5	50
Organosols	2	3–4	45–60
Polyester wood lacquers	1·5	3–4	45–60
Polyurethane lacquers	1–1·5	3·5–4	50–60

* The diameter is greater for pressure-feed paint containers.

† The pressure gauge should be located at the spraying booth, not at the air tank in order to prevent wrong readings. A pressure gauge is desirable for every spraying booth. Pressures for electrostatic spraying e.g. Ransburg-Process No. 1 are lower, usually around 1·25 atm (about 15–20 lb/in^2).

‡ For gravity bucket or syphon cup feed.

§ Or larger nozzles with correspondingly higher air pressures for larger surfaces when using pressure-feed equipment. The spraying distance for water-bound paints is greater than usual, i.e. of the order of 2 ft (61 cm).

‖ The pressure varies with the composition of the polyester lacquers.

Notes

[1] Anon. (1975), *Ind. Fin. Surf. Coatings,* **27**, No. 323, 17.
[2] Hughes, R. (1974), *Paint. Decorat. J.,* **94**, No. 119, 8.
[3] Schmitz, F. (1975), *Ind. Lack.-Betrieb.,* **41**, 235.
[4] Dunn, J. (1970), *Product Fin.,* **34**, No. 11, 41.
[5] Private communication.

2.5 Electrostatic spraying difficulties

The group of electrostatic spraying processes fit well into fully automatic production lines, producing even coats on less accessible parts of work pieces and their rear sides with little loss and many savings.

Only short reference can be made to the functioning of the plants.

Coatings are metered in controlled quantities and conducted, according to the process, either to a nozzle from which they are sprayed, e.g. with the low pressure of 15 to 20 lb/in^2 into an electric field, or to a sharp-edged bell, blade, reciprocating disc, etc., which is under high tension (corona effect). From the latter, in some cases with the assistance of centrifugal action, lacquer is discharged in very small electrically charged particles along the electromagnetic field towards the earthed work pieces.

Bells, etc., can be arranged in staggered positions so that all areas of the work pieces can be completely covered by sprays. 'Hand guns' work on a somewhat similar principle as the bell system.

Well-designed electrostatic systems will accept most paints with only minor changes. Normally, it is only necessary to reduce the material with solvent of the correct conductivity.

Table 2.2 lists the conductivities and flash points of common solvents which may be used as electrostatic thinners. It is important to check that the paint system is soluble in the solvent before it is used to modify the conductivity.

Special cases

Conductive paints
These may be sprayed through automatic equipment by insulating the pressure feed supply container. Manual equipment may be modified in a similar manner but highly inflammable paints should be avoided since a capacitance effect may generate excessive power in the feed line which could cause high intensity sparking and ignition.

Water thinnable paints
There is little difficulty in spraying these materials with automatic electrostatic equipment. It is important, however, to have correctly designed atomizers, discs and bells because of the high surface tension of these paints. If the pressure feed

Table 2.2

Non-polar or non-conductive solvents	Resistivity	Flash point
White spirits	More than 10^{12} ohm/cm	min 34°C (93°F)
Solvent naphtha	More than 10^{12} ohm/cm	min 32°C (90°F)
Trichloroethylene	8 x 10^{9} ohm/cm	non-inflammable
Toluol	More than 10^{12} ohm/cm	4·5°C (40°F)
Xylol	More than 10^{12} ohm/cm	23°C (73°F)
Polar or conductive solvents		
Industrial methylated spirits	10^{5} ohm/cm	14°C (57°F)
Isopropanol	1.2 x 10^{6} ohm/cm	12°C (53°F)
n-butyl alcohol	8 x 10^{5} ohm/cm	46°C (116°F)
Diacetone alcohol	1.2 x 10^{6} ohm/cm	40°C (104°F)
Acetone	2 x 10^{6} ohm/cm	−16°C (2°F)
M.E.K.	5 x 10^{6} ohm/cm	−7°C (19°F)
M.I.B.K.	4 x 10^{6} ohm/cm	23°C (73°F)
Ethyl acetate	8 x 10^{7} ohm/cm	−4°C (25°F)
Butyl acetate	8 x 10^{8} ohm/cm	39°C (102°F)
Cellosolve	2 x 10^{5} ohm/cm	40°C (104°F)

system is isolated there is no problem with manual applications for these products are non-inflammable.

Metallic paints

It is normally possible to spray metallic paints with metal contents up to 7% if the paint manufacturer has used coated metal, as near spherical as possible, and a resin solvent blend of ideal resistivity. Aluminium should be, preferably, of the non-leafing type. Hammer finishes can also be applied if correctly formulated and they should be sprayed at viscosities used with conventional equipment. It will be noticed that the hammer pattern is slower to form with electrostatic spray application.

Certain difficulties are specific for the electrostatic spraying methods.

(a) Uneven deposition can occur in all processes by:

(i) Unsuitable jigging, particularly in the original 'Ransburg' No. 1 process.

(ii) Bad overlapping adjustment of bells in the bell processes, resulting in striping. This can also lead to differences in film thickness, and will become the more visible, say on sheet metal, if the coatings have not the best obliterating power, such as creams.

(iii) Repeated circulation through the plant of some work pieces, thus obtaining thicker film deposits than others. This seems to happen most frequently in the reciprocating disc type plant.

(iv) Lack of cleanliness. Discs, bells, blades, etc., must be wiped clean after each interruption.

(v) Unsuitable thinners containing solvents having evaporation rates which do not allow sufficient wrap round. In the case of 'hand guns' these can lead to the formation of comparatively large 'doughnuts' (uncovered areas in the centre) over the surfaces. It must be borne in mind that the travel time of electrically charged particles is considerably slower than of those which are moved by compressed air.

(vi) Uneven temperatures in the plant, especially where the reciprocating disc types are employed. Once such a plant has been set at a temperature this must be maintained over the whole operating period, e.g. by air conditioning.

(vii) Unsuitable composition of the medium. The resistivity of paints can to some extent be modified by additives based on ethoxylated amines (Cataphor) [1] and by solvents (see above).

(viii) A high metal content, as in bronze-containing coatings, prevents successful results. Well-formulated polychromatic lacquers can, however, be used. Difficulties have also been experienced with certain pigments, possibly only under specific conditions. Anatase titanium dioxide has been reported to work with less trouble than rutile.

(ix) The article to be coated must be well earthed to ensure an even distribution of paint.

(x) Imprecise tuning of conveyor movement and speed may result in uneven coverage and film thickness.

Cures are mostly self-explanatory. Atomized particles should remain wet longer than when sprayed with compressed air, so that they can flow out after arrival at their destination.

(b) Flotation trouble is more frequent in electrostatic than the orthodox spraying processes.

(c) The application of the processes to non-conductors offers difficulties. In the case of spraying of polyester lacquers to wood this has been solved by adjusting the critical moisture content of its surface layer to minimum 8 to 10%. Polyester lacquers are put on in two coats, wet on wet within 6 to 8 min or so. If the surfaces can be kept sufficiently moist a somewhat similar procedure should be possible with fibre board and some plastic materials. An alternative up-to-date method especially on wood consists of the application of a very thin conducting coat immediately prior to the spraying of a polyester lacquer, both coatings being compatible.

When, however, several layers of nitrocellulose lacquers are applied electrostatically to wood, the conductivity is reduced from coat to coat, and the third or fourth coats become doubtful.

If nitrocellulose lacquers must be used those of high solids content may in some cases give satisfaction with two coats.

(d) Problems encountered with the electrostatic spray application of powder coatings.

(i) Blocking of powder may occur in the injector nozzles feeding the spray gun.
This may be caused by high moisture content in the powder; water in the air line; or the presence of fine particles which agglomerate in the narrow aperture of the injector. With fully automated powder spray units it is essential to remove fine particles below 15 microns.

(ii) Poor deposition efficiency of the powder coating.
The problem may result from poor powder flow due to the presence of fines; excessively dry atmospheres; poor earthing of the coated object; excessively high electrical field causing back ionization and repulsion of like charged particles; or, poor charge acceptance due to low intensity electrical fields or low specific resisitivity of the powder particle. The ideal specific resistivity of powder particles is greater than 10^{13} ohm/m.

Note

[1] ABM Chemicals Ltd.

2.6 Dry spray and cobwebbing

The dry spray defect may not only be noticeable by sandy looking surfaces but in various other forms, e.g. a kind of orange peel effect. In all cases the film strength is reduced. Transparent lacquer films may only show cloudiness in slight cases, but pigmented coatings, in bad cases, show discontinuous films; in slight cases a dulling effect. As the adhesion of such films is reduced, they may tend to scaling.

Cellulose, synthetic enamels, lacquers, some high-build coatings and zinc-rich paints may be subject to this failure.

Dry-sprayed films can sometimes be improved by rubbing down with water-resistant abrasive paper of finest grain, and re-spraying.

Viscous lacquers containing highly polymerized synthetic products, e.g. vinyl esters (p.v.a., p.v.c., etc., and their copolymers), acrylic and methacrylic resins (including their various copolymers), or polystyrene resins, raw or chlorinated rubber of medium or high viscosity are, in presence of very volatile solvents, often difficult to spray and may exhibit cobwebbing, instead of dry spray. This effect may be used to advantage in the production of decorative effects.

(a) Too great a spraying distance, too wide a spray fan and excessive air pressure in relation to the size of the nozzle, also excessive temperatures in the spraying shop or the air line causes the coatings to lose part of their most volatile solvents during the spraying process before reaching the objects and to impinge on the sprayed surface in a partly dried state, producing in the worst cases a more or less 'gritty' film.

Hot spray lacquers are similarly susceptible to excessive atomizing or top air pressure and to overheating (dry spray, bubbling and whitening).

In the electrostatic spray process too long a passage of the droplets before deposition can have a similar effect, particularly if the solvents present evaporate too fast [see sub-section (c)].

Cobwebbing of the lacquers referred to in the introduction to this section becomes more pronounced as the spraying distance is increased. For normal work, this tendency to stringiness makes the application and the obtaining of smooth continuous films most difficult (but generally has little effect on the brightness); sometimes it lessens the adhesion of the layers.

The speed of evaporation of solvents increases as much as 3% for every rise in temperature of 0.5°C (about 1°F).

Air conduits may sometimes best be led along (or in) the floor.

In the case of spirit varnishes additions of products containing terpene alcohols are recommended by O. Jordan.

(b) Contrary to dry spraying, too close a spraying distance may result in an uneven film surface with ripples or small grooves, caused by the high velocity of the air/coating mixture. The heavier a film is put on this way and especially if the surface is vertical, the greater the probability of failure (including sagging). It is more difficult to rub down to a smooth surface than if an orange peel [see above, sub-section (a)] has to be removed.

Suitable counter-measures are: Correction of the spraying distance, pressure or room temperature, or of all three factors, if required, until the fault is eliminated, i.e. until a uniformly spreading wet film surfaces is produced. In the case of small and medium-sized objects a suitable spraying distance, e.g. for nitrocellulose lacquers, is 8 to 10 in (20 to 25 cm). For larger objects the lacquers should be made up in such a way that they possess enough 'reserve flow' to allow a greater play with regard to spraying distance. This is possible by the selection of an orifice where the size of the spray can easily be regulated and perhaps by addition of less volatile thinners. Thus failures which are occasioned by the fatigue of the operator at the end of his working day or by the changing distances of the different surfaces of the object to be sprayed are reduced to a minimum.

If spraying is necessary at low room temperatures, the air should be suitably pre-heated.

It must be borne in mind that the air for pressure is dependent on seasonable changes of temperature and is sometimes taken from the adjacent rooms, the atmosphere of which can be objectionable.

(c) If due to large percentages of low-boiling solvents the lacquer dries very rapidly, a correspondingly large percentage of the solvent, even in a normal working spraying plant, may evaporate on its way to the sprayed surface, whilst in extreme cases (depending on the spraying pressures) the lacquer particles may reach it in form of 'dry dust', producing a flat and brittle layer as described above. This is important also for electrostatic spraying, see above.

When hot spray lacquers, the solvent content of which is comparatively low, are overheated on application similar failures are possible, because an excessive proportion of volatile matter could prematurely evaporate whilst travelling from nozzle to work piece. In clear lacquers a whitening effect could be produced.

Higher-boiling ingredients must be selected.

To avoid cobwebbing it is particularly important to use correctly formulated thinners in the right dilution ratios. Generally the volatile portion coming off last should always be a good solvent for the medium contained in a coating. The addition of small quantities of dipentene or other high-boiling solvents is sometimes helpful.

(d) Unsuitable size of the fluid nozzle. Since highly viscous lacquers do not easily pass a small nozzle, this condition may also lead to dry-sprayed films. The spraying globules then do not always flow out to an even film. The use of too small a nozzle, or one partly blocked by skin or dirt, means waste of energy. On the other hand, the use of too wide an aperture can lead to excessive consumption of coatings by overspray, to runs and other disturbances of flow; similarly, if the diameter of a small nozzle has been increased by improper cleaning.

Use a larger nozzle and clean daily with proper solvents and not with hard mechanical tools, but,for example, pipe cleaners.

(e) Too low top or atomizing pressure effects the flow badly and produces orange peel.

(f) Badly functioning air suction of the exhaust fans may cause precipitation of dry lacquer dust throughout the entire shop and on freshly lacquered surfaces.

Overspray should be caught by baffles or paint arresters. Airless spraying methods hardly necessitate any exhaust. Superheated steam spraying reduces overspray. Electrostatic spraying cuts out overspray for all practical purposes.

(g) If the surface to be sprayed is not held at a correct angle toward the jet, even thin-oil base enamels may induce dry spray. The cone of the jet passes laterally, but the coarser particles located around its cone impinge on the imperfectly sprayed portions of the surface, forming spotty, finely granulated areas.

Avoid dead angles. Big objects with many angles and edges such as cars and coaches want the use of lacquers as described above in sub-section (a).

(h) It is also possible, especially in the case of dilute enamels, that if the opposite sides of the objects are sprayed alternately (drying the one side in the air or in stoves before applying on the other side) the back of flat objects may show the images of the spraying supports. This can be due to spraying at incorrect angles (too flat, for instance, by holding the spraying objects too high) so that part of the jet drizzles upward through the free spaces between the sprayed objects and the supports.

Proceed as directed in sub-section (g) above. Spray in a more downward direction. The sprayed parts should be placed on to trays sufficiently open to prevent blocking of the fine droplets of the jet.

2.7 Formation of spraying mist, and exhaust faults

Apart from the economic disadvantages induced by excessive consumption of paint materials and by dry-spraying, the formation of paint mists should be prevented for hygienic reasons, since these mists contain more pigment and vehicle particles than pure solvent vapours and on account of their high degree of cohesion are only removed with difficulty by the exhaust fan.

Overspray can also cause defective work, e.g. cratering and bleeding.

(a) The formation of mist is promoted by

(i) strong cooling;
(ii) the presence of moisture particularly when the relative humidity is high;
(iii) existing condensating nuclei (spraying dust); and
(iv) overthinning.

These difficulties can be eliminated by the following measures: Reduce the spraying pressure, filter and preheat the compressed air. Make certain of dry shop atmospheres and dry compressed air, and use only solvents and pigments free from moisture.

The lateral spattering of paint particles from the spraying cone or fan must be prevented by using clean nozzles free from burrs and of uniform operation. The trigger control and the needle valves should operate smoothly and evenly.

The gun must not be held too far from the articles to be sprayed.

'Mistless' spray guns are designed so that an air cone surrounds the spraying jet.

On the other hand, electrostatic spraying practically eliminates excessive spraying mist, as also do the 'airless' spraying processes.

(b) Faulty functioning of the exhaust.

A normal spray gun in connection with a normal spraying booth is claimed to require about 5300 ft^3 (150 m^3) of vapour-air-spraying mixture and atmosphere per uninterrupted working hour. If about half of this time is assumed to represent actual operating time (half for pauses), only about 2650 ft^3 (75 m^3) of air need be removed per hour. The speed of the air current in the exhaust pipe would then amount to about 33–50 ft/s (10–15 m/s) and at the spraying post about 5 ft/s (1·5 m/s) at the most. See also Ignition of spraying dust, Fire danger, and Health hazards.

Shape and size of the material to be sprayed should govern the type of construction and placing of the spraying booth. Water curtains or paint arresters may be advisable. Another important factor for consideration is the mode of application.

2.8 Dipping, flowcoating and curtain coating

Dipping is one of the cheapest, if not *the* cheapest mass finishing process. All round, uniform coating of the objects is achieved, with low losses of paint and low labour costs. The articles must, however, be suitable. The processes are widely used in the production of motor cars, agricultural machinery, components for tractors and trucks, steel furniture, tools and steel sections.

There are numerous variants of the process: ladling and hand dipping, automatic slow extraction and conveyor dipping have been in use for a considerable time. In the more recent gel dipping, using certain cellulose ester lacquers, the latter are liquefied at approx. 50°C (approx. 120°F) and set on cooling to a firm gel of up to 10 thou (250 μm) thickness per dip. Dry dipping processes include both conventional and electrostatic fluidized bed application of thermoplastic and thermosetting powder coatings.

Large objects having long draining areas, smooth surfaces, and of compact build and which are all to have the same colour can be finished by flowcoating [1], a process which is a highly mechanized development of the rather primitive ladling. The coating flows under some pressure through hoses and multiple stationary spray nozzles. Surplus paint is collected and re-used after filtration and adjustment of viscosity. Comparatively small quantities of paint are in the system at any time. Flat, large panels or tubes can be flowcoated without runners, but the loss of thinners can be high. The process fits into conveyor systems, where the jigging of the work has already been studied for spray cleaning and phosphating. The avoidance of runners and air bubbles depends on the good functioning of the plant. Flowcoating of large objects having the same colour is normally done under 1–1·5 atm (15–25 lb/in^2). The multiple stationary spray nozzles as used in conveyor work are usually fan-shaped.

During the last decade trichlorethylene thinned paints have become accepted in dipping baths [2]. The advantages of trichlorethylene are its non-flammability, its excellent solvent power for most polymers and resins (nitrocellulose being a notable exception) and its good compatibility with other solvents. The toxicity of trichlorethylene must be rated as a disadvantage, but good plant overcomes this drawback. Trichlorethylene dipcoatings can be operated in hot baths at about 80°C close to the boiling point of trichlorethylene which is 87.2°C. Hot baths require preheating of the objects to be coated to 65°C, and the capital outlay is greater due to the necessity of incorporating a solvent recovery plant. Plants operating at ambient temperatures (20°C to 25°C) are therefore more popular. The evaporation rate of trichlorethylene is 3.8, compared with xylol (14) and white spirit (50). This promotes rapid drying, permits much faster operation of the production line and gives less trouble with ‘fat’ edges and drips.

Some points made in the following sub-sections are also applicable to flowcoating.

(a) The varnish or paint material in open dipping tanks is subject to gradual thickening, depending on the rate of evaporation of the solvents and thinners contained.

In flowcoating plants the drainage area for the actual flowcoating section should be sloped, preferably in all directions to a drainpoint with sufficient fall. The drain connection to the paint-settling tank should be of adequate size [3].

The tanks must remain well covered while out of use. The temperature in and surrounding the plant is also of interest as affecting the rate of evaporation and should therefore be carefully controlled. Keep the tank's surface as small as possible.

In all dipping, flowcoating and curtain coating processes, maintenance of the correct paint viscosity is of great importance and daily viscosity checks are essential. Some modern plants incorporate viscosity control.

The success of a flowcoating unit, and some dipping plants, depends to a great extent on the solvent concentration which can be built up in the drain section immediately after the point of application, and for this reason the drain area is completely enclosed except for the ends. If a high solvent concentration is obtainable the paint will remain open long enough for efficient draining and flow out. A good solvent vapour seal is necessary for economic running of the plant [3, 4].

The consistency of the paint material should be checked at regular intervals, e.g. by determining the specific gravity using a hydrometer, and corrected by adding suitable quantities of thinners. See also below, (b) and (e).

(b) If the dipping material contains pigments, possible settling tendencies (governed by specific gravities, etc.) and the corresponding deviations of colour, covering power, gloss and so on, must also be considered. Water-thinnable coatings tend particularly to aeration. They are the more troublesome if they leave a 'track'.

Constant automatic circulation and slow agitations is useful in large plants. This in turn can promote aeration and accelerate oxidation and fattening of the medium. For dipping large panels in heavily pigmented primers, wavemakers are used on the surface of the tank which cause the primer to flap against the objects as they are withdrawn and thus drag off the wet film. This promotes even films free of sags and runs. Aeration can be countered by the addition of surface active agents, or specific silicone oils in minute quantities.

The adjustment of viscosity of water-thinnable materials according to instructions is of great importance for the avoidance of bubble trouble. In some cases air bubbles can be removed with the aid of a weir in the dipping tank. The circulating pump should be of a non-aerating type. See also the advantages of flow coating in the Introduction.

Allow impurities to settle below a fine wire mesh at the bottom of the vat.

Before filling up any paint dipping vats, it is advisable to control the specific gravity of the remnants in the tank in order to make certain that these have not become faulty due to excessive concentration of pigment. [See above sub-section (a)]. (H. Wolff once observed a concentration of more than 50% above the original value.) In such cases thin down, preferably with thinning varnish.

At the completion of a shift the main circulating system is normally changed over to solvent circulation to enable the whole of the pipe work and coating area to be washed down, the contaminated solvent draining back to the solvent tank.

(c) The higher the drying speed of the dipping material [as to evaporation of solvents see above sub-section (a)], the more slowly should the dipped objects be withdrawn from the dipping bath in order to prevent the formation of faults such as runners, fat edges, etc.

It is advisable when using quick-drying dipping lacquers such as acetate cellulose or nitrocellulose lacquers for small parts (such as pencils, penholders, wires, cables, etc.) to employ accurately operating slow extraction dipping devices, in which the speed of withdrawal is not quite as high as that of the excess lacquer running off the surface of the objects. In the case of high-temperature stoving varnishes, evaporation takes place some time after all the excess varnish has run off the parts. It is easily possible with dipping material of this type to carry out the dipping, flowing or pouring-on operation rapidly by hand or by conveyor.

(d) On dipping some thermo-hardening coatings, particularly those containing crosslinking or urea or melamine resins, the flow may be somewhat slow during the initial air drying period with sometimes a tendency to silking and to the formation of runners. Such coatings should be specially made for dipping.

Gelling can also be found especially of polyurethane coatings due to aeration and accelerated hydration or excess of catalyst. Such conditions are possible, e.g., on curtain coating.

The addition of a percentage of solvents with a low rate of evaporation as suggested by the manufacturers may be of benefit.

(e) Too thick dipping material also favours bubble trouble and runner formation particularly if it sets rapidly or is subjected to stoving shortly after the dip, the more so if unsuitably jigged or hung. A typical winter complaint.

Warm the paint for dipping, if possible, or thin down. Sometimes adjustment of the draining angle by different hanging helps. Slower or faster thinners have to be selected according to the length of the work pieces. In trichlorethylene dipping the film thickens should not exceed 35 μm [2].

Soft undercoatings can also be lifted, causing, e.g. rippling. This can be more pronounced the longer the stay in the dipping vat. See also above (a) and (c).

(f) Rapidly drying oil base coatings containing wood oil and other drying oils and/or much drier, especially cobalt, tend to form skins of varying thicknesses in contact with air. On stirring up carelessly, skins are divided up into small bits which make the paints, etc., unsuitable for dipping. The early addition of antiskinning agents, e.g. oximes, may be of help, but previous consultation with the manufacturers would be necessary.

There are, however, materials which sometimes form skin independent of oxidation. To these belong unstable chlorinated rubber and also shellac lacquers. Skins of the latter become frequently insoluble. It starts with partial evaporation of solvents. Coatings containing cross-linking resins, e.g., amino resin–alkyd lacquers can form skin-like gels when for some reason, possibly old age or unsuitable solvents, polymerization continues. This sort of skinning may develop into an irreversible gel of the whole paint or lacquer mass.

The above defects can also occur in partially filled or open cans.

Cellulose and non-shellac spirit lacquers and most coatings which dry by evaporation do not form insoluble skins. In the case of non-drying oil modified stoving synthetic resin varnishes, skinning is rare. The behaviour of straight urea and asphalt varnish (including the air-drying types) or coatings containing little or no drying oil (for instance, those containing dammar resin) is similar. Some urea type finishes, especially if pigmented, may offer other difficulties on dipping as set out above in sub-section (d).

(g) If a dipping bath is used for a considerable length of time it is gradually contaminated with the impurities carried into the bath by the objects to be lacquered or varnished, and often also by numerous air bubbles. The impurities are particularly objectionable on glossy films. Contant aeration can upset the stability of coatings.

The paint material should be filtered through very fine mesh strainers, tiffany or gauze. Large quantities are cleaned by circulation through filter pumps. Some distance over the bottom of the dipping tanks sieves should be fixed to allow sediments and sludge to settle out and to be drawn off. See, however, above subsection (b). Air bubbles should be skimmed off over a weir.

(h) Rope-figured parallel streaks (ropiness, groove formation, silkiness) can sometimes disturb the shine of dipped coatings. This is likely to be due to one of the reasons given for afterthickening, excessive viscosity or silkiness.

Dip wooden objects, if possible, with the end grain at the lowest point because this tends to absorb all or some of the lacquer surplus.

Curtain coating is a technique widely used in the furniture industry for coating large panels with lacquers or paints. The principle underlying the method has been used in other industries, for instance chocolate manufacture, long before it was adapted for surface coatings [5]. The machine consists essentially of a head tank which contains a slit from which the 'curtain' flows. The objects to be coated pass on a conveyor through the 'curtain.' A micrometer adjustment permits varying the gap from 0 to 2 mm. A typical setting could be 0.5 mm. The coating material is pumped to the head tank and is forced under pressure through the gap. Width of gap, pump pressure and viscosity of the material determine the film thickness deposited on the objects. The coated articles pass on to a second conveyor for drying; surplus material is pumped back to the head tank. Conveyor speeds of up to 135 m/min (450 ft/min) are possible.

(i) Uneven curtain and break in the curtain.

This can be due to partial blockage of the gap. It is essential to include fine sieves in the pump circuit to remove particles of skin or foreign matter from the paint. Trapping of air in the curtain must be avoided and an airbleeding device on the top of the head tank should rectify this fault.

(j) Uneven film thickness.

Uniformity of the gap along its whole length should be ensured by compensating any distortion by means of a tensioning bar fitted to the outside of the head tank.

Notes

[1] Page, R.G. (1968), *Ind. Fin (U.S.)*, **44**, No. 11, 24.
[2] Giesgen, H.R. (1973), *Ind. Lack.-Betrieb*, **41**, 32.
[3] Stordy, J.J. and Appleton, W.G. (1956), *JOCCA*, **39**, 565.
[4] Pursche, G. and Cain, H.D. (1968), *Plaste and Kautschuk*, **15**, 297.
[5] Tatton, W.H. and Drew, E.W. (1962), *Product Fin.*, No. 8.

2.9 Electrocoating and fluidized bed coating

Electrocoating

A number of advantages have been claimed for the electrocoating (electrophoretic deposition) technique [1] of dipping over the traditional methods. These are briefly:

(i) a more even deposition, i.e. no runs or sags, and uniform film thickness, but maximum film thickness is limited to about 20 μm
(ii) covering of sharp edges
(iii) no flash-off time required due to the low solvent content of the deposited film
(iv) no solvent wash can occur before or during stoving.

Normally this technique lends itself only to the application of a single coat of paint to metallic articles. This is due to the fact that the cured paint film will present too large a resistance to allow a satisfactory current to flow in an electrophoretic circuit. However, if a conductive paint is deposited as a first coat, a second coat can be applied electrophoretically.

The process is complex and depends on the transport of charged colloidal particles through a polar medium to an electrode of opposite charge. The article to be coated is made one electrode (usually the anode), the other electrode (the cathode) being the tank or a metal probe inserted into the paint. The most frequently employed binders are carboxyl group containing polymers derived from acrylates or phenolic resins and maleinized oils. Aqueous paints are used and they must be diluted with deionized water. Inorganic ions occluded in the film could reduce its corrosion resistance materially. By ultrafiltration of the effluent from the tank, savings of deionized water can be effected and paint wastage reduced. The solids content of the bath and the pigment/binder ratio of the paint must be monitored.

Electrodeposition has introduced a number of specific film defects which are outlined below:

(a) The final appearance of the deposited film will depend on the length of time of immersion and on the strength of the electric current (amperes). A coarse deposit is usually an indication of a short immersion time at too high a voltage.

Longer immersion times at smaller voltage or current values will produce smoother films.

(b) Poor deposition, foaming, discontinuity and lack of opacity of the deposited film may be due to several factors.

(i) Poor deposition can be caused by a bad contact between the article to be coated and the anodic jig. It may also be due to settling of pigment in the bath.
Settling can be prevented by the use of circulating pumps, poor contact by adjustment.

(ii) Foaming and blistering is caused by the formation of bubbles (oxygen) at the electrodes due to electrolysis.

Its effect can be minimized by either a reduction in the voltage or current values or by adequate circulation.

(iii) Inability to coat shielded portions of the article, indicating lack of 'throwing power'.

Improve by increasing the rate of deposition and hence rapidly increasing the resistance of the deposited film. However, this could create other difficulties, see above, (a). A better remedy would be a paint which deposits a thin film of high reistance.

(iv) Badly pre-treated metal surfaces; oxide layers making them bad conductors.

(c) Precipitation of the medium on the bottom of the tank due to change in pH. During the operation of an electrophoretic dipping process the more acidic portion of the medium will be attracted to the anode, the alkaline, e.g. amine, portion to the cathode. This will eventually lead to a build up of alkaline material in the bath which would cause a considerable pH drift. Too high a pH can also lead to porous films and to re-solution of some of the deposited film.

Accurate pH control must be excercised at frequent intervals during the working period of a bath.

It has been suggested that a semi-permeable membrane placed around the cathode would allow the alkaline materials to collect in a confined volume from where they could be removed with ease. A further method would be to pass the paint through an ion exchange resin. The pH must be maintained according to the instructions of the paint suppliers.

(d) Temperature can affect the efficiency of deposition. Higher temperatures may cause loss by evaporation which would require adjustment of pH and solute.

It is advisable to run the process at 16–26°C (60–78°F).

(e) Runs, thick edges and blistering on stoving may be caused by a thin layer of paint being deposited on top of electrodeposited films by normal dipping techniques. This usually occurs when the electric current is switched off prior to the removal of the article from the bath though this can happen to a lesser extent if the current is not switched off if the supernatant layer had not been rinsed away.

Surplus paint can be removed by immersion in a tank of water or by passing the dipped object through a curtain of water. The latter is preferable as the former can create water marks, (see also Snail track blistering). There should be no residual waterdrops left on the newly deposited film when it is placed into the oven.

(f) Marring of the very soft deposited film before stoving; or under-curing resulting in poor adhesion and resistance to stove-chipping [2].

(g) Variations in phosphate pre-treatment of metals, crystal size, coating weight and composition of final rinse are critical for the successful electrodeposition of water-soluble paints, though apparently not of emulsion paints.

(h) According to British Patent No. 1, 027, 716 (Pressed Steel Co.), pinholing and high-current surges can be avoided by applying the coating in two stages. In the first stage a low voltage is used and this is increased in the second stage to produce the desired film thickness.

(i) Tanks containing electropaint are prone to bacterial infection, and in South Africa this has been found to be by far the worst problem encountered in electropainting [3]. Some fifteen varieties of organisms have been identified, *Proteus vulgaris* being the most common one. Biocides must be used, but as bacteria soon develop resistant strains, two or three biocides of known effectiveness must be used in sequence. Infection can stem from many sources; one is the ion-exchange resin beds of the water demineralizing plant. This can be overcome by treating the demineralized water with ultra-violet radiation.

Good housekeeping and frequent cleaning of the tanks diminishes the problem.

Powder coatings applied by fluidized bed

(a) The fluidized bed dip process: this principle was developed as a method of dipcoating metallic articles by research workers at the German firm of Knapsack-Griesheim.

At the bottom of the fluidizing tank there is a plenum chamber separated from the tank proper by a porous tile wall. The plenum chamber is fed with low pressure air and the powder is fluidized by the air diffusing uniformly up through the porous wall, which is made of sintered bronze, 2 mm to 5 mm thick, with an average pore size of 12 μm.

Fluidization is a term used to describe a process in which a rapid gas current, normally air, passes through a finely powdered substance in such a manner that the powder particles become separated and airborne without being carried away by the air stream. Articles to be coated by this dip method are heated to a temperature exceeding the softening point of the powder coating. The particles melt and adhere to the surface of the articles when they are immersed in the fluidized bed.

It is essential for the correct application of coatings that the bed operates smoothly without any settling or surging of the powder. The stable state in which the pressure drop across the bed is just sufficient to support the weight of the particles is referred to as incipient fluidization.

Channelling is a term used to describe uneven air flow through the fluidized bed which is often caused by the presence of very fine particles or cohesion between the particles, resulting in agglomeration. This defect can be reduced by vibrating the fluidized bed or by the introduction of some form of mechanical vibration.

Slugging or spouting is caused by the surge of a large gas bubble through the bed which carries away powder from the fluidized container. This problem is often resolved by increasing the diameter of the fluidized bed.

For prolonged smooth operation, it is essential that the gas used is dry and free from oil contamination.

(b) The electrostatic fluidized bed was developed by S.A.M.E.S. in France to improve the efficiency of the fluidized bed and to enable articles of simple shape to be coated without the need to immerse the article in the bed of powder. The main advantages of this new method were the improvements in cleanliness of the process and the considerably reduced volume of powder kept in stock within the fluidized bed.

The bottom of the tank consists of a ceramic or porous plastic plate through which air passes to fluidize the powder. Electrodes connected to a high voltage generator are distributed over the porous plate. When an earthed article passes over the powder an intense electrical field is created between the article and the electrodes. Powder particles receive an electrostatic charge and they are conveyed to the earthed article by the transporting influence of the electrical field.

It is almost impossible to coat intricate articles by this method but it has proved to be successful with small objects, continuous wire and webbed material. Film thicknesses up to 500 μm may be obtained and it is not necessary to preheat the article to cause the powder to adhere, although the objects must be heated after coating to flux the powder layer and produce a continuous and fully adherent film.

This type of application equipment should not be operated without stringent safeguards.

Notes

[1] Fest, M. (1974), *Ind. Lack.-Betrieb* **42**, 380.
[2] Walker, I. (1964), *Int. Metal Finishing Conference,* London, 1964.
[3] Nicolay, R.L. (1977), *JOCCA* **60**, 169.

2.10 Coil-coating

Pre-coated metal sheets, produced by the continuous coating of a coiled strip of metal by reverse roller application have in recent years been used for purposes as varied as facia panels for motor cars, filing cabinets, curtain walling, shelving and garage doors, to name but a few. Many different coatings are in use, such as alkyd/amino stoving finishes, acrylics and solution vinyls but p.v.c. organosols and plastisols have been the most popular. The latest polymers applied in this field are oil-free polyesters, polyimides and fluorocarbons.

In the United States, coil-coating has been practised for about 30 years and large quantities of pre-coated steel and aluminium have been used as 'sidings' in building construction. Often exaggerated claims for durability have been made.

For decorative purposes p.v.c. calendered films on steel are available in a variety of textures [1].

In the UK, the commonest coating has been p.v.c. plastisol applied at a thickness of 125–200 μm. This is used to a considerable extent as external cladding in the building industry. Its performance depends on the formulation of the plastisol and on the climatic conditions of exposure, and variable results have been reported.

The principal defects are:

(i) lack of corrosion resistance; this can be overcome by galvanizing the strip before coating.
(ii) dirt pick-up and retention; this is due to the softness of the coating and can to some extent be remedied by topcoating with a mar-resistant lacquer.
(iii) poor light-fastness. Light fastness must be of a very high order, as the main claim put forward for pre-coated metal sheeting is a long period of service before maintenance is needed.

It has been reported that under severe climatic conditions near the sea in southern England, p.v.c. plastisol and organosol precoated steel has failed frequently through change of colour to black or brown, blistering, corrosion and loss of adhesion. Polyvinyl fluoride coated steel and aluminium has shown excellent durability, with no chalking and little colour change after prolonged exposure, but some susceptibility to mechanical damage has been noted [2].

In the manufacture of polyvinyl fluoride coated sheeting it is important to control the fluxing operation most carefully at the correct temperature to avoid incomplete fusion of the polymer particles. (See Fig. 2.2.)

The main problem posed in the use of pre-coated steel or aluminium for fabricating various articles is the difficulty of finding a coating which combines all the desired characteristics. p.v.c. based coatings are much favoured, but close co-operation between coil coater and coating manufacturer is essential.

Notes

[1] Newton, D.S. (1973), *JOCCA* **56**, 566.
[2] Whiteley, P. and Rothwell, G.W. (1975), *JOCCA* **58**, 453.

2.11 Foaming of coatings

Foaming is the formation of a stable gas-in-liquid dispersion, in which the bubbles do not coalesce with each other or with the continuous gas phase (air above the surface of the liquid) [1].

(a) A disagreeable fault in, e.g. the roller-coating of paper (labels, hardboard, film pictures, etc.) with spirit and other roller-coating lacquers.

Antifoam preparations have been developed for the purpose of overcoming the foaming trouble. The defect can be reduced by lowering the speed of the rollers. The viscosity of the varnish has also to be watched.

(b) Water-thinnable coatings (emulsions and solutions) foam easily and persistently on transport, stirring or dipping.

To reduce surface tension small additions of surface-active agents, e.g. silicone anti-foam compounds or higher alcohols, can be added according to the advice of the manufacturers. These must be used with great care. Dipping tanks should be fitted with a suitable weir for the removal of bubbles and dirt.

Increase in pigmentation, decrease in viscosity and an increase in alcoholic solvent content as well as the addition of 1% of aliphatic hydrocarbon, boiling between 90° and 105°C have been suggested in connection with the dipping of paints containing specific melamine and phenolic resins [2].

(c) Plastisols are prone to aeration.

Vacuum de-aeration is effective.

Surface-active agents [3] (for the reduction of interfacial tension between air and liquid coating) help to disintegrate air bubbles more easily. Small percentages on the non-volatile content should, if possible, be incorporated at the manufacturing stage of the plastisol, though additions which are difficult to incorporate may often still be useful if made to the finished product.

Notes

[1] Keozthelyi, S.E. (1974), *J. Paint Tech.*, No. 598, 31.
[2] North, A.G. (1961), *JOCCA* **44**, 132.
[3] Surface active agents used for this purpose include (for example) polyethylene glycol derivatives. Some anionic surfactants may aggravate foaming.

2.12 Sticking together of enamelled parts in a tumbling machine

(a) Light articles, especially if their shape is flat or they have protruding sections, are apt to stick together on lacquering and drying in a tumbling machine. Tumble coating is most suitable for small parts such as buckles, buttons, rivets, etc.

Insert (together with the goods to be enamelled), shot, the diameter and weight of which has to be selected according to the circumstances, or small blocks (tumble jacks). Use just enough coating without surplus. Mix articles of different sizes and shapes so that the smaller pieces fill the spaces between the larger ones [1]. Alternatively a rotary paddlelike wheel, called a 'breaker' is fitted in the tumbling barrel to prevent sticking [2].

It has also been suggested 'to add a very small amount of pine oil after the coating is evenly distributed. With shellac, a little mineral oil may be added and later removed by tumbling with a few rags to absorb the excess oil' [1]. To avoid fat edges and drops, objects with sharp edges (e.g. nuts, bolts) are successfully coated by centrifuging. The latter process is faster and also causes no waste. It is suited for large quantities of small articles.

Excessive tumbling may roughen the surface and affect the lustre. Too great speed also promotes nesting and sticking.

Recommended speed: 20–40 rev/min [1].

(b) Too viscous and sticky coating.

Tumble warm if the covering or hiding power requires it and provided that the enamel, etc., is suited for this; otherwise thin down.

Notes

[1] Wampler, R.H. (1946), *Modern Organic Finishes,* pp. 102–115.
[2] Beach, W. (1972), *Metal Fin.* **70**, No. 5, 90.

3

Difficulties mainly due to faults in the liquid paints

3.1 Excessively thin consistency (as received)

(a) This can lead to settlement of pigment during storage. Containers should be turned over from time to time. Before use the full container should be shaken or vibrated on a paint conditioner and the contents then stirred thoroughly.

(b) On application, excessively thin consistency will result in the formation of runs on vertical surfaces and pulling away from sharp edges. Opacity and film build will also be poor.

(c) The paint material is too warm, for instance, due to storing too near a stove of radiator.

20°C (70°F) is the preferable temperature for storage. During hot seasons, with temperatures up to 30°C (85°F), the viscosity of paint materials is greatly reduced. In the same way, paints of a viscosity correct for temperate climates may be far too thin for application in tropical districts.

(d) Some solvents contained in thinners may reduce the viscosity of a paint more than others. Alcohols generally have such an effect. Aromatic solvents, when used in oleoresinous or alkyd type paints, reduce the viscosity more than aliphatic hydrocarbons used in the same proportion.

Ketones are very powerful solvents and reduce viscosity of paints very rapidly. They should be used very sparingly. Common ketone solvents are methyl ethyl ketone (MEK), methyl isobutyl ketone (MIBK) and cyclohexanone ('Sextone').

(e) With long storage, increased wetting of the pigment may occur with consequent reduction in viscosity.

(f) Excessively thin emulsion paints may contain insufficient water-soluble thickener or the latter may have suffered decomposition by bacterial action as a result of insufficient preservative. Low consistency resulting from pigment settlement is uncommon in emulsion paint.

3.2 Faults due to excessive addition of thinners or additives

Most paints are supplied at the correct consistency for the prescribed method of application and no addition of thinners or additives should be made. There are, however, special cases in which addition of thinners is permissible, for example, to a primer when it is necessary to seal a highly absorbent surface. In such cases the correct thinner should be obtained from the manufacturer and his instructions on the quantity to be used should be followed.

When thinners are used they should be added gradually with stirring and the stirring continued for a while after all thinner is added. This ensures a homogeneous mix.

(a) Failure to observe these precautions can lead to the following defects: loss of gloss and opacity, poor build, and sagging – all the outcome of excessive thinning. Thinners should be measured or weighed and not added 'by touch'. The latter frequently leads to over-thinning.

(b) The consistency of a paint is influenced considerably by temperature and consequently less thinner is required in summer than in winter periods.

(c) Excessive thinning can in extreme cases produce bad adhesion, greater moisture permeability and precipitation, e.g. of the resinous component. In the first stage the medium becomes cloudy (only visible when unpigmented, of course), the clouding then intensifies until an insoluble deposit settles out which may become hard or sticky and which cannot be stirred in again. The same can happen if perfectly compatible thinners are poured into, say, a lacquer, in one go instead of adding them steadily under stirring. Innumerable complaints are due to such carelessness, particularly during the cold season. In the presence of much pigment the film may show seeding. Such defects occur, e.g. when oil-base varnishes and enamels are thinned mainly with white spirit.

Water-thinnable coatings as used for stoving have a critical point at which water causes no great change in viscosity. Beyond this point, however, even small further additions of water can thin excessively. Throw-outs may develop. The stability of aqueous solutions is greater than that of aqueous dispersions.

As explained above, thinning has to be done with great care. In the case of oil-base coatings terpene compounds such as dipentene are useful. In many cases xylol or naphthas are superior to white spirit. Throw-outs can hardly ever be brought back into solution, with the exception of cellulose lacquers, where this may be done with the aid of strong solvents, e.g. of the glycol ester or cyclohexanone ('Sextone') type.

(d) Unsuitable thinners can cause flocculation of pigment or even resin precipitation. This often happens when a mixed solvent contains a high-boiling component which is a non-solvent for the resin.

No thinners containing hydroxyl groups (alcohols, glycol ethers, etc.) or water should be added to polyurethanes as these would react prematurely with the isocyanate and cause gelling. For this reason thinners for polyurethane must not be left unsealed or moisture is likely to be absorbed.

(e) Reduction of durability. The physical film properties—adhesion, flexibility, tensile strength and elongation—can vary considerably according to the thinners used (or solvent combination contained), especially of cellulose and other lacquers which dry by evaporation. Excess thinning may change even long-oil paint materials so that they will behave like short-oil paints with a corresponding alteration of period of life [1]. This is due to the fact that:

(i) Films of this type may be so thin so that the decomposition of the film substances through oxidation, etc., proceeds much more rapidly than that of a normally thick film, increasing the speed of embrittling correspondingly.
(ii) In the case of pigmented paints the thickness of the films in relation to the size of the pigment particles is reduced to such an extent that the latter are not sufficiently surrounded by the film material, facilitating the penetration of moisture and gases (oxygen and carbon dioxide), thus facilitating attack on the film and base.

Experiments carried out with boat varnishes in water always showed that those with the largest percentage of varnish body are more resistant.

(f) Excessive additions of water or prepared aqueous diluents to water thinnable paints may have results such as streaking, pigment flotation [2], instability of emulsions, curtaining, sagging, etc.

Flotation and colour separation can occur in emulsion paints, water paints and distempers by additions of excessive amounts of water or aqueous diluents. This fault is more accentuated in strong colours and pigment combinations which include carbon blacks. Phthalocyanine blues are particularly susceptible.

The effect is observed in brush application as either a patchy appearance of colour or a darker or modified colour area at the extreme edges of the brush marks. When overlapping an edge painted some minutes previously, light or dark brush marks due to the overlap may be observed as a colour patchiness. Streaking occurs when aqueous paints have relatively poor flow and the colour separation is shown usually as darker or more intensely coloured lines corresponding to the 'valleys' of the brushmarks. This effect is invariably made worse where high porosity of the surface creates a suction which reduces the wet-edge time of the paint.

Application at a higher viscosity, i.e. with less thinner additions, can often cure flotation of this type. This can be due to the reduction in mobility of the pigment particles at the higher viscosity, although brush marking may be increased. The additions of solutions of anionic and non-ionic wetting agents can sometimes rectify the defect where this is due to insufficient wetting at the first stages of manufacture. Carbon blacks and many dyestuff pigments can show this defect due to inadequate wetting. Care in the selection of wetting agents is very important, as the wrong type or the use of excess can result in more serious flotation defects. Manufacturers should be consulted. Additions of thickeners can sometimes give satisfactory results due to improved flow properties at increased viscosity. Again care in selection is important.

(g) Certain coloured aqueous paints can give emulsion instability on excessive dilution when hydrophobic pigments are present (e.g. arylamide yellows, azo reds, phthalocyanine blues, etc.), although a similar paint formulation in white, or containing hydrophilic pigments as tinters, can be stable.

Poor dilution tolerance of emulsion-based paints of all kinds can be accentuated when the emulsion is intolerant of hard water. The presence of soluble salts in thinning water can be responsible for flocculation of the pigment dispersion. This causes a change in the balance of dispersing agents present, often resulting in pigment/binder coagulation either as dispersed rubbery particles or in severe cases as a rubbery mass.

Emulsion instability can rarely be rectified. Where the breakdown has not resulted in complete coagulation of the disperse phase (in O/W–oil in water emulsions and vinyl polymer dispersions), a temporary refinement or reconditioning can sometimes be effected by passing the paint through a fine-mesh sieve. (Refinement by milling is undesirable, since excess shear may result in complete breakdown.)

Notes

[1] Gardner, H.A. *Circular of Am. P. & V. Man. Assn.,* No. 260.
[2] Bell, S.H., (1952), *JOCCA.,* **29**, 373.

3.3 Curtaining, running and sagging on vertical surfaces; pulling away from sharp edges and corners; bad flow

Curtaining and sagging

A downward movement of a paint film between the times of application and setting, resulting in an uneven coating having a thick lower edge. The resulting sag is usually restricted to a local area of a vertical surface and may have the characteristic appearance of a draped curtain, hence the synonymous term 'curtaining'. See Fig. 3.1, B.S. 2015 [1].

Runs

Narrow downward movements of a paint or varnish film; may be caused by the collection of excess quantities of paint at irregularities in the surface, e.g. cracks, holes, etc., the excess material continuing to flow after the surrounding surface has set. See Fig. 3.2, B.S. 2015 [1].

Tears are small, characteristically shaped runs.

These may result from the following causes:

(a) Excessively heavy application generally. Conventional brushing paints should be spread evenly, special attention being paid to mouldings and corners. Heavy coats will tend to sag on flat vertical areas and will drain into corners and cause 'runs'.

Brushes should not be charged heavily except with thixotropic paints when heavy coats can be applied without risk of running or sagging even on sharp edges and corners.

With spray paints generally, it is necessary to select the correct nozzle size and to adjust the air pressure so that a film of the correct thickness is obtained. Spray strokes should overlap only to the extent necessary to give uniform film thickness.

Air-inhibited polyester lacquers, i.e. those containing wax, offer difficulties when sprayed to vertical surfaces too generously. The wax migrates to the surface and causes runners; it cannot penetrate horizontally and exclude air, thus function as it should by promoting setting.

Draught can have a similar effect.

Higher additions of catalyst can speed up setting and curing, though reducing pot life of polyester lacquers.

(b) The spraying shop is too warm and the varnish is a slow-drying one. The coating, e.g. a low-melting bitumen solution, usually solidifies after the evaporation of the solvents, but under the above condition remains soft.

The shop temperature should be about 20°C (70°F). Spraying trays or trolleys charged with freshly sprayed material should not be located too near to radiators, ovens, in direct sunlight, etc. If in excessively warm shops the objects are painted by brushing on quick-drying coatings, brush marks frequently remain visible.

(c) Excessive thinning with rapidly evaporating solvents. Such coatings may be touch-dry before it is possible for them to flow out smoothly, especially if the room temperature is too high, as pointed out in sub-section (b) and (e). Too strong an exhaust at the spraying booth may cause somewhat similar failures. In some cases fast solvents can prevent sagging, but on the other hand too slowly evaporating thinners can cause it.

Besides running, silking [2] can also arise, e.g. with some thermo-hardening synthetic coatings, if unsuitably thinned for spraying.

Use the recommended thinners only and with care.

(d) Insufficient spraying distance or not holding the gun at the right angle to the work piece may lead to runners, ripples and small grooves.

Try out the correct spraying distance and angle by experiment. In the case of small surfaces a distance of 20 to 25 cm (8 to 10 in) is satisfactory. Spraying nozzles must always be kept clean.

(e) Incomplete brushing out, especially of quick drying materials. Application of quick drying paints and lacquers should be confined to small areas only, otherwise the film sets and becomes tacky before it can be spread uniformly. In addition, brush marks will not flow out. A warm atmosphere accentuates the defects.

(f) Fresh paint applied over an old but still glossy paint may sag. Old paint should be well rubbed down to remove all gloss. Wet paper should be used to avoid dust which might contain lead. The surface must be allowed to dry before painting.

(g) Enamels and paints for brush application may sag or curtain if evaporation of the solvent component is slow or delayed, as in the case of poor ventilation conditions, or the incorporation of excessive amounts of high-boiling thinners.

Check material under good drying conditions, change the thinners to faster-evaporating type where practicable. Use only recommended thinners.

(h) Sagging and curtaining of emulsion paints and distempers, etc., can occur on new plaster, when applied in the early drying stages of the plaster. The presence of excess water in the plaster surface prevents the absorption of the water content of the coating, and evaporation is delayed due to the high humidity at the surface. The paint remains wet for an excessive period and may become even further diluted by the water contained in the plaster.

Delay painting until conditions are more suitable. Improve ventilation in order to speed the drying-out of the plaster.

Emulsion paints can also show sagging on vertical surfaces when containing excess amounts of certain stabilizers or thickening agents, e.g. polyacrylate solution.

Notes

[1] B.S. 2015 (1965), *Glossary of Paint Terms.*

[2] Silking (or streakiness) is due to orientation of pigments in the direction of drainage. *See also* Kresse, P. (1966), *JOCCA.*, **49**, 868.

3.4 Formation of drops on dipping of well-bodied coatings

Similar defects are possible on flowcoating or ladling.

Where the film is very thick, drying through is usually delayed.

At the lower ends and at projecting points of the surfaces the coating tends to collect and to dry superficially before it has dripped off (fatty edges).

(a) Excess paint, etc., must be made to flow down smoothly by hanging or jigging the dipped objects in an angle most conducive to drainage, or be taken off with a dry brush, electrostatic de-tearing, blowing of air or by some centrifugal device before drying commences.

(b) Stoving varnishes should be permitted to drip off sufficiently, see (a), before charging the material into the stove. Conveyor work can also be made to pass an electrostatic field of high voltage to remove the drops from a draining point of not more than 0.25 in in radius and an electrode distance of 8 in. This method is considered economical, and yields films of even thickness without a trace of drips [1]. This process is applicable to metal, wood and plastics [2].

(c) The consistency of the varnish, etc., should be reduced by suitable thinning or by preheating, e.g. to about 80°C (175°F), or to any other temperatures suited to the paint materials. The right consistency may vary with the seasons.

(d) If screw-holes, nuts, etc., cause runs coupled at times with the formation of bubbles, a careful addition of an antifoam solution (normally a silicone compound) may be an advantage.

Consultation with the lacquer manufacturers is essential.

Centrifuging creates even films on mass-produced, possibly intricately shaped objects, e.g. buttons, nails, screws, where jigging or racking would not be economical or tumbling too slow. Centrifuging is a fast process without waste. It is usually followed by stoving.

Rotational dipping (Rotadip), by which even hidden areas may be uniformly covered, can be considered for suitable bigger articles, e.g. priming of motor-car bodies.

Attention should be paid to the grooving tendencies exhibited by cellulose dipping lacquers.

Notes

[1] Miller, E.P. (1946), Electrostatic Detearing, *Ind. Fin. (Amer.),* June.
[2] Wampler, R.H. (1946), *Modern Organic Finishes,* p. 93.

3.5 Difficulties of application due to cold winter temperatures

The British Standard Code of Practice BS 5493 (1977), formerly CP 2008 recommends that no external painting should be done when:

(i) Atmospheric temperature falls below 4°C.
(ii) Relative humidity rises above 90%.
(iii) Inclement weather (rain, fog, snow, etc.) prevails.
(iv) The temperature is liable to fall below the dewpoint of the ambient air.

Extensive investigation of meteorological data by the Paint Research Association [1] shows that, based on the above criteria, the number of days in the year on which painting can be satisfactorily carried out in the U.K. is remarkably similar for locations as far apart as London, Manchester and Prestwick. It was found that conditions are worst in the early morning and that no job should be started before 10 a.m. Over a ten year period in London, acceptable conditions held for 12 hours on 29% of all days. It was also found that, on the whole, most days were either fairly good or fairly bad for satisfactory painting. As a result of this work a technique has been developed for assessing the expected painting efficiency in any month for specific sites. A contractor can call in the PRA to carry out a meteorological survey for any particular area provided, of course, that the size of the contract justifies the cost. Otherwise weather forecasts should be watched.

Before discussing difficulties of application in detail, attention is called to the fact that surprises are possible in well-finished films which have been subjected to low temperatures. Coatings containing wood oil and nitrocellulose lacquers if they have been exposed to frost too early after application, e.g. to coated leather, artificial leather and oil cloth, may crack. This applies also to stoved stamping, forming, and drawing enamels whilst films, of, e.g. linseed oil modified alkyds may suffer in damp cold by showing matt patches.

(a) If the paint materials are stored in cold paint shops or storerooms or in the open (at low outside temperatures), they tend to thicken considerably. Consequences: faulty drying, difficult application, especially brushing, formation of runners. Separation may also be caused which cannot be corrected.

The consistency of the paint material usually becomes satisfactory again as soon as it has returned to normal temperature.

Hot spraying of course, eliminates such difficulties.

(b) Similar difficulties are encountered if cold surfaces, for instance metals, are to be coated, e.g. with oil paints or cellulose lacquers. In the latter case blushing or flotation are a possibility.

(c) Formation of pinholes due to excessively cold mixtures, cold spraying air, etc.

(d) Unsuitable consistency: if for one purpose or another the paint material is heated or subsequently thinned, it must be borne in mind that the quantity of thinners employed must be adapted to the season, i.e. different quantities are required in summer and winter. Less solvent evaporates in winter owing to quicker cooling.

(e) The drying rates of paints are considerably reduced in winter due to the lower temperatures prevailing, the higher humidities, the lower light values and the reduced hours of daylight compared with conditions during summer months. The rate of evaporation of the solvent components of all types of paint is reduced by the lower temperatures and high humidity, as well as the fact that ventilation is also often poor, due to closed rooms, resulting in the atmosphere becoming saturated with solvent vapour. This may in severe cases lead to runners, curtaining, sagging, etc.

Air-drying paints which dry by oxidation and polymerization of a drying oil or oleoresinous or synthetic medium (e.g. alkyd) modified by drying or semi-drying oils, can have drying rates severely affected by winter conditions. The speed of oxidation and polymerization can be much reduced by low temperatures, e.g. 5–13°C (40–56°F) and by the poorer actinic value and shorter hours of the daylight. Drying times can be more than doubled under these conditions. Alkyd-based paints, particularly those modified with semi-drying oils, can be checked at a sensitive stage of the drying process and take a considerable time to recover. Under extremely severe conditions the tacky stage may be prolonged into days due to the absorption of water in the paint film when humidity was high and subsequent drying conditions were poor. The slower-drying long-oil alkyd enamels can be greatly affected by the cold, damp winter conditions due to their slower set-off period involving a sensitive stage of probably 1–4 hours. Although the quicker-drying enamels have a much shorter set-off time, they can still be affected in the same way if the cold, damp conditions coincide with the shorter sensitive period.

(f) Chemically curing coatings such as amine or polyamide cured epoxies will not cure satisfactorily at temperatures below 10°C. At low temperatures the curing process will be incomplete and tackiness, dullness, various other surface imperfections and lack of chemical resistance will result.

(g) Whenever possible, good ventilation and daylight should be given as painting proceeds and during the drying process. The application of heat, e.g. by the turning on of central heating, etc., in a large building such as a school, hall, etc., is not necessarily beneficial, as severe condensation may result when the temperature falls during the night, causing a check in the drying or a matting of the film in the case of gloss finishes. Painting should be carried out in the early part of the day, when the maximum drying time under the best conditions can be given. Where a paint film has remained tacky without hardening satisfactorily there is no other choice but to allow it to harden slowly (rub down lightly and re-coat under good drying conditions if and where the surface has become damaged during the prolonged drying process). The use of a quicker-drying paint is often recommended for winter conditions, or the manufacturer may supply an 'accelerator' in the form of a balanced drier solution to be added to the paint in fixed proportions in order to speed up the initial touch-dry stage.

It is possible that under certain conditions the films undergo an apparent drying process which does not preclude re-softening and stickiness of the films as soon as the temperature has risen again. The subsequent drying at normal temperatures may also be considerably delayed thereby.

(h) During winter and the cooler seasons rapid falls in temperature occur, giving rise to deposition of moisture or frost formation. Fog also accounts for damp surfaces, and it is therefore important to ensure that exterior surfaces are thoroughly dry before painting is commenced. The formation of emulsions, or the absorption of the surface water by the paint, may give rise to poor drying or subsequent loss of adhesion.

Do not commence painting in the early morning after fog, frost or heavy dew has been deposited on exterior surfaces. Allow surfaces to dry out thoroughly before applying paint or varnish, e.g. dry off with a chamois leather to remove surplus water and then allow to dry out.

Note

[1] McKelvie, A.N. (1970, 1971), *Analysis of External Painting Weather,* PRA. (RS/T/77/70, RS/T/28/71).

3.6 Brushmarks, bad brushability, wet edge time

The formation of brushmarks is to a large extent caused by the bristles of the brushes creating furrows in the paint film during application. Careful work has shown that this is not the only mechanism at work, but there is insufficient evidence for other factors [1].

Brushmarks are always formed in the first instance, whatever the type of coating, but their occurrence is only objectionable if they do not flow out before the coating dries. The force at work within the film which determines the ultimate degree of levelling is surface tension. It is opposed by viscosity which increases as solvent evaporates from the film. In other words the flow properties are influenced by the recoverable strain in the system which represents the proportion of elastic deformation in relation to the total deformation under the pressure of the brush [2].

Bishop [3] has determined the shear rates in brush application by means of a brush connected to a strain gauge. He reached the conclusion that normal brushing produces a shear rate of approximately 10^4 sec^{-1}. The shear rate is calculated by dividing the speed of application by the film thickness. At a shear rate of 10^4 sec^{-1} the viscosity of a decorative finish applied at 62–75 μm wet film thickness should be 2.0–2.5 P, and that of a protective finish applied at 87–100 μm wet, 3.0–3.5P. The viscosity measurements were carried out with the ICI cone and plate viscometer as described in BS 3900, Part A7. This type of instrument is also used by other workers in this field, e.g. Ehrlich *et al.* [4], who have pointed out that single point viscosity measurement are meaningless in predicting brushing and levelling properties of solvent based paints. An automated instrument for measuring brushmarks and flow in emulsion paints has been described by Lehmann and Salzwedel [5], who measured the profiles of paint films with an optical scanning method and an oscilloscope. Brushability is usually judged subjectively by the operator [6], but it can be measured with several instruments, e.g. the B.K. Brushability Recorder or the Electronic Brushability Meter developed in Australia [7]. This measures the force necessary to overcome brush drag by means of an electric resistance strain gauge.

Various additives have been recommended to improve the flow and levelling properties of paint. Some of these agents are claimed to improve the wetting of the substrate and to reduce the formation of pinholes and craters and minimize frothing tendencies [8]. The 'wet edge time' of a brush applied film can be defined as the maximum time that the film can be left and still permit satisfactory joining up with freshly applied paint. It has been shown [9] that this property depends on the wet film thickness and the character of the solvent system employed. This should have the best solvent power for the resin used as film former in the paint and be as slow as possible in its evaporation rate. Glycol ethers often confer a significant improvement on the solvent power. To illustrate the influence of film thickness: a wet film 50 μm thick will have a wet edge time of 7 minutes, whereas the same paint at 75 μm will stay open for 12 min.

Brushmarks may have been present in the preceding primer or undercoat, unnoticed due to their lower gloss. A glossy finishing coat will show them up.

(a) Due to low temperatures.

Add a little more thinner, if required, or warm paint, etc., slightly.

(b) Unsuitable brushes (coarse, short-haired or dirty) cause difficulties, especially in covering large areas and also if the paint or varnish is of the quickly drying type.

Soft brushes made of split bristles, camel or badger hair should be selected. Small, specially shaped brushes should be used for narrow and not easily accessible sections, corners, cornices etc.

(c) Irregular or too thick application. The quicker a coating dries, the more susceptible it is to flow marks. Emulsion paints are no exception to this rule. In the case of brushing cellulose lacquers these faults are frequently also caused by the use of bristle brushes. Transparent gold lacquers particularly should be applied in thin films only, otherwise gritty, patchy-coloured films are easily produced.

The art of brushing is not particularly difficult in the application of most types of paints and lacquers, but it becomes increasingly problematical with rapidly drying materials. Easing with a little 'slow' thinner often helps.

For applying transparent cellulose, lacquers, etc., on polished or finely rubbed-down surfaces it is necessary to use soft-hair brushes. Attention is called to the suitability of the dipping process especially by slow extraction for obtaining smooth, well flowing out and wetting cellulose lacquers films. Centrifuging of small parts with thin lacquers is also worth consideration for mass production.

Cotton wool pads have occasionally been used for rapidly drying coatings, e.g. thin cellulose and spirit lacquers for metal, while for castings and porous brass, bronze surfaces, etc., bristle brushes can be employed (see above, (b)). Much depends on the method of applying the brush, which should be dipped into the lacquer slowly and left there until all air bubbles have risen to the lacquer surface. All excess lacquer should then be wiped off and the brush conducted smoothly over the surface to be lacquered, but not too quickly. Small areas are better suited to this treatment than larger ones.

(d) Excessively thick consistency, e.g. due to long storage or to chilling of coatings resulting in changes in the colloid dispersion.

Highly pigmented primers are prone to leave brush marks.

(e) Brushing of quick-drying lacquers, etc., on to large areas frequently leads to bad wet-edge times. Medium and short oil length alkyd enamels belong here.

Quick working is absolutely necessary in applying lacquers of this type and over-brushing must be avoided once the paint has begun to become tacky, as this is bound to aggravate the marks.

Soft, flat brushes have to be used. Sometimes slowly evaporating thinners are added, but great care must be exercised if pool distillate is added to decorative paints, as, whilst improving wet-edge time, it delays drying and re-coatability, and may promote blooming, because of its content of higher fractions. Better results are obtained with the spraying process.

(f) Contamination of oil-base materials by water increases their viscosity by the agglomeration of pigment particles at the oil-water interface, upsetting the flow as indicated by brush marks. Water in oil emulsions are formed, accompanied by flocculation of the pigments. Rough and consequently flat surfaces may also be caused thereby.

It is necessary to protect paints, etc., against contact with water.

Notes

[1] Smith, N.D.P. *et al.* (1961), *JOCCA.*, **44**, 618.

[2] Khanna, R.K. (1969), *Paint Tech.*, **33**, No. 1, 23.
[3] Bishop, D.M. (1969), *Paint Tech.*, **33**, No. 4, 16.
[4] Ehrlich, A. *et al.* (1973), *J. Paint Tech.*, **45**, 58.
[5] Lehmann, H. and Salzwedel, M. (1972), *Fatipec Congress Book*, 575.
[6] Fink-Jensen, P.H. and Ronschou Nielsen, H.K. (1971), *J. Paint. Tech.*, **43**, 60.
[7] Jones, K.L. (1961), *JOCCA*, **44**, 637.
[8] Paton, C. (1969), *Paint Var. Prod.*, **59**, No. 7, 57.
[9] Cray Valley Products Ltd., *Resinotes*, Summer 1966.

3.7 Bad flow of varnishes, lacquers and powder coatings. Effect of surface irregularities

Orange peel of sprayed films, an appearance-resembling the skin of an orange, and similar effects are in most cases due to too rapid evaporation of the most volatile solvents and to failure of the film to flow out to a level surface. Rapid evaporation generates strong convection currents in the still liquid film. According to observations by Bartell and van Loo [1], these currents break up the film into cell units with raised walls and indented centres which cannot disappear if the film becomes prematurely too rigid to allow a second flow. These films consequently show a rough structure.

Wave formation and sags, however, are due to excess flow caused by excessive film thickness coupled with slow setting, particularly on vertical surfaces. Glossy finishes obviously show imperfections of film much more than do matt films.

These defects once formed, can where possible be corrected only by rubbing down with emery paper, etc., followed by over-spraying with a solvent coat or thin lacquer, or by strong polishing of the surface with a grinding and polishing paste. This applies particularly to cellulose lacquers. When re-surfacing faulty lacquer films by a wet method, the treated surfaces must be dry before the next coat is applied. Orange peel is more easily rubbed off than wave formation. Too fast drying paints have also too short a wet-edge time if they are brushed.

(a) The liquid films set too quickly before they have spread satisfactorily, because:

(i) Unsuitable thinners containing insufficient quantities of actual solvents or excessive amounts of highly volatile constituents have been added (especially in the case of cellulose lacquers). The lower boiling xylol may be a better solvent than the heavier naphthas for thermohardening urea or melamine alkyd media. Similar difficulties may also be met if cellulose lacquers contain too much ethyl alcohol in presence of insufficiently alcohol-soluble nitrocellulose.

It is advisable to use only the thinners specified by the manufacturers and to beware of adding too large quantities. The more pronounced the orange peel or the other faults mentioned, the more rubbing-down work with water-resistant emergy paper, mechanically or by hand, possibly eased by a thin solution of soft soap, is necessary until the surface is smooth again. Paper grades used are, for preliminary cutting down 180–220, for finishing 320–400. Polishing may also be required.

(ii) The varnish or lacquer is sprayed at too high a pressure through too small a nozzle, especially in the case of viscous material, so that no wet coat can be deposited. Too low a pressure may also cause bad flow.

(iii) The varnishing shop or the spraying air or the lacquer or even the article itself is not at the right temperature (see sub-section (d) below).

A temperature about 20°C (70°F) should be maintained; substantially higher or lower temperatures may lead to faulty results except for hot spraying.

(iv) Nitrocellulose brushing and other rapid-drying lacquers are applied with unsuitable brushes (mottled appearance if transparent and dyed).

Use soft-hair brushes.

(v) Sprayed articles remain too long after application near the spray booth, the air current of which may cause acceleration of the evaporation of the solvents, as may the draught as such if excessive. Larger articles, for instance, automobiles, which have to remain in or pass through the spray booths for a considerable time and may begin to dry before the spraying of the whole object is finished, are obviously much more prone to the danger of too long exposure to air currents or to forced draught caused by the exhaust equipment.

Remove the articles from the spray booth as quickly as possible after coating or use a lacquer which is adapted to the conditions by having a higher content of less volatile solvents.

A similar effect might be caused when spraying takes place on automatic conveyors which pass the spray booth too slowly.

(b) Excessively thick application. This fault is particularly pronounced in the case of the rapid-setting materials, such as some synthetic resin, rubber compound and spirit varnishes as well as cellulose lacquers.

Spray cross-wise at lower pressures and with smaller nozzles. One coat hot-sprayed lacquer equals about the film thickness of one and a half cold-sprayed coats. In addition, the flow is usually better, and the tendency to runners reduced.

(c) Excessively rapid heating or too fast charging into too hot an oven of some stoving coatings, or overloading of stoves.

Many, though not all, coatings can be adjusted for such extreme working conditions by the manufacturers.

(d) Excessively low temperature of spraying air.

The air can be warmed before entering the spray gun.

(e) The priming coat or filler is sandy or has not been sufficiently rubbed down. The top film then is not smooth and covered with specks showing small bright dots.

Rub down carefully with pumice stone or with water-resistant emery paper of medium fineness and re-lacquer.

(f) If the lacquered or varnished surfaces are not entirely plane (as, for instance, metal sheets), films of high gloss are frequently uneven. Some decorative lacquers, e.g. ripples, then exhibit brighter stripes at the raised edges.

Transport the articles carefully to keep them exactly straight and plane. Straighten or fill the panels for lacquering, if necessary.

(g) Dry-spraying of, e.g. nitrocellulose lacquer, causes the deposit of fine-grained spraying dust. Overspray from other booths can cause the same defect. Airless and electrostatic (including the 'hand gun') spraying create practically no overspray.

If the shape of the articles which are to be coated is complicated it is often difficult to spray all parts of the surface at the proper distance or angle, thus causing some overspray.

Avoid dry spraying as much as possible. Also see to it that no spraying dust precipitates on unfinished, etc., surfaces. The failure would show up similarly as described in above sub-section. Ensure perfect exhaust at all times.

(h) The flow of paints containing wax is often poor.

As far as possible apply with a stipple brush.

(i) Lining can occur with roller coating or coil coating varnishes. This defect can often be cured by the use of a slower solvent or reduction of film thickness. The fault can also be apparent when roller coated varnishes are applied wet-on-wet over certain inks. Once again the problem may be resolved by the correct choice of solvent, i.e., replacing aromatic hydrocarbons by alcohols.

A special case occurs with roller coating varnishes cured by ultra violet irradiation where curing is effected before the film has had an opportunity to flow out. The problem should be referred to the manufacturer.

(j) Powder coatings may exhibit pronounced orange peel in the cured film (Fig. 3.3). Possible causes are; poor storage stability of the powder if the batch has been in stock longer than six months; the entrapment of air bubbles below the film surface; the presence of too many fines in the powder giving rise to agglomerates. The last two faults can only be remedied by the manufacturer.

Note

[1] Bartell, F.E. and Van Loo, M. (1925), *Ind. Eng. Chem.*, **17**, 925.

3.8 Pinholing, pockmarking, cratering, cracked blisters [1]

Pinholing is the formation of minute holes in a paint film during application and drying, sometimes due to air or gas bubbles in the wet film which burst, forming small craters that fail to flow out before the film sets (Fig. 3.4).

Cratering describes the formation of small bowl-shaped depressions in a paint or varnish film.

Pockmarking is a defect in the shape of irregular depressions formed during the drying of the film. Pockmarking in sprayed films often resembles the skin of an orange and this defect is also known as 'orange peel' (see Section 3.9).

Pinholing occurs more often or at least becomes more visible, in unpigmented varnishes or in such paint or enamel films which contain a comparatively low percentage of pigments such as carbon black, Prussian blue, and especially when or where they dry as a thin layer, i.e. when applied by dipping, spraying or rolling. The differing explanations result from varying causes of pinholing.

In many cases the failure is due to the content of incompletely dissolved or minute thrown out gel particles in the vehicle, in others to insufficient wetting. Many theories have been put forward, but today it is considered that the most common cause of pinholing in all resin and varnish systems is the presence of high molecular weight fractions and it has been shown that even incompatible resin mixtures do not exhibit pinholing if they have a similar spread of molecular weight.

Epoxy resins are particularly prone to pinholing problems and this is probably due to high molecular weight fractions or, in the case of epoxy esters to the presence of gel particles. Modern epoxy resins manufactured by the 'advancement process' are less susceptible to pinholing.

Now for the conditions under which the failures are likely to occur:

(a) Excessive after-thinning frequently causes some of the properly dissolved varnish body, etc., to throw out again, but for epoxy resin varnishes, etc., some of which are prone to pinholing, a rather high dilution with the right thinners may be necessary. This also applies to water-thinnable stoving paints intended for spraying.

Thin carefully with suitable thinners. Manufacturers may in some cases of high polymer resin based lacquers suggest the addition of a flow agent as a remedy.

(b) The addition of normal quantities of thinners, if unsuitable for the medium, also leads to pinholes, particularly on stoving. For instance, if stearin pitch varnishes are thinned with kerosene, the pitch constituents, under unfavourable conditions, may separate completely. The pitting of such films by stoving is accelerated, especially if very low- and very high-boiling solvents, e.g. tetralin, are present simultaneously. However, an excess of low-boiling thinners can also lead to the formation of pores and other troubles in oil- and pitch-type vehicles, because there is less chance of a second flow taking place.

The addition of medium-boiling solvents, of outstanding solvent power, sometimes causes elimination of the failure.

The solvent mixtures contained ought to have a balanced rate of evaporation, especially for stovings. Warming may be helpful.

(c) Insufficient mixing of the thinners with comparatively viscous pigmented or unpigmented media induces the formation of volatile occlusions in the latter. The thicker the paint, etc., on thinning, the more likely is the appearance of the defect. Quick evaporation of occluded solvents, e.g. by stoving, leaves only the respective holes in the film. This failure arises especially during the cold season when coatings are thicker and therefore less miscible. (See Fig. 3.5). It happens even with fairly heavily pigmented enamels.

Remnants of cleaning spirits in the spray gun which has been filled with paint can cause pinholes, especially at the beginning of the spraying, if the first sprays are used for application.

(d) Under conditions where dry spray is bound to happen, e.g. if a highly viscous coating is sprayed through a relatively small nozzle, and if the room temperature is low or the substrate is cold, (especially if the paint is quick drying), pinholes can appear.

Overspray of one paint upon another film is also considered as a possible source of the defect [2].

Further thinning, increase of air pressure or of the size of the nozzle, or even warming of the objects to be sprayed, or of the coating itself, are remedies. The room temperature has to be watched.

(e) Mixing of different or even identical varnishes or paints of high viscosity, perhaps due only to low temperature, can have the same detrimental effect as the admixture of cold thinners, etc.

Warming of the paint, etc., to be treated, as near as possible to the boiling point of its solvents or thinners, and intense mixing of the two components, has to take place. Be careful to avoid fire. Oil-base varnishes, etc., with a flash point of above 32°C (90°F) can be heated up to 80°C (175°F) since the solvents contained will boil at temperatures above approximately 148°C (300°F). The varnish, etc., can also be heated in a clean pot by means of a pre-heated water-bath or an electric heater by which open fires are avoided. Solvents and thinners which are to be used for cold cut processes should be stored so that they are at room temperature when they are needed. Never add hot thinners to cold or thick varnishes, etc.

(f) The cold season is treacherous if paints or varnishes have to be stored or to be transported over a comparatively long period. If they have to be mixed, or to be thinned, miscibility and solubility are reduced and the viscosity is increased. The temperature of work pieces has also to be watched.

Separation phenomena are to be expected.

Do not store paint materials or thinners in the yard. The ideal storing temperature is about 20°C (70°F). If necessary, the materials or objects should be slightly warmed before they are used, as already stated in sub-section (e) above. Use the oldest stock first!

(g) If the application is by spraying, cold air can cause failures similar to those described above.

An air compressor should, because of the cold seasons, not suck the air directly from the open. The air chamber and the pipe lines, at least, should be located in heated rooms, if possible, near heating pipes or radiators. Alternatively, cold compressed air should be pre-heated in some suitable manner. The air must always be stripped of oil and moisture by passing through filters (see below (i)).

The standard temperature of the spraying air should be maintained at about 20°C (70°F).

(h) The mixing of varnishes or lacquers which are incompatible with one another can cause separations, etc. [3].

In such cases no large-scale applications should be carried out without preceding experiments. Paints should be procured ready mixed and of the desired shade and features from the manufacturer or they should be selected for easy miscibility.

(i) Heavy mineral oil, wax, silicon compounds or water, which contaminate the paint or varnish, or are present as impurities or remainders of cutting oils or mould release agents on the surface to be coated, may lead to pinholes, or even (as in the case of water resulting from the spraying plant) to blisters (see Fig. 3.6). If an undercoat or an old paint film which contained a silicon compound is repainted, pinholing or cissing frequently occur. Similar defects occur with wax-containing polyester lacquers. If stovings, such as stamping enamels, contain incompletely dissolved wax particles the same defect is liable to appear.

Oil and water extractors of spraying plants should be drained regularly and their fillings (coke, glass wool, etc.) renewed in the required intervals. Do not allow condensed water to form in open cans, vats, or in pipe lines. Dipping tanks must be cleaned out at regular intervals, as silicon compounds can accumulate around the sides, causing complaints which would not arise with the normal low contents in dipping paints. Keep the containers of paints well closed. Water is heavier than varnish and is bound to settle slowly at the bottom of the containers. To prevent pinholing in coatings containing wax the addition of a solution of ammonium linoleate has been proposed.

(j) Films contaminated by dirt, fibrous or gelled particles and pieces of skin may also show a pinhole-like cissing effect, as is well known. This is partly due to air occlusions on such foreign bodies [4].

Leave the material to settle quietly if sufficient time is available. Otherwise filter through fine-meshed muslin, if necessary, after warming the varnish, etc., to a sufficiently liquid state. Cork stoppers should always be surrounded by suitable pieces of goatskin, oil-paper, etc.

Painting shops must be kept as free from dust as possible. Great care must be taken if rags are used for cleaning.

(k) Where pinholing was solely due to great surface tension of the wet film towards the substrate the nucleus of the pinhole consists of a remainder of the film-forming coating. This form of the failure is not dependent on stoving.

The addition of small quantities of wetting agents and/or metal driers, such as solutions of zinc naphthenate, may be helpful.

(l) Paint films subjected to a sudden rise in temperature on stoving, etc., particularly if containing low-boiling solvents, can show pinholes.

Bubbles are created by the boiling of too volatile solvents during the fusion process within some organosol films.

Water-thinnable coatings tend to bubbling if stoved without sufficient flash-off time.

Where possible, coatings should be placed in a not-too-hot oven, and the temperature raised gradually. If speed of production requires it, manufacturers of coatings should be approached with a view to having the low-boiling solvents replaced by higher-boiling types. Alternatively, a different medium may have to be chosen.

Water-based coatings may need about half an hour at room temperature, or less at gradually increasing temperatures.

(m) If oven temperatures have been raised too high, and possibly also too rapidly, or if coated goods have been loaded into too-hot ovens, craters and bubbles can be created.

Thus boiling of solvents in primer-adhesives for plastisols, applied on to metal strip in continuous roller- or knife-coating can be promoted. Solvents should be driven off, however, before the plastisol is applied.

If plastisols of the p.v.c. type are overcured in the fusion process they begin to decompose, show brownish discoloration and form bubbles. This defect would be enhanced if insufficient of the right stabilizers had been present in the formulation to neutralize the released hydrochloric acid.

Large plastisol mouldings need to be manufactured under vacuum for deaeration before being put into the mould, to avoid cratering.

(n) Failures suspected to be pinholes are often cracked (blowholes) or uncracked blisters, or are not at all connected with the paint, etc., itself.

Air bubbles or blisters can arise in rather viscous, even sparsely pigmented coatings, due to aeration on intense stirring, e.g. when dipping too fast, on flow coating as a consequence of the 'jetting' action, or in the case of roller application by their too quick rotation, not necessarily connected with too thick a consistency. Water-thinnable coatings easily form air bubbles. Plastisols and silk-screen pastes frequently tend to form blow-holes.

Prolonged standing often eliminates the trouble. Where applicable the viscosity should be corrected by addition of suitable thinners to improve flowing out, or by slight warming.

Anti-bubbling silicone oil solutions in very small doses are also in use. For flow coating the functioning of the vapour tunnel must be satisfactory.

Slow extraction dipping creates no air bubbles. Evacuation of all air is necessary before applying plastisols in particular by rotational moulding, but also by dipping, slush moulding, calendering, extrusion- or knife-coating.

(o) Many of the lacquers containing such components as urea or melamine formaldehyde resin, shellac, non-oil modified phenol or maleic resins, stove rapidly. Their films, and those of silk-screen pastes, if heavy, very often show a special tendency to form bubbles and blowholes. Extreme cases have been observed where the formation of countless tiny bubbles, etc., have caused the impression of mattness of a glossy film. The defects may hardly be visible to the naked eye.

Opportunity to escape smoothly must be given to the solvent vapours and products of oxidation and condensation of such quick-setting stovings before their films become surface dry or rigid, thus allowing a second flow to take place [3]. It is therefore important that the consistency of such coatings is adjusted, and the application is made so as to avoid too heavy film layers being formed. It is preferable to raise the temperature gradually. In some cases the addition of some anti-foaming agent is worth trying.

(p) Films of polyester lacquers, particularly if pigmented and of the non-air inhibited, i.e. wax-free, type, are at times blemished by blowholes, because they had been sprayed with too high air pressure, or from too close a distance.

Unsuitable wood stains can also produce in these lacquers blowholes and cratering and cause bad adhesion or prevent drying.

(q) If lacquered alloys of low melting point, e.g. containing tin, or if objects containing soldered seams are subjected to comparatively high temperatures, the metal may melt beneath the film and cause the formation of a blister-like effect. Infra-red (radiant) heating may encourage this failure, eddy current heating may be suitable.

In cases like this it is necessary to apply coatings drying at low temperatures, usually not higher than 80–100°C (175–212°F). Radiant heat used should be at comparable temperatures.

(r) Attention should be called to the fact that pinholing may also arise in insufficiently matured air drying or stoving oil base varnishes if they contain excessive amounts of driers. On the other hand, pinholing of alkyd resin type stovings which may not contain driers is said to be checked by the addition of driers in modest quantities.

If, as in the case of excessive content of driers, microscopic fine particles cause the pinholing, a remedy is the addition of a suitable solvent which will hold these particles in solution until the film has set.

Additions of small amounts of calcium naphthenate solutions have also been suggested to counter throw-outs of lead driers, particularly in alkyd media.

This failure ought to be avoided in the manufacturing of the varnish by careful filtration, and also by avoiding excessive additions of lead compounds. Turbidity of clear media may, of course, be encountered simultaneously.

(s) Andeś and Stock described as a possible cause of pinholing the application of an oil varnish over another coat of oil varnish which was insufficiently dry or still 'sweating'. This may well be the consequence of bad wetting.

(t) In the case of application of varnishes, etc., by a roller coating machine, pinhole-like impressions or streaky patterns can spoil the smoothness of a film when the surface of the roller is uneven or contains pores.

Apart from the use of smooth rollers, a good machine should be equipped with a good working distributor roller with lateral motion (vibrator, inker) in order to equalize the film.

(u) Overspray has been found to be responsible for cratering and pinholing if differently composed coatings are sprayed on the same line, and extraneous spraying dust has settled on an uncured film [6]. Water-thinnable stoving lacquers and emulsions can be susceptible to overspray from non-aqueous coatings.

Clean working, good ventilation, correct spraying pressure, distances, viscosity, solids, also prevailing humidity and temperature or a combination of all these circumstances need to be watched. Air conditioning and water curtain paint arrestors should be of great help.

(v) Powder coating films may show pinholing due to contamination with small amounts of chemically different powder types; polyester or acrylic powders will affect epoxy powder coatings. There is no simple way to overcome this defect and the problem should be referred to the manufacturer.

Notes

[1] Parts of this section from M. Hess, (1943), *Pinholing and Cissing of Coatings,* paper given before a meeting of the Birmingham Paint, Varnish and Lacquer Club, *Oil Col. Tr. J.,* **103**, 664.
[2] Harvey, A.A.B. (1953), *Ind. Fin.,* **4**, 625.
[3] Bartell, F.E. and van Loo, M. (1925), *Ind. and Eng. Chemistry,* **17**, 925, 1051.
[4] Schieber, J. (1952), *Farbe Lack,* **5**, 221.
[5] ICI Ltd., *Effect of Driers on the Properties of Alkyd Resin Finishes,* Dyestuffs Division.
[6] Murdock, J.D. and Segall, G.H. (1961), *Off. Dig.,* **33**, 713. Observation on various alkyd type coatings over a thermosetting acrylic copolymer substituted with styrene. By using the more compatible vinyl toluene instead, and altering quantity and type of acrylates, the defect did not occur.

3.9 Cissing and beading

Cissing is the partial creeping back or contraction of the wet film into 'beads', craters (usually with raised rims), islands or pinholes whereby the substrate becomes visible (Fig. 3.7). It is in a broad sense due to bad wetting, i.e. if the liquid coating forms a high angle of contact to the surface. As a consequence the coating tends to recede into drops instead of forming a homogeneous film. Frequently it is an aggravated form of pinholing [1]. Unpigmented varnishes show cissing more frequently than pigmented enamels or paints, and the latter mainly if they have a comparatively low pigment content. The defect is more frequent in thinner than in thicker films such as are obtained by brushing. Greasy or damp surfaces promote bad wetting. But coatings of the same qualitative but not quantitative composition can behave quite differently towards grease, because the composition of the medium or its diluents or solvents may dissolve the grease according to their solvent power.

As foreign contaminating matter may be responsible for the defect and a surface be re-contaminated within a short time, intervals between cleaning and painting should be kept at a minimum.

The surface on which cissing takes place may consist of either a metal, plastic, ceramic or other smooth material or of a varnish or paint film.

Where retrogression causes a partial increase of the film thickness, further unpleasant consequences such as tackiness are to be expected.

Crawling is a term used to describe an extreme case of cissing where a coating recedes from large areas of a surface.

(a) The application of a gloss finish over an old hard glossy surface, as in re-decoration, can lead to cissing due to the existence of high inter-facial tension between the new paint and the surface. If the old paint is not removed, it must be rubbed down with abrasive paper to a matt finish in order to provide a key for the new paint. Even when rubbing down has been carried out it is good practice to apply an undercoat before the gloss finish since undercoats key better to old surfaces.

(b) Surfaces may carry films of dirt, grease, sweat residues, moisture or possibly incompletely hardened or 'sweating' varnish, giving high interfacial tension which may lead to pitting and cissing. Some stoving finishes are very intolerant toward contamination. Great care is necessary to remove grease from porous castings. Water-based paints are particularly sensitive to grease.

Silicone compounds, often present in polishes or as slip-promoting agents in paints, are particularly active in promoting cissing and are difficult to remove by normal methods. When these are present the existing paint film should be stripped or burned off.

Polyester and other mouldings are often contaminated with mould-release agents which can also cause cissing. Often these are water-soluble and hence not removed by normal degreasing.

Coatings have been developed which are not very sensitive to grease (important for special purposes), but it is advisable in all cases to use a satisfactory degreaser. Wiping the surface of the material with rag soaked in petrol or some other solvent, or rubbing with clean rags only, is not entirely reliable. Degreasing plants show excellent results and are in general use. Heating of metal, e.g. by radiation, to above 170°C (338°F), followed by wiping, can be effective. Attention is also called to the fact that a dirty base may reduce the life of films. The solvent methods are not reliable for the removal of perspiration and water-soluble residues.

(c) Damp or cold base, or cold humid atmosphere in the varnishing shop or outdoors (fog). Spirit varnish is especially sensitive towards moist deposits on the surface to be varnished.

(d) Due to oven conditions: the external dampers may have been closed too far, possibly in order to save fuel. The vapours, which are mainly solvents, thus concentrated within the stove may locally remove part of the moist film (solvent wash). Too close packing can have the same consequences.

This fault is frequently encountered and is occasionally accompanied by a considerable prolongation of the drying period. In case of doubt with regard to the correct degree of opening of the dampers it is preferable to err on the high side.

(e) Overcharging the stoves: consequences like those described in the previous sub-section ensue.

Economy in the wrong place! Even the strictly financial side of the matter is miscalculated, as cooling down of the stoved charges requires a correspondingly longer time. The prolongation of the drying period may also induce embrittling and excessive matting of the finishes. The density of the packing of the charge for each and every purpose depends on the character of the parts in question and on the state of dryness previous to stoving. Usually the danger of defects is passed as soon as the film has become tacky, as a large proportion of the solvents has evaporated by that time.

(f) Abrupt heating: unsuitable solvents for this drying method are driven off too rapidly, and the disturbed surface cannot return to its original smooth state due to advanced drying before the second flow can take place.

But if a finish cisses on normal heat treatment it can safely be assumed that its wetting properties are insufficient. There are cases where coatings do not show any cissing while air drying, because their slow solvents possess good wetting properties. This hides the bad wetting of its non-volatile components which have been given a chance to build up a film structure before the solvents have completely vaporized.

Raise the stoving temperature more gradually. In continuous ovens reduce the speed of the conveyor. If these manipulations are not possible, another type of coating should be selected for the job. In the case of alkyd stovings small additions of butyl alcohol and/or dipentene may be useful.

(g) When cissing is due to high surface tension inherent in the coating, specific additives are frequently used e.g. cellulose acetobutyrate (actually a flow-promoting agent) for polyurethane lacquers, and anionic or non-ionic surfactants for water-thinnable (especially emulsion) coatings. Water-based coatings are very sensitive to surface contamination especially on metals where the cleaning must be very thorough.

Note

[1] OCCA. (1976), *Introduction to Paint Technology.* 4th Edn, p. 154.

4

Faults related to drying and curing

4.1 Excessive drying periods – air drying

The majority of liquid paints (a notable exception being the solvent-free epoxies) contain volatile solvent which acts solely as an aid to application. When a paint film has been applied to a surface the following sequence of changes takes place.

Loss of volatile solvent

If the material is a non-convertible type such as chlorinated rubber or cellulose lacquer, loss of solvent leaves the film in its final form and no further change takes place. The film remains soluble in the solvents used in the liquid paint. The rate at which the solvent leaves the film depends on a number of factors: the vapour pressure or volatility of the solvent (e.g. xylol evaporates much more rapidly than white spirit); the relative humidity of the air (low humidity favours evaporation); temperature (low temperatures delay evaporation); the movement of air to remove the solvent vapour and the solvent-releasing properties of the film [1]. The majority of non-convertible coatings contain very volatile solvents and the problem of slow drying is not often encountered.

Vinyl [2] and acrylic [3] emulsion paints also qualify for the description 'non-convertible' coatings. Loss of water from the wet film leaves a matrix of polymer particles or globules together with pigment and additives. Provided the temperature of the film is above the Glass Transition Temperature (T_g) the polymer particles will coalesce and incorporate the pigment into a film. Film formation is purely a physical process, no chemical reactions being involved. If the temperature of the film is below the T_g of the polymer then a discontinuous film is formed. With copolymer types of emulsion the T_g of the polymer can be lowered by the use of coalescing solvents. When emulsion paints are applied to a porous substrate absorption takes place and this can appear to reduce the drying time. Such substrates require a specially formulated sealer, otherwise the absorption will result in an underbound film.

Loss of volatile solvent followed by chemical reaction in the film

Materials of this class are known as 'convertible' coatings and can be divided into two types: (1) coatings which dry or harden by absorption of atmospheric oxygen, the process being catalysed by certain metallic compounds (notably those of cobalt, manganese, lead and zirconium); and (2) coatings which cure by stoving or harden by chemical reaction in which atmospheric oxygen is not involved. Among the latter are the two-pack systems and include epoxies, polyurethanes, acid-cured alkyd-amino and some polyester systems.

Type (1) includes oleoresinous media, alkyds (including the modified types), epoxy esters and urethane oils. Drying commences as soon as most of the solvent has left the surface and involves several stages:

(i) Film set.
(ii) The 'tacky' condition.
(iii) The 'surface dry' or 'touch dry' condition.
(iv) The 'through dry' or 'hard dry' condition.
(v) The state of fitness for overcoating.

The binders mentioned all contain esters of unsaturated fatty acids and the initial reaction is absorption of oxygen at points adjacent to the unsaturated linkages, a reaction catalysed principally by cobalt and to some extent by the other drier metals [4]. This absorption of oxygen takes place through the film surface and the role of cobalt is described as 'surface drying'. Hydroperoxides are formed at the unsaturated centres and decompose to give 'free radicals' which, in turn, induce polymerization of the mass of the binder. This polymerization is catalysed by lead, manganese and zirconium which are described as 'through' driers and the reaction 'through drying'. It is important to maintain a correct balance between the quantities of 'surface' and 'through' driers used. If these are out of balance, drying of the film will not take place uniformly.

The drier metals are usually added as solutions of their salts with organic acids, the commonest being naphthenate and ethyl hexoate. The drying process does not appear to be influenced by the nature of the acid radical.

Since the drying process involves fatty acids it is influenced by the amount of fatty acid present in the binder (often referred to as the 'oil length').

The 'oil length' of a varnish is the ratio of oil to resin by weight. Varnishes are described as 'long oil', 'medium oil' or 'short oil' and these correspond roughly to the ratios of 3/1 and upwards, 2/1–3/1 and 1/3–2/1 respectively. The properties are controlled to a large extent by the oil length, the long oil types being slow drying and durable while the short oil types are rapid drying and not suited to exterior exposure. The term 'oil length' is also applied to alkyds and refers to the percentage of fatty acid present. There are long oil, medium oil and short oil alkyds whose properties also vary with the fatty acid content.

In the U.S.A. oil length is expressed as gallons (U.S.) of oil per 100 pounds of resin. Thus a '30-gallon oil length' varnish would contain oil and resin in the ratio of 30 gallons to 100 lb.

The short oil types dry rapidly (often with cobalt alone) and this is aided by the fact that they usually contain very volatile solvents such as xylol. The long oil types are generally thinned with white spirit and dry more slowly. Both cobalt and a through drier are usually required.

Temperature plays an important part in the drying reactions. The time required to reach the 'hard dry' stage can vary considerably between summer and winter. It is unwise to apply these paints at temperatures of 4°C or below.

Humidity also affects the drying rate, the higher the relative humidity the longer the drying time. At very high relative humidities (above 90%) painting should be suspended in view of the risk of condensation.

Type (2). Epoxy resins are cured by amines, amine adducts or polyamides. The latter are the most convenient to handle but the curing reaction is slower than with amines or adducts; the cured film is also slightly inferior in respect of chemical and solvent resistance. Although the time required to reach the hard dry stage varies from 12 to 24 hours, in each case the film requires about seven days to reach the fully cured condition. The rate of cure at each stage is increased by rise in temperature and 'forced' drying is often an advantage. Temperature also influences the 'pot life' of the mixed components; this can vary from eight hours under warm conditions to sixteen or twenty hours in winter.

Polyurethanes are produced by reaction between polyols and/or polyesters with aromatic or aliphatic isocyanates. The type of polyol or polyester can be selected to give films of any desired degree of elasticity or hardness. Aromatic isocyanates give films with a pronounced tendency to yellowing on exposure but the tendency is considerably less with the aliphatic compounds. Water reacts with isocyanates and the presence of moisture during curing can retard the process. Use is made of the reaction with water in the so-called 'moisture curing' polyurethanes. A polyol which reacts with isocyanates more slowly than do the normal types is mixed with an isocyanate solution. The mixture is stable until a film is exposed to the air when atmospheric moisture causes reaction and curing to take place.

Polyester resins are produced by reaction between certain polyesters and styrene. The reaction depends on the presence of 'free radicals' which are produced by decomposition of a peroxide in the presence of cobalt. Solutions of peroxide ('initiator') and cobalt ('catalyst') must not be mixed under any circumstances; they are added separately to the polyester-styrene mixture. The curing of polyester films in air can be inhibited by oxygen. One method of overcoming this difficulty consists in dissolving wax in the resin so that it forms a skin on the surface of the film. When the film is dry the wax is removed by polishing.

The processes of drying have been described in detail because of the fact that the statements made with regard to permissible drying periods and temperatures as well as to the hardening-through period, are frequently very inaccurate, giving rise to numerous complaints which could easily have been eliminated if the drying processes were better understood. It is clear that the requirements for the final conditions of drying of a bridge paint differ very much from those of a much-handled tray enamel.

Causes of slow air drying

(a) A comparison is possibly being made between paints of different drying characteristics. Consumers used to the application of, e.g. nitrocellulose lacquers frequently complain if they have to work even with rapidly drying oil or synthetic resin varnishes, but as the latter cannot always be replaced by nitrocellulose lacquers their characteristic properties must be accepted. It is absolutely essential when ordering a coating to submit detailed information concerning the type required and the drying period possible, in order to avoid misunderstandings between the makers and consumers.

(b) The film has been applied too thickly – one of the most frequent causes of trouble. Many consumers of paint materials are not aware of the pronounced influence exerted by the thickness of films on the drying process particularly of oleoresinous, alkyd and other films which dry by air oxidation. If a certain maximum film thickness is exceeded then oxygen will diffuse to the lower layers only very slowly. The surface layers will dry but through drying will be extremely slow and, with long oil types, wrinkling is likely to occur. On vertical surfaces excessive film thickness becomes evident through sagging, runs, etc.

The drying difficulties with thick oxidizing films set a limit to the thickenss of film which can be applied with 'high build' oleoresinous paints. For example, a high build oleoresinous paint pigmented with micaceous iron oxide can be applied by airless spray to give a dry film thickness of 75 to 100 μm and it is unwise to exceed this figure. Films which cure by chemical reaction not involving oxygen e.g. epoxies cured by amines or polyamides, can be applied at appreciably greater thicknesses. High build chlorinated rubber and vinyl paints are non-convertible coatings but require considerably longer for complete loss of solvent than do their conventional counterparts. In normal decorative work it is a mistake to fill cracks, small holes, etc. by applying extra thick paint films. The cracks and holes should be cut out where possible, primed and then stopped up with a filler followed, when dry, by rubbing down and sealing.

(c) The viscosity of the paint is too high due to storage under cold conditions or as supplied. Gentle warming (caution!) may remedy the first but any addition of thinners should be made in accordance with the manufacturer's recommendations.

(d) The ambient temperature is too low. The setting of most coatings which air dry by cross-linking is the more delayed, the colder the ambient temperature. This applies, e.g. to cold-setting amino resin/alkyd, amine-curing epoxy and to polyester resin based materials. The curing of the latter comes practically to a standstill below 20°C (68°F). Polyurethane-based lacquers set, however, even at low temperatures, though the reaction is slowed down. The drying of non-convertible coatings is also slowed down, but to a far lesser degree. Chlorinated rubber paints, for example, will dry at temperatures below 0°C [5].

(e) Too much high boiling thinner has been added.

(f) The previous coat was not completely dry. With conventional decorative paints an interval of 16 hours (which is, in practice, overnight) is necessary before overcoating. For other types of paints the overcoating times are usually stated in manufacturer's technical information sheets.

(g) Indifferent cleaning of the surface. Traces of wax (from polishes or paint remover) will prolong the drying time very considerably. The effect may be apparent in isolated areas only and drying in patches is a certain indication of wax deposits. To ensure the complete removal of wax the surface should be swabbed thoroughly with white spirit. A blow lamp can be used on wood. Any paint remover used should be of the waxless type and preferably water-rinsable. Special attention should be paid to areas which have been fingered e.g. round the handles of doors.

Oil or grease on metal surfaces can slow down the drying rate of paints applied over them as well as softening the film and impairing the adhesion. On wood the use of certain types of wood stain can slow up the hardening of both conventional paints and polyesters.

(h) Rain, fog and frost are detrimental to a newly painted surface in many respects, bad drying being one of the worst features. It is particularly important that the paint be applied to a perfectly moisture-free base especially after previous rubbing down. A high degree of humidity in the air can delay the process of drying of oxidizable films such as alkyds and oleoresinous paints.

Sufficient fresh air is essential for satisfactory drying of interior paintwork. Rooms are often closed to ensure freedom from dust, and drying may suffer from this cause. Again, surfaces closely packed together after painting (as often happens when numerous portable cabinets are stacked together for drying) may exhibit slow drying on the sides not open to free air circulation.

Painting must be effected on perfectly dry bases only, and exterior painting gives satisfactory results only in fair weather. It is necessary after continuous rain or wet fog, for instance, to wait until the surfaces to be covered or the priming coat are perfectly dry. Wiping the surface with rags is not enough for this purpose, and sufficient time must be given to ensure complete dryness. Moisture within the pores cannot be removed mechanically and would prove very harmful later on if not entirely removed by evaporation.

Paints are made containing water-displacing additive (e.g. ammonium linoleate, stearine amine, etc.) which are reasonably satisfactory when applied to moist surfaces, but a perfectly dry surface is still preferable. Ensuring of a warm dry surface for painting is one of the advantages of flame cleaning in the case of steel.

Attention is called to the fact that a relative humidity of about 50% is normally required for perfect drying and durability of the film. The oxygen necessary for indoor painting can be obtained by thorough aeration and ventilation of rooms, workshops and drying chambers, although care must be taken to exclude dust, etc.

Big objects, i.e. motor cars, are dried artificially in special plants. While for oil-base applications a certain minimum degree of humidity is required as mentioned above, drying of cellulose, shellac, vinylpolymer, etc., lacquer films is best carried out in as dry an atmosphere as possible in order to ensure satisfactory flow and films free of blush.

(i) Long storage of paints containing certain pigments can result in very prolonged drying times. The most troublesome pigment is carbon black, but the defect can also occur with titanium dioxide, red oxide of iron and, to a lesser degree, with blanc fixe and china clay. Alkyd media are particularly sensitive to the defect which is the result of adsorption of drier metals by the pigment. The adsorbed metals appear to be loosely bound to the pigment surface and, by re-milling the paint, 70 to 80% of the drying properties are often recovered. Proprietary materials (e.g. Nuact [6]) are available for checking drier adsorption. These materials are added at the dispersion stage and presumably satisfy the adsorption requirements of the pigment. If facilities for re-milling are not available, additional driers may be added, preferably after consultation with the manufacturer.

(j) Poor drying may occur with u.v. cured coatings, especially pigmented finishes where the pigment, e.g. titanium dioxide, absorbs in the same wave band as the emission source. This fault can only be resolved by the manufacturers use of a different photo-initiator [7]. Incomplete reaction of the monomer can also result in slow drying.

Notes

[1] Hansen, C.M. (1965), *Off. Dig.,* **37**, 57.
idem (1968) *JOCCA,* **51**, 27.
[2] Cogan, H.D. and Mantz, A.L. (1968), in *Treatise on Coatings,* (R.R. Myers and Long, J.S., eds), Vol. 1, Part 2, p. 381, Edward Arnold, London.
[3] Allyn, G., *ibid,* Vol. 1, Part 1, p.1.
[4] Love, D.J., (1977), *JOCCA*, **60**, 214.
[5] ICI (Mond Division) Leaflet *'Alloprene' Chlorinated Rubber.*
[6] Durham Raw Materials Ltd.
[7] Parrish, M.A. (1977), *JOCCA*. **60**, 474.

4.2 Excessive stoving or heat conversion times

The chemical and physical reactions which take place on air drying or cure of a film have been explained in the introduction to the previous section, of which this section is a continuation. The very important polymerization processes improve the mechanical resistance as well as the water and weather resistance of the films. A certain stage of polymerization of the vehicle may already have been effected during the making by the cooking process and materials of this type must be regarded as particularly valuable. Of course, the slight stoving treatment employed to accelerate the drying processes based on evaporation of e.g., chlorinated rubber, vinyl polymers, acrylates and cellulose lacquers, and spirit varnishes, etc. (force drying) does not affect the molecular structure.

Thermohardening coatings (without catalysts) have a critical minimum temperature below which the curing is very slow. Generally the rate of reaction is doubled with every 10°C (approx. 18°F) rise in temperature. In the case of thermohardening epoxy resin systems, whether solvent based or in the form of powder coatings, there is a minimum temperature below which the curing of the coating is inhibited. No amount of overstoving will complete the cure of these inhibited films.

Epoxy based coatings will not suffer adversely from overstoving but it is essential that they are not understoved. Frequently such films are subjected to a 'solvent test' as a check for complete curing. This is, however, not infallible, as some cross-linking media are less solvent resistant than others. If a solvent resistance test is employed as a check for efficiency of film curing then it is advisable to compare the performance with a cured standard film of a previous acceptable batch. The manufacturer should be consulted about the value of the solvent test as a means of assessing optimum film properties.

When stoving in the usual ovens (the so-called convection method) the temperature, as shown by suitably fitted thermometers, actually indicates the temperature of the hot air surrounding the coated article.

Whether this or the paint film itself will really attain this temperature depends on several factors, such as stoving time, heat conductivity, absorption of heat, etc., and in the case of infra-red radiation on a number of additional factors.

'In a batch type oven the major heat transfer to articles near to the walls has been by radiation. Articles shielded from the walls are heated by convection and by radiation from work nearer the walls as this becomes heated more rapidly. If a forced convection conveyor oven can be used in such a way that articles passing through it are not shielded from direct radiation from the walls and the conveyor system accelerated, it has become an infra-red oven with the added advantage of convected heat transfer.' (L.W. Anderson and E.A.C. Chamberlain [1]). Thus the old-fashioned convector oven also effects drying partly by radiant heat.

Forced movement of the air leads to greatly increased heat transfer, but may promote surface hardening. In a forced convector oven rates of heating vary with the air velocity. All objects should be swept by air of reasonably uniform velocity to ensure uniform heating. Modern convection ovens are most effectively heated by hot air circulation.

The stoving of coatings does not only help in the flow of production and results in shock-resistant finishes, but promotes also the development of continuous non-permeable films at a thickness considerably less than is possible with coatings formed at lower temperatures.

(a) The stove is charged too densely; the products of evaporation cannot escape and thus delay the normal drying process by prohibiting contact with the necessary oxygen, etc.

In addition the film may at this stage still be susceptible to action of solvent vapours and could be washed away. This occurs mainly on the upper portions of work pieces (solvent wash).

Coatings drying by oxidation at low temperatures, i.e. up to about 80°C (175°F) can be more quickly dried by an artificial supply of oxygen (for instance, with the aid of an ozone plant).

By far the best results are obtained in an oven fitted with an efficient air circulation equipment (forced convection) which

(i) prevents overheating and insufficient local temperatures,
(ii) continuously carries fresh atmospheric oxygen to the drying film surfaces,
(iii) ensures a uniform product, and
(iv) induces efficient utilization of the available heat.

If sufficient humidity (25–35% humidity) is present, wooden objects can be force dried in chambers at up to 78°C (160°F) without the wood showing signs of cracking or the glue being embrittled.

Paper goods withstand a maximum temperature of approximately up to 110°C (230°F) for several minutes; hardboard is much less susceptible, and half an hour at 120°C (approx. 250°F) is frequently applied.

(b) The vapour exhaust flap is completely open, and the heat is not perfectly utilized.

Air seals can be fitted at the ends of conveyor ovens to save heat.

(c) The temperature indicated by the thermometers does not agree with the inside temperatures of the stove.

In most instances the thermometers indicate too low temperatures because of incorrect location, for instance, near the oven door. Test (maximum) thermometers should be placed from time to time for control at various points of the stoves to check up, or a thermocouple (also available with recording instruments, e.g. for use in conveyors) connected to the drying goods.

(d) Thick-walled and heavy metal objects require different drying periods [3] from thinner plates, sheets, etc., as it takes a longer time to heat the material to a uniform temperature. On the other hand, the cooling down takes correspondingly longer. Differences in heat conductivity of different bases have to be borne in mind.

It is necessary either to subject heavier materials to longer drying periods or to increase the temperatures.

For metal parts having differing metal thicknesses the best-fitting average time/temperature cycle should be established.

(e) Various types of metals and alloys also require different drying periods.

(f) The high curing temperature and/or long time is characteristic for some synthetic media, e.g. silicone resin lacquers.

The addition of 0·5% of lead in form of a drier, calculated on resin solids, one hour before application has been recommended for speeding up this process [3].

(g) Uneven heating, e.g. due to unequal adjustment of infra-red radiation [4], possible also, e.g. in box ovens can cause partial under- and over-stoving of the same batch. Circulating air, the temperature of which is not below that of the oven walls [4], has an equalizing action. Stoving of electro-insulating varnishes is dealt with in Section 7.29.

(h) Powder coatings are free from volatile by-products and they do not need the high rates of air exchange which are necessary for stoving solvent or water based coatings. This factor is an economic advantage when using powder coatings and it also reduces the problem of dust contamination.

Notes

[1] Andrew, L.W. and Chamberlain, E.A.C. (1943), *Radiant Heating for Industrial Purposes,* The Institute of Fuel.
[2] *The Journal of the Institution of Electrical Engineers,* 1945, p. 395.
[3] Cahn, H.L. (1956), *Ind. Fin. (Amer.),* **1**, 42.
[4] Anon. (1951), *Ind. Fin. (Brit.),* **10**, 181.

4.3 Tackiness, softness, insufficient hardening of oxidizable air-drying paints

The above faults are frequently accompanied by rapid collection of dirt on the fresh films, especially in the open. They may be due to one or other of the following causes:

(a) A deficiency of cobalt drier can result in air drying oxidizable films remaining tacky for some time. Any addition of cobalt drier must be made with care since an excess of cobalt can result in a soft film with a tendency to smear; wrinkling can take place also, especially with tung oil varnishes. Lead/manganese should be used in this case.

(b) Excessively thick films resulting in solvent retention over several days. Runs on vertical surfaces as well as rivelling can occur.

(c) The use of unsuitable high boiling thinners such as distillate or paraffin. In addition to causing prolonged softness and stickiness through retention of thinner, incompatibility of thinner and resin can give rise to seediness in the film.

(d) Use of other unsuitable, low volatile thinners such as dekalin or tetralin, or of semi-drying or non-drying oils. If in contrast, thinning of oil-base coating is carried out with turpentine, especially if this is fresh, with pine oil, or dipentine, which all form peroxides, acceleration in drying times has been observed immediately or within some months after thinning, depending on the peroxide content. This content may vary with different supplies of the same solvent. The higher it is the more the drying time is reduced.

(e) The primer, filler or undercoating was not sufficiently dry before applying the next coat. With air-drying alkyds the top films applied over insufficiently hard undercoats can remain sensitive to pressure for a long time, depending on the oil length. On the other hand, some types of finishes show poor adhesion if the undercoat is too hard.

(f) Softness can result from the presence of certain residues on the surface which tend to penetrate the paint film. These include waxes, mineral and cutting oils, fats, badly washed off or incompletely neutralized cleanser such as caustic soda or other alkaline liquors including alkaline paint removers. The latter can cause saponification. Wax residues usually originate in organic solvent-type paint remover. Attention must be paid especially if furniture and woodwork are to be re-decorated after having been treated with polishes containing wax or silicone compounds. These faults can, of course, be avoided if care is taken to ensure that all surfaces are clean and dry before painting. Wax, oils and fats can be removed by swabbing with white spirit; alkaline residues by swabbing with water containing a little acetic acid (or vinegar).

(g) Excessively rich or hot pre-oiling of wood with boiled linseed oil or use of too long oil priming paints, especially for hard wood of low absorption capacity or for smoothly finished wood, preventing thorough drying of the films. Stickiness as well as resoftening are particularly pronounced if a short oil varnish is applied on wood thus prepared.

In the case of wood which is rich in rosin, oiling is better omitted in order to avoid ensuing stickiness of paint films.

(h) The material of which the base is made may cause re-softening of some coatings. This is often observed when plastic articles are coated, particularly cellulose ester plastics, polyvinyl chloride or acetate, or polystyrene, and is caused by migration of the plasticizer in the base into the coating film. This can take place within a few days or weeks, and is more severe in tropical climates.

The number of plastic materials and of plasticizers is now so numerous that no golden rule can be given for avoidance of this defect. A lacquer based on the plastic material (e.g. a polystyrene lacquer on polystyrene) may sometimes assist, or an insulating coat in the solvents in which the plasticizer is insoluble. Frequently it is necessary to experiment with a range of different coatings.

(i) Soft and sticky spots or small areas (often discoloured) can result from bleeding of resinous compounds from knots in wood. These should be sealed by knotting or other oil-free sealer before painting. In repair work the affected areas should be stripped and the wood cleaned with white spirit. Knotting should then be applied before touching in with the paint system.

(j) Application at low temperatures can result in films which remain sticky and soft for some time. This applies particularly to some chemically-cured systems such as epoxies and polyesters where the chemical curing reactions take place very slowly, if at all, at low temperatures.

(k) Freshly lacquered or varnished objects should not be wrapped in paper, especially of the fibrous kind. Neither should fresh films be subjected to heavy loads by high piling or loose stowing.

If necessary, the drying process should be accelerated, e.g. by stoving or force drying. Products to be packed and shipped as soon as possible after finishing can be lacquered with suitable cellulose lacquers.

Layers of greaseproof paper may sometimes be useful.

4.4 Loss of gloss, dulling, matting, differences in gloss of stoved coatings [1]

(a) The base is too absorptive or has been sand-blasted, etc., too strongly. Pre-treatment with chemicals, such as rust inhibitive treatment, frequently acts in the same way. This defect is often corrected by recoating.

(b) Insufficient hardening through of the previous coating. Restoving and recoating will often correct this fault.

(c) Excessive overthinning may coagulate and precipitate the binder resin or varnish. A small amount of true solvent may correct this fault, but it is best to consult the paint manufacturer.

(d) Gas checking in wood-oil type oils (conjugated double bonds) containing coatings can cause matting as the consequence of extremely fine rivelling which can only be verified under a microscope.

The defect is sometimes visible as fine ripple or a homogeneously fine matt surface, particularly if the work pieces had been left too long before being charged into the oven, or if they had been put into a recently lit cold convection oven. Articles left inside a conveyorized oven which has been closed down between shifts, often show loss of gloss due to gas checking when the curing cycle is eventually completed.

(e) Addition of too much pigment. This fault can only be corrected by the paint manufacturer.

(f) Drying for too long a time or at excessively high temperature [1] possibly on account of wrong indications of the thermometer or of unequal heat distribution within the stove.

Uneven metal thickness can, apart from variable hardening through, also produce areas which possess lower gloss. The consequences of heterogeneous heat radiation or distribution could be similar.

It has been observed that excessive heat on a heavy layer of black pitch enamel on steel caused a netlike structure in the film (no cracks) which was the more pronounced the thicker the film. Still more extreme conditions would probably have caused cracks.

(g) All iron plates and sheets made of soft iron develop the so-called 'temper colours' [2] when heated to elevated temperatures. Besides, numerous manufacturers subject some of their iron plate and sheet products to special processes in order to reduce the disagreeable colouring of the edges, and most of these processes produce characteristic effects causing the plates or sheets to assume a violet to blue colouration on being heated to temperatures of about 150°C (300°F). Thin films of, e.g. black enamels, which are dried on such sheet material at correspondingly high temperatures, can become partly or completely matted, so that different results are obtained with the same paint material on different lots of sheets.

If it is impossible to dry the films at lower temperatures, it is necessary, of course, to use sheets less susceptible to the showing of annealing colours. Much can also be done by avoiding non-uniform distribution of heat, by installing suitable air-circulation equipment. If these ways are not possible the fault can be eliminated by applying a number of coats so as to cover the discoloration.

(h) Finished flat thin metal sheets sometimes show the pattern of the trays where they lie flat and closely on the latter during stoving, the films being somewhat more matt at these points due to more intensive heating.

This phenomenon can be avoided by suitable jigging or suspension of the coated articles or use of wider mesh trays.

(i) Different metal bases influence the appearance of stoved films the more so the thinner the film. Some coatings on brass, for instance, will yield films with a greater degree of mattness than on iron or aluminium, particularly since brass coatings require longer stoving periods. Differences in the surface structure of the same or different kind of metals have also to be considered.

The self colour of surfaces, e.g. copper alloys, may affect the colour of insufficiently opaque coats.

(j) Even traces of mineral acid fumes in the oven or a fouled atmosphere may cause flatting or 'gas-checking' of certain thermo-hardening synthetic coatings (especially of the alkyd plasticized melamine or urea resin type). Acid fumes or chlorinated hydrocarbon solvents (e.g. trichlorethylene) used in close proximity, for instance, in a degreasing plant, a nearby plating shop, or a paint-remover tank, are commonly the source of such faults. But acid catalysts sometimes contained in synthetic resin varnishes may also be the cause if they induce too rapid surface drying. This failure is really one of the 'gas-checking' phenomena, see above (d).

It is obvious that the remedy consists in the elimination of the acid or noxious fumes. Watch that the oven dampers are sufficiently open.

(k) Heavy layers of glossy enamels (especially those containing amino-resins), where the water formed during the thermosetting process (polymerization and condensation) is expelled rapidly by fairly high stoving temperatures, may not only show blowholes but also give a homogeneous matt film without pores necessarily being visible to the naked eye.

As the same material may give perfect results at the same temperature, if only the film thickness is normal, attained either by thinning or by less generous application, no reduction of the stoving temperature appears to be necessary.

(l) Prohibition of a proper second flow by too rapid stoving leads to celled surfaces (Bénard cells) and accumulation of portions of the pigments or of an immiscible phase at the air interface not covered by 'oil', as a consequence of the vortex-reaction. This shows up as surface dulling.

The 'oil body' ought to flow over the cell walls at the stage of the secondary flow leading to the eradication of the cell configurations and to surfaces of high reflection. (F.E. Bartell and M. van Loo.)

(m) Loss of gloss during the stoving of powder coatings is often due to the presence of small quantities of other types of powder coating, i.e. anhydride cured epoxy powder coatings will reduce the gloss of amine cured systems [3].

Notes

[1] Similar failures appearing by using infra-red drying plants are dealt with in Section 4.12.

[2] The temper colours and temperatures of 'iron' are as follows:

Light yellow	220°C (430°F)
Brown-yellow	250°C (480°F)
Reddish-brown	265°C (510°F)
Violet	285°C (545°F)
Light blue to cornflower blue	310°C (590°F)
Grey to green	330°C (630°F)

Prolonged action of stoving temperatures causes the colours to appear gradually at lower temperatures than those indicated, the changes being more marked with thin and bright sheets. It is possible that the composition of the atmosphere of the stove exerts a certain influence, as the presence of much hydrogen, of steam, carbon monoxide and oxygen causes changes in the temper colours. In practice it is only the blue coloration which may give rise to troubles. It may appear under the above conditions at temperatures as low as 150°C (300°F). The temper colours render iron somewhat more resistant to corrosion. E. Hedges explains that these colours are due to the interference of light by a thin film of oxide.

[3] Harris, S.T. (1976), *Technology of Powder Coatings,* Chapter IV, Portcullis Press, London.

4.5 Flooding and flotation; mottling; silking; striations

Flooding is a defect which appears soon after application in many gloss finishes containing two or more pigments differing appreciably in particle size, specific gravity (there are exceptions to this) and possibly wetting characteristics [1]. Typical pigment mixtures giving rise to the defect are titanium dioxide/phthalocyanine blue [2], titanium dioxide/carbon black and lead chrome/Prussian blue (chrome greens). The effect is a gradual and uniform deepening of the colour of the surface which dries to a deeper shade than the bulk of the paint [3].

Flotation (or floating) resembles flooding in that one pigment appears to 'float' to the surface, but the latter assumes a mottled appearance and close examination reveals the existence of hexagonal Bénard cells (Figs. 4.1 and 4.2). The mottled appearance is due to differences in pigment concentration between the edges and centres of the cells. As a consequence of the differences in concentration, strains are likely to be set up in the film, particularly on weathering. It is possible that certain forms of checking (e.g. 'crowsfoot') may originate in this way.

Silking is an effect observed on dipped articles where, as a result of drainage, the Bénard cells have become elongated giving the surface a silky appearance.

Striations (or 'streaks') are produced on brush or dip application of paints where one of the component pigments (usually the stainer) has flocculated. The flocculates break down under the shear produced by brushing or drainage and produce striations or streaks of deeper colour.

During the process of solvent evaporation from a wet paint film, convection currents are set up while the film is still fluid, and these carry some of the pigment particles to the surface. The particles are usually the finest and often (but not invariably – see below) those of lowest specific gravity. Any tendency to flocculation by these particles appears to assist the process.

The type of solvent evaporation exerts a profound effect on the type of surface pattern produced. Under equilibrium conditions solvent evaporation is by a diffusive process and results in flooding or uniform distribution of pigment in the surface. When the solvent vapour is removed rapidly by air circulation, evaporation becomes ebullioscopic and takes place from points in the film. This leads to Bénard cell formation and a mottled appearance on the surface (flotation).

Flotation can take place with single pigments but is not often evident. The Bénard cells (many of which will be elongated) can be seen clearly if a droplet of an alkyd paint, pigmented with toluidine red, is placed on a flat surface and examined under low magnification. In thixotropic paint systems pigment mobility is very restricted and therefore flotation is least likely to occur.

The addition of certain silicone oils (1 part to about 40 000 parts of paint) will often convert flotation into the less objectionable flooding. Certain proprietary materials (e.g. Troykyd Anti-Float), at 2 to 2·5% on the weight of pigment, will sometimes reduce both floating and flooding significantly. The effect of this additive varies from one medium to another.

The incorporation of finely divided extenders (e.g. Aerosil, Calofort S, Calofort T) can reduce floating considerably [4]. However, in view of the pronounced flatting effect exerted on gloss finishes by extenders of this type, care must be taken to control the quantity used.

The type of flotation can be influenced by the relative humidity of the air. In coloured paints containing titanium dioxide, floating of the hydrophilic titanium dioxide has been observed [5] at high humidities whereas under drier conditions the coloured pigment has floated.

Electrostatic effects (charges on pigment particles) appear to influence flotation. An increased tendency for this defect to appear has been observed when paint films have been applied by electrostatic spray.

Green paints produced by blending Prussian blue with lead chromes (chrome or Brunswick greens) are more prone to blueing, i.e. floating or flooding of the blue component, than those made by using the co-precipitated 'struck' greens.

Modern methods of high speed dispersion, together with basic improvements in the surface preparation techniques of organic pigment manufacture, have considerably reduced the problems of flooding and floating [6].

Frequent causes of these faults are:

(a) Insufficient mixing of paints before use. Fully automated spraying equipment is fed from paint reservoirs which are stirred continually.

(b) Charging of parts freshly sprayed with stoving lacquers into the hot stoves. (Specially sensitive to this fault are, e.g., the thermo-hardening enamels, based on phenol resins.)

The freshly applied film should if practicable be subjected to a preliminary air-drying (flash-off) period of 15 to 30 min, or should be charged into a stove of low temperature, up to 20–30°C (70–85°F). Quick-drying coatings are more endangered.

(c) Too close packing in stoves of the enamelled articles possibly in conjunction with too rapid temperature increase.

Sufficient space must always be left, particularly where larger parts are stoved.

(d) Excessively thick films due to heavy application or to too high viscosity, especially on spraying. The faulty appearance of the films is particularly in evidence where the film layer is heaviest, e.g. if a full layer of a thin paint or lacquer is sprayed on or dried whilst the objects are not lying horizontally so that an uneven film thickness is formed (Fig. 4.3).

Sometimes hexagonal Bénard cell structures are formed coupled with bubbling (solvent escape prevented by skinning over), particularly if hot sprayed to a cold surface. Winter mornings are especially dangerous.

Faults due to surface tension, insufficient wetting, flotation and different specific gravity of a paint are less visible with thin films.

(e) Formation of emulsions in paints due to contamination with water. When pigments are mixed with medium the former may already be moist.

Pigments should be stored as dry as possible

(f) Existence of an extremely matt, absorbent base.

(g) Unsuitable solvents or thinners. If they are fast evaporating they favour flotation. Polar solvents, e.g. acetone, butyl alcohol, etc., should not be present in large quantities.

(h) Metal soap formation *in situ*, e.g. by reaction of zinc driers with alkyd resins resulting in the formation of zinc phthalate, can produce flotation.

(i) Brushing with a too coarse bristled brush. The opinion that cracking along the brush marks, that is where the paint film is thinnest, is possible under great stress cannot be denied.

Employ split bristle or hair brushes, depending on the purposes in question. Use a more careful brushing technique, employing the brush-tip, i.e. the tapered parts of the bristles, for 'laying off'.

(j) Use of brushing method in connection with too rapid drying paint materials or in excessively warm surroundings.

(i) Employ another method of application such as, for instance, spraying, where possible, or use another more easily and satisfactorily brushable paint material, or

(ii) Add some heavy thinners, e.g. white spirit, turpentine where oil-base coatings are concerned, otherwise, e.g. one of the glycol esters. Make sure, however, that these are compatible with the medium of the paint.

(iii) In the case of water-based and emulsion paints where brush marking occurs due to excessively warm conditions or very porous substrates, additional thinning may be desirable or the use of a prepared solution instead of water in order to give improved wet-edge time i.e. use aqueous thinners containing colloidal thickeners, emulsion, etc.).

(k) Using paint materials too stout for brushing, e.g. semi-prepared. This may cause streaked or checkered appearances of the film similar to those caused by spraying and dipping.

Pronounced colour differences can be observed between paint surfaces which have been brushed on in different directions.

Either thin the too thick material to suitable viscosity or employ a paint which is better qualified for brushing.

(l) Coatings which contain a high percentage of low-boiling solvents or thinners may show streaky films, especially if application is coating by roller. The movement of the rollers produces air-currents, thus quicker evaporation of the highly volatile constituents resulting in uneven films. Dirt on the scraper (leveller) which is sometimes used shows up on repeated rotation in the form of streaks (reverse brush marks).

However, coatings produced by reverse roller coating give more uniform results than knife coating.

Sometimes uneven absorption of e.g., fibrous material is to blame.

Addition of high-boiling thinners may be helpful. The room temperature should be well watched.

Mechanical knife-coating may have its advantages for cardboard, fabrics and paper, venetian blinds etc., when stout coatings are to be applied with low speed. Curtain coating machines have found entrance for the even application of nitrocellulose lacquers and polyesters to flat wood panels etc. as relatively thick, mostly unpigmented films.

Notes

[1] Ferguson, I. (1959), *JOCCA*, **42**, 529.
[2] Crowl, V.T. (1967), *JOCCA*, **50**, 1023.
[3] Parfitt, G.D. and Lewis, K.E. (1966), *JOCCA*, **49**, 92.
[4] Laporte Technical Bulletin, No. 6/62. *Flooding and Floating in Paints.*
[5] Marwedel, G. (1960), *Farbe Lack*, **66**, 379.
[6] Smith, F.M. (1975), *JOCCA*, **58**, 205.

Proprietary Materials

'Troykid Antifloat' – Troy Chemical Co.
'Aerosil' – Degussa.
'Calofort S and T' – J. and E. Sturge Ltd.

4.6 Wrinkling, rivelling, webbing, frosting and gas-checking of oil-base paints, synthetic resin and bituminous enamels

Wrinkling, also known as *rivelling* or *crinkling,* is the development of wrinkles in a film during drying, usually due to the formation of a surface skin.

Webbing, is the development of wrinkles, often in a well-defined pattern, in the surface of a coating during drying. This may be aggravated by gas fumes in a gas heated stoving oven and is then called *gas-checking.*

Frosting is the formation of a translucent 'crystalline' surface pattern consisting of fine wrinkles, and frequently happens during stoving in gas ovens. It is characteristic of paints and varnishes containing oils which have been incompletely polymerized. Frosting, webbing and gas-checking are to some extent interchangeable terms.

Wrinkle finishes are coatings in which a regular pattern of wrinkles is deliberately induced. Such paints are often used to hide imperfections in the substrate.

Wrinkling occurs if the upper film layer dries quicker and expands more than the lower layer. This can be due to several reasons. Wrinkled paint films, especially on wood, will under the influence of penetrating humidity preferentially increase the volume of the creases, on the ridges of which cracks are formed when dry, if under strain. The rapid breakdown of the whole film is then only a matter of time.

Causes

(a) Excessively thick application, perhaps in addition at a temperature above or below room temperature, especially of:

(i) Materials with high oil content, such as anti-corrosive and oil paints, long-oil varnishes, etc.

(ii) Varnishes, even of the shorter oil types containing wood oil or similar oils and too much cobalt drier (not to be mistaken for the gas checking mentioned in following sub-section).

(iii) Synthetic resin varnishes (containing alkyd or phenol resins with wood oil or similar oil), e.g. tung, oiticica, dehydrated castor oil.

Thick films of many an enamel based on alkyd resins are prone to rivelling because they become rapidly surface dry, but take considerably longer to harden through. The thicker the films, the coarser the wrinkling. It is claimed that thixotropic alkyd paints tend less to this defect due to their somewhat slower drying properties [1].

(iv) Black enamels containing bitumen-like constituents.

Wrinkling as well as stickiness of the films may be the consequence. The water impermeability (important in connection with anti-corrosive paints) and life may also be impaired thereby. It is an old and reliable rule that a number of thin consecutive films of traditional surface coatings gives better results than one thick one. This applies to application by brushing, spraying, dipping, etc. If the paint material itself is too viscous, it must be thinned, of course. Partly emptied containers should not be stored for a prolonged time. The containers should always be stored well closed and at room temperature (20°C = 70°F). If these precautions are not taken, the viscosity of the paint may change noticeably within a few days.

Small additions of solutions of zinc naphthenate, resinate or octoate, etc., calcium naphthenate, etc., or of turpentine, may improve the performance. A most useful additive to cure these film defects is a solution of rosin.

(b) Stoving bituminous enamels drying at high temperatures and applied too thickly tend sometimes to wrinkling, especially if the vapour exhaust is too strong.

Reduce the vapour exhaust, for instance by partial (but not complete) closing of the oven dampers, which also saves fuel.

(c) For rippling and 'gas-checking' up to total partial apparent matting of films on stoving, the oven conditions are generally blamed. While the presence of wood-oil type oils would make films more susceptible, the phenomenon is shown also by films of amino resin [2] containing non-drying oil modified (i.e. saturated) alkyd or epoxy resin media, if the conditions favour its formation. Already R.S. Morrell [3] noticed 'gas checking' of vanishes containing perilla oil. He further observed the failure under the effect of darkness and by absence of ultra-violet rays. A number of papers [4] deal with possible reasons. It is suggested that the contributory factors are as follows:

(i) Gaseous products of combustion are being re-circulated inside the oven owing to the setting of the out-draught, or to atmospheric conditions. Incompletely burned gases containing soot (H. Dannenburg) make matters worse. Carbon dioxide, sulphur dioxide, oxides of nitrogen and ozone [4] are either acid catalysts or oxidants, both of which promote surface drying. Close packing of the ovens might accentuate the defect.

A catalysing or oxidizing effect caused by contamination from outside, even another factory building, cannot be ruled out, particularly on a foggy day. It could, e.g. be acid vapours, such as hydrochloric acid emanating from trichlorethylene decomposed by moisture, sunlight or hot oven surfaces.

(ii) Cold and damp ovens, particularly on autumn and winter mornings and on foggy days.

(iii) If the flash-off time after application and before stoving is extended for several hours, or even overnight, most coatings will by then be surface dry and tend to 'gas-check' on stoving [5]. This defect sometimes produces the matt films already mentioned, where rippling can only be verified under a magnifying-glass. There is reason to believe that the ripple type of this defect occurs in thin films containing a comparatively low proportion of alkyd to amino resin, whilst the webbing ('crystalline') type is to be found more where the proportion of oil-modified alkyd resin is high.

(iv) The presence of fast curing, e.g. amino resins [2] could under suitable conditions also be a promoter of the defect.

The above faults can frequently be avoided by good ventilation of the painting shop, or by fully oxidizing or eliminating gases of combustion from ovens. These should not use up all the atmospheric oxygen. Thin films are frequently, but not always, more resistant to the defect.

If 'gas checking' is not inherent in the medium, small additions of the following solutions will be worth trying: hundred per cent phenolic resin, anti-oxidants or amines. In some cases the placing of ammonium carbonate inside ovens has had a beneficial effect by neutralizing acid products of combustion. For epoxy resin based coatings the following agents have been suggested: if amino resin, modified epoxy esters, anti-oxidants and basic products, such as triethanolamine, hydroquinone, 2, 6-di (tertiary butyl) 4-methyl phenol and combinations; for straight epoxy resin modifications (Epikote 1007) with urea formaldehyde resin the morpholine salt of p-toluene sulphonic acid [6].

(d) Excessive additions of liquid driers (Fig. 4.4) to air drying or stoving paints, particularly driers containing cobalt. This metal mainly promotes surface drying.

If drier additions are required at all, they should be added sparingly, especially to alkyd media. Manganese driers usually promote wrinkling and should be kept to a minimum. Combinations of lead, cobalt and calcium naphthenates often counteract wrinkling.

(e) In the case of stoving varnishes, too sudden an increase of temperature may also cause trouble, especially with alkyd resin or oil varnishes containing wood oil or similar oils.

The stoving temperatures must be increased gradually if possible.

(f) Normally the painting of a short-oil over a longer oil coating leads to alligatoring and cracking. If, however, the short-oil paint is applied before the long-oil paint film has dried through, rippling can occur. This seems to be mainly the case if the films are applied to a porous absorbent base such as wood or cardboard.

Crocodiling or *alligatoring* is a drastic type of crazing (see checking, p. 262). producing a pattern resembling the hide of a crocodile.

(g) Wrinkling of powder coatings can occur when rapid cure systems begin to age on storage.

Notes

[1] North, A.G. (1956), *JOCCA*, **39**, 864.
[2] Urea and melamine formaldehyde resins.
[3] Morrell, *Varnishes and their Components,* p. 280.
[4] Dannenburg, H. *et al.* (1950), *Ind. Eng. Chem.,* **42**, 1594.
Rogers, L.R. (1953), *JOCCA,* **36**, 95.
Phillips, I.H. (1957), *JOCCA,* **40**, 271.
[5] Hess, M. (1957), *Off. Dig.,* **29**, 681.
[6] Swift, G. and Stolton, R.E. (1957), *Off. Dig.,* **29**, 664.

4.7 Sandy, dirty and bitty films, seeding

A '*bitty*' paint or varnish is one containing bits of skin, gel, flocculated material or foreign particles, which project above the surface when the paint or varnish is applied.

Seediness is a defect in a clear varnish or lacquer caused by small particles sometimes visible in transmitted light. After application, varnished or lacquered surfaces may present a bitty or sandy appearance due to this defect. The term 'seedy' specifically denotes bits which have developed in a paint or varnish during storage.

(a) Precipitation of irregularly distributed dust particles (long in shape, like tiny hairs) before or during application over the freshly varnished surfaces caused by:

(i) Untidy, cluttered varnishing shops with, possibly, rough and unfinished walls, etc., where dust may collect and be scattered by draught.

Shops should be provided with smooth walls covered with light-coloured oil paints and equipped with a minimum of furniture. Regular shop cleaning. No superfluous equipment should be permitted in these rooms. High-quality articles should be dusted before painting, with a clean brush or a dusting gun. Tack-ragging especially before applying the last coat is advisable.

The finishing shops should be kept separate from the other departments, particularly where slow-drying varnishes are applied. Keep floors oiled.

(ii) Excessive artificial air pressure above that of the atmosphere in spraying shops which is sometimes created in order to prevent penetration of dust into the shops and to eliminate eddying. Excessive suction by the exhaust of the spraying post.

Too high pressure is bound to increase the dust danger, as the dust is forced toward the exhaust equipment at the spraying posts where it escapes together with the excess spraying particles after passing the freshly varnished wet parts.

Spraying booths can be fitted with water wash. Air filters including the electrostatic type can be built in. Keep floors oiled.

(iii) Close proximity to driving belts, conveyors, etc., results in fluffy pimples.

Moving mechanical parts should not be tolerated within the varnishing shops or must be encased so as to prevent the formation and raising of dust. Driving belts also grind off fine leather dust etc., which is partly distributed throughout the air.

(iv) Leaving freshly finished parts with coatings, which set slowly or which are being coated wet on wet with an interval outside the stoves previous to drying, for too long a time, or maintaining excessively low room temperatures when applying such materials.

If the hurdles or trolleys carrying the freshly varnished parts into the stoves must be kept waiting for any considerable time, they should be temporarily covered against dust and transferred into the stoves as soon as possible. The painting shops should be maintained at a temperature of about 20°C (70°F).

(v) Proximity of grinding shops, or insufficient removal of pumice or emery powder by rinsing after rubbing down.

Dry rubbing should be effected in front of a suitable exhaust, using a branch tube of the spraying exhaust if necessary (advisable also for sanitary reasons). The grinding shops should be an entirely separate unit.

(vi) Unsuitable working clothes of the operators or use of fluffy rags for cleaning near the spraying booth while the fan is working.

The working apparel of the operators, especially their overalls, must be maintained in an externally clean state at all times, but they should also be beaten thoroughly at short intervals so as to keep them free of dust. Fuzzy or frayed clothing should not be worn. The hair of the operators should be covered with a cap or hood.

(vii) Suction of dirty air into the spray gun.

Where steel and galvanized iron pipes are in use this can be caused by rusty particles found inside the conduits between air filters and guns. The air intake should be at a point where the air is not contaminated by dust, etc. Build in air-filters.

(viii) Omission of thorough dusting by hand or spray gun prior to painting, especially finishing.

(ix) Careless washing after one of the pretreatments.

(b) Storage of varnishes at too low a temperature. This causes seediness, as does varnishing in too cold an atmosphere. Sudden changes of temperature during the varnishing operation are also to be avoided.

(c) Stirred-in skin, found in tins of oleoresinous and alkyd resin air-drying varnishes. (Indented, flat or lump-shaped bits in films.) The introduction of anti-skinning agents into paint manufacture has greatly reduced the occurrence of skin formation.

Do not store containers filled with rapidly drying paint materials unstoppered or half-filled. A small additional air-space beyond that required for the expansion of the contents is sufficient to supply enough oxygen for skinning. Full tins or cans, e.g. of white gloss paint, have for a long time past been stored upside down for the sake of avoiding the entrance of air through ill-fitting lids, etc.

Filter the varnish or lacquer through fine-mesh vibratory strainers, silk gauze, or finest muslin in addition to wire strainers depending on requirements. Where the requirements are exceptionally severe, the muslin can be covered with a layer of long-fibred cotton wool.

In some instances (particularly in connection with finely ground enamels) such materials can be rendered usable again, at least for a time, by repeated grinding. Filtration may also be tried to remove the immiscible particles. Impurities can be due to too new or too cold oil varnishes.

(d) Stirred-up and insufficiently dispersed pigments in paints containing relatively dense or old sediment. (The shape of the little lumps thus formed in the films is round and voluminous.) Some of the synthetic resin enamels tend to form such sediments within much shorter time than the oil-base enamels.

Frequently paints have been stored in their containers for a time without shaking or rolling, etc. Stir thoroughly and disperse the soft parts with a rod before application. Rather risk a difference in shade than disturb really hard sediment.

(e) Paint, etc., drippings have been burned to fine cinder in the stoves, and contaminated freshly coated parts.

An oven must be maintained absolutely clean at all times. Place dripping pans at suitable points and clean them at proper intervals.

(f) Freshly stirred paints and varnishes which are viscous or contain much pigment show numerous air bubbles which remain in longer the colder the material. On premature application they will form tiny pores or craters in the film or small concealed blisters.

Similar defects can occur by excessive circulation or by leaky glands on the suction side of circulating filter pumps [1].

Store such paint materials in warm surroundings or thin down. Previous to placing the painted parts into the stoves, the bubbles must have vanished. Strain if necessary through a sieve.

(g) A frequent cause of trouble is the agglomeration of pigment particles (for instance, in finely ground black enamels) [2] around solids, e.g. tiny skin or gelled particles, metallic soaps, copal or resin particles due to cold storing, or caused by the separation of certain thickening agents. Every one of these particles then forms a nucleus for the formation of small grains or even lumps of occasionally considerable size.

Some colours which have been forced into solution by hot storage may be thrown out of the medium fairly coarsely at normal temperatures [3].

(h) Insufficient wetting of the pigment on grinding either due to insufficient grinding, unsuitable mill charges, or too low an 'oil' content of the medium. Grinding promotes stable suspensions of the pigments, etc., because the pressure of the rollers, etc., aids the normal but insufficient forces of capillarity to ensure thorough wetting.

Regrinding may have to be considered. Addition of wetting agents can be useful. The latter may, however, sometimes increase the defect.

(i) The right selection of dye-stuffs and solvents is one of the responsibilities of the paint manufacturer. It is, therefore, advisable to buy the lacquers already coloured. Sometimes the addition of suitable solvents may be of help.

(j) A sandy surface may be caused by the use of water stains on wood or composition board, or by bleaching of wood, which tend to raise the grain, particularly if the finishing operation follows on without sufficient interval for air or force drying.

Spirit or naphtha stains do not raise the grain of wood. Normally and only occasionally on damp days (alcohol is hygroscopic) fix loose wood fibres by a 5–8% shellac solution (wash coat). Then rub down and fill the pores [3].

(k) Coarse substrates, e.g. metals subjected to badly controlled phosphate treatments, require a greater number of coats to hide their rough surface, otherwise the 'points' stick through the film and are frequently thought to be the result of 'bitty' paint.

Pretreatment processes must be closely supervised.

(l) If paints which contain strong solvents such as are typical for cellulose lacquers are filled into lacquered or used cans previously holding, say, oil base paint, residues of it may be partly dissolved or swollen, thus spoiling the new material by coarse particles.

(m) Dirty brushes are a frequent cause of bitty finishes.

The remedy for bitty surfaces is to rub down gently and carefully with pumice stone and felting pad and apply another coat of paint or varnish.

Notes

[1] Mason, H.J. (1947), *JOCCA*, **30**, 470.

[2] An explanation as to why carbon blacks tend to agglommerate so much in enamels has been given, viz. that pure carbon black particles are repelled by most vehicles and attracted by each other. This has been successfully prevented by surrounding the ultra-microscopic carbon particles by a phase, e.g. a carbon oxygen complex, which allows them to be attracted by the vehicle (U.S.A. Patents 2 013 774 and 2 013 775). Bartell and van Loo make 'greasy layers' round carbon black responsible for the difficulty to wet them; solvents should therefore dissolve those first so that the oil medium can effect the wetting.

[3] Wampler, R.H. (1946), *Modern Organic Finishes,* pp. 210–394.

4.8 Lifting of paint films

Lifting describes the attack by a coating on an old or new paint film, usually due to solvent action, often resulting in blisters, blowholes and/or the formation of a wrinkled film (Fig.4.5). Among films on which this failure occurs are the long oil alkyd types which are sensitive to solvents and water during the drying period and for some time after. Contact with solvents results in swelling and blister formation, the effect varying with the type of solvent and the age of the film. Weak solvents such as white spirit are the least harmful, aromatics are more active whilst strong solvents such as ketones and esters attack the film rapidly. Ageing of the film reduces the sensitivity to solvents and a similar effect is produced by stoving. Similar defects occur in emulsion paint films when water is present but these develop their maximum water resistance more rapidly than alkyds although the final resistance may be lower.

The defects may be due also to the following:

(a) Application of spirit varnishes or cellulose lacquers over films of oleoresinous paints or fillers. Remedies are:

(i) Use of a barrier between the cellulose and the earlier coat.

(ii) Complete removal of the existing system followed by the use of hard drying primer and filler.

(iii) Replacement of cellulose lacquer by one-pack polyurethane lacquer.

(b) Application of a second coat of cellulose lacquer before the first has dried out thoroughly. Spraying at too high a viscosity can lead to pinholing; this may also be caused by too high a proportion of low boiling solvent. The film tends to set up before the entrapped air bubbles have escaped. The defect is eliminated by the use of a higher proportion of medium boiling solvent.

(c) Lifting due to swelling is possible if cellulose lacquers containing plasticizers based on alkyds are overcoated with the same lacquers after the alkyd has set but before it has oxidized sufficiently to resist the solvents of succeeding coats. Often the times for satisfactory recoating are within 3 to 4 hours or after a period of several days. Similar difficulties are encountered in the case of multiple coats of short-oil alkyds or vinyl modified alkyds in solvents such as xylol. Alkyds in white spirit are usually quite safe.

It is necessary to apply the various films as long as the lower film, although tack-free, is still sufficiently soluble for the next film to adhere. Such films can occasionally not be re-coated for several months because of their tendency to swell in contact with solvents.

Both films unite at least partly to a more or less single film because of an exchange of solvents and plasticizers, hardening through in one operation. Cellulose combination lacquers containing non-drying (non-oxidizing) alkyds have no tendency to lifting, and are therefore suitable, e.g. for touching up of lacquers of this kind.

(d) In the finishing of furniture by spraying with nitrocellulose lacquers, blisters are a frequent occurrence; their causes are set out below. Some of them can also apply to other coating materials, such as polyester lacquers, though there can be additional causes.

(i) Whilst pores of wood may show up in the finish through a thin layer of lacquer, occluded air can be responsible for blisters, if the surface drying of the lacquer over unfilled or incompletely filled pores was too rapid due to draught, or to high shop temperature; also if its viscosity was too high. Fluffy types of wood are usually covered with numerous small blisters if not filled. Air bubbles may not be visible in the first coat, but due to re-dissolving (softening) come out on subsequent coats.

Such bubbles remain at least partly stuck in the 'middle' of the film layer. Somewhat similar conditions or even very much worse ones can be found in polyester glass fibre mouldings.

Open-grained wood and other substrates need filling. To avoid swelling and lifting the pore filler must be completely dry before sealer or lacquer is applied.

(ii) Residues of hydrogen peroxide and other bleaching agents, if sealed in, can cause gas bubbles to disturb the film. Too severe bleaching can make the wood 'spongy', in which condition air is retained. The bubbles can, also in this case, become noticeable in the final stages of lacquering.

After bleaching operations a generous time interval should be allowed between after-treatment of the bleach and coating. Fluffy wood requires repeated spraying with diluted lacquer to fix the fibres.

(iii) If the lacquer is too thick, it cannot penetrate the pores, and air is held even by wood which would not necessarily require pore filling.
(iv) If blisters appear over hard annual rings, apply a thinner film.
(v) On hot spraying, bubbles can be caused, if (1) the flow rate, e.g. in litres per minute is too high, building up too thick films too rapidly; (2) the air pressure of the pot is too high (minute bubbles can be trapped in the film, giving a whitish appearance at times); (3) the temperature of the lacquer is so high that it approaches the boiling point of solvents contained; (4) the temperature of the lacquer is too low, and the lacquer thus too thick; (5) the thinners are too fast, i.e. too low boiling.

(e) Chlorinated rubber lacquers which usually contain benzene homologues as solvents are according to their composition more or less sensitive to lifting if applied in several layers, and the more so if brushed. Do not overbrush.

In this case lifting is not accompanied by wrinkle formation, but results mainly in a relatively high degree of brush marking. Mottling can also appear if the colours of successive coats differ.

Try out best re-coating time by experiment before carrying out large-scale work.

(f) Some short-oil styrenated, or similarly modified, alkyd paints, etc., dry very rapidly, but are prone to lifting on re-coating, although attaining a certain degree of hardness. One can even be deceived in this respect after short stoving times. The oxidation process of the drying-oil content proceeds for some time after. Heavy coats are more prone to the defect, so is application by brush or dip as compared with spraying.

Re-coat either within, say, two hours or after, say, three days. The best times have to be verified by an application test, and the temperature has to be borne in mind.

(g) Understoved amino resin-alkyd lacquers can fail by lifting even when the following coat was of the same composition.

Rather try wet-on-wet application first, than understove. Some degree of understoving is sometimes practised to obtain better intercoat adhesion between films of thermosetting lacquers.

(h) Small blisters recognizable only with the aid of a good magnifying-glass, and white dots on the surface of weathered nitrocellulose lacquer films represent definite signs of incipient breakdown.

It is advisable where possible to treat exposed nitrocellulose lacquer films with wax or similar pastes at regular intervals to maintain their resistance to weathering. This procedure has been found to be most satisfactory in the case of automobile lacquers. Otherwise it is very difficult to produce nitrocellulose lacquers resistant to weathering for a long period.

(i) Latex emulsion paints should not be recoated before the first coat has dried thoroughly. At normal temperatures and provided the plaster is dry they can be recoated safely in one and a half to two hours.

(j) On recoating polyurethane films, blistering has been observed if the intervals between coats is too short; the top coat was redissolved by trapped solvent in the first coat.

4.9 Bleeding of dyes and pigments contained in coloured coatings

Bleeding, in general, refers to the diffusion of a colouring matter from a substrate and also to the discoloration resulting from such diffusion [1].

The use of dyes in coatings is confined to a relatively small number of materials. Chief among these are spirit and water stains for woodwork and dyes for lacquers, both air drying and stoving.

Dyes used in water are generally of the acid type and these have little tendency to bleed into organic coatings applied over them. Spirit soluble dyes are basic and soluble in many organic solvents. They tend to bleed into paints applied to the woodwork but the bleeding is often confined to the primer. If stained woodwork is to be painted it is wise to rub it with a rag soaked in solvent to ascertain the tendency to bleed. If this is positive, the surface should be given a coat of 'white knotting' (conventional knotting pigmented with aluminium) to act as a sealer.

Pigments vary considerably in their tendency to bleed, i.e. in their solubility in solvents and media. Inorganic pigments (with the exception of certain chromates in which slight water solubility confers their corrosion-inhibiting properties) are insoluble in both aqueous and organic media and so show no tendency to bleed.

Organic pigments vary considerably. They are generally insoluble in cold water and, with two notable exceptions (Para Red and o-Nitraniline Orange) they show little tendency to bleed in linseed oil. Many show a pronounced tendency to bleed when in contact with strong solvents, resins and waxes; this tendency is greater with the pigment dyestuffs than with the toners [2]. The solubility is not, usually, of a high order and it is not a serious fault if it is confined to the paint in which it is used. When a paint containing a pigment such as Para Red is used in the normal way, no problems of bleeding arise. If, however, white stencilling is applied bleeding of the red takes place resulting in discoloration (pink). This bleeding is due largely to the solvent in the second coat and is much more intensive when strong solvents such as xylol or ketones are used than with white spirit. Non-bleeding pigments are chosen if there is the least contact with another material or surface into which the colour may bleed. This is important in printing inks used on packages for food-stuffs.

Bleeding can be troublesome or can be accentuated under the following conditions:

(a) The application of a light coloured coating containing strong solvents e.g. a cellulose nitrate lacquer or chlorinated rubber paint over a paint containing certain organic pigment dyestuffs (as described above). The colour can bleed through more than one coat (see Fig. 4.6). A similar result is obtained on a plastic material containing such a pigment. The remedy is to seal the surface with a spirit lacquer into which the pigment will not bleed.

(b) Heat increases the solubility of the pigment dyestuffs in organic media and so increases bleeding. In some stoving finishes where a pigment dyestuff has been used as the major pigment, the temperature of stoving causes an appreciable amount of the pigment to dissolve in the medium. On cooling, much of this separates and appears as a 'bloom' on the surface [3] (see Fig. 4.7).

(c) If traces of lacquers containing organic dyes have been left in spray guns, etc., differently coloured spots such as, for instance, red or lilac bleeds may appear on white nitrocellulose lacquer films, etc. Faults of a similar nature may be caused by the deposit of spraying dust on raw or lacquered goods due to insufficient ventilation.

Red, brown and black nitrocellulose lacquers are particularly obnoxious in this respect.

The counter-measures are obvious. Spray guns must be cleaned out thoroughly after every spraying and suitable cleaning solutions should be used for this purpose. The painting shops must be maintained in a high state of cleanliness. 'Dry spraying' of lacquers should be avoided, and perfect ventilation is an absolutely necessary prerequisite of modern finishing shops. It may sometimes be advisable to increase the speed of the exhaust. Dry aniline dyes, stains, etc., should never be used or handled within the painting rooms.

(d) Undercuring of thermosetting lacquers containing colouring matter which would perform satisfactorily if properly cured can result in objectionable bleeding, e.g. after white lines had been applied on cycle frames.

Raise stoving temperature or increase curing time. Alternatively the addition of a catalyst may help without the necessity to change the drying schedule. For this consultation with the manufacturer is advisable.

(e) Marking of conventional paint films by felt-tipped pens or by 'aerosol' sprays (graffiti). Most of these contain organic dyes which stain the paintwork. Blacks are pigmented with carbon black. If the paintwork is sufficiently hard much of the colour can be removed by solvent. In any case it is necessary to rub down and to apply a flat obliterating coat followed by two-pack polyurethane which may be clear or pigmented. Polyurethane coatings are resistant to staining by dyes and subsequent graffiti are more easily removed by solvent. The flat obliterating coat must be suitable for use under two-pack polyurethane and in view of the strong solvents used in the latter it must be specially formulated. Graffiti stains on emulsion paints or exterior wall surfaces are virtually impossible to remove and it is necessary to re-coat the surfaces with, preferably, the system mentioned above.

Notes

[1] See also B.S. 2015, 1965. *Glossary of Paint Terms.*
[2] Geissler, G. (1973), *Defazet.* **10**, 462.
[3] Geissler, G., *ibid.*

4.10 Yellowing and discoloration of coatings during convection stoving or at elevated temperatures

The more pigmented the enamels, etc., the less their films are affected by discoloration of the vehicles. (See, however, (k) below.) Most natural resins and rosin derivatives as well as drying-oil modified alkyds tend to discoloration under the effect of heat and also light. Owing to great progress in the manufacture of a variety of synthetic resins the failures under discussion are less frequent now.

(a) The films were baked too long or at too high a temperature for the coating used. Vehicles as well as pigments, including whites, vary in sensitivity towards elevated temperatures. Some media are subject to rapid yellowing and darkening, e.g. phenol formaldehyde resins, others, e.g. the acrylic resin family, remain water white at temperatures hitherto never reached with organic coatings–some up to 177°C (350°F)–and for longer times. They are only surpassed by white silicone resin lacquers, at 200°C (400°F).

White and pale shades of polyurethane lacquers yellow at elevated temperatures according to type, though improved resins have been produced. Stoving temperature should be kept well below 120°C (250°F). In fact, stoving of lacquers of this type could accentuate their tendency to discolour under the effect of light.

Overstoved thermosetting alkyd/amino resin based whites, some of which can be cured at up to 135°C (275°F), frequently bleach out on exposure to light.

Some white straight high molecular epoxy amino resin lacquers can safely be stoved at 180°C (360°F) maximum. When catalysed with a curing agent the temperature may have to be dropped to 160°C (320°F), as otherwise not only yellowing but also severe embrittling could be observed. Epoxy ester lacquers produced with suitable oils, such as coconut or dehydrated castor oil, and modified with amino resin can be stoved at 120–130°C (250°F).

Polyvinyl acetate, chloride, their copolymers, and some other vinyl resins, when stabilized are safe up to 120°C (250°F). Overheated they can even turn black.

White silicone enamels can keep their colour over fairly long periods of exposure.

Conventional oleoresinous coatings cannot be taken much higher than 60–80°C (140–175°F), some whites, e.g. those containing dammar, not beyond 90°C (195°F).

Medium oil length alkyd enamels withstand some increase of stoving temperature, above 120°C (250°F) according to the oils with which they have been modified.

Even bleached shellac lacquers discolour gradually with increasing temperature from about 120°C (250°F) when they begin to crosslink, though gaining in hardness.

Polyvinylbutyral lacquers perform similarly. Polystyrene lacquers are good up to the region of 150°C (300°F).

Straight styrene-butadiene copolymer stoving lacquers either in solvent or in water require for optimum curing 175–205°C (350–400°F), which makes them unsuitable for whites or delicate shades. Modified with amino resins, the temperature can be reduced to 95–120°C (200–250°F), at which no discoloration should take place.

Outstanding is the colour retention of cellulose acetate butyrate lacquers, which show hardly any change at 180°C (360°F). It must be borne in mind that with the duration of heat exposure the film weight gradually diminishes until in gross cases nothing, except the inorganic pigments, remains, resulting in chalking. The behaviour of the various binders in this respect differs greatly.

The above maximum stoving temperatures refer to normal convection-type ovens, and for stoving times which suffice to obtain hard, or cured films. Efficient air circulation, very short baking at high temperatures, infra-red drying and other radiation methods require modification of these figures.

Greys are almost as susceptible to elevated temperatures as whites.

It must be borne in mind that loss of gloss and embrittling can also result from overstoving.

As to the part played by some pigments, see sub-section (k) below.

(b) Water-thinnable stoving lacquers behave, as regards discoloration, comparably to their equivalents in organic solvents. Some of them require, however, fairly high stoving temperatures, i.e. 150°–175°C (300–350°F). Some of these, e.g. styrene butadiene copolymer and melamine alkyd based lacquers, show fair discoloration; the phenolic type, of course, shows more.

(c) Polyvinyl chloride lacquers, organosols, and plastisols can discolour, owing to decomposition (dehydrochlorination) by u.v. light, if overheated in the fusing process, or by action of a catalyst. Film thickness, time, and temperature (particular for each resin) play their parts in the process. If the coatings had been properly stabilized, discoloration should neither take place at normal fusion temperature nor under exposure to light. The defect produces colour changes to yellow, orange, brown and even black (complete degradation). This may possibly occur only in spots, accompanied by embrittlement, and can also be accompanied by sweating out of plasticizer.

Even white polyvinyl organosols, etc., should withstand the customary fusing temperature of 163–180°C (325–365°F) for the necessary duration, which depends on the film thickness. If, however, the material had not been stabilized against photolytic or thermal breakdown, discoloration must be expected, in the latter case around 100°C (212°F).

(d) Local overheating in certain parts of the drying stoves caused by irregular distribution of heat inducing partially faulty results.

A.A.B. Harvey thinks that the most accurate control of convection tunnel ovens is by steam heating [1].

See that flow and distribution of heat in the stoves is rendered as uniform as possible, preferably by installing automatic temperature control, as well as an air circulation device ensuring uniform temperatures in all sections of these stoves, and also by giving attention to the vapour flaps. As to temperature control, see sub-section (g) below.

For heat-resistant coatings see above, (a).

(e) In objects of closed design, or fitted with covers, which are painted in light colours, such as bread bins, the heat inside is not subject to sufficient air circulation, and may be retained for a longer time, causing discoloration.

See sub-sections (a)–(c) above. Open the covers, etc, also increase air circulation.

(f) Entrance of the gases of combustion into the drying space of directly heated stoves or through leaks in the bottom, etc., of cased ovens. These gases may discolour the fresh lacquer and varnish films.

(g) Thermometer indicates too low temperatures. A thermometer might be in itself a highly accurate recording instrument. Its 'feeler' can, however, only register the temperature with which it is in contact, or by which it is surrounded. A thermocouple attached to the article in the stove reads the article's true surface temperature. Such instruments can also be used to travel and record on a conveyor line.

The thermometers of ovens must be checked at regular intervals, e.g. by means of a standard thermometer, e.g. in an oil-bath. The temperatures in the various parts of the stoves may vary and can be checked by means of maximum thermometers. Never heat the stoves to the maximum temperature permissible for a coating, but try to remain 5 or 10°C (10 or 20°F) beneath, unless certain of accurate temperature indication.

(h) The thermometers may possibly be located in wrong places, for instance too near the doors, where they could be cooled below the real stove temperature by cold outside air.

Frequent opening and closing of the oven doors can also interfere with the keeping of a constant temperature.

Fair readings are usually obtained with thermometers introduced through apertures bored laterally through the sides of the stoves.

See above (g).

(i) Unsuitable additions, especially of driers or varnishes.

(j) Thinning of stovings with, for example, crude solvent naphtha, which contains sulphur compounds or which tends to discolour gradually on account of constituents of the indene resin type. L. Blumer reported that cyclohexanol acetate when used in white stoving enamels as a solvent caused yellowing above 110°C.

Suitable thinners should be procured directly from the manufacturers. Iron containers are often unsuitable for thinners, due to formation of rust or iron compounds, or possibly catalytic action.

(k) Several inorganic pigments and dyestuffs are very sensitive to elevated temperatures. A number of these, such as some lead chromates, are sensitive to light as well. Pure vermilion also is very sensitive to heat, and its tendency to turn dark, and finally black, with rising temperatures has been made use of in the manufacture of the so-called 'warning colours'. Vermilion also is somewhat sensitive to light. Nearly all white pigments, with the exception of titanium dioxide and oxide of antimony, are sensitive to elevated stoving temperatures, but in some cases regain their full colour on exposure to light, see above sub-section (a). Some iron oxides can be discoloured at high temperatures. It has been observed that the fastness to light of phthalocyanine greens can be degraded by formaldehyde, i.e. in any finish releasing the latter at high temperatures, e.g. on stoving of urea or melamine formaldehyde resins. It should, therefore, not be used for obtaining bright, fast-to-light pastel shades [2]. Instead mixtures of good chrome yellow or better, cadmium yellow, with phthalocyanine blue, are recommended [2]. For cadmium pigments heat resistance to temperatures over 350°C (660°F) is claimed.

Some vinyl resins, especially those containing polyvinyl chloride, are thermally decomposed by zinc and iron pigments (iron oxides expected), even in presence of heat stabilizers, see above sub-section (a)–(c).

(l) Certain synthetic resins, in particular polyvinyl resins, discolour and decompose on stoving, even when stabilized against the effect of heat, if applied directly over surfaces of iron and iron compounds, zinc, copper or tin [3]. Compare with sub-section (k) above. Stabilized polyvinyl chloride coatings can, however, be stoved on phosphated iron.

Discoloration can also be caused if the surfaces were dirty before application. This would be the case whether the coatings were plastisols or organosols.

The application first of a suitable primer, including an etching primer, or of a phosphating process recommends itself. Phosphated iron has no discolouring effect. Heat stabilization is essential.

See sub-sections (c) and (k) above.

(m) Epoxy based powder coatings cured with amino-amide type hardeners, e.g. dicyandiamide and substituted dicyandiamides, yellow more readily on overstoving than epoxy powder coatings cured with anhydrides. Amidine and polyamide curing agents are severely yellowed even during normal stoving schedules.

Small additions, e.g. 0.01% of dioxazine violet, will partially compensate the yellowing in white pigmented systems.

Notes

[1] Harvey, A.A. (1953), *Ind. Fin. (Brit.)*, p. 632.
[2] I.C.I., Dyehouse Dept. Circ. No. 197, 12.5. 1954.
[3] Chatfield, H.W. (1953), *Varnish Constituents*, p. 371.

4.11 Blooming of stoving enamels (mainly blacks) in gas ovens

Bloom is a deposit, resembling the bloom on a grape, which sometimes forms on glossy paint films, reducing the gloss and dulling the colour. Some forms of bloom may be removed by wiping with a damp cloth. The term 'hazing' is often used to describe a permanent form of blooming.

Black enamels with a stoving temperature range of about 60–140°C (140–285°F), oil based as well as containing synthetic, e.g. urea resins, are mainly affected.

Although the failure has been observed occasionally on white synthetic stoving enamels, it seems that black enamels, which are pigmented with carbon black only, cause the main trouble. Possibly the different carbon blacks react differently with regard to the occlusion of incompletely burnt products of combustion. It may also be that rapid setting finishes counteract such occlusions. On the other hand, it is feasible that a slowly surface-drying film, especially if sufficiently thick, may incorporate these fume occlusions due to its convection currents before solidification.

(a) Incompletely burnt gases of combustion penetrate into the drying chamber of the stove, causing blueing of a black film. This blueing is usually removable by washing with water. Certain unfavourable local conditions, such as, for instance, using spraying air from near the exit opening of a chimney, also favour the blueing reaction. Conveyor ovens, e.g. of the hump-back type, in which the hot gases of combustion very often are circulated together with hot air, may be similarly troublesome. In the same gas-heated conveyor oven where black enamel films showed blueing, the film of a white stoving enamel containing a synthetic dicarboxylic acid resin produced iridescence.

Leaks in the bottom or walls of an indirectly gas-fired drying chamber of a stove must be closed at once, possibly by pressing asbestos string into them. Tighten all bolts, open the exhaust damper sufficiently. Indirect heating is an absolute prerequisite of success for such type of enamelling. Black alkyd resin enamels containing oil appear to be less sensitive to the action of gases of combustion than the gum-oil types.

Dampers of conveyor or other ovens with circulating air for the stoving of sensitive coatings should be adjusted in such a way that the quantity of gases of combustion sucked into the drying channel is reduced, or the plant should be adjusted so as to allow indirect heating.

(b) Baking in open (direct) gas-fired stoves (officially prohibited in some countries) is a frequent source of trouble.

Gas stoves should be fitted with separate drying chambers whenever possible, which are hermetically sealed from the fires and the flues (triple-cased stoves), provided that no open gas flames are required to develop, say, a crystalline effect. The flames should at least be covered with a metal sheet.

(c) Mixing of gases of combustion and vapours derived from the drying of paints (prohibited in some countries). Close neighbourhood of the exit openings of the chimneys leading from the firing flues and the drying chambers respectively. The same applies if a number of drying chambers are connected up to one chimney, while one of these chambers, e.g. for development of crystal figuration, is directly heated. In fog, low clouds or hot sunny weather, the flue gases may then be forced back into another, perhaps colder, drying chamber, causing the kind of bloom described in above sub-section (a).

The counter-measures are self-evident.

(d) The iron sheets themselves commence to colour in the heat and show up through a thin film.

(c) The defect is due to close proximity of a trichlorethylene degreasing plant (white or yellow, partly iridescent bloom) or of a galvanizing plant.

Fumes of chlorinated solvents on coming in contact with sources of heat liberate hydrochloric acid and other chlorine compounds. Even phosgene may be developed under special conditions. Gas flames are particularly troublesome in this respect. Hot sun or glowing wires are other potential causes of decomposition.

Decomposition of trichlorethylene is favoured by the presence of water and light; in particular u.v. rays greatly accelerate it. The presence of metals, such as nickel, stainless steel, aluminium and iron further increase the rate of decomposition.

Recommended counter-measures are:

(i) Removal of the degreasing or galvanizing plants to another location. The working of the condensing coils, etc., need at least checking. Vapour or liquor degreased parts should be dry and not carry liquid trichlorethylene into the surface-coating processes.
(ii) Where (i) is impossible, change the composition of the degreasing agent.
(iii) If spraying air is drawn by the compressor from the surroundings of such plants, this may cause the difficulties in question. Procure the fresh air from a less contaminated source.
(iv) Dimethylguanidine and hexamethylene tetramine are recommended as stabilizers.

4.12 Failures in connection with the peculiarities of drying by infra-red radiation

The terms radiant heat and infra-red heat are used to define the same effect, although radiant heat more precisely describes the operation of heaters at 800°C and above, while infra-red heat relates to black heaters operating at 400–500°C. A radiant heater is much more effective for curing paint films at the higher temperatures than the infra-red heater, since comparable heat inputs produce considerably more radiation.

A radiant heat burner emits the highest proportion of its energy in the medium wavelength, while the black heater emits in long wavelength. There are also electrical high temperature heaters which emit short wave radiation. Paint films preferentially absorb radiation at some wavelengths and reject radiation at others so that the medium wavelength radiation does offer the best compromise.

For the sake of better understanding of the differences between drying by radiant heat and by convection heat, it is recalled that in the latter case, the air surrounding the coated articles has to be heated first before the heat is transferred to them for the drying process, the air thus having a retarding effect. Radiant heat is independent of the temperature of the surrounding air, just as the sun's rays reaching our bodies on a cold winter day make us feel their warmth though the atmosphere is cold. Similarly, near infra-red and visible wavelength rays are absorbed by the paint film and/or the base, and the absorbed rays instantly transformed into heat within the film or base [1]. The radiation energy will partly be reflected and transmitted. Whilst absorption of a high percentage of radiated energy is essential, reflection is usually present and is a feature of the visible spectrum. For a given material the magnitude of each effect varies with wavelength throughout the electromagnetic spectrum [2]. The temperature of an infra-red plant rises in a more or less steep curve immediately after being switched on if electricity, or lighted if gas, etc. is the source of energy. The material receiving heat is not necessarily the absorbing substance, e.g. in the case when the coated material is irradiated from the back, which will be sufficient in many cases, or of a transparent lacquer which may be heated by absorption in the underlying substance. Absorption during transmission is likely to give the most successful results.

Great care must be taken to ensure that the intensity of radiation is substantially uniform over the whole surface [3]. Uneven radiation is the source of many failures e.g. the lower portion of a work piece could have remained soft whilst the upper end may have been overcured, showing perhaps yellowing, embrittlement or loss of gloss.

The main functions of a radiant heat plant will be to accelerate evaporation of the solvents and condensation and polymerization of the medium. In the case of paints drying mainly by oxidation the time saving may not be economical.

Its preferential application appears to be for the drying of large numbers of similar simple and suitably shaped articles on a conveyor with approximately the same surface to weight ratio, or on flat banks. The process has also been applied to wooden panels and even furniture. Articles with re-entrant angles require special care, so that the heating of the shadowed portions by conduction will proceed at approximately the same rate as the parts exposed to direct radiation. Any process which can be carried out in a short time at a high temperature instead of in a long time at a low temperature is particularly suited to the application of radiant heat [4]. More uniform heating of paint surfaces is possible when using the i.r. radiation emitted by an IRASER (infra-red amplification by stimulated emission of radiation) [5]. Tunnel drying with gas-heated black emitter panels has been praised by J.H. Nelson [3] for its very uniform results. Ceramic cup gas burners are also in use. Of the electric sources of radiation robust tubular sheathed electric elements provide a higher energy concentration than lamps and should have a longer life. Heating panels can be arranged according to the contour of the work pieces. The merits of infra-red radiation for drying and curing paints, generated by electricity and emitted through fused quartz flat plates have been reviewed by Cohnmann [6].

Handy and easily movable small units, e.g. for repair work or even for purposes of house decorating, have been developed, some of which use butane-type gas.

In a properly working arrangement drying by irradiation should occur so that all solvents and condensation, etc., products can evaporate without disturbing the flow of the film surface. It has to be decided, therefore, according to the circumstances whether the irradiation should be applied from the film side or (if technically possible) from the uncoated side.

It may be observed that properly cured thermohardened films show a considerably higher corrosion protection if treated by radiant heat than when they have been subjected to stoving by convection heat at the time-temperature cycles regarded as economical, for instance at 150°C (250–300°F). The beneficial polymerization process is taken much further in the shorter exposure time to radiant heat.

In selecting a drying plant the many factors which determine its suitability have to be weighed up carefully. Whilst normally a great saving in floor space can be achieved with infra-red plants, this is not always possible. 'In tinplate lacquering, for instance, it is found that, although electric lamp equipment enables stoving times to be reduced by as much as ninety per cent as compared with convection times, it actually necessitates the use of longer plant owing to the need for full exposure to radiation of each individual sheet' [3]. As to the economics of using radiant heat, it is claimed that 'equally fast stoving times can be obtained with forced air circulation, if the circulating air temperature is correspondingly increased, the highly turbulent air flow being aimed to obtain efficient transfer of heat from the circulating air to the work piece. Where a direct-fired system using forced circulation heating can be employed, for a given output the running cost will be approximately half that of a radiant type' (owing to great heat losses). 'The triple-cased (convection) unit may be regarded as the earliest form of a radiant unit, since almost the whole of the heat is transferred by radiation from the inner sheet-metal casing' [7]. To increase efficiency of drying by radiant heat additional devices, such as circulating air, combinations of infra-red heating with ultra-violet radiation and use of air enriched by ozone have been introduced [8].

(a) Excessively long drying times are hardly to be expected if the right type of coating has been selected. Oil-base finishes for the drying of which oxidation plays a great part will, (though not without exception), be much less suitable than thermo-hardening types. Clear lacquers or varnishes over a bright, reflecting base may dry more quickly in a convection oven. Heavy objects need longer drying time.

'A coating over a polished surface will require up to 20 per cent longer baking' [9]. Reflection of some of the radiant heat can take place even through an opaque enamel.

'Silver, gold, copper and aluminium maintain a high reflectivity through the infra-red region of wavelength' [2]. Ferrous metals, including polished steel, absorb the radiation faster than chromium [9] and the former metals. This is likely to affect the drying times of transparent films more than the opaque ones.

Transparent and black enamels on polished brass can be successfully dried by the infra-red process. The time-temperature cycle, however, may have to be adjusted generously.

Differences of temperature in the higher and lower parts of the plant depend largely on convection currents of cool air passing over the lower sections of the objects, resulting in insufficient hardness. This is particularly noticeable if the coating is of the alkyd type [10] but occurs, too, with the thermosetting lacquers of, e.g. low alkyd and high amino resin (e.g. urea formaldehyde resin) content.

'At the top of the infra-red heated tunnel the air temperature may be as much as 45°C (approx. 80°F) higher than at the bottom. This is compensated for, to some extent, e.g. by placing the lamps closer together at the bottom or using larger-wattage heat bulbs there' [9]. Panel types of heat sources are very adaptable. See also sub-sections (d) and (f) below, Heat losses.

Mass (thickness) and specific heat [11] of the base (e.g. of different metals) or of the film depending on their chemical or physical characters, the surface condition of either (whether matt or glossy) and the colour and transparency of the coating are important. Glossy surfaces and lighter colours need longer drying and vice versa. Thus drying times of different shades of an otherwise similar paint can vary somewhat. A white synthetic enamel absorbed 50% of near infra-red radiation, but absorption may be as high as 80% for various colours, approaching 100% for black [2]. Matt surfaces dry especially quickly, causing little reflection. White pigments, especially titanium dioxide [12] in contrast, reflect a great percentage of the incident heat and consequently need longer heating. Chromes and cadmium orange are also strongly reflecting [13]. Green, blue and red are slower than black in that order [14]. The difference in temperature between black and white panels reached a steady temperature is given as usually to be 43–50°C (70–90°F), but only 11–17°C (20–30°F) with gas-heated dull emitter panels [3]. However, as the bulk of the wavelength at which tubular or sheathed electric wire elements operate is long, differences are practically eliminated in such plants (except for some metallic paints), as these long wavelengths are absorbed equally well by all colours from glossy white to matt black [12].

If delay in drying is caused by reflecting bases, the substitution by a non-reflecting surface, e.g. by a matting treatment, should be considered.

Coatings on heavy articles, e.g. shells, can be dried in high intensity radiant-heating furnaces with forced convection cooling. Even thin-walled metal objects have been treated successfully in suitable plants within seconds at very high temperatures (up to approx. 500°C (900°F)) without noticeable discoloration of even touchy colours and without substantial loss of gloss. Paints used must be adjusted for this treatment.

Thermometers cannot be used for temperature control, but, e.g. thermo-couples, or as a guide special paints or crayons, the changing colours of which indicate the temperature.

(b) Overheating, causing excessive dulling, embrittling, also sometimes yellowing or underbaking resulting in incomplete polymerization of the film. Such failures may happen before the most favourable working conditions have been established for a new article, particularly if it is one of differing thickness. The transition from a perfectly dried to a damaged film usually takes only a very short time owing to the short drying periods involved, and because the rate of reaction increases rapidly with the increase in temperature. It must be watched that radiation from both sides (especially in the case of flat articles) may have a detrimental effect. Matt surfaces cause much less re-radiation than glossy or polished ones.

Films of thick-walled objects may be spoiled more rapidly.

Control temperature with a thermocouple preferably on the actual work piece.

Reduce either drying time, e.g. by increase of the conveyor speed or, in electric plants, switch off part of the units or tunnel sections or reduce voltage (if provided for in the plant); in gas-fired plants the gas rate of a whole unit, or of a section only, can be lowered. Local changes in emissivity, as recommended by L.W. Andrew and E.A.C. Chamberlain [4] are achieved by painting the normally black surface of the irradiating panel with heat-resisting aluminium paint. Conveyor speed must always be strictly controlled. The fitting to the conveyor of starwheels and trips which turn the articles steadily will assist to avoid overbaking of parts nearer to the hottest area.

(c) The flow may not be satisfactory owing to:

(i) Rapid setting.

Allow if possible the coated articles to flow out and set in a cool area or pass them through gently rising temperatures before exposure to full heat. Try irradiation from the other side. The addition of some heavy boiling thinner or a plasticizer may help. Act in accordance with sub-section (b) above.

(ii) Too much closing of the external baffles (many plants are, however, open) causes drying times to be exceeded, the flow to be disturbed and heat wasted because the volatiles cannot escape steadily.

Although circulation of air should be kept low to reduce heat losses, it must be sufficient to remove the vapours.

(d) Irregular results are possible if the sizes and shapes and thickness of the bases (especially metals), colour, surface conditions, film thickness vary or if parts of the coated objects are shielded from direct radiation.

Avoid cool air current by better enclosure. Devices which turn the articles steadily are recommended.

It must be considered whether the use of convection heat (e.g. box ovens) would not be more advantageous. Shielded areas can only harden through in a radiant drying plant if the heat conductivity of the, e.g. metal base is good, because radiant energy waves travel in straight lines.

(e) Thermohardening coatings have less tendency to blistering when dried by radiant heat, because of their drying from the bottom upwards. However, it must be borne in mind that heavy coats of such finishes are particularly prone to blistering, owing to the release of water in the condensation process and the rapid setting which tends to seal some of this water prematurely.

(f) The plant must be designed to provide for the maximum flux density in the area where the drying takes place. Heat losses on infra-red drying can be due to:

(i) Rays which do not strike the objects to be dried or which are reflected.

Surround plant with reflecting metal coated or white covers. Trough-type plants and flat banks of lamps or other emitters for flat objects are also available, e.g. for drying aircraft wings.

(ii) Air currents caused by the conversion of radiant energy into heat 'may carry away heat by convection' [9].

Enclose heating unit as far as possible.

Notes

[1] In the drying processes which use radiant energy, thermal energy is generated by absorption and conversion of electromagnetic waves of 770 to 18 000 nm within the body.
[2] Maxted, R. (1942), *Trans. Illum. Eng. Soc. (Lond.)*, No. 1, 1–14.
[3] Connell, L.C., Humphries, O.W. and Rycroft, J.L. (1945), *J. Inst. Elec. Eng.*, **92**, Part II. 388.
Andrew, L.W. and Chamberlain, E.A.C. (1943), *The Institute of Fuel*, July 21st. July 21 st.
[5] Vianova Kunstharz A.G., Austrian Patent No. 329 717; *Paint Resin Pat.* 1976, **13**, 220.
[6] *Can. Paint Fin.* **50**, 11, 1976.
[7] Stordy, J.J. and Appleton, W.G.J. (1956), *JOCCA*, **39**, 573.
[8] Daimler-Benz, A.G., German Patent No. 1 054 884.
[9] Wampler, R.H. (1946), *Modern Organic Finishes*, p. 177, New York.
[10] Frost, R.L. (1946), *Pt. Man.*, **16**, 358.
[11] The specific heat of a substance is the number of joules needed to raise the temperature of one kilogram of the substance by 1°C. The specific heat then, is the thermal capacity of unit mass of the substance.
[12] Dovey, D.M. (1958), *Ind. Fin.*, **1**, 36.
[13] Taylor, D. (1958), *JOCCA*, **41**, 707. The author has classified the various pigments into optical groups.
[14] Quarendon, R. (1942), *Paint Manuf.*, **12**, 51.

4.13 Comparison of the various artificial drying methods

A number of up-to-date drying processes, which may be suitable for specialized work of paint and varnish drying, are covered by this section. They open up ways for the application of forced drying to fields hitherto not considered suitable, and in addition in some cases lead to improvements of quality and production schedules.

The available curing methods are convection (natural and forced), radiation heating (e.g. by gas, by electric lamps and by electric resistor equipment in furnaces of high intensity), dielectric and eddy-current (low and high frequencies) flame curing and radiation by ultraviolet rays and electron beams.

The various defects or disadvantages connected with the better known drying methods, i.e. by means of *convection* (transfer of heat in liquids and gases from hotter to colder bodies by the actual movement of the liquids or air) in box or conveyor ovens with or without forced air circulation or by means of radiation have been treated in special sections. The introduction of fan-assisted natural heat seals on convection ovens has made extremely rapid stoving schedules possible. Such ovens have an open base, but a closed top to prevent escape of heat. The walls of such ovens are not hotter than the circulated air and therefore cannot radiate heat to the work pieces [1]. This makes for uniform heat transfer. Ovens of this type are also being used for so-called 'flash baking' for short time schedules at high temperatures.

Flame curing was used in Continental wine cellarages on wood vats with shellac solutions long ago. It has recently been introduced in mechanized set-ups for its rapidity of drying and curing, e.g. of acrylic resin lacquers. Films pass under a source of heat of say, 650°C (approx. 1200°F) for, e.g. 3 seconds. As the usual hydrocarbon solvents would leave soot, they must be replaced by other more expensive organic solvents. Flame curing of wet coats of certain styrene butadiene copolymers has also been recommended.

Flame spraying of epoxy powders was developed in Japan about twenty years ago and has been used in industry for the last ten years. The powders are conveyed to the front of the gun at a rate of 40–50 m/s by compressed air. The powder particles are melted in the high temperature flame (about 2000°C) and strongly adhere to the substrate. A spray capacity of 5–6 kg per hour is possible.

Artificial drying of coatings can also be carried out by radiation curing, which is a term used to describe two distinct processes. Ultra-violet radiation and accelerated beams of electrons [2] can polymerize lacquers which do not contain organic solvents, in a similar manner to the curing of unsaturated polyester resins by peroxides. The curing process is, however, more rapidly effected by radiation and the following advantages are claimed for these drying techniques:

(i) The level of pollution is very low
(ii) Considerable space savings
(iii) Low energy consumption
(iv) Little or no heat to affect the substrate
(v) Lower wastage of material compared to conventional coatings
(vi) Process operates immediately it is switched on
(vii) Coated articles can be passed immediately to the next processing stage.

Electron beam curing employs high energy levels in the region of $2–5 \times 10^5$ eV and the accelerated electrons generate ions by breaking the chemical bonds (molecular fragmentation). These ions become free radicals and initiate photosensitizers in electron beam coatings although these are essential in ultra-violet curing compositions. The electron beam is so penetrating that pigmented coatings may be readily cured.

The major obstacle to the widespread industrial use of electron beam cure is the very high capital and maintenance cost, and it is necessary to use an inert atmosphere in the curing zone to displace oxygen which would inhibit the rate of cure.

A power unit steps up the line voltage and rectifies it to direct current. An accelerator unit comprising a beam tube, electron gun and scanner produces an electron stream when the high voltage energizes a tungsten filament at one end of the beam tube. These electrons are concentrated by a Pierce lens and accelerated to full velocity by the electron gun. Electromagnets produce a side to side deflection of the beam as it leaves the accelerator through the scanner which is covered at the wide end by a nickel aluminium alloy or titanium metal foil window. This device permits the electrons to pass out of the accelerator without releasing the high vacuum inside.

A typical commercial accelerator produces 3×10^5 eV using a current of 25–200 mA and gives a dose rate of 1–10 Mrad/s. X-rays are produced as a side effect of the electron beam and shielding with lead or concrete is an essential safety measure. Ancillary equipment includes a vacuum pump, a cooling device to prevent melting of the metal foil window in the scanner, and an inert gas generator.

Coatings have been successfully cured by electron beam methods on plastic, metal, wood, leather, concrete, glass and asbestos sheet.

Development work is proceeding to produce an accelerator with improved efficiency so that less energy will be required. This would considerably reduce the capital costs and the need for excessive shielding. Improvements in the scanning device could produce a wider beam increasing the area covered per unit. Coatings developments are directed towards compositions which are not inhibited by air, to remove the need for inert gas atmospheres.

High energy ultra-violet light will cure correctly formulated coatings at ambient temperatures in the radiation range 3–10 eV, which is considerably lower than the energy levels employed in electron beam curing. It is necessary to incorporate a photoinitiator into the coatings and in the case of pigmented finishes, specialized chemical additives are effective.

Ultra-violet energy is obtained by the use of a high intensity mercury vapour tube emitting u.v. light in the long wave end of the u.v. range i.e. 315–400 nm. The curing unit is a relatively simple installation consisting of multiple lamps arranged in parallel within reflectors. A typical unit would consist of 4–12 lamps of 1.3 m (4 ft) length operating under high pressure (5–50 atm) or medium pressure (1–2 atm).

Although the power required is of the order of 80 W/cm of tube the u.v. efficiency is not more than 15–20% with some emission of infra red. The efficiency is higher at internal tube temperatures of 280–300°C which can raise the temperature of the cured coating to 40°–50°C [3].

Curing times vary between five seconds and two minutes depending upon the type of coating, the film thickness and the u.v. intensity. While these curing times are slower than those of the electron beam process they are considerably better than those of convection oven. A comparison of the three processes is given in Table 4.1 [4].

Electric insulators, when subjected to rapidly alternating electric fields, absorb power with a consequent increase in temperature. 'The rate of the rise in temperature depends on the strength and frequency of alternation of the field and the dielectric constant and its power factor' [5]. *Dielectric* (electrostatic high frequency) heating secures the rapid heating of materials which have low thermal conductivities [6]. It employs frequencies above 1 000 000 Hz.

When an electrically conducting material is placed in an alternating magnetic field, eddy currents are induced in it, 'power is absorbed and heat is generated' [6]. In *eddy-current* (*induction*) heating the frequencies employed are usually below

Table 4.1 Comparison of curing processes

	Electron beam	Ultra-violet	Convection oven
Capital cost	High	Low	Medium
Overall length	10–30 ft (3–10 m)	10–30 ft (3–10 m)	90–300 ft (30–100 m)
Switch on operation	Yes	Yes	No
Inert atmosphere	Yes	No	No
Cure temperature (°C)	Ambient	43–55	60 +
Cure time	1 s or less	Seconds	Minutes
Cure coatings on plastic	Yes	Yes	No
Cure pigmented coating	Yes	Limited types	Yes
Catalyst required	No	Yes	Sometimes
Coatings solids	90–100	90–100	35–65
Coating coverage (ft^2/25 μm thickness)	1000–1400	1000–1400	400–500
Coating cost	High	High	Low
Hazard	X-ray	u.v.	Heat

1 000 000 Hz. Following are several factors of practical importance in connection with eddy-current heating:

(i) The thickness of the layer in which heat is generated may be kept exceedingly small by using sufficiently high frequencies. (This allows heating of the surface of metallic articles without appreciably heating the centre, e.g. important in the case of soldered articles and certain light metals.
(ii) The length of an article over which heat is generated may be controlled by confining the magnetic field to the required region.
(iii) On account of their higher resistivity and permeability, iron and steel may be heated more readily than metals such as silver and copper. Higher frequencies must be used for non-magnetic than for magnetic materials, in order to secure the same power input per unit area. (The metal will have to be selected accordingly.)
(iv) In order to obtain a given power input per unit area, a higher frequency must be used, the smaller the diameter of the article to be heated [6].
(v) A fruitful field for the application of eddy-current drying appears to be that of impregnating varnishes on electric coils, the finishing of tube furniture, conduits and the like. The shape of the objects must be suitable for the passage through the inductor coils, i.e. it should be cylindrical.
(vi) Coatings applied to such metallic articles which are fitted with non-metallic objects which are susceptible to heat, e.g. wood, may be dried without harming the non-conductor.

It must be left to the users to select the heating method or a combination of methods to suit their requirements.

The final degree of gloss or of mattness is only settled after practically complete evaporation of the solvents and thinners. A decision as to the exact degree of gloss developed by a film can frequently only be given several hours after drying, at the earliest.

Notes

[1] Anon. (1959), *Ind. Fin.* **10**, 181.
[2] Deninger, W. and Patheiger, M. (1969), *JOCCA,* **52**, 930.
[3] Young, S.E. (1976), *Progr. Org. Coat.,* **4**, 225.
[4] Levinson, S.B. (1972), *J. Paint Tech.* **44**, No. 571, 35.
[5] Connell, L.J.C., Humphreys, O.W. and Rycroft, J.L. (1945), *J. Inst. Electrical Eng.* **92**, No. 29, 385.
[6] Zlade, H.P. (1944), *Plastics,* **8**, 100.

4.14 Damage caused by the stoving of coated pure aluminium, aluminium alloy and magnesium alloy

The stoving schedules given in Table 4.2 refer to traditional convector box ovens. The short or mild methods of force drying may, however, be applied analogously, according to the case. Radiant heat is not likely to have a detrimental effect even on copper- or magnesium-bearing alloys (see sub-sections (c) and (d) owing to the 'rapid soaking with heat' in this process, except on extremely thin metal sheets of, say, 20 S.W.G. (0.91 mm).

(a) Alloys which derive their strength from some cold working treatment, such as cold pressing, rolling, etc., or to a hardening treatment [1] are endangered by heating. This may induce losses of strength properties, and also cause serious distortion.

Baking varnishes must never be employed in connection with aluminium-magnesium alloys, which have been subjected to plastic deformation (see below sub-section (d)).

Thin-walled parts made of magnesium alloys (e.g. Electron metal) are sensitive to high temperatures since these may cause distortion. Thin-walled sections of pure aluminium should not be heated to temperatures exceeding 120°C (250°F). Insufficient stoving on such metals may, however, impair the anticorrosive action of otherwise good coatings.

Magnesium-rich alloys can easily be subjected to stoving temperatures of up to 150°C (300°F) for as long as five hours.

Use a zinc chromate-alkyd resin primer for air-drying purposes, followed by nitrocellulose lacquers.

(b) Heavy-walled sections and castings of aluminium, and many aluminium alloy parts, except the age-hardening alloys [1] can usually be baked in box ovens at temperatures of up to about 180°C (350°F) for a maximum period of two hours.

Table 4.2 Temperature limits for the baking of coatings on some aluminium and magnesium alloys. The main metals only are mentioned

Main metals	Al thin pressed and rolled	Al thick and castings	Al-Mg thin pressed and rolled	Al-Mg castings	Al-Cu-Mg age-hardened	Al-Cu-Ni including castings age-hardened	Al-Cu including castings age-hardened
Temperature 20°C	Air drying		Air drying		Air drying		
80°C					The corrosion resistance diminishes above 80°C. This is, however, not so in the case of plated metals of this group.		
100°C							
120°C	120°C max.			120°C max. 2h	120°C begins losing strength		
150–180°C		180°C max. 2 h			150°C* Several hours. No appreciable effect on the strength. Heated about 2 h at 180°C. Strength is already somewhat reduced. Temperature limit therefore depends on intended use.†		160°C* max. suitable in connection with the annealing process.
Any up to annealing temp. of max:							

Al-Si-Mg and Al-Mg-Si including castings age-hardened	Al-Si-Mg-Mn	Al-Mg-Mn semi-hard and hard metal including castings	Mg Elektron types	Al-Mn	Al-Zn-Cu castings	Al-Si including castings	Al-Si-Cu castings
100°C. Some begin losing strength							
		120°C* max. 2 h				120°C* max. 2 h	
160°C* max. suitable in connection with the annealing process.	150°C* max. 2 h		150°C* max. 5 h				
		180°C* max. without noticeable loss of strength		180°C* max. 2 h	180°C max. 2 h	180°C* max. 2 h	180°C* max. 2 h
		350°C	220°C max. 1 h	350°C		castings: 350°C	

* The compositions of the alloys vary within the same group and with it the temperature limits. In addition the thickness and pre-treatment of the metal has a profound effect. It is therefore advisable to make sure of the directions of the makers.

† Magnesium Data Book, 1961, p. 40.

Such aluminium-magnesium-silicon alloys, though they may be age-hardened, can be heated up to approximately 160°C (320°F), while aluminium-magnesium castings, if baked at all, should not be heated above 120°C (250°F) for more than two hours.

Use coatings drying at low temperatures, for instance, at 120°C (250°F). Varnishes suitable for aluminium can, nearly always, also be used for aluminium-magnesium alloys. Maintain uniform temperatures throughout the box ovens by means of air circulation, so that local over-heating is rendered impossible.

(c) Baking can hardly ever be applied to the following age-hardening [1] groups of alloys containing copper:

Aluminium-copper-nickel, aluminium-copper-magnesium, to which belong the important Duralumin alloys, and also the corresponding cast alloys. Whilst these alloys can be heated to temperatures of up to 150°C (300°F) for several hours without incurring serious deterioration of strength and elongation, their corrosion resistance, however, is noticeably diminished if heated above 80°C (175°F). A reduction of corrosion resistance on heating the parts above 80°C (175°F) is not induced in the case of plated materials of the aluminium-copper-magnesium group of alloys. The age-hardening alloys of the aluminium-copper group, and of the aluminium-magnesium-silicon group, can be heated as those in above sub-section (b), but not above 160°C (320°F), and it is advisable in this case to combine stoving with annealing.

As to air-drying coatings, see sub-section (a) above. The maximum baking temperatures should, in the case of the Duralumin alloys, be limited, depending on the contemplated use.

(d) Investigations conducted by K.O. Schmidt indicate that in the case of certain types of light metals, films of coatings showed apparent brittleness, which in reality had been caused because the structure of the metal itself had possibly suffered certain changes already under the influence of the comparatively low temperature of stoving. It is obvious that in cases of this type the lack of heat resistance does not lie with the coating.

Note

[1] Age-hardening of aluminium alloys of suitable composition is effected by annealing at temperatures of about 500°C (930°F) and quenching in water, followed by ageing the alloy at ordinary temperature, when the hardening takes place automatically.

5

Defects apparent shortly after application

5.1 Formation of blisters and blowholes in filler coats on porous substrates

For pictorial standards see Fig. 5.1.

A definition of blistering is given in Section 5.2.

The enclosure of ultimately volatile or even of gaseous products such as air within thick film layers is the cause. These can sometimes emanate from the substrate.

(a) Excessively thick films. The solvents and thinners cannot evaporate completely during the period of plasticity of the coating to permit the latter to flow out and form a continuous film closing the pinholes originated by the evaporation. Pores of, e.g. pinhead size are thus developed in air-drying and in particular stoving films. The retained solvent residues force their way out of the more or less superficially hardened film.

(b) Rapid charging into initially too hot stoves. The solvent vapours are forced out of the film with great pressure, while the filler film solidifies at the same time. The defect hardly ever occurs where conveyors are used.

(c) The baking temperatures for the last top coats may be too high for previously applied stopper and filler coats which may have been stoved at lower temperatures. The entire structure of the system may thereby be ruined, and faulty areas may show matting through several finishing coats.

Normally thick stoving stopper or filler coats should be left in the air sufficiently long (at least one hour) until superficial drying has set in. They are then placed into the stove and the heat gradually increased. The maximum temperatures for the first applications must always be considered in conjunction with the baking of the last films. It is absolutely necessary, of course, to select all the paint materials so that their technological properties are in agreement with one another.

Blowholes or pores are filled with a knife stopper. The work is then carefully stoved, rubbed down, and a sufficient number of finishing coats are applied.

(d) Stoving a top coat over insufficiently filled polyester glass fibre mouldings or porous cast iron or aluminium sprayed metal coats and other rough areas which thus retain pockets of air. The defect is worse in unduly hot areas of ovens, and the less penetrating the filler has been.

After fettling, if a rough casting, or other preparation to smooth the metal (or other) surface, fill properly, and let the filler coat dry slowly to permit enclosed air to escape while it is still permeable.

(e) Porous castings may be pretreated with a thin coat application of vinyl or phenolic butyral primer after preheating the specimen or better still applying the coating under vacuum which has removed the air from the pores.

(f) Pores of the wood may show up in the finish as a consequence of incomplete filling with a pore filler.

It is recommended to brush or rub in a sufficient quantity of pore filler and to apply with the grain as well as across the grain of the wood followed by sanding. No dust, emery powder, or such-like particles should be allowed to bar the entrance of the filler into the pores.

(g) Premature application of the superimposed films, i.e. before the previous heavy layers of, e.g. fillers have had a chance to dry thoroughly. Solvent residues penetrate through the covering films, which can also cause swelling and lifting.

(h) Some roller-coated tinplate lacquers, especially those based on epoxy resins, have been reported to exhibit blistering, when the strip or sheets had been cooled down with water [1].

Phenolic and vinyl resin based lacquers have been found to be 'quite superior' in this respect. Removal of the water is necessary, as even traces of it can cause also rusting and spotting on coiled or stacked tinplate [1].

(i) Powder coatings have better filling properties than solvent based fillers but they often require the application of a solvent-borne epoxy phenolic primer when the castings are very porous.

Note

[1] Bendix, G.H. and Dempsey, J.N. Jr. (1961), *Paint Industry Mag.*, 2, 11.

5.2 Blistering, sometimes followed by destruction and lifting

For pictorial standards see Fig. 5.1.

Blistering is the formation of dome-shaped projections in paint and varnish films by local loss of adhesion and lifting of the film from the underlying surface (Fig. 5.2). Blisters may contain liquid, vapour, gas or crystals.

Blistering is a common phenomenon and one of the most frequently observed paint film defects. In most cases it is due to liquids or gases trapped in or underneath the film. The causes and the mechanism have been investigated by many workers and there is no doubt that complex physico-chemical processes [1, 2] such as electro-osmosis (on metal substrates), osmosis, diffusion of water, swelling and loss of adhesion [3], temperature rise, and deformation of the film are involved. Of these processes, osmosis seems to be the most universal [4]; most paint films are semi-permeable membranes which permit passage of moisture through the film. Soluble salts in the film are dissolved by the water, generating the high osmotic pressure which leads to blistering. In heavily pigmented films above the critical pigment volume concentration (CPVC – see Section 7.1) blisters should normally not occur unless the adhesion of the film is poor. Such films have high permeability and allow water vapour and gases to escape without blister formation.

Whilst water is the most common cause of blistering and associated failures [5] such as cracking and peeling, even very small quantities are often sufficient to cause blistering particularly on wooden substrates. Underhaug [6] has made a study of blistering on wooden structures and concluded that dampness promotes rapid hydrolysis of linseed oil in primers and can give rise to breakdown products which become liquid in warm weather, destroy the adhesion of the paint film and the expanding air blows up blisters. This phenomenon has particular significance in Scandinavian countries where wood is often oiled with linseed oil before painting and low PVC linseed oil primers are frequently used. This defect does not occur with lead based primers.

Other workers [7] have concluded that blistering on timber structures is most commonly due to extractable constituents of the wood which diffuse to the paint/wood interface under the influence of water and heat. In the U.S.A., where wooden houses predominate, blistering of house paints is often a problem and much work has been done in an effort to overcome it [8]. Suggestions have been made to formulate paints with a greatly reduced tendency to blister by:

(i) Minimizing the presence of unesterified carboxyl and hydroxyl groups in alkyd resins.
(ii) Blowing oxidizing type vehicles, thus preoxidizing them.
(iii) Avoiding alkaline pigments.
(iv) Modifying alkyds by partial replacement of semi-drying oil fatty acids by alcoholized tung oil or eleaosteric acid.

Gans [9] has devised an accelerated blister test for house paints which shows good correlation with field experience. Painted panels are soaked for 7 days in water and then heated in an oven at 150°C (300°F) for up to 30 minutes. Observation at 5 minute intervals permits assessment of the degree of blistering to be expected in service.

Blistering on metallic substrates is often due to the presence of residual corrosion products. On galvanized metal the zinc coating may be porous, particularly after weathering, and the pores are liable to retain soluble zinc salts. Water, which permeates through the superimposed paint film, is absorbed by the zinc salt (chloride or carbonate), thus initiating the blistering process [10].

As paint films age and are subject to stresses, due to further water absorption and thermal expansion and contraction, blisters will tend to crack. As a result the protection of the substrate is diminished.

(a) Influence of direct sun rays or artificial sources of heat on wet or comparatively fresh paint and varnish films (particularly on slow-drying long-oil paint materials). Favoured by insufficiently dried priming paint films, greasy impurities and contamination on the lacquered or varnished surfaces and too amply pre-oiled wood, etc.

If an entire surface need not be recoated, rub the affected sections down to the base, touch up and repaint carefully. If the direct radiation of the sun cannot be avoided, the addition of suitable quantities of a slowly evaporating solvent such as high boiling petroleum derivatives, hydrogenated naphthalene, dipentene, turpentine, etc., is sometimes helpful.

(b) Rain-drops on incompletely dried films (especially of long-oil paint materials, including alkyds) cause dull, spongy or uneven surfaces like cracked blisters, faults by mechanical displacement (and consequent whitening) of the coating, emulsification of the water in the medium or, occasionally showing concentration of the pigment particles at the water-oil interface, visible by the formation of whitish patches.

But even on some dry films rain-drops or condensed water can cause circular, usually lighter coloured spots, e.g. on shellac-containing films. These may not only remain permanently, but also the gloss can suffer, and at times blistering and lifting can occur.

The continuous action of steam or water vapour influences soft, water-sensitive films in a similar manner.

Coatings which have a high linseed type oil content are especially susceptible to moisture.

Painting of timber under adverse weather conditions has been investigated by J. Boxall [11]. The results of this work indicate that many systems have good tolerance to poor application conditions, yet show inadequate outdoor weathering performance. It is suggested that suitable primers possessing good flexibility are most important [12] and that the recent British Standard 5358:1976 (low lead primers for wood) will contribute to the improvement of lead free primers on the market. The use of high performance lead primers is declining owing to the toxicity of lead compounds.

(c) Under exceptionally warm and humid conditions films of nitrocellulose and some synthetic resin lacquers, especially dark colours, have been found to blister. This has occurred in Great Britain; the risk is, however, much greater in the hot and humid areas of the tropics, because of heat absorption by dark colour.

Seeing to it that the coatings are thoroughly dry before exposure, if necessary by applying a force-drying method, will greatly help.

(d) Blisters containing aqueous liquid are formed if coatings are applied to green or moist wood surfaces (due to rain, fog, moisture left from rubbing down, bleaching processes, alkaline paint removers, including solutions of ammonium chloride, which penetrate deeply into the wood) or to moist plaster possibly covered with soluble exudations, or to undercoats still moist from flatting down, etc. A minimum interval of forty-eight hours before any painting is recommended.

Some paints in addition to blistering may not dry properly and show a very patchy appearance if they are applied to a moist base which exudes traces of salts. These salts may consist of efflorescence from the brickwork of magnesium chloride solution, frost-protective salts added to the mortar during the building, residues of sodium chloride resulting from sea sand or untidy work with brine, previous use of fire extinguishers, etc.

Joinery wood which has been piled in a marine atmosphere or transported on decks of ships may contain sea salt in the surface cells. Under the effect of moisture 'local pores, crops of small pimples and efflorescence' may consequently develop on a paint surface.

Blisters are also caused by rain-water entering cracks and crevices of the wood in exposed position after the drying of the film. Painted or varnished wooden parts of new buildings such as flooring, lining, panels, etc., frequently develop blisters caused by the moisture exuded by the fresh brickwork, ceilings, etc. Panels delay the evaporation so that the drying of the walls takes much longer than usual. Cracking, alligatoring and scaling may also develop, frequently leading to peeling of the entire films if the houses are built during damp seasons and if the paint material is applied too soon after the completion of the masonry work. Excess moisture generally moves along the easiest channels, such as resin ducts and medullary rays. The thin-walled cells of spring wood carry water more easily than the heavy-walled, darker summer wood cells. The position of blisters is determined principally by the wood structure, particularly by local variations such as knots. But mechanical changes due to sweating and shrinking and modifications of the wood cells due to pressure from the planing knife have also to be considered. When a paint film is only slightly permeable, lifting and flaking can occur.

Relatively rapid increase in temperature is usually also involved.

As to the counter-measures suggested, see sub-section (a) above.

A good deal of early paint failures through blistering can be avoided by sound design of buildings and steel structures. The aim should always be to minimize condensation by good ventilation, to minimize cold surfaces where condensation can occur and to encourage condensation to move away from situations susceptible to damage [13]. Poor building practice, mainly unsatisfactory storage of components on site, allowing the surfaces both of steel and timber components to deteriorate, is another major contributor to early paint failure.

Moisture remaining from the removal of alkaline paint removers can be avoided by employing one of the organic products, though some of them can also be washed off with water.

The application of a sealing coat over those parts of a wall which exuded salts has chances of lasting success only if the wall was first given enough time to dry out, and, secondly, does not become damp again.

To remove contaminations of salt in wood, break the surface cells by rubbing down with sandpaper, followed by thorough washing with water and complete drying before painting. It is good policy to apply a primer before such exposure. But even then the salt will have to be removed by washing, drying and gentle rubbing prior to painting. Wood should not be painted until it is seasoned.

Wooden parts of buildings should be suitably primed and painted at least at the most exposed sides. Built-in sides of doors and window frames should always be given a priming coat, as they will otherwise absorb large quantities of moisture after being fitted into new walls. End grains must receive two priming coats, because they are usually more exposed and absorb water most easily and carry it by the capillaries over considerable distances.

White lead/red lead primers give excellent performance and are popular in the U.K., but their use is declining on grounds of toxicity. Shop coats of emulsion paints appear to work well and acrylic emulsion primers are now frequently used for factory priming of timber components, often by a dipping process.

All woodwork and the masonry with which it comes in contact must be permitted to dry thoroughly before painting. The maximum water content of well-seasoned building timber in equilibrium with the atmosphere is about 16 to 20% of the dry weight under ordinary exterior conditions. If that water is permitted to escape slowly, the varnish or paint films are much less endangered than by rapid evaporation of, say, 10% of the moisture. The safe content of well-seasoned joinery wood for work in Great Britain is 10 to 12% and about 11·5 to 12% for use in furniture manufacture, of the dry weight; in the U.S., however, only 6–8% for the latter purpose.

In countries with an extraordinarily damp atmosphere it is possible that builders' wood absorbs as much as 30 to 40% of moisture. Such conditions are bound to cause paint failures by blistering and subsequent flaking if very great care is not taken.

When all free water is lost relatively rapidly, wood arrives at the so-called 'fibre saturation point'. Subsequent to this the drying takes place more slowly, so that an equilibrium can only be achieved after several weeks or months. Wood left about in wet situations, as during building operations, undergoes rapid changes of condition. This is more pronounced at the ends of timbers. A dry surface does not prove that the bulk of the wood is dry.

The presence of moisture can be determined by various means, e.g. one of the moisture meters measuring dielectric constant or electrical resistance. Priming coats on wood should dry for about one or two months before finishing. This helps to prevent the penetration of moisture, but if moisture should be present due to condensation, it should be permitted to escape before finishing the job.

(e) Blistering under temperature gradient.

This occurs when a heat flux passes through a coating, particularly a low water absorption coating, to a colder surface as in chimney linings or on cold water pipes exposed to a higher temperature [14]. It has been ascribed to osmotic pressure build-up under the coating and to compressive stresses due to swelling and differing coefficients of expansion of coating and substrate. The phenomenon has been confirmed experimentally. A very gradual raising of the temperature is recommended to avoid this failure.

(f) Gaseous emanations from wood which is rich in essential oils or contains resinous occlusions are liable to cause blistering. In the case of resinous occlusions the first tiny blisters grow and develop finally into cracks. Under the influence of warmth the rosin flows out through them and hardens on the top of the paint film.

Remove the paint with a paint remover, and wash the wood, etc.

Fine blisters can be caused by timber which has been treated with a solvent based wood preservative prior to painting. Such timber must be allowed to dry thoroughly before coating. The preservative treatment is beneficial.

(g) Fast air drying films, if applied too thickly sometimes exhibit bad flow, leading to the formation of numerous tiny blisters. Relatively rapid temperature increase favours the failure.

Thick airdrying films sometimes blister due to entrapment of solvent if insufficient time has been allowed before overcoating.

Other synthetic resin lacquers, especially of the thermosetting types, which contain, e.g. urea or polyurethane resins or shellac varnishes, are particularly sensitive to this fault, especially on stoving. Coatings of this type stoved by radiant heat show less tendency to blistering.

E. Assheton reports that too rapid polymerization on air drying or stoving is responsible. In many cases surface drying or the water which is set free by the condensation process of thermosetting synthetic resins is to blame. In the case of some synthetic resins, e.g. some styrenated alkyds, the time elapsing between successive applications must not be too long if lifting of the film is to be avoided. Chlorinated rubber lacquer films may also exhibit this defect.

(h) Bubbling, especially of low solids p.v.c. organosols, is caused by application of too thick films and/or too sudden exposure to the gelling (fusing) temperature. The diluents used may boil up because they are too volatile, or they could have been sealed in. The gelling at 165–180°C (325–356°F) is an operation which is necessary for obtaining continuous films of all p.v.c. type organosols and plastisols for solvating of the resin particles by the plasticizers.

In the case of plastisols which are to be applied in a single coat in a film thickness surpassing 250 μm (approx. 10 thou) de-aeration of the paste is necessary to avoid the appearance of permanent air bubbles in the film.

The required higher film thickness should rather be obtained by building it up in several coats. It may be more advantageous to work with a high solids organosol or a plastisol. The latter needs no pre-setting at any film thickness before fusing, nor do low solids organosols. High solids organosols, the more frequently used types, need however, pre-setting at approx. 150–165°C (300–325°F), either prior to re-coating, or if the time lapse before fusing is longer than 10 min. Otherwise cracking (mud cracking) could occur. Thick one-coat films of over approx. 150 μm (6 thou) of high solids organosols do not require pre-setting before fusing within 10 min.

(i) Air blisters which sometimes remain on the surface of films produced from very thick paint materials usually represent rather harmless blemishes. They can, however, give the impression that pinholing had taken place.

(j) Blisters due to the formation of chlorine gas have been observed as a result of the action of water rich in salts, especially of sea-water on red lead paint films. However, the continuous action of steam may also induce the formation of blisters in films insufficiently resistant to this action. [See sub-section (b) and (c).]

Employ suitable varnishes or paints for purposes of this type. Good results are frequently obtained with the highly resistant phenolic resin varnishes or chlorinated rubber lacquers. The latter should not be exposed to temperatures above 60°C (140°F) at which it decomposes and yellows, except when it contains heat stabilizer with which it has been claimed to withstand up to forty-eight hours at 100°C (212°F).

(k) Traces of oil from the compressors or moisture from condensation, especially on days of high relative humidity, in the lines of spraying plants, or if the oil and water separators have not been cleaned for some time, may cause the formation of large blisters, although these appear only rarely and in small numbers.

A number of other cases of blistering are mentioned in several sections (see Index).

(l) Water-thinnable stoving paints blister on stoving if the flash-off time allowed was too short. Understandably this can somewhat vary in extent according to the relative humidity prevailing at the time of application.

Such bubbles may collapse partly, leaving a circular image or crater behind.

Apparently the residual water 'boils off' through the film. This observation has been made on water dispersions of polyvinyl acetates, acrylics, polystyrene and their copolymers and styrene-butadiene, as well as on water solutions of aqueous phenolics, melamines and treated oils.

Apart from giving sufficient flash-off time, hot spraying at 50–70°C (120–160°F) has been recommended at least for the curing types as a means of reducing the flash-off time. Not only should this method help to avoid bubbles, but also allow to build up films of greater thickness, as spraying at higher viscosity is possible. To avoid overheating, non-recirculating apparatus is recommended. It has also been suggested that flash-off time in water-thinned alkyd/amino paints can be reduced and 'boiling' eliminated by neutralizing the alkyd portion with dimethyl-ethanol-amine. (Bayer Technical Information sheet KL44173, 1974.) Solvent-thinned stoving finishes are also liable to 'boiling' if insufficient flash-off time is allowed.

(m) Blistering of 'oil-free' saturated polyester coatings. This is due to air entrapped in the sprayed coatings. It is claimed that it can be avoided by addition of 1–3% isophorone or Shellsol R and/or proprietary antifoaming agents (e.g. Byk Antifoam SL) [15].

(n) Internal blistering in bituminous coatings used in lining hydrophores or water conduit pipes [16].

This fault has been traced to expansion of dissolved gas under varying pressure conditions which lead to formation of gas bubbles as in soda water when the pressure is released. The blisters appear in regions of relatively great thickness and not where the coating is thin.

Notes

[1] Kresse, P. (1966), *Farbe Lack,* **72**, 1179.
[2] van der Meer-Lerk, L.A. and Heertjes, P.M. (1975), *JOCCA,* **58**, 79.
[3] Brunt, N.A. (1964), *JOCCA,* **47**, 31.
[4] Prosser, J.L. and Bullett, T.R. (1962), *JOCCA,* **45**, 836.
[5] Whiteley, P., Rothwell, G.W. and Kennedy, J. (1975), *JOCCA,* **58**, 200.
[6] Underhaug, A. (1970), *Fatipec Congress Book,* p. 83.
[7] Ulfvarson, U. and Pattyranie, C., (1972), *JOCCA,* **55**, 207.
[8] Long, J.S. *et al.* (1967), *J. Paint Tech.,* **39**, 169.
[9] Gans, D.M. (1972), *J. Paint Tech.,* **44**, 71.
[10] Etheredge, D.E. (1965), *Decorator,* **64**, Nos. 760, 761 and 762.
[11] Boxall, J. (1977), *JOCCA,* **60**, 43.
[12] Whiteley, P. and Rothwell, G.W. (1975), *JOCCA,* **58**, 90.
[13] McKelvie, A.N. and Sherwood, A.F. (1975), *JOCCA,* **58**, 345.
[14] de Vries, Th. (1970), *Anticorrosion,* **17**, 11.
[15] Bayer, A.G., Technical Leaflet: *Alkynol 1160 Polyester.*
[16] Bilven, P. *et al.* (1972), *J. Appl. Chem. Biotech.,* **22**, 165.

5.3 Lack of opacity or hiding or obliterating power

Opacity can be defined as the ability of a paint film, when applied at a given thickness, to hide or obliterate the surface or undercoat. It should not be confused with 'covering power' which indicates the area of surface over which a given volume of paint can be spread.

The measurement or assessment of opacity must be carried out at a known film thickness and the standard method [1] generally employed specifies the use of steel shims and doctor blade for casting the films. The average dry film thickness of a single coat of conventional paint is of the order of 25 to 37 μm (1 to 1·5 thou) but high-build types can give dry films of 125 μm (5 thou) and upwards.

In addition to film thickness, the opacity of a paint will depend on the type of pigment used and its absorption and reflectance characteristics. Blacks and dark colours e.g. browns and dark greens absorb a high percentage of the incident light and confer high opacity. Titanium dioxide has a high refractive index and reflects or scatters most of the incident light and so paints based on this pigment also show good opacity. A number or organic pigments, when dispersed in organic media, give semi-transparent films due to the proximity of their refractive indices to those of the media. In this case the opacity cannot be improved significantly either by increasing the film thickness or the amount of pigment in the paint.

It should be noted that paint films which dry to a matt finish usually show good opacity as a result of the scattering of the incident light by the uneven film surface.

It has been shown [2] that the opacity of an alkyd/titanium dioxide film is related to the composition of the adsorbed resin layer at the pigment surface, viz. the acid value and molecular weight of the adsorbed resin.

Poor opacity can be due to one or more of the following causes:

(a) The paint has been applied too thinly or too much is expected from a paint. Gloss finishes are generally lightly pigmented to give maximum gloss and pale shades cannot be expected to obliterate dark shades in one coat. Undercoats need to be used and the shade should approximate to that of the finish.

Thixotropic or 'non-drip' paints possess very little 'brush drag' and, in consequence, there is a tendency (which must be resisted) to spread them too far. The resulting films show poor opacity.

(b) The paint may have been overthinned. This results in a thin dry film and a tendency to run and sag on vertical surfaces.

(c) The paint may have been in stock for a long period and settled pigment may not have been thoroughly re-dispersed.

(d) The pigment content of the paint is too low. This can occur in finishes but more often in undercoats or low gloss paints. In the latter it is usually due to an excess of extender and too little opaque pigment. Extenders contribute little to opacity since their refractive indices are close to those of organic media.

(e) The pigments used may have refractive indices close to those of the paint media. This applies to many organic and a few inorganic pigments.

(f) In thin films of white paint (as, for example, those produced in roller coating) the opacity is influenced by the reflectance characteristics at the paint/substrate boundary [3].

(g) The pigment may have flocculated during the drying or curing process. Pigment flocculation has been encountered during the curing of white and pale shades of certain acid-cured alkyd/amino finishes. Here it appears to be associated with reaction between the acid and the basic oxide coating of the titanium dioxide.

(h) Pigment flocculation can also occur in emulsion paints resulting in loss of opacity. The inclusion of a small proportion of finely divided silica (e.g. Gasil [4]) helps to keep the pigment in a dispersed condition. Methods for improving the opacity of emulsion paints by modification of pigment composition, pigment content (PVC) and particle size of dispersed polymer have been suggested by P.D. Aronson [5].

Notes

[1] Specification B.S. 3900, Part D4.
[2] Goldsburgh, K. and Peacock, (1971), *JOCCA,* **54**, 506.
[3] Tunstall, D.F. (1967), *JOCCA,* **50**, 989.

[4] Joseph Crosfield and Sons Ltd, Warrington
[5] Aronson, P.D. (1974), *JOCCA*, **57**, 59.

5.4 Lack of filling power (or film build)

Filling power can be defined as the ability of a coating to fill pores, slight depressions and irregularities in a surface. The application of very thick films should not be attempted as a means of overcoming lack of filling power in a conventional paint.

Filling power is a property usually associated with primers and undercoats rather than finishes. In industrial finishing operations, particularly in the painting of metal articles, the surfaces are primed and then sprayed or brushed with a specially formulated filler. This is rubbed down before subsequent coats are applied and under these circumstances little trouble is encountered.

In some types of work the use of filler can be avoided by the use of high-build coatings which yield films sufficiently thick to fill the irregularities. Chlorinated rubber, two-pack epoxies and vinyls are produced in high-build form. These high-build coatings are more expensive than the conventional types but this is offset by the fewer coats required to achieve a specified film thickness. Even thicker coatings can be obtained by the use of solvent-free epoxies, some polyesters and epoxy/pitch compositions.

Whilst it may be convenient to use filler on manufactured articles, filling ability of primers and undercoats is often more vital in maintenance work where patching may be necessary.

On metals, poor filling power can be evidenced in several ways involving the shining-through of metallic surface defects, crevices, pores etc. The causes may lie

(a) Too few coats of primer or undercoat have been applied (this fault may lie in the specification).

Apply either a suitable stopper or a well-filling undercoat. If a high degree of resistance to shock, etc., is required, an increasing number of paint films must be applied (possibly without the use of a filler). The differences of filling power of the various types of paint materials are due to the nature of the vehicle and to the kind and quantity of pigment contained. It is obvious, of course, that the latter factor is limited to a greater or lesser extent by the degree of gloss or flatness required, as well as by the necessary proportion and the type of vehicle used. In making comparisons, attention should be paid to the fact that a flat film appears to be of greater filling power than a glossy one, due to the lesser reflecting capacity of the film with regard to the defects of the base, e.g. grinding scratches.

(b) The parts to be coated may show excessively coarse grinding marks or roughness from the filler or from phosphating.

In view of the fact that fine-grinding, or polishing, of say, the metal base is normally more expensive than the building up of some films of coatings, deep marks and cracks caused by mechanical treatment of the metals (such as drawing operations) will usually have to be filled by the use of an efficient stopper. If the metallic surface is buffed with a fine-grained emery wheel, costs are increased accordingly, but the finishing treatment of the parts is simplified in proportion.

Attention must be paid to the fact that ground metallic parts should be stored in dry rooms laid out with wooden or other soft flooring, or they should be piled up on cardboard in order to prevent scratching by sand, etc. It is possible in some instances to substitute the fine-grinding treatment of metallic surfaces by special fillers or primer fillers, which have good adhesion on bare metal, and exhibit the filling capacity of a stopper without its relative degree of brittleness.

Painting must be executed at the earliest possible stage.

The condition of pre-treatment baths must be checked in regular intervals.

(c) The applied films are too thin as a result of poor application, or the paint having been overthinned.

(d) Paint material unsuitable for one single or a small number of coats.

For the sake of economy it is usually attempted to manage with as few coatings as possible. It is true that in many instances excessive requirements are made on the filling capacity of paint materials which sometimes can be fulfilled only at the expense of other important technical properties.

(e) Wood usually absorbs an appreciable quantity of a coating, depending on the type and hardness of the wood, causing patchy results on parts or over the whole surface. Softwoods possess a large pore structure and have high absorption; the small pore hardwoods have relatively little absorption and the large pore hardwoods occupy an intermediate position.

Much of the joinery timber used in building construction is softwood but a single plank can comprise both dense heartwood and more absorbent sapwood. Examination of a prepared wood surface under magnification will reveal a large number of slight depressions of various sizes (cells which have been cut through) and the fibrous cellulose structure. These depressions in softwood are seldom sufficiently large to justify the use of fillers but in high grade work the primer is often followed by two or more coats of undercoat, with rubbing down between coats.

Primers for softwoods are generally based on oleoresinous media, alkyds or one-pack polyurethanes. Pigments used are titanium dioxide, aluminium (mainly but not exclusively on hardwoods) and calcium plumbate. Acrylic emulsions are increasingly used as primers for wood. The traditional 'pink' primer, based on white and red leads in linseed oil, is unsurpassed for protection but is now little used on account of toxicity.

Primers should possess good hiding power and adhesion to the wood surface; at the same time they should seal the surface so that on application of further coats there is no patchiness between adjacent areas of heartwood and sapwood. This is particularly important in exterior joinery where moisture penetration, which is greater in the open grain sapwood than in the denser heartwood, can lead to blistering and flaking [1]. Where hardwoods are used, for example, for exterior window sills, hard drying aluminium primers are frequently used. Good adhesion is essential. Large pore hardwoods require special treatment (see below).

Lack of filling power on wood can lead to the following results:

(f) Patchy areas which may be localized or distributed over the entire surface. The wood is absorbent. More than one coat of primer is necessary. Alternatively a more heavily bodied primer can be used.

(g) Wooden pores were insufficiently filled with pore filler, etc., or insufficiently rubbed down or the grinding dust was subsequently not completely removed. In the latter case checking above the pores is also possible.

For cellulose and shellac lacquers it is advisable to apply a thin coat of lacquer before the pores are filled. This may prevent the whole built-up film system from losing gloss and to lack in filling power showing indentations over the pores. The wood should be brushed clean before the pores are filled.

In the case of porous woods and especially large-pore hardwoods such as oak, it is necessary in order to get maximum smoothness of finish and therefore least dirt retention and maximum durability, to fill all large pores. This cannot be achieved by a single application of priming paint, and it is usual to use a filler after the priming coat. (In the U.S.A. painting practice is to use a filler followed by primer. British practice is to use primer first, then to fill with an oil-type knife or brush filler.) Inadequate filling results in early breakdown at the sharp edges of the large pores [2].

(h) Plaster surfaces, particularly where repairs have been carried out, require special attention, otherwise patchiness will result. Rough plaster must be rubbed down and, if possible, given a 'skim coat' of fine plaster. All new plaster surfaces must be adequately sealed to prevent patchiness. Vinyl emulsion sealer is effective but two coats are usually necessary. Where new and old plasters are to be joined up, the application of a coat of vinyl emulsion sealer to the edge of the old plaster will reduce the absorption of water from the new material and result in a more satisfactory bond. Plaster surfaces are usually finished with low sheen emulsion paints, low sheen alkyd-type paints or wallpaper. With such coatings complaints of lack of filling power are infrequent. When they occur in the two former types the cause is usually bad sealing of the surface resulting in uneven absorption. The application of highly pigmented finishes over an inefficiently sealed surface will not seal it even if several coats are applied. The only remedy is one or two coats of appropriate sealer over the entire surface before re-finishing.

Notes

[1] OCCA (1976), *Introduction to Paint Technology,* 4th Edn, p. 156.

[2] van Loon, J. (1966), *JOCCA,* **49**, 844.

5.5 Patchiness in the form of lighter or duller areas in films on newly finished or refinished motor cars, etc.

(a) Filler lacquered over before completely dry.

(b) The filler or knife stopper has been insufficiently or unevenly rubbed down, particularly along the edges, so that no satisfactory surface can be obtained with the superimposed, e.g. thin cellulose lacquer films, causing a patchy film surface.

It is absolutely necessary for the production of a perfect finish over fillers, etc., to apply the latter in sufficient film thickness in one or several layers with a knife, brush or by spraying according to the circumstances and to rub down each coat smoothly with water-resistant silicon carbide abrasive paper, i.e. first with a coarse grade (e.g. 180), then a finer one (400). Oil-base and alkyd resin varnishes are less sensitive in this respect than the very much thinner cellulose lacquers. The application of a surfacer over the filler is advisable.

(c) The last traces of water from washing films of fillers, primers or old paints, e.g. after rubbing down, sanding, etc., must be removed by leathering before they dry. This is particularly important in connection with the finishing of car bodies. Hard water is worse than soft in having 'a tendency for lime rings to be produced, and rapid breakdown of the paint film may result' [1]. There is a similarity with the blistering of paint films on phosphated metal over residual water-soluble salts, possibly deposited from the rinse water.

Swabbing over with a chamois leather removes the bulk of the water, leaving only a film to be evaporated in the oven [1].

(d) Application of an insufficient number of coats or of, e.g., a cellulose enamel of insufficient filling power on patched-up objects.

It is obvious that additional coats are required, and the technique of hot spraying should be considered as an alternative to conventional spraying.

(e) Insufficient film thickness of the finishing coats over coloured, e.g. nitrocellulose undercoats, or the filler, so that thin and translucent areas are caused on polishing (but no peeling).

The layers of stopper, etc., must be rubbed down carefully, so that thin lacquer films in contrast to the thicker oil-base, etc., films are not rubbed-through so easily.

(f) Unevenly thick or not full enough applied coats, e.g. because they were not sprayed on crosswise, look patchy due to thin film areas allowing the differently coloured base to grin through.

(g) Patches may also be formed if in repairing lacquered surfaces the stopper used exhibits a porosity different from that of the old filler, or, in the case of stoving, the temperature given to the repair coat is higher than that applied to the previous film system. Blowholes may appear at the same time.

Select a more suitable filler. The application of a layer of sealer or surfacer may have to be considered.

Note

[1] Harvey, A.A.B. (1953), *Ind. Fin.*, **4**, 632.

5.6 Spotted, cloudy and/or greyish wood lacquer films

Some of the following failures may be noticeable immediately or within days or months. A number of them can occur not only with cellulose lacquers but also in connection with cold curing urea-alkyd, polyurethane or polyester lacquers.

(a) Defects located beneath the lacquer films: Greyish pores, or spots and streaks in a polished cellulose lacquer film of a colour varying from grey to white or a cloudy effect in the film may appear on wood, especially if stained in a dark colour.

(i) This may be due, e.g. to the penetration to the pores of glue used in the manufacture of the veneer wood. It is a sign that the excess of glue had not been thoroughly washed off.
(ii) Another possibility is that dust from grinding of the veneer or from rubbing down the pore filler had not been completely removed, the former by washing or brushing, the latter by brushing or wiping. This is particularly to be expected if an excess of or unsuitable filling powder has been added to the pore filler.
(iii) The colour of the pore filler has, of course, to be compatible with that of the wooden base. If the pores are sealed by a coat before the filler could dry thoroughly, greying and also lifting are possible.

The use of veneer which is bonded with synthetic resin may be preferable. Application of a wash coat is advisable.

(b) Defects appearing within cellulose wood lacquer films: Excessive pre-oiling of the wood may cause cloudy films, especially if mineral oil was used.

The too early and too lavish use of polishing oils, etc., for levelling and final polishing is likely to lead to the retention of some of the oil in the upper face of cellulose lacquer films.

(c) Defects noticeable on a cellulose film surface: Greyish streaks and patches on the polished film surface visible after complete hardening through may result from the lavish use of pumice powder for polishing if the film was not dry enough for this operation or if the powder was not properly removed before the application of the re-softening levelling coat, as usually carried out with a solvent mixture. Similar trouble can arise if the dust ground off on flatting the films was not completely removed before the levelling operation or before another coat was put on.

Scratches on the film surface and frequently unsatisfactory gloss are the result of premature sanding and polishing whilst the film is still too soft.

In two-component lacquers softness can have been caused by wrong dosing of, or even omission to add, the catalyst or accelerator. Alternatively the catalyst, etc., may have been added and reacted prematurely. Also the shop temperature could have been below the minimum allowed.

5.7 Blushing, milkiness and dulling, especially of clear lacquer

Blushing is a milky opalescence which sometimes develops as a film of lacquer dries.

It is usually confined to lacquers which dry solely by evaporation of solvent. Most commonly, blushing is caused by the action of water from the atmosphere under conditions of high humidity (water blush [1]), which is taken up by the drying lacquer to such an extent that the dilution ratio of the solvent for water may be exceeded [2]. Thus precipitation of resins or cellulose is induced.

'The final temperature reached on drying may be substantially below the dew point of the atmosphere and minute amounts of water are deposited into the lacquer. This leads to the formation of an emulsion, producing a semi-opaque, whitish appearance. When the water has evaporated the space of the water globules is filled by air leaving a porous (matt) milky film [3]. The decorative effect and mechanical resistance of such films is reduced if milkiness remains more or less permanently visible.'

(a) High relative atmospheric humidity of above 80%, causing the condensation of water due to the low temperature, which may be below ambient temperature, produced by the evaporation after application of the solvents and thinners, [3] particularly on damp hot summer days. This can occur in well-conditioned workshops and even in bright sunshine. The humidity of the air can change considerably even within half an hour. If the parts to be lacquered are cooler than the surrounding atmosphere, this failure will, of course, be aggravated accordingly. In the case of spirit, vinyl resin varnishes or of such (e.g.) nitrocellulose lacquers which, owing to their content of alcohol soluble nitro cotton contain a larger percentage of high-grade methylated spirit, the failure can be promoted as a consequence of hygroscopicity. Cellulose acetate lacquers show for somewhat similar reasons a great tendency to blushing.

When the evaporation of water from the liquid film is faster than that of its solvents, only temporary blush will occur. The proportion of water decreases as the drying proceeds until the blush completely disappears. This happens frequently with shellac lacquer films. If, however, the evaporation of the water is restricted by the high humidity of the air 'the relative proportion of water in the film will increase rapidly by the evaporation of the solvents. Under such circumstances the intensity of the blushed condition will increase rather than decrease as the drying of the film proceeds [3].

In order to combat the adverse influence of moisture one endeavours to use, in general, water-repellent solvents which do not cause a big drop in temperature due to evaporation or which form azeotropic mixtures with water, but have a convenient evaporation rate, e.g. butyl alcohol [2].

It is preferable to check the atmospheric humidity with the aid of a hygrometer. This fault can be observed particularly in transparent films, but it may also appear in rapidly drying pigmented films, although in these instances it is not as easily recognizable; as a rule, pigmented lacquer films show the consequences of atmospheric moisture only as more or less noticeable dulling, while under very unfavourable conditions whitish disoloration may be caused. The recognition of this trouble requires some experience.

(i) Add slowly evaporating thinners, known as anti-blush thinner, procured from the manufacturers. This step will assist the water to evaporate before part of the solvents.

(ii) Warm the parts before lacquering to a temperature of about 40°C (110°F), or force dry the fresh films slightly during cooler spring or autumn days when heating of the entire shop may prove too expensive. This, however, does not always help. For films of shellac lacquers temperatures of up to 60–140°C (140–284°F) have been recommended.

The warming disperses the condensed water vapour. The lacquers themselves should be at room temperature when applied. If they are to be warmed, this should be done with great care owing to the presence of volatile solvents, etc.

Hot-spray lacquers, provided they are used at the correct temperature (for nitrocellulose maximum 75°C (167°F)) do not suffer from 'water blush', most likely because the temperature of the substrate never drops below room temperature whilst the film is forming. Some resins, in particular toluene sulphonamide, as used in nitrocellulose lacquers are very susceptible to overheating, which can more easily occur in circulating hot spray units than in those where heating takes place only immediately before spraying. Overheating causes a mechanically weak film.

(iii) In case of spraying reduce the air pressure. The brighter the metal base, the more visible are the faults and the thinner the gauge the more sensitive are the films.

If the blushing occurs on dipping, especially on the sharp edges of thin mass-produced metal articles, dipping into the same lacquer should be repeated carefully (re-solution), avoiding draughts. Re-spraying of such articles is less risky. Blushed films of pigmented lacquers are sometimes improved by re-coating under normal conditions with a suitable, if possible transparent, lacquer.

(iv) Spray a mist coat of thinners containing anti-blush solvents over the milky film.

(b) Very rapidly drying lacquers containing a large percentage of low-boiling solvents and thinners. Such films, e.g. of shellac or vinyl polymer lacquers, may apart from blushing, even be porous.

The whitening of shellac films disappears rapidly, if the atmospheric conditions are not too abnormal.

In order to ensure compatibility until all the volatiles have evaporated, thinners used should contain a proportion (tail) of solvent of low volatility and good solvent power.

Wood, cardboard, plastic masses, etc., which have much lower heat conductivities, are not so prone to these faults as are metallic surfaces.

(c) Water has contaminated the lacquer. The following precautions are recommended:

(i) Empty the water and oil separator of the spraying plant regularly.

See below, sub-section (f). Unsuitable thinning with methylated spirit of too high a water content.

(ii) Avoid the formation of condensed water in the air piping, compressed air receiver, compressor, containers, etc., by fitting and keeping them in comparatively warm places.

(d) Unsuitable thinners such as, for instance, those containing large percentages of petrol or if the solvents contained evaporate faster than the thinners. A high content of a water-soluble solvent or diluent, see above sub-section (a), even if it evaporates slowly, may also be detrimental. Even thinners supplied by manufacturers which are adapted to the proper lacquers must be added with care. They must be compounded in such a way that they dissolve and keep in solution, e.g. the resins, cellulose and 'oils' of a lacquer, until the film is set, as otherwise throwing out of one or more of the solid substances becomes visible immediately after application as 'gum or cotton blushing' [1]. If it happens in the containers, deposits or skinlike separations may be formed.

Thinners must always be added carefully, and not in one go, to avoid throw-outs. They also ought to be stirred in thoroughly.

(e) Cold and draughty finishing shops, especially during the cold seasons.

The temperatures prevailing in painting shops are frequently overestimated, particularly during spring or autumn. The most satisfactory operating temperature is about 20°C (70°F). Heating of these shops should be indirect, due to the fire danger.

(f) Excessive thinning or incomplete mixing, say, of spirit varnishes and nitrocellulose lacquers made from alcohol-soluble nitro cotton with methylated spirits. The same quantity of different thinners may dilute differently.

Do not overdo the thinning of spirit varnishes. If nitrocellulose lacquers of the type in question must be thinned considerably, special thinners should be procured from the manufacturers. It is always necessary to mix the thinners with the lacquers as intimately as possible so as to obtain a perfectly homogeneous mixture. Non-dewaxed shellac varnishes, however, always look cloudy in their liquid state.

Employ high-grade spirit denatured with methyl alcohol.

(g) The nitrocellulose lacquer films may cover wood containing vegetable glue of alkaline reaction. This glue might possibly absorb water and sooner or later cause hazing. Similar troubles may be caused by residues of oils or fats, especially on metal bases.

Use animal glue for any purposes of this type, and do not apply the lacquers until the glue is absolutely bone dry. (F. Zimmer.) Degrease carefully.

(h) Some transparent or pigmented (e.g. cellulose acetate) lacquers which were originally in good condition can deteriorate on long storage when acidity is developed (e.g. in white glass containers, due to the influence of light on the cellulose or plasticizer). Not only does blushing of the film take place, but also some reaction with, say, a metallic base (on, e.g. an iron base brownish or yellowish spots or patches). In the case of iron, under-rusting may develop. Similar failures are possible with air-drying crystal lacquers containing nitrocellulose and acid crystals.

Such lacquers can hardly be reconditioned, and if faulty should not be used on a metal base. In some cases stabilizing could have avoided the defect.

Notes

[1] Kraus, A. *Handbuch der Nitrozelluloselacke,* Teil I. Lösungsmittel, p. 62.
[2] Jordan, O. (1937), *Technology of Solvents.*
[3] Bogin, Ch. F. in Matiello, J.J. (1941), *Protective and Decorative Coatings,* p. 689.

5.8 Differences of shade in paints by comparison with a standard

These faults make their appearance during their application as well as later on, the difference in shade being noticeable on comparison with that selected from some colour chart or with a customer's standard. Colour comparisons should always be carried out under strictly standardized conditions. For example, the British Standard (Automobile Series) Au 148, part 1, 1969, specifies the type and size of the panels (150 x 100 mm), surface preparation, the coating method (BS 3900) and the type of illumination under which the painted panel should be compared with a panel of a standard shade. The light source must emit light with a colour temperature of 6500 K and produce not less than 1000 lumen/m^2. The interior of the booth in which the comparison is carried out must be painted with a specified matt grey. Only if these or other similarly precise standards are observed, can reproducible results be expected. It is most important always to use a well defined light source as many colours will appear quite different if viewed under different illuminants. This phenomenon is known as 'metamerism'. In works practice it is quite usual to compare colours with the standard shade in diffused light near a window facing north.

(a) If for any reason the degree of gloss or of flatness of a coating differs due either to the properties of a pigmented or of a clear last coat, the colour shade when dry will appear to differ, even if it is correct in the wet condition. Matt films will look whiter, glossy films darker.

In comparing colour shades of pigmented paint materials, it is necessary that the films shall be of the same degree of gloss or flatness. It is also necessary in all these cases to determine whether identical conditions prevail, including the condition of the surface of the base, lighting, etc.

(b) The surface condition of the base or undercoating as well as its colour, influence the colour shades of the last coat.

Only the colours of extremely well-pigmented layers of paint are practically uninfluenced by a coloured base or undercoating.

A white base, for instance, renders the films of all but dark enamels more bright and lively, while dark bases, such as iron and steel plates, exert a discolouring influence. The same white undercoating can be employed for differently coloured last coats, in many cases. The colour of the fillers should also be considered in relation to that of the last coat.

(c) It is also important that separation of mixed pigments having different specific gravities should be prevented in containers stored for a considerable time. It is obvious that in case of separation of these pigments and if thorough stirring is not successful the colour shades of the resulting films cannot be the same as those of the formerly homogeneous paint. It is even possible in instances of this type to obtain very differently coloured films from the same container.

(d) When investigating the fading, discoloration or yellowing of films, attention must also be paid to the different behaviour of transparent last coats towards pigmented undercoats in so far as transmission of light or ultra-violet rays are concerned.

(e) In the case of coatings containing certain dyes it may happen that their films, sometimes shortly after drying (generally by stoving) or much later change their colour, showing some sort of blooming, bronzing or very fine efflorescence. (See also Bleeding.)

This may be due to sublimation or migration of recrystallized particles of colour to the surface of the film, involving some loss of gloss. Heat would promote and accelerate the trouble, e.g. for toluidene reds at temperatures above 150°C (approx. 300°F). Such coatings are also inferior as to light fastness.

Other pigment dyestuffs which could be affected in this way include the phthalocyanine pigments which may show a reddish bronzing, and some of the azoic reds and maroons which may show a bluish bronzing. The right selection of solvents should be made by the paint maker.

A coat of clear lacquer or varnish containing suitable solvents can often remedy the above type of bronzing if applied after cleaning the surface with an appropriate solvent, with which the dyestuff will not bleed, in particular white spirit.

(f) The colour of coatings applied to surfaces which are mainly excluded from daylight may darken or yellow in comparison with a standard and with other surfaces of the same material on which good daylight falls. This effect is pronounced in air-drying gloss enamels, particularly pastel colours such as light blues, greens, etc. Situations where this can occur are numerous including the inside of cupboard doors, garage doors, etc., and in particular colour standards! This type of colour change is caused mainly by the yellowing of the oil contained in the varnish or medium.

Re-exposure to daylight will gradually reconstitute the colour match. Similarly the standard may have changed, e.g. through having been kept in a drawer.

(g) Changes of colour in paints containing Prussian blue or phthalocyanine blue occur on storage.

(h) Colour differences on woods are caused by their varying self-colours. Only a few white woods such as birch, hornbeam and maple are exceptions to this rule.

Elimination of the differences should be as far as possible attempted by bleaching prior to any staining.

(i) Instrumental colour matching, although involving expensive equipment, is becoming more generally accepted in the paint industry. The Paint R.A. has developed a colorimeter [1] which is particularly suited to accurate measurement of the colour of surfaces. It is known as the 'fibre optics colorimeter'. It differs from more conventional colorimeters in that the viewing head, containing illuminating and viewing elements, is separated from the body of the instrument by fibre optics light guides. These consist of bundles of thousands of small diameter glass fibres, which conduct light without any loss to the detector head. Great versatility can be achieved as the flexible fibre light guides can be several metres long and several viewing heads can be used with areas of illumination varying from 1 mm^2 to 100 cm^2. The instrument is also suitable for comparing the colour of bulk batches of wet paint and can therefore be used for colour matching during paint manufacture.

Note

[1] Ishak, I.G.H. (1971), *JOCCA*, **54**, 129.
British Standard Colours: B.S. 381C (1964), *Colours for specific purposes.*
B.S. 4800 (1972), *Paint colours for building purposes.*

5.9 Iridescence, rainbow effects, on films of transparent lacquers

(a) Application of excessively thin films of about 1 μm of nitrocellulose, or synthetic resin lacquers, for instance, on polished silver, brass, glass, etc., may lead to iridescence. Iridescent areas have also been observed on films of emulsions. This may be due to similar causes. The 'colours of thin films' are due to the interference of light [1].

Use less thinner. In the case of nitrocellulose solutions the phenomenon makes its appearance as soon as the cellulose content drops to approximately 2%; some synthetic resin solutions show it with about 10% solid content.

Transparent films which have been exposed to heat for a long period can become so thin that they produce iridescence because part of their substance has volatilized.

(b) Application of thin cellulose brushing lacquer with insufficiently filled brush or cotton pad causing lack of wetting of the lacquered surface.

Use a soft hair brush or a cotton pad to soak up the lacquer fully and brush over lightly.

(c) The lacquer is applied on an incompletely degreased surface.

Especially for clear lacquers the base must be freed from even the slightest trace of fat, oil or grease. Exceedingly slight traces of fatty substances will cause iridescence of the lacquer films, and it is a typical indication of incomplete degreasing; besides a pronounced lack of adhesion is possible.

Note

[1] Such films are only a few light wave-lengths thick. The phenomenon is seen as a set of coloured bands, the contour of each band marking a change in film thickness, each colour representing a particular thickness.

5.10 Various degrees of gloss and appearance of black enamels

Gloss is the degree to which a painted surface reflects the light in a mirror-like manner (specular reflection). The following stages in increasing order of gloss are normally recognized:

Flat (*or matt*). A surface free from sheen, even when viewed from an oblique angle.
Eggshell flat. A surface showing very low sheen.
Eggshell gloss (Silkfinish).
Semi-gloss. A surface possessing moderate reflecting properties.
Full gloss. Smooth and almost mirror-like surface when viewed from any angle.

A number of instruments are available for measuring gloss by comparing the percentage of light specularly reflected from a surface at 45° or 60° compared with a standard high gloss tile (see Section 5.18).

Gloss is not a single sensation, but a complex of at least three simpler sensations. These were found to be: sharpness of mirror image, variations in the brightness of the surface when viewed at different angles, and the parallactic effect in which we seem to be looking at one surface through another. Gloss is a combination of all three, but not in fixed proportions. When a beam of light strikes a matt surface the incident light is scattered in all directions, and little or no trace of a specularly reflected beam can be seen. This is known as 'diffuse reflection [1]'.

There is a definite relationship between gloss and surface roughness (see Fig. 5.3). Finer grinding increases gloss by reducing the size of pigment grains protruding from the film [2]. The size of these humps must probably be several times the wavelength of light in linear dimension in a completely matt surface. 'Surface features much smaller than the wavelength of light may cause the surface to be veiled', i.e. to appear matt, when viewed at normal incidence. Larger-scale features affect gloss over a wider range of angles of incidence, but some grazing incidence sheen remains until the features are many times larger than the wavelength of light. Semi-gloss or matt paint films obtained by using high pigmentation or metallic soap additions usually exhibit surface structure of a relatively larger scale, but the surface is undulating rather than rough [3]. See Fig. 5.4.

'A surface imperfection, in order to cause a measurable reduction in gloss, must be of the order of magnitude of 0·5 μm. Since the pigments which are commonly employed in the manufacture of gloss paints or enamels possess average particle diameters considerably smaller than this, it follows that only aggregates, agglomerates or flocculates of quite a few particles are able to affect gloss adversely. It also explains why relatively coarse inerts must be employed if a high degree of flatness is desired' [4].

The degree of mattness of a coating depends partly on type of pigment and its percentage in the dry film. Heavily pigmented films can contain pigment particles or agglomerates ranging in diameter from less than 1 μm up to 100 μm. This can cause much light scattering at the air/coating interface and almost eliminate specular gloss [5].

The sensitivity of measurements to the surface character of the sample (a paint film) makes it important that all colour measurements should relate only to films of the same degree of gloss. There is no way by any single colour measurement to differentiate between decreased gloss and increased whiteness, e.g. through fading [3].

The reason for too matt or grey an appearance of a black film, therefore, may be:

(a) Black stoving enamels usually turn more matt by stoving at too high temperatures or for too long a time, although they may be perfectly glossy if stoved under suitable conditions. A few, e.g. bituminous, coatings stay glossy even though the optimum stoving temperatures and periods are considerably exceeded.

(b) A black enamel selected in too matt a shade. This happens easily if the match from which the selection was made had been on a rounded surface, whilst the application was made on to a flat surface.

Select a glossier composition.

(c) Enamels losing their gloss for other reasons, e.g. by absorption of vehicle by the base.

(d) Paints and lacquers yielding films absolutely free from reflection can be produced with black flock. Such coatings are required for some optical instruments. Although it is impossible to make such coatings completely smooth, the roughness can hardly be felt, but they are susceptible to fingerprints and rubbing.

Notes

[1] Harrison, V.G.W. (1945), *Definition and Measurement of Gloss,* Printing and Allied Trades Research Association, London.
[2] Hess, M. (1956), *JOCCA,* **39**, 185.
[3] Tilleard, D.L. and Bullett, T.R. (1953), *JOCCA,* **36**, 545.
[4] Elm, A.C. (1959), *Off. Dig.* **31**, 733.
[5] Bruxelles, G.N. and Mahlman, B.H. (1954), *Off. Dig.* **26**, 299.

5.11 Difficulties on rubbing down of fillers and undercoatings

(a) *Grinding through* is due frequently to the rubbing down of filler or undercoats, etc., before they have dried through sufficiently for this purpose. If the films are too thick and not sufficiently dried below, the already hardened top layer is easily torn off by the friction.

It is absolutely necessary to permit the layers to dry thoroughly down to their base before subjecting them to any rubbing or cutting treatments. Wet sanding results in smoother and more thoroughly rubbed down films and gives satisfactory results on hard undercoats, e.g. based on such thermo-hardening resins, where dry rubbing would be unsatisfactory.

Rubbing down operations can either be carried out by means of:

(i) Abrasive powders, such as pumice, tripoli or diatomaceous silica, and felt.

(ii) Grinding blocks.
(iii) Abrasive papers or cloth.
(iv) Fine steel wool.

They are available in various grades. Application of the powders may be dry or wet, of blocks preferably wet, of sand papers dry, of waterproof (wet or dry) abrasive papers or cloth dry or wet, and of steel wool dry. Instead of water, soap solutions or thin mineral oils are used as lubricants. The choice depends on the job. One starts with a coarse grade, e.g. Nos. 180–220 and finishes with a fine grade, e.g. Nos. 320–400.

Sand or abrasive sheets must be stored at room temperature, preferably in a cabinet. Too cold storage conditions embrittle the binder of the sand, silicon carbide, etc., too hot storage can dry it (e.g. glue) and also result in brittle paper.

Always rub parallel to any grain in the surface.

(b) *Corning* occurs when bits of abrasive, sticky film particles and sometimes also lubricants or polishing compound join to form hard little balls. Nasty scratches in the film result, which involve a great deal of cutting down work for their removal. Corning is possible when rubbing down takes place:

(i) Before thorough hardening of the film, particularly of fat edges, which as a rule are still soft, when the rest of the film is hard enough for rubbing down.
(ii) With a lubricant, when this greatly softens the film. Normally a lubricant tends to prevent corning owing to its cooling effect.
Scrape the corns from the felt, etc.
(iii) If a lubricant, e.g. xylol, is not used on sanding a polyester lacquer film of the wax-containing type, damage is likely.

(c) The weight or excessive use of sanding machines can cause rubbing through on small strips, e.g. edges. This risk is great when, as in car body finishing, large areas have to be rapidly cut with sanders hung from a counterbalance.

Prevent the full weight from being unevenly distributed. Rub edges by hand if possible.

5.12 Difficult polishing of films, smearing on polishing

Polishing is a refined grinding process using a lubricating carrier medium, e.g. wax, for the very mild abrasive power. In slightly defective work fine scratches are produced; in more serious cases matt or smeary films result, and if it comes to the worst the film may be rubbed through. This happens when the film system is still too soft, or its constituents have too low a softening point as heat is generated by the high pressures or speed of friction. In fact, a comparison could be drawn with metal polishing (Beilby layer).

Somewhat similar effects as with polishing are obtained by various other methods, e.g. by the application of pull-over solutions after rubbing down, or by the use of blades. Both methods are popular with wood finishes.

(a) Use of too long oil varnishes, etc., or fillers.

See (f) and (i) below.

(b) Polishing too soon after application.

Polishing should not be carried out until the films have dried through. In case of lack of time, artificial heat should be used to accelerate the drying process.

(c) Excessively thick films cannot harden through satisfactorily. The lower layers of, e.g. oleoresinous films remain soft for weeks after the surface has become sufficiently hard for polishing. It is possible in the case of nitrocellulose lacquers that even if the polished surface exhibits a satisfactory appearance for some time it will become flat after a while, or in some cases it could develop cracks.

(d) Insufficiently hard baking of stovings, e.g. cycle enamels.

The films should be dried for longer periods or at higher temperatures until absolutely scratch-hard. The better dried through the easier the polishing. Actions which may delay the hardening through in an oven have also to be considered.

(e) Cellulose and other lacquer films rich in pigment.

Add clear lacquer if necessary, or cover the film of pigmented lacquer with a clear lacquer and polish the latter.

(f) Application of excessively soft cellulose lacquer unable to withstand the friction of the sanders or high-speed buffing bands or wheels. Apart from the rubbing-through defect, corning with subsequent scratching of the film is likely to happen. Cracking of such films has also been observed. Too high speed of the wheel, etc., produces too much friction and too great heat which would damage the soundest film.

Give doubtful films sufficient time to dry thoroughly, or apply a forced drying method. Use polishing machines, etc., with reason and apply suitable paste, etc., to the wheels regularly. Avoid fat edges. The linear speed of a buffing wheel is approx. 1000 to 1500 metres (approx. 3500 to 5000 feet), per minute on the circumference, which equals 1500 to 2000 revolutions per minute on a 25 cm (10 in) wheel.

(g) A consequence of the orange-peel effect on films. Also dry spray of, e.g. cellulose lacquer may have caused precipitation of spraying dust of extremely fine grain.

The more uniformly a film is sprayed, the easier it is to polish, and the lower the costs.

(h) Excessive application of polishing pastes may cause superficial smearing of the film. The same applies to polishing pastes containing too much wax.

(i) Coats of thermoplastic synthetic and natural resin finishes are susceptible to the heat of friction. Many of them are unsuitable for polishing, since it is practically impossible to avoid a certain increase of temperature due to the frictional heat induced by the polishing process. The same applies to soft (e.g. overplasticized) cellulose lacquers. Solvent retention, e.g. of acrylic resin lacquers, aggravates this shortcoming.

Examples of soft natural resins are rosin and dammar.

Genuine shellac varnishes if rubbed show no ‘gum’ on the sandpaper etc., whereas many shellac substitues and adulterants (such as rosin) do.

Longer drying periods or the use of increased drying temperatures counter solvent retention. Bearing this in mind, nitrocellulose-acrylic lacquer etc. can be polished.

See sub-sections (a), (c) and (f) above.

(j) If polyester lacquers have been applied at too low a temperature they may never get hard enough for successful sanding or polishing. The minimum room temperature generally recommended is about 20°C (68°F). The relative humidity should be under 85%.

Note

[1] Wampler, R.H. (1946), *Modern Organic Finishes,* p. 390.

5.13 Difficulties encountered in the forming of coated sheets

It is important that properly gauged and annealed metals are used, because otherwise the metal surface itself may crack on forming or deep drawing. In most cases application is by roller coating. Curtain and knife coating and automatic spraying are also practised for covering large surfaces.

(a) Varnishes and enamels, the films of which slide smoothly along the forming (stamping) dies, are in general use, but greasing is also practised here and there. Modern coatings eliminate the necessity of lubrication, since they contain, e.g. a wax or silicone oil which acts as a lubricant. The presence of these lubricants in the film is often shown as a distinct 'bloom' which is removed in the forming operation. There is a disadvantage to the technique of including a lubricant in the film applied to flat tinplate sheets, for the coated sheets slide easily during transportation and this can be hazardous.

Oleoresinous stamping enamels can soften, if greased. A more thorough stoving may make the coating less susceptible to softening, but the more to forming.

(b) Cracking or peeling-off of films, which are elastic enough under normal conditions, particularly when no overbaking had taken place, can happen on stamping if the sheets are too cold.

This defect resembles the flaws on kinking of cold oil-base varnished American cloth.

(c) Damage to the films at the edges and corners on forming may have occurred for one of the following reasons:

(i) Too short drying or excessive stoving. With the fast-drying coatings now available, such as polyvinyl resin lacquers, stoving times have been reduced to a fraction of what they used to be.
(ii) Application of too thick film layers.

(d) To prevent damage to polished sheets during forming, strippable protective temporary coatings, mostly on a vinyl resin base and applied by spraying, roller coating, curtain coating or spreading, are in use. They guard the metals also on transport, in storage, on a site or on forming.

(e) The success of post-forming operations with pre-coated steel and aluminium depends on:

(i) Selection of the optimal curing temperature [1] which will not only give complete fusion, but also the correct impact resistance as determined by (e.g.) the Gardner Variable Impact Tester, and the appropriate surface hardness measured by a Pendulum method (Persoz or Konig).

(ii) Forming the coated sheet at sufficiently high temperature to avoid the formation of cracks in the coating during the process.

Note

[1] Schmid, E.V. and Helmen, T. (1975), *Farbe Lack,* **81**, 118.

5.14 Faults in vacuum evaporated metal-lacquer-film systems

We are concerned here solely with the vacuum deposition of aluminium for the production of silver-coloured, sometimes dyed, decorative effects or chromium plating imitations on metals, e.g. reflectors, or plastics including sheeting.

Practically any surface can be vacuum metallized; all types of plastics can be used. However, thermoplastic materials, such as acrylates and polystyrene, give smoother surfaces than thermosetting amino- or phenolformaldehyde resins, which would need scouring prior to lacquering. (L. Holland [1].) This would make the application of several lacquer coats necessary to obtain a specular surface. Of metals, iron and zinc, as zinc die castings, are commonly used.

The process normally consists of cleaning, application of a base coat lacquer, vacuum metallizing and application of a finishing lacquer to protect the metal coating. No surface defects on or contamination of the substrates are permissible. Plastic articles are cleaned in water containing a detergent, and the water must be dried off by heat before the next stage of processing. For polystyrene, ethyl alcohol is frequently used. Also mould release agents, lubricants, etc., must be removed. Similarly metals must be cleaned and degreased by one of the well-known pretreatment processes. Operators should wear gloves, to exclude finger marks and air must be filtered, to avoid imperfections due to dust, lint, etc. Lacquers should be re-filtered before re-use.

Some plastics, in particular polyethylene, require a surface pretreatment to improve adhesion of the base coat. This can be either treatment with a hot oxidizing flame or an acid dip or a brief immersion in a hot solvent such as perchlorethylene [2].

The reflectivity of vacuum-deposited pure aluminium is of the high order of 89.5%, its thickness below 0·05 μm, 0.454 kg (1 lb) covering 2323 m^2 (25 000 ft^2) [1].

In order to evaporate the metal to the best advantage, i.e. obtain a specular reflecting surface, the base is levelled out with at least one coat of suitable lacquer. After the base lacquer has been completely dried or cured, the vacuum deposition of the aluminium can proceed. After melting, the evaporating metal atoms at 1300–1500°C leave the cathode and are deposited on the articles to be coated having travelled in a straight line. Evaporated metal films are invariably porous and contain microscopic pinholes, due to contaminating particles accumulating on the surface during deposition, and are therefore never completely impermeable to gases and vapours.

Highest specular reflectivity is obtained by slow-extraction dipping of the base coat in the correct viscosity. Full coats are thus obtainable without drips and runs. According to the condition of the surface one or two coats will be required. Spraying of base coats would give less even coats; besides, many more may be required. It is, however, sometimes needed for intricate work. Flow-coating followed by centrifuging off excess lacquer is occasionally used for lower quality work.

Lacquer systems include acrylics, amino resin/alkyd combinations, urethane alkyds and nitrocellulose or two-pack epoxies. Base coats and finishes must possess good adhesion, water resistance, heat resistance, hardness and weather resistance [3]. The solvent balance in all these systems is of great importance and solvents which could attack the plastic substrate must not be included; otherwise accentuation of stress marks, blistering and increased pumping time in the metallizing stage could result [4]. In extreme cases crazing and cracking will take place. Force-drying and stoving temperatures must not exceed 90°C.

Vacuum-deposited metal films provide no corrosion protection for metal substrates, but a good base lacquer can contribute to this. As well as having good adhesion, build and flow, a base coat must not, after curing, give off residual solvents under vacuum, nor plasticizers or adsorbed water. On the contrary, if the substrate consists of a plastic the base lacquer should seal its surface against outgassing, i.e. emanating volatiles, e.g. plasticizers, at low pressures [1]. In fact, it should accept a certain amount migrating into it without disadvantage. Outgassing may impede the smooth running of the plant and contaminate the vacuum system. The usually waterwhite, clear top lacquer is sometimes applied by dipping, but spraying is preferable. The top lacquer must adhere strongly to the deposited metal film, have no self-colour, except when intentionally tinted, e.g. gold, nor develop one on ageing. It also should be water resistant and impervious to aggressive vapours, so that the brightness of the system is preserved in use. Mechanical protection, too, must be provided by it.

Attractive results are obtainable by vacuumizing directly on the reverse, ornamented side of transparent, e.g. acrylate plastics. A backing lacquer takes the place of the top lacquer. This is called 'second surface' metallizing.

(a) Cissing, pinholing and crater formation are usually encountered for similar reasons as on other substrates. In particular mould release agents can be troublesome; they may even have been incorporated into the plastic. Zinc stearate, waxes and silicone oils are used as lubricants.

The design of articles can often be such that strain-free moulding without the use of a release agent is possible. Alternatively remove lubricants, e.g. stearates, by thorough wiping with suitable solvents. Proprietary brands claim to be successful even for the removal of silicone types.

(b) Dust and flying lint cause the same type of trouble as in other processes, i.e. bittiness. The brilliance of the metal film, however, exaggerates each unevenness. Finger marks must be avoided, they may show through.

Use filtered air. Operators should always wear gloves.

(c) Faulty adhesion of lacquers and deposited metal can have the same causes as described above, i.e. 'greasy surfaces', or it could be caused by the type of lacquer in use. It can also be due to incomplete curing of the base coat in a badly ventilated oven, to excessive curing of the base coat or to the presence of moisture [5].

'Backstreaming' (oil from the diffusion pump seeping into the vacuum chamber) causes loss of adhesion, iridescence and contamination of the article being metallized [5].

(d) Prevention of the formation of an opaque and brilliant metal film over one of the moulded plastics such as cellophane, cellulose acetate or -aceto butyrate, or some phenolic resin compounds can be due to outgassing, i.e. evaporation of plasticizer and other additives from the substrate. This effect has also been described as 'blackening' or 'burning'. Blisters, too, can arise under such conditions, which in turn affect adhesion.

During evaporation of aluminium from tungsten the temperature of the surface of the metal film being formed can rise to well over 100°C (212°F). This may cause solvents (from the base lacquer) or plasticizers (from a plastic substrate and/or base lacquer), also residual moisture, to evaporate. These vapours oppose the on-coming metal vapour. Dulling of the metal coating can also occur when a low molecular weight plasticizer migrates into the basecoat during heating [5].

Thermosetting plastics outgas very little, depending on the type of filler used [6].

Thorough drying or curing and degreasing are essential treatments before coating.

(e) Starfracture (crow foot cracking or 'dropthrough') with a pinhole at the centre, crazing, rippling, crinkling, accompanied by loss of brillance are due to solvent attack emanating from the top (protective) lacquer penetrating through the pinholes and pores of the metal layer. Solvent absorption causes the base lacquer and/or a plastic substrate surface to swell so that the aluminium film fractures under stress. This may show itself 'as a minute starfracture of the metallic film with a pinhole at the centre, frosted to the naked eye. Thermosetting (base) lacquers are not easily attacked, when completely cured', if not cured the same defect can occur. 'Disruption of the metal film may become even more pronounced when the protective (top) lacquer coating is stoved over an uncured base lacquer, because the latter contracts as stoving proceeds' [1]. The slightest distortion in the base lacquer ruptures a deposited aluminium film, and this can happen either when the top lacquer is applied or stoved [1]. See Fig. 5.5.

(f) Striation of the dip applied base coat can affect the brightness of the vacuum metallized film. This is particularly noticeable in the coating of head-lamp reflectors.

The problem is invariably caused by the solvent balance used in dipping which must be slow enough to prevent striation yet sufficiently volatile to be removed from the film during the stoving cycle. Naphtha solvents should be replaced by xylene or trimethyl benzene, the choice depending upon the ambient temperature in the finishing shop.

(g) A number of failures can have various causes.

(i) Iridescence is sometimes observed. If this appears after the top coat has been applied, it is most likely due to its own excessive dilution resulting in interference colours.

On the other hand, too heavy or possibly wavy, uneven films can be the result of too rapid dipping, also leaving runs and drops. Application of a lacquer of too high viscosity would affect the flow similarly, except if applied by a slow extraction process.

Iridescence can also be caused by high stoving temperatures as explained above (e), if the base coat is thermoplastic or was undercured and is subjected to higher temperatures during or after the deposition of the metal film causing distortion of the metal layer.

The surface temperature of an aluminium film whilst being formed by vacuum evaporation can rise to well over 100°C (212°F). The defect can further originate in use, e.g. by tropical heat. As the sharpness of the image suffers in such cases the failure has also been described as blushing, hazing and fogging. Blushing can also be due to reasons which have no connection with the vacuum deposition process, but solely with the property of a lacquer, its solvents or thinners and with humidity. It can prevent the deposited metal film from being smooth. Increasing the flash-off time allowed for the topcoat and stoving the topcoat at a lower temperature than the base-coat may be helpful. Iridescence is also shown when the thermal coefficient of expansion of the lacquers differs much from that of the plastic article being coated.

Humidity control of the application room will prevent blushing.

Correctly carried out, slow extraction dipping contributes to obtaining a full enough base coat in one operation on suitable surfaces.

Base-coated plastics must not be too close to the filaments in the unit [6].

Selection of suitable lacquers and oven control are important.

Distortion of mouldings can arise either from excessive stoving temperatures or through excessive tension from bad jigging. Stoving temperatures must be carefully watched and not exceed the specified limit. Distortion gives rise to iridescence.

(ii) Discoloration of the top coat is either the consequence of too high stoving temperatures for the lacquer used, or the lacquer's colour is at fault.

(iii) Incorrect jigging is responsible for 'shadowing', i.e. preventing the metallization of some areas.

(iv) 'Soak-in' describes uneven sealing of a plastic substrate by the base coat, caused by variation of porosity in the moulding.

(v) Granular appearance is due to an overlong metallization process.

Notes

[1] Holland, L. (1956), *Vacuum Deposition of Thin Films.*
[2] Kut, S. (1971), *Ind. Lack. Betrieb,* **39**, 192, 251, 454.
[3] Francis, S.E. (1971), *Ind. Lack. Betrieb.,* **39**, 409.
[4] Moore, V.A. (1971), *Paint Tech.* **35**, No. 3, 22.
[5] Kut, S. (1967), *Trans. J. Plastics Inst.,* **35**, 621.
[6] Narcus, H. (1960), *Metallizing of Plastics,* London and New York.

5.15 Faults characteristic of metallic, hammer and polychromatic paints

(a) Insufficient covering power of metallic paints: Different metallic bases, e.g. steel, aluminium, tin, require paints of different metal powder content, in particular if the pigment is aluminium powder. The additions required of copper bronze powder or pastes are about twice as high by weight as those of aluminium.

(i) Inform the manufacturer of the type of metals to be painted.
(ii) If the bronze paint is mixed in the workshops, experiment first with different concentrations, but weigh every component.
(iii) Use aluminium in paste form in preference to the use of powders. Wet out the paste with a little of the varnish to obtain a creamy smooth dispersion before adding the remainder of the varnish. Do not wet out with solvents.

(b) Lack of adhesion.

The adhesion of aluminium and bronze powder paints is obviously better on the rougher-surfaced metallic bases, such as sheet iron, than on smooth and bright metallic surfaces, such as, for instance, polished aluminium sheets or on certain types of glass or plastic.

For surfaces of this type it is advisable to apply a coat of etch primer first, or to roughen the smooth surfaces by some suitable process.

(c) Excessively coarse grain due to:

(i) Coarse metal powder particles. These give, however, best protection and gloss if of the leafing type.
Employ a finer grade of powder or paste (which is usually more expensive) of, e.g. aluminium.
(ii) Addition of too large a percentage of metal powder.
Reduce the powder content of the paint.
(iii) Use of unsuitable vehicles tending to gel and formation of tiny lumps during prolonged storage in mixture with fine metal powders (this fault sometimes is accompanied by the troubles described below). This may also happen if the medium contains chlorinated rubber.

If the painter admixes the powder or paste, it is advisable to prepare immediately before the work starts not more than that necessary for a single day or so. If the paint is procured ready mixed from the manufacturer it is advisable to enquire whether it will keep during a reasonable time. It is sometimes possible to recondition paint turned gritty by adding small quantities of, e.g. fatty acid or fatty acid soaps.

(iv) Wrong method of mixing metal powder or paste with medium. Bad dispersion can also lead to pinholing of sprayed, thin aluminium lacquer films. (See above).

Any thinners have to be put in at the end. Ageing of the mix for, say, two days, helps to obtain homogeneous results, and to reduce the risk of pinholing.

(d) Greening of copper bronze powder, gelling of copper and aluminium bronze powder. Due to unsuitable vehicles, especially on prolonged storing of metal powder nitrocellulose lacquer mixtures, possibly in conjunction with sandiness and change of colour of the medium.

Tarnish-resistant grades of bronze powders are now available.

(e) Greying, leadiness and reduction of gloss of aluminium paints and lacquers is due to loss of leafing capacity of the aluminium powder contained. This may not only affect straight aluminium paints, but also hammer and polychromatic lacquers.

The smaller the number of metal flakes present, the more of them can orientate themselves parallel to the film surface (leafing), and thus produce the greatest specular reflection. Enough flakes must, of course, be available to form a continuous metal film.

Reflectance is reduced if flakes can stand on edge. The bigger the flakes in proportion to the film thickness, the brighter the result.

Causes of loss of leafing power are as follows:

(i) Vehicles of low viscosity, which dry sufficiently slowly to allow orientation of the flakes, contribute to paints of very high reflectivity. According to the nature of the vehicle, however, prolonged keeping in a mixed condition destroys leafing sooner or later. This is largely a function of the acidity of the medium. Also the stearine with which the flakes are coated in order to create high surface tension in the media may be dissolved, possibly due to the presence of strong solvents or thinners, or even the binder itself. If the acid value of the latter is high, the aluminium metal itself would possibly be attacked and hydrogen gas evolved.

Repeated re-circulation or overstirring of aluminium paints has an effect similar to prolonged keeping. Leafing in these cases is detroyed by more contact with air, and the solution of the stearine coat is speeded up.

(ii) The choice of paint media is most important. Generally speaking, non-polar resins are chosen to promote good leafing of the aluminium, i.e. coumarone-indene resins, styrenated long oil alkyds. Highly polar resins such as epoxy, melamine formaldehyde or hydroxyl terminated acrylics bring about rapid deleafing of aluminium flakes.

(iii) Grinding instead of mixing of aluminium pigment with vehicle, or in the case of one of the fancy finishes with pigment and vehicle, though this does not concern all of the latter. It applies, however, also to protective primers and paints in which a percentage of aluminium flakes, which could be non-leafing, is incorporated to increase durability by thus increasing the reflection of destructive light rays.

(iv) Entry of moisture into the paint. Up to 0·2% of the total paint is permissible. Otherwise not only greying but also gassing (hydrogen development) will follow, causing at times a minor 'explosion' of the container. Such paint will also gel and become completely useless.

Containers must be kept firmly closed. This applies particularly to those of the aluminium pigment, the splashing of which can lead to spontaneous ignition.

(v) Use of too recently manufactured aluminium powder or paste.

Aluminium bronzes mature with age and improve in brightness after approximately four months, according to storage conditions.

(vi) Use of lead driers. Cobalt driers are recommended instead.

(vii) Use of non-leafing aluminium pigment instead of one of the leafing types.

(f) Premature destruction of the films.

An aluminium paint applied to a new cement or plaster surface is subject to premature destruction, as it is sensitive to alkaline reaction. The cement or plaster surfaces should be permitted to dry for at least one month, although a period of as much as one year is frequently allowed before applying metallic paints.

The base should be bone-dry throughout, but it is advisable to interpose in addition an insulating coat of a suitable fluosilicate compound. The washing of the bases to be coated with a solution of zinc sulphate sometimes causes strong efflorescence.

(g) Metal paints which contain leafing pigments, in particular aluminium, differ in brightness according to the method of application. They are more glossy when brushed or roller-coated than when sprayed as a thin film. Special grades of aluminium paste have been developed for use in water borne finishes including electrophoretic coatings. But streaky and dark lined aluminium paint films result from too much 'working' with the brush.

Dipping and flow-coating, especially of aluminium paints, can give excellent results, but the disadvantages of standing, mechanical stirring or re-circulation have to be considered.

Sprayed films should be sufficiently thick to allow the flakes to align themselves. When brushing lay the paint on lightly, brush in one direction only, and apply not too heavy coats.

(h) Blooming of aluminium paints has been reported to be caused by temperatures under 10°C (50°F) prevailing on application. Whitish bloom caused by epoxy resin media which are catalysed with an excess of amine soon shows up particularly badly, on air drying, if the pigment used is aluminium.

(i) Fading of bronze-coloured metal- and fancy-finishes can be observed when the colour has been obtained by using dyed bronzes or by adding dyes to, e.g. aluminium bronze, which were not fast enough to light for the purpose used. Similarly heat can affect such colours, if the dyestuffs, etc., used are not fast enough to heat.

(j) Pronounced pinholing in hammer finishes can be due to building up on spraying, if the type used obtains its hammer effect from the presence of silicone oil. The amounts are small but very critical. Such pinholing may please one observer but displease another. For corrosion protection apply first a coat of etch primer where necessary. Pinholing in silicone-containing hammer finishes may be minimized by high speed stirring of the paint. A small addition of cellulose acetobutyrate solution may also be effective. Try a thinner film.

(k) Cratering of hammer finishes based on amine cured epoxy media can sometimes be remedied by the addition of small quantities of a ketone resin.

(l) Variations in shade of polychromatic lacquers and in shade and pattern of hammer finishes depend very much on application. It is very difficult to obtain exactly the same result with different operators or different makes.

Maintaining the same film thickness helps greatly.

5.16 Defects developed on wrinkle (ripple, rivel or crinkle) finishes

A definition of wrinkle finish is given in Section 4.6. Films for this kind of decorative coat must be fairly heavy. The necessary surface skinning takes place whilst the lower layer of the film is still wet. The skin extends and thus forms ripples. As soon as this stage has been achieved the figuration is finished and the stoving can be proceeded with. Usually the wrinkle-type finishes are based on a content of wood-oil type oils or alkyd resins containing conjugated oils, like dehydrated castor oil.

When the ripple effect has developed as a defect it is usually termed 'gaschecking'.

(a) Some wrinkle finishes are susceptible to the entrance of combustion gases into the drying chambers. They develop then a crystalline finish or matt areas in the film.

Only indirectly heated ovens should be used. If a triple-cased stove is employed, leakages of fuel gases must be repaired.

(b) Partial or complete lack of figuration due to charging the coated articles in an oven which was too cool or which was already too hot. In such cases the films cannot form the surface skin. This is frequent on drying by an infra-red process.

For the latter it might be necessary to figure first at lower temperature before full radiation is applied. Ventilation must draw off vapours from lower parts of the objects to avoid interference at the upper parts.

The figuration temperatures range between 60°C (140°F) and 120°C (250°F) for convection stoving. Rippling can be assisted by an extended flash-off time at least if the medium contains a dehydrated castor-oil alkyd.

(c) Faulty patches appear also if the lacquer was applied in part too thinly or where it had run off, i.e. near the upper edges of vertical surfaces.

The consistency of the ripple coating ought to be kept thicker if a heavier application is of no avail. The incorporation of quickly evaporating thinners is useful to avoid running on vertical objects, but precautions connected with the dropping of the flash point below 23°C (73°F) have to be taken.

The thickness of the films on vertical areas can be built up by going over the surface with the spraygun several times, each time spraying on a thin layer and allowing some of the low volatile thinners to flash off before the next spray. This heavy coat is then dried.

(d) Irregular figuration, due to:

(i) Uneven temperature or hot air currents within the heating chamber. This happens mainly with large objects.

Examine the stove and check the temperature in its various sections.

(ii) Formation of shiny patches or stripes on the film surface over comparatively deep irregularities of the base.

Ripple finishes are usually selected to cover slight irregularities. If no filling and priming of rough bases can be carried out, only fairly smooth articles can be finished in one coat.

(iii) Uneven application also results in patchy films.

The films must neither be too thick nor too thin. The spraying process should be used wherever possible. The heavier the film, the coarser the figuration.

(iv) Uneven wrinkle pattern may occur on castings due to different wall thicknesses of metal.

(e) Fat edges due to excessively thick application.

Some finishes tend more, others less, to this failure.

(f) Complete failure sometimes results if the coating has been subjected to frost for a considerable time.

(g) Unsuitable additions can kill the effect. To these belong vehicles containing phenolic or melamine formaldehyde resins or antioxidants.

(h) Wood oil based wrinkle finishes have a marked tendency to yellow during stoving and this effect becomes pronounced when the finish is applied to castings with different thicknesses of metal.

(i) Non-oil based wrinkle finishes formulated with polyacrylates or unsaturated polyesters have been described. They are cured by irradiation [1]. Certain liquid, free isocyanate blocked, urethane compositions, containing a small amount of the zinc salt of a fatty acid, are claimed to give a wrinkle finish [2].

Notes

[1] PPG Industries, U.S. Patent No. 3 783 004, (1974).
[2] Textron Inc., U.S. Patent No. 3 645 943, (1972).

5.17 Some lacquer defects occurring on plastic articles

(a) The Silver-spray method. The 'Mirror-spray' or 'Silver-spray' method as used for decorative purposes on plastics requires stainless steel dual spray guns and stainless equipment, including the spray booth. Exceptionally the dipping process is used. In this method a quasi normal (genuine) silvering process is carried out *in situ* on the articles over one or several base lacquer coat(s) after preceding sensitizing. The ammoniacal silver nitrate solution is sprayed on simultaneously with the reducing chemicals, in presence of wetting agents, through separate nozzles. The build-up of the film system is comparable to that of the vacuumizing process, and a number of faults can be traced to similar causes. Base coats help to seal the plastic surface and to obtain good lustre and adhesion of the silver. Cellulose and force drying cold-curing lacquers of traditional compositions are in use. The clear protective top coat, which can also be tinted, e.g. gold, is even more important than the one used in the vacuumizing process because it must prevent blackening (silver sulphide formation) in addition to giving mechanical protection.

(i) Apart from using daily freshly prepared and accurately made up solutions, working conditions must be exceedingly clean to prevent adhesion troubles.
Thoroughly rinse with distilled water between each process and coat.

(ii) In order to obtain the most brilliant silvering no dust should fall on the work pieces, and excessive humidity must be avoided, the latter also to prevent blushing of lacquers.
Filter air and install air-conditioners.

(b) The coating of phenolic plastic mouldings with two-pack polyurethane finishes. It is much more economical to coat a dark coloured phenolic moulding with a white pigmented or pastel shade paint finish than to use a melamine formaldehyde light coloured moulding. This practice is widely used in the manufacture of domestic iron handles. Application is carried out by electrostatic spray methods to iron handles which have been cleaned to remove traces of mould lubricant, followed by force drying at 140–160°C for ten minutes.

(c) Cissing of paint film: This may be due to inadequate cleaning. Another common cause of cissing is inadequate mixing of the two components, e.g. polyester with an aromatic, or better still, an aliphatic di-isocyanate. The blend of base and hardener should be allowed to stabilize for a least thirty minutes after mixing.

Use of incorrect thinner. There should be no departure from the manufacturers' recommendations.

(d) Sagging of film or running away from edges.

Check viscosity i.e. approximately 15–17″ B.S. Flow cup No. 4 at 20°C. Apply two thin coats, 'wet-on-wet', with a flash-off period between coats.

(e) Wrinkling on edges: Too rapid heat-up during stoving.

(f) Poor deposition of paint: the non-conducting plastic may not provide a suitable earth. In this case, a pretreatment with a dilute solution of a polyamine compound should improve the efficiency of deposition.

(g) The coating of phenolic plastic mouldings with powder coatings: this technique is a one-coat electrostatic spray application to the warm plastic moulding following the pretreatment process. Advantages are shown by good edge coverage and freedom from sagging. Rivelling of the coating can occur with very rapid cure epoxy powders.

5.18 Loss of gloss and matting of air-drying finishes

The gloss of a surface is a visual impression and indicates the degree of specular reflectance, i.e. the light reflected from the surface at an angle equal to the angle of incidence. By contrast, from a surface which is devoid of gloss, the light is reflected at miriads of different angles. Such reflectance is known as 'diffuse'. There is a gradual decrease from high gloss to flat and the following degrees of gloss are generally recognized:

(i) Full gloss.
(ii) Semi-gloss.
(iii) Eggshell gloss (Silk finish). Eggshell flat.
(iv) Flat or matt.

There are, however, no established standards.

A qualitative comparison of gloss can be made in the following manner. The paint in question and the comparison standard are applied, side by side on a plane surface such as a metal or glass panel and allowed to dry. The panel is then held horizontally about 6 to 7 feet from a window with stout glazing bars. The image of the glazing bars is viewed at grazing angle; the sharpness of the image indicates the reflectance ability of the film.

It is usual to quote the gloss of a surface as a percentage of the gloss of a standard glossy black tile. The measurement is carried out by means of a gloss head attached to a galvanometer. The gloss head consists of a lamp and photoelectric cell mounted at 90° to each other so that the beam of light is incident upon and reflected from the surface at 45°. The method, which is described in detail in B.S. 3900, Part D2 [1], is satifactory for gloss readings down to about 60%. Below this the variable gloss head is used. In this instrument the angles of incidence (and reflectance) of the light can be varied and are set at low values for paints of low gloss. In stating gloss figures the angle of measurement should be quoted.

Causes of loss of gloss

This fault may appear a few hours or some months after application.

It is generally accepted that a glossy paint film contains a layer of lightly pigmented binder above a more heavily pigmented layer [2]. The surface of the film reflects the incident light in accordance with the rules of specular reflection. Any disturbance of the surface smoothness will tend to scatter the incident light. Fine rivelling or profiles of pigments (or pigment aggregates) will cause scattering and so replace part or all of the specular reflectance by diffuse reflectance. In an eggshell finish, for example, the level of pigmentation is fairly close to the critical pigment volume concentration. During the first few weeks of exposure the film often shrinks and the gloss falls. This is due to an increase in the number of pigment profiles in the surface causing an increase in the diffuse reflectance. In a matt paint film, although there is a very thin layer of binder on the surface, the pigment (or extender) profiles are so close together that there is practically no specular reflection.

The following are the common causes of loss of gloss:

(a) The material to be painted is highly absorbent, e.g. new plaster, wood, hardboard, cardboard or Gyproc. The surface must be sealed effectively before application of gloss finish otherwise sinkage, which is never uniform, will lead to variations in gloss. Highly absorbent surfaces require at least two coats of sealer. When repair work has been carried out on walls, the new plaster is likely to be more absorbent than the old surface and needs special care in sealing.

Emulsion paints and low sheen wall paints can also dry to patchy or 'sheary' finishes over substrates of uneven porosity. Variations in colour can also result. When patchiness occurs it is futile to continue to apply further coats of highly pigmented material. The only remedy is one or two coats of a good sealer over the entire surface. The finish can then be applied with relative safety. For solvent-type wall paints a varnish type sealer should be used; for emulsion paints a suitable sealer is clear glaze (unpigmented emulsion) or a very dilute solution of p.v.a.

(b) Underbound (i.e. over-pigmented) primers, fillers, stoppers and undercoats will lead to sinkage of the finishing coat and consequent loss of gloss. The best treatment is removal of the whole system since underbound films are likely to lead to premature failure. If this is not possible a coat of good quality sealer should be applied before the finish but this system is not likely to give very good service.

(c) Application of the finish before the undercoat has dried through can lead to visual defects. The solvents of the finish will tend to dissolve and so disturb the undercoat film. This can lead to excessive brushmarking of the top coat and impaired gloss. Overcoating should not be carried out before the undercoat is in a suitable condition. A film may not have oxidized sufficiently although it may be 'hard dry'.

(d) An excessively rough surface e.g. of castings, heavily blast cleaned metal, etc. can give unsightly profiles in the paint surface and consequent poor gloss. This topic was discussed by C.M. Hansen [3], who considered performance in relation to roughness profiles. Surface roughness was measured by a Talysurf 4, but no mathematical correlation between roughness figures and degree of gloss was demonstrated.

An undercoat in which the pigment aggregates have not been dispersed sufficiently can result in profiles appearing in a gloss finish, particularly if this is of low build.

(e) Poor flow and levelling in undercoats resulting in excessive brush marking can also result in an uneven surface and poor gloss in the finish. The remedy in the above cases is to rub down with wet abrasive paper and allow to dry before applying the finishing coat.

(f) Loss of gloss in the form of hazing is sometimes seen in alkyd gloss finishes pigmented with titanium dioxide and zinc oxide [4]. This appears to be due to chemical reactions in the film. Seeding also is often apparent.

(g) In the production of low sheen finishes, care must be taken in the choice of pigments and extenders particularly if replacements are to be made. The sheen of the paint is influenced by both the oil absorption and the pigment volume contribution and replacements should be made on the basis of equal volumes and similar oil absorptions.

The gloss of a flat or matt finish can be improved by the obvious method of addition of medium.

When the gloss is too high it can be reduced by addition of matting agents which are often manufactured silicas of very fine particle size. For matting agents there is an optimum degree of dispersion to produce the maximum number of surface profiles, i.e. maximum reduction in gloss [5].

(h) Films of long-oil alkyd varnishes may show as a consequence of condensation big matt areas after having been exposed to a very cold atmosphere saturated with moisture. This defect is not unlike that of wood oil checking. The life of such defective films is obviously more or less reduced, so are its protective qualities. The observation of F.B. Gardner that where high-gloss linseed-oil varnishes came in touch with cold draught, the films became silky is undoubtedly connected with the above case.

The majority of air-drying gloss finishes may be affected to a greater or lesser degree by the condensation of moisture on the paint surface at some time during the drying process. The loss of gloss may show in patches, e.g. where a cold air stream has touched the surface, or may be present on all areas, e.g. when a rapid fall in temperature has resulted in dew formation, or when rain has settled on the surface, e.g. on window sills. The conditions necessary for this defect are that the varnish or paint shall be at a sensitive stage of the drying process when moisture deposition takes place. In the case of the slower-drying gloss finishes, e.g. long-oil alkyd based, the sensitive period is much longer than in the case of fast-drying alkyd or oleo-resinous-based paints and therefore their critical period is longer. To the naked eye the loss of gloss of the above types of coatings appears as a permanent bloom, but microscopic examination usually shows it to be a minute rivel or orange-peel effect. Although generally encountered under winter conditions the loss of gloss due to deposited moisture may occur at other times of the year, when either a change in temperature causes dew formation or rain becomes deposited on the surface, or in interior conditions especially in kitchens, bathrooms, etc., when steam is condensed on the surface, all these occurrences coinciding with the early drying stages of the film.

Special care has to be taken if outside painting is done in early spring. Draught must always be kept away from fresh films, if only to avoid contamination with dust, etc.

Where the film has matted completely check that the surface is dry and hard, rub down with a fine grade of glass paper or flat down with fine carborundum paper (e.g. 240 grade) preferably used wet, allow to dry and then under favourable drying conditions apply a further coat of gloss paint. In the case of localized matt areas, flat down as above and re-coat the affected section with gloss paint.

Cationic surfactants have been recommended to be added to paint to permit it to absorb collected moisture, allowing painting to continue under damp conditions or on wet surfaces, thus permitting water absorption with subsequent evaporation, resulting in no loss of gloss under conditions of high humidity and condensation.

(i) Painting of too moist wood, or presence of traces of moisture, e.g. in wood due to rain or, for instance, from the washing off of residues of alkaline paint removers.

Exterior woodwork should not be painted when the temperature is below 10°C (50°F) or in urgent cases 4°C (40°F), and never when the temperature is expected to drop by 10°C (approx. 20°F) to well below 10°C (50°F), nor after such a drop. The work should be stopped early enough in the afternoon to allow the paint to set before a sudden drop may occur. At least one week of 'dry sunny weather' should precede the painting of green or very wet timber.

(j) Moisture in pigments. The gloss, brushability and keeping properties of alkyd finishes can be adversely affected by moisture in the pigments. A matting effect can result with both alkyd and oleoresinous finishes if moist zinc oxide is present. Paints which are glossy on manufacture may give matt or semi-gloss films after storage.

It is necessary in any case to avoid even the slighest contamination of paint materials with water (for instance, by moist storage of dry colours). This applies not only to the pigments above mentioned, but to all other pigments as well. Small additions of water-miscible solvents such as butyl alcohol, glycol ethers, etc., may sometimes assist in absorbing the moisture present and reducing the defect.

(k) Unsuitable solvent mixtures. When a mixture of solvents is used it it essential for the resin to remain soluble in the slowest evaporating component, otherwise 'throw out' may ensue, resulting in poor gloss and a weak film. Diluents, for example, must evaporate completely before the true solvent. An excess of pure solvent can also result in loss of gloss.

(l) Incompatibility of resins. Hazing or matting can result if two resins, which appear to be compatible in a solvent mixture, become incompatible when one or all the solvents evaporate. This can occur with certain alkyd/vinyl and alkyd/chlorinated rubber mixtures.

(m) Under tropical conditions loss of gloss occurs more rapidly. Alkyd media are the most favoured type. Whites and tints are inferior to the dark colours but the latter retain their gloss for only one to two years. This is not, however regarded as a defect but is due to intense radiation in the tropics.

Notes

[1] B.S. 3900, Part D2. *Gloss (specular reflection value)*.

[2] An electron microscope examination of this layer was carried out by Murley, R.D. and Smith, H. (1970), *JOCCA*, **53**, 292.
[3] Hansen, C.M. (1972), *J. Paint Tech.* **44**, 61.
[4] O'Neill, L.A. and Brett, R.A. (1969), *JOCCA*, **52**, 1054; Bullett, T.R. (1971), **54**, 849.
[5] Colling, J.H., Craker, W.E., Smith, M.C. and Dunderdale, J. (1971), *JOCCA*, **54**, 1057.

5.19 Difficulties in coating plastics or wood impregnated or bonded with plastic [1]

Most plastics are derived from the same or similar materials to those used in the various synthetic coatings applied by the usual methods of painting or spraying to numerous bases, such as metals or wood. These coatings are then air dried or stoved to form hard films.

The materials from which the plastics are made are often powders which have to be moulded, extruded, etc., by heat and pressure to give the solid article required. Table 5.1 shows some characteristics of high polymers, most of which can be used both in plastics and coatings.

Some of the troubles experienced in the coating of plastics may have to do with their content of catalysts consisting of acids or salts.

As in the field of synthetic coatings, a division is made in the field of plastics, viz.

(i) thermoplastic (two dimensional), which soften on warming, undergoing no chemical change, and
(ii) thermosetting (three dimensional), which become insoluble and infusible on heating or by the action of catalysts, due to cross-linking, and cannot be re-formed by further heating.

There is no absolute division of the two groups, as the properties of each may be modified by the incorporation of drying, or non-drying oils, resins, etc.

Apart from the usual film defects, there are other troubles which are characteristic for plastics:

(a) Surface disturbances such as cracking or crazing are frequent on lacquered polystyrene if this has not been annealed [2] to remove strains as far as possible.

The excessive softening of coatings over plastics consisting of vinyl polymers and 'some of the softer cellulose acetates' is usually caused by the migration of unduly great quantities of plasticizer from the plastic into the coating [3]. The behaviour of cellulose acetobutyrate as to painting has been described as similar to that of cellulose acetate [4].

Compounds used for lubricating the moulds (release agents), e.g. castor oil, silicone fluids or zinc stearate [5] may have similar effects, and impede adhesion if not carefully removed.

Migration may also result in wrinkling or crazing, as observed, e.g. over polystyrene plastics. Some may distort on stoving owing to their low softening point. Crazing of methyl methacrylate plastics is also possible.

Lacquers used for thermoplastics should contain as little and as mild solvents and thinners as possible, e.g. alcohols 'to avoid excessive softening of the surface, resulting in wrinkling' [3].

Some moulding materials, e.g. polyethylene, most of all polystyrene, are easily charged with static electricity, which attracts dust, forming patterns. As destaticizers the following have been recommended for temporary protection: detergent solutions, buffing compounds, de-ionising salt sprays, waxes, etc.; for permanent protection; lacquer coats, e.g. cellulose acetobutyrate lacquers [6]. Ionising air guns for destaticizing have also been applied.

Lubricating agents should only be used when unavoidable, and must be carefully removed before coating.

Sealing coats may be necessary in some cases. Internal stresses and irregular thickness of work pieces cause crazing. This is accelerated by solvent action [4]. Strain patterns in transparent plastics can readily be observed in transmitted polarized light.

As a test for strains in polystyrene articles, their dipping into *n*-heptane is suggested [7]; if crazing takes place annealing will be necessary [2].

(b) Distortion of thermoplastics can be caused by stoving at too high temperatures if their softening point is low.

(c) Plastics in sheet form or celluloid can also be affected by the action of unsuitable solvents or thinners in the coatings.

Celluloid, etc., printed on one side only, tends to bend after spraying of lacquer on to that side. This tendency increases with decreasing thickness of the sheets. The superficial dissolution of the surface depends on the plastic's composition and on the kind of solvents used in the lacquer.

Where lacquers containing solvents for celluloid, etc., are applied it is obvious that a certain action on the surface will take place. On evaporation of the solvents to give the new film, stresses may be set up causing distortion of the base (see above (a)).

A coating of minimum solvent action should be employed, or the sheets be lacquered on both sides. If, for example, merely letters are sprayed by means of stencils, the trouble does not usually arise.

(d) One of the most frequent defects is bad adhesion. Thermo-hardened plastics, e.g. of the phenolic and urea formaldehyde type, which have become insoluble after their curing and which have generally quite smooth surfaces offer very little key. Generally speaking, non-polar plastics like polyethylene, polypropylene and polystyrene are poor substrates for obtaining good adhesion, whereas polar plastics like acrylics, polyesters etc. accept coatings more readily [9].

The surface of the non-polar polyethylene (and similar compounds) requires pretreatment by oxidation for promotion of adhesion of surface coatings including printing inks.

The adhesion of commercial oil-base paints on the types mentioned above may be good enough for house decorating, they may in addition protect these plastics at least from weathering, but hardly resist rough handling.

Plastics should, if possible, be degreased before coating with mild solvents, such as alcohols and aliphatic solvents.

Thermohardened plastics can be slightly roughened by fine sand blasting or water sanding to ease the bonding of a coating.

On phenol formaldehyde resins a first coat of etch primer can sometimes improve adhesion. Nitrocellulose and urea alkyd lacquers are in use for finishing this type of plastic. For amino resin plastics, alkyd enamels and vinyl copolymer lacquers have been recommended [8]. It may also be worth while to try the adhesion of a polyamide epoxy (two can) lacquer.

Plastic mouldings based on phenol formaldehyde should be degreased in a warm sprayed alkaline solution, dried and coated whilst the article is still warm from the drying chamber with a one coat application of electrostatic sprayed powder coatings or a two coat wet-on-wet application of a two pack polyurethane followed by low temperature curing. Excellent adhesion and a lower reject rate is shown by the powder coating.

Some thermoplastic-type plastics may receive a mist coat of strong solvent [3] in order to redissolve their surface immediately before the application of a lacquer, etc.

Oxidation pretreatments for polyethylene after degreasing are by:

(i) Oxidizing gas flame.
(ii) Low voltage discharge in a vacuum or by applying high voltage (10 000 to 30 000 volts) stress or a corona discharge [10].
(iii) Chemical treatment using oxidizing agents, ozone, potassium permanganate, halogen-gas (e.g. Cl) together with u.v. light, and a chromic acid process.
(iv) Pretreatment of polypropylene with boiling trichloroethylene has been recommended [11].

Method (iii) is said to be costlier and more difficult than (i) or (ii). The thus pretreated articles should remain untouched until coated or printed. In certain cases it was found that vinyl butyrate lacquers adhered satisfactorily without any of the above pretreatments. See also Table 5.1.

If such preparation does no improve the adhesion, the paint manufacturers may be able to offer a suitable coating for a particular material.

It is likely that coatings which are based on a compound similar or related to that of the plastic are most successful, e.g. p.v.a. or p.v.c. lacquers on p.v.c. plastics.

(e) Overcuring of a plastic is possible if its maximum curing time is exceeded by the stoving operation of a finish. For polyethylene 15 to 20 minutes at 75°C (158–167°F) has been suggested [9].

The use of an air-drying coating or one needing shorter stoving time or lower oven temperature must be considered. Once again a two pack polyurethane is the recommended coating.

Table 5.1 Some characteristics of high polymers

Plastic articles are made from casein and from cellulose esters and ethers as well as from the class of materials generally understood by the term 'synthetic resins'. The latter can be subdivided into the linear or thermoplastic polymers, and the crosslinked or thermosetting* polymers. The linear polymers can include copolymers of two or more monomers, as well as the homopolymers which involve only one monomer. In addition there are further classes of linear polymers in which the arrangement of the monomer units can be orientated in a stereospecific way, giving rise to the isotactic or syndiotactic forms. For example, ethylene can be polymerized under high pressure to give the normal form of polyethylene, whereas at low pressures and using a Ziegler catalyst, the crystalline or high density form is produced.

The following list, while not exhaustive, includes THE MOST COMMON HIGH POLYMERS, with some indication of the type of coating likely to produce satisfactory results.

Crosslinking plastics: Polymers	*Coatings: Recommended*
Polyester (usually reinforced with glass fibre)	Etch primers Alkyd enamels Nitrocellulose-alkyd combination lacquers Polyvinylbutylether-based primings Vinyl polymer lacquers
Phenol-formaldehyde	Etch primers Nitrocellulose lacquers Alkyds Epoxy powder coatings Polyurethanes
Epoxy resins	Most coatings suitable
Urea-formaldehyde and melamine-form-maldehyde	Alkyd and urea alkyd enamels Stoving vinyl resin lacquers Polyamide epoxy resin lacquers
Polyurethanes	Alkyds Polyurethane lacquers
Alkyd resins	Alkyds Nitrocellulose lacquers
Linear polymers	
Polyvinyl chloride	Post-chlorinated polyvinyl chloride lacquers Polyvinyl acetate and copolymer lacquers
Polyvinyl acetate	Vinyl polymer lacquers
Polyvinyl acetate-chloride	Vinyl polymer lacquers

Table 5.1 Some characteristics of high polymers (*continued*)

Crosslinking plastics: *Polymers*	*Coatings*: *Recommended*
Polyacrylates	Nitrocellulose lacquers Cellulose acetate butyrate lacquers
Polystyrene	Ethyl cellulose lacquers (using alcohol solvents) Polyvinyl butyral and other vinyl polymer lacquers
Polyethylene, high density polyethylene	Surface preparation by flaming, exposure to gaseous chlorine or ozone or ionizing radiation, etc. Many coatings then suitable
Polypropylene	As above
Polyhexamethylene adipamide (Nylon)	Polyester-polyisocyanate lacquers
Polyethylene glycol terephthalate (Terylene)	Polyester-polyisocyanate lacquers
Polyvinyl butyral	Vinyl polymer lacquers
Styrene-butadiene†	–
Acrylonitrile-butadiene†	–
Neoprene†	Polyurethane lacquers Polyacrylate lacquers
Other Polymers	
Cellulose acetate	Cellulose acetate butyrate lacquers Cellulose acetate lacquers
Cellulose acetate-butyrate	Cellulose acetate lacquers Cellulose acetate butyrate lacquers
Ethyl cellulose	Nitrocellulose lacquers Ethyl cellulose lacquers
Nitrocellulose, Celluloid	Nitrocellulose and other cellulose ester lacquers
Casein	Most coatings suitable

* Cross-linking may take place without heat in some instances as with, e.g. polyesters (free radical polymerization with peroxide catalysts) or polyurethanes (reaction of isocyanate radicals with other groupings to give cross-linking by chemical rearrangement).

† These synthetic rubbers may be cross-linked by ‘vulcanization’.

(f) Plastics and rubbers which contain dyes which are soluble in the solvents, etc., of their coatings can bleed.

(g) The content of sulphur in vulcanized or synthetic rubber can cause discoloration of a coating if reacting with those pigments which form coloured sulphides.

Remove thoroughly by abrasion, etc., before lacquering, any bloom, surface skin or mould lubricant. Change in pigmentation or application of a sealing coat may be advisable.

Notes

[1] Mainly from Hess, M. (1947), Difficulties in coating plastics, *Verfroniek,* **20**, No. 8, 167.
[2] Bateson, S. (1953), *Vacuum,* **3**, 35.
[3] Wampler, H.R. (1946), *Modern Organic Finishes,* p. 272.
[4] *Encyclopédie Francaise des Matières Plastiques, numéro spécial 1955 de la Revue Industrie des Plastiques Modernes,* p. 235.
[5] Brooks, D.E. (1952), *Prod. Fin.,* **11**, 88.
[6] Monsanto Plastics Ltd, Tech. Serv. Bulletin No. P6/3.
[7] Parks, L.E. (1958), *Mod. Plastics;* **102**, 105.
[8] Rabat, J.L. (1956), *Paint Pigm., Vernis,* **2**, 121.
[9] Lindberg, B. (1975), *JOCCA,* **58**, 408.
[10] Pye, C.R. (1966), *JOCCA*, **49**, 719.
[11] *Hercules Chemist,* No. 43, 14, (1961).

5.20 Various difficulties connected with the varnishing of paper, fibreboard or metal

(a) Cracking of the resin film on varnish-impregnated paper, e.g. for transparencies. This fault is identical with the cracking of short-oil varnish films on long-oil undercoatings. The resin layer really takes the place of a very short-oil varnish film.

Label varnishes dry quickly without penetrating into the paper, while on account of its oil content, and also of two-sided application, transparent varnish can penetrate completely.

(b) Faulty drying of varnish, drying by oxidation, over printed paper, metal, etc.

Some pigments contained in printing inks retard the drying of, e.g. roller coated clear varnish. Procure more suitable inks. Drying can be accelerated by one of the safe heating processes.

(c) Loss of gloss of lacquered fibreboard, etc., due to a strongly absorptive or an insufficiently pretreated base (sizing or priming). This may apply to cellulose, stoving synthetic, polyester lacquers, etc.

Procure fibreboard, etc., better suited for the purpose or apply several priming coats. Fibreboard and paper destined for this sort of finishing must always be well sized, resin impregnated, etc. Application by hot spraying might be helpful.

(d) If the varnished printed surface shows numerous matt patches, the printing inks used may be unsuitable and too absorptive. Sometimes they do not accept superimposition of a lacquer at all ('crystallizing'). The same applies to printed metal sheets.

Special printing inks or overprinting varnish may be required.

(e) If the lacquering is streaky or cloudy, cold or draughty working rooms are indicated, affecting flow. Alternatively, if application is by roller coating the rollers may need attention.

Spirit and cellulose lacquer films may also show cloudiness if susceptible to high moisture content of the atmosphere.

The temperature for lacquering or varnishing should be about 25–30°C (75–85°F). Good results have also been obtained by slight warming of the coating on the machine.

Rollers should be even, clean and correctly set.

5.21 Sweating out, patchiness and dulling of various lacquers, due to oil, wax etc.

This is a defect which does not disappear even after complete drying.

(a) The fault is caused by penetration of remnants emanating from the substrate of mineral oil, paraffin wax (derived from paint removers), fat, mould release agents or similar substances, and can be accompanied by other disadvantages, such as partial re-softening of the film.

Residues should be carefully removed.

(b) The polishing wax or oil employed contains too large amounts of high-boiling solvents. Traces of smeary waxes, etc., remain, causing faults, exhibiting the appearance of blooming or sweating.

If a polishing medium is not suited for a lacquer film, solvents contained in the former can superficially dissolve its surface. These finishes, of which the highest gloss is required, are sometimes extremely susceptible to such treatment and not only develop matt patches, but also bluish or greyish-looking surface effects in places. This defect is more pronounced where too heavy coats had been applied, e.g. by hot spraying.

A more suitable polishing medium should be used. Blueing may be due to the action of vegetable glue on nitrocellulose lacquer films.

(c) The wax contained in polyester lacquers of the air-inhibited type can cause defects:

(i) It will crystallize if the interval between coats has been too long. This will not only result in unsightly whitish films, but also in bad intercoat adhesion and cissing troubles in spite of sanding.

(ii) Force drying can result in films which never harden.
(iii) Low ambient temperature, i.e. below 20°C (68°F), slows down hardening through of both wax and non-wax polyester lacquers.
(iv) The wetting becomes unsatisfactory towards the end of pot life, causing cissing and pinholing.

Several difficulties can be circumvented by using two-feed spraying guns or the 'active ground coat' method, in which peroxide and accelerator are incorporated in two different coats, which are applied one over the other. The 'metering' in the first method must be exact all through.

Re-coating of wax types presupposes careful and thorough removal of wax residues from the 'old' film.

The relative humidity on applying and drying all polyester lacquers should be kept below 80%. Too high shop temperatures will shorten the pot life.

(d) It is important in connection with the finishing of leather and artificial leather that a sweating effect may be due to the presence in coatings of relatively large amounts of free fatty acids or in nitrocellulose lacquers of semi-drying oils (oily exudation), especially of castor oil, which is no solvent for, e.g. nitrocellulose, or to the presence of unconverted crystalline phthalic anhydride in the alkyd (white solid exudation).

Excess plasticizers in synthetic resin lacquers can be sweated out. Solid plasticizers, e.g. triphenylphosphate, may cause crystallization in cellulose acetate lacquer films.

(e) Non-solvent plasticizers in, e.g. cellulose lacquers may be partially expressed by the application of pressure, or sweated out of the film by heat. The danger increases with increasing amounts of plasticizer. Reduce amount of plasticizer or select a more suitable one. Careful attention must also be paid to the balance between true solvents and diluents in the formulation of lacquers.

(f) Softening and tackiness of a nitrocellulose furniture lacquer was caused by contact with vinyl covered upholstery cushions. The plasticizer in the vinyl covering acted as a solvent for the nitrocellulose finish. The cushions showed serious staining. Changing the furniture lacquer to a cellulose acetobutyrate base solved the problem [1].

(g) 'Sweating-up' is the technical term applied to an oily gloss which appears on varnished surfaces after pumicing. Do not allow too long an interval between pumicing and application of the next coat.

Note

[1] *P.O.C.J.* **152**, 291, (1967).

6

Defects related mainly to poor adhesion

6.1 Brittleness, poor adhesion, flaking, peeling, etc., in solvent type paints

Brittleness in a film can arise in a number of ways. The following are probably the most common:

(i) by the use of a medium of too short oil length (the effects are apparent very early);
(ii) by oxidation and weathering of externally exposed films;
(iii) by the use of reactive pigments such as zinc oxide – formation of zinc soaps is accompanied by embrittlement of the film;
(iv) the use of insufficient or unsuitable plasticizer in non-convertible films such as cellulose nitrate or chlorinated rubber.

Flaking of a film is usually preceded by cracking; both faults are the result of internal stresses set up in the film during the initial contraction when drying [1]. On a rigid substrate the film can remain intact but movement of the substrate, such as dimensional changes in wood due to moisture, can result in cracking and adhesion failure giving rise to flaking. If the wood is completely sealed so that moisture cannot enter, then the paint film should remain intact [2].

Causes of brittleness, poor adhesion and flaking:

(a) The use of an excessively short oil medium (possibly to meet a short drying time requirement) will result in a brittle film. If this has been applied over a longer oil undercoat, cracking or crazing will probably take place. The only remedy is removal of the coating and replacement by another type of medium.

(b) The consequences of applying fillers, putties, etc. directly to an unprimed substrate should be mentioned here. These contain a relatively small amount of binder and absorption of a proportion of this into the substrate will leave an underbound filler or putty with poor adhesion. Premature detachment from the surface is likely. All surfaces, with the possible exception of metals, should be primed before the application of putty, filler, etc.

(c) Absorption of moisture. Some media, particularly alkyds and some oleoresinous varnishes, absorb significant amounts of water if they are immersed or in contact with water for some time. The film swells appreciably and in the wet condition the adhesion to the substrate falls to a very low value [3] and the film is readily removed. On drying out, much of the adhesion (except under blisters) is recovered.

(d) Moisture on a surface will impair the wetting of the surface by a paint and poor adhesion is likely to result. In consequence of this many painting specifications forbid application under conditions of high relative humidity (above 90%) or low temperature (at or below 4°C).

(e) Paint should not be applied to a loose, friable surface and the use of sealers on such surfaces is fraught with risk. Loose and friable areas of plaster should be removed and renewed but new plaster should not be applied over old paint films. Powdery or flaking paint should be removed so that a firm base is established for the new paint system.

(f) Lack of inter-coat adhesion as a consequence of bad wetting is observed if paints are applied without preparation to old hard and glossy paint films. Such surfaces should be rubbed down with fine abrasive paper (preferably wet) to provide a key for the new paint. Excellent inter-coat adhesion is obtained with non-convertible coatings such as chlorinated rubber where the solvent dissolves the surface layer of the preceding coat. The 'gloss on gloss' process used with some paint systems is successful provided the second coat is applied before the first coat has become too hard to be softened slightly by the solvent of the second coat. The time limit is usually 24 hours for alkyds, epoxies and polyurethanes. Failing this the film must be allowed to become thoroughly hard and is then rubbed down before recoating.

The results of an investigation into the adhesion of multicoat systems on weathering has been described by P. Walker [4]. The conclusion drawn is that many typical paint systems, correctly applied, are unlikely to fail by loss of adhesion.

Under-eave peeling can occur in spite of the protected situation of the film. This can be due to the formation of water-soluble products during progressive oxidation of alkyd or oleoresinous binders. These products form a water-soluble film and deposition of dew and/or frost results in this film becoming saturated with water, leading to loss of adhesion and peeling. Zinc-containing paints are particularly prone to this defect.

Before repainting the water-soluble film must be removed by thorough washing and rubbing down.

Old alkyd or oleoresinous paint films, after rubbing down, can be repainted with similar material or with chlorinated rubber paint. If the latter is to be used it is wise to apply a small test area to check that no lifting is likely to occur.

Chlorinated rubber paint can be overcoated with the same material and, as mentioned earlier, inter-coat adhesion is excellent.

Films of thermosetting alkyd/amino type enamels can show poor inter-coat adhesion after recoating as a consequence of over-curing. This is probably due to over-stoving of the first coat; a shorter stoving schedule for the first coat or 'wet-on-wet' application greatly assists in these systems. With cold curing alkyd/amino enamels, excessive additions of some types of acid curing agent can lead to embrittlement and loss of adhesion.

(g) The use of amorphous types of zinc oxide causes embrittlement with media in which zinc soaps can be formed. Progressive embrittlement in paints exposed to the weather leads to cracking and finally flaking of the film. A considerable improvement results from the use of the acicular form of zinc oxide; this type finds considerable use in the U.S.A.

(h) In cold weather the temperature of exposed paint films can fall below the glass transition temperature (T_g) and the film is in no condition to withstand flexing or extension. On rigid substrates no danger is likely but if flexing or impact occurs, cracking (and possibly flaking) can result.

(i) Grease and wax residues can have a detrimental effect on the adhesion of surface coatings. Polyester lacquers are also affected if an unsuitable wood stain has been used, or the excess of an otherwise perfect stain has not been sponged off. The same cause can also promote drying difficulties and cratering and blistering.

Contaminated surfaces must be thoroughly cleaned before painting, e.g. in a degreasing plant, or, at least by thorough wiping with solvent, usually white spirit or naphtha.

For polyester systems water stains appear to work satisfactorily as a rule. A test is, however, recommended before large-scale application. Under no circumstances must stains be used which are soluble in the monomer used for crosslinking, usually styrene, or which would be oxidized by a peroxide catalyst.

Notes

[1] König, W. (1962), *Fatipec Congress Book,* p. 424.
[2] Saarnak, A., Nilsson, E. and Kornum, L.O. (1976), *JOCCA,* **59**, 427.
[3] Walker, P. (1967), *Paint Tech.,* **31**, No. 8, 22; No. 9, 15.
[4] Walker, P. (1974), *JOCCA,* **57**, 241.
[5] Schurr, G.G. and van Loo, M. (1967), *J. Paint Tech.,* **39**, 128.

6.2 Unsatisfactory adhesion, brittleness, flaking off, etc., on coated metals

In the case of baking varnishes these failures generally become apparent directly after drying, except on such metals as zinc and brass.

Adhesion failures are most commonly related to faults in stoving ovens and deterioration due to overstoving is usually accompanied by severe discoloration of the thermally unstable components.

'Organic coatings consist of a complex mixture of diverse kinds, sizes and shapes of molecules. It is this variation in molecular architecture of both substrate surface and the paint medium which provides the source of bonding forces and the resultant tendencies to develop adhesion' [1]. Much depends on the presence of hydroxyl and, as far as no detrimental corrosion is produced, carboxyl groups of the vehicle solids, which favour adhesion. K. Broockmann [2] reports that media containing these groups produce on aluminium a measurable, thicker layer of aluminium oxide than those which possess poor adhesion.

'It is now generally accepted that adhesion between surface coatings and substrates results mainly from the so-called dispersion or van der Waals forces between the molecules of the coating and the atomic lattice of the substrate. If close contact (within a few Ångström units) is maintained over the surface, these forces, even for a non-polar coating, can be ample for strong adhesion. With polar coatings the forces are greater for the same contact area, and the margin of safety is therefore bigger.' (T.R. Bullett and A.T.S. Rudram [3].)

Poor adhesion arising from greasy or dirty surfaces is only to be expected [4].

Broadly speaking, good adhesion on a very smooth, maybe polished, surface, is more difficult to attain than on a rough one. Roughening of a smooth surface, i.e. the production of interstices, for instance, by sandblasting or by chemical treatment, therefore assists the adhesion of the film on account of the anchorage provided even on 'difficult' metals. 'Roughened surfaces provide more actual surface area per unit of macroscopic area. In the case of a depth of oxide coating of aluminium of 10^{-3}, the estimated surface increase would be approximately twelve and a half times the original surface area' [5]. In cases of plated metals such a surface can be produced with an even, consistently fine rough metal deposit by directing the plating process accordingly.

Heating increases the adhesion of films even of the cellulose type provided polar groups have been introduced by the addition of plasticizers and/or resins. The strength of adhesion is closely connected with the wetting properties of a coating on a surface, the degree of wetting depending on the surface tension involved. What happens in the fresh liquid or semi-liquid surface has also a bearing on the final film behaviour. It is during that time that the molecules of the coating orient themselves.

Sometimes chemical reactions take place between free carboxyl groups of the coating and the metal base [6].

To avoid adhesion difficulties on metals several factors have to be considered together, viz., right selection of the metal or alloy for the job, of the coating for the base and for the job, of the best pretreatment of the base (if necessary), and of the coating system in so far as type and number of coats and method of drying are concerned, not omitting, when stoving is required, that for certain metals limiting stoving temperatures have to be observed. (See Table 4.2, pp. 106–107.)

As to the various possibilities of failures:

(a) The coating used is not suitable for the metal in question, i.e. for aluminium, cadmium, chromium, brass, etc. [7].

Sometimes the adhesion of the same coatings varies because a recuperated metal of changed composition is being used, e.g. the copper content in an aluminium alloy may have been unduly increased.

Alkali-cyanide residues of the plating electrolyte present in the electro-deposits which are difficult to remove by simple washing, such as in cadmium-plated steel, are probably responsible for flaking of paint films.

The majority of the usual works or laboratory tests are carried out on steel, tin or aluminium. Whilst it is true that the adhesion of most coatings, for instance on tin base, offers no great difficulties, it is worth mentioning that one can also find media which possess good adhesion on, for instance, the somewhat difficult 'smooth' cadmium base or on brass, whilst they do not necessarily give satisfactory results on tin [9].

The linear expansion of most coatings is of the order of 0·0001 of the original length per degree centigrade, that is about ten times that of steel and four times that of aluminium [10].

There is no coating of which it can be said that because of good adhesion on one of the more 'difficult' metals it will show good adhesion on other or even all bases, nor is there a metal or other base of which it can be said that if the adhesion of a typical coating on it is satisfactory it will be a good base for other coatings. It is necessary to beware of generalization [9].

In dealing with this sub-section it is assumed that the degreasing [4] of the metal has been carefully carried out. Suitable special varnishes, etc., have been developed for almost every base. However, almost any varnish or lacquer can be used for nearly every metal surface by proper preliminary treatment of the latter, i.e. by sanding, rubbing with emery paper or by chemical etching, although the latter is not so highly recommended.

The application of a matt first coat or an etch primer is being used successfully to allow a finishing coat to key on to a smooth base.

To clean and key on, e.g. a cadmium-plated surface, immersion for a few minutes in an aqueous solution of about 1% of chromic acid and about 0·1% of sulphuric acid whereby a thin surface film of oxide is formed, has been recommended [8].

(b) Result of overbaking, due to extreme drying periods or temperatures, especially of the first coat which has to key the whole film system to the base [7]. Local overheating is even possible in box or conveyor type convection ovens with circulating air.

Whilst these observations apply to almost all types of coatings there are a few exceptions to the rule: many of the cross-linking epoxy resin based and the polyurethane lacquers may be brittle when undercured but become more elastic at higher temperatures or longer baking, though, naturally, up to a limit.

Modern synthetic thermosetting finishes must be cured in accordance with the manufacturers instructions and problems are more likely to occur through undercuring than over-stoving. It must not be assumed that an undercured film can be rectified by subsequent re-stoving at the correct temperature, for the film will attain a partially crosslinked structure which prevents further molecular mobility to complete the desired degree of crosslinking. This fault is most likely to occur with phenolic varnishes, epoxy phenolics and epoxy amino blends when stoved below 330°F.

Undercuring can often be confirmed by a simple solvent resistance test in comparison with a correctly cured standard panel.

Control and regulate stoving ovens by use of a thermographic recording device in preference to a maximum thermometer. This method of assessing correct temperature control is particularly important when standardizing conveyorized ovens. Automatic temperature regulation in an ordinary box oven does not lower the chance of divergences, as it indicates only the temperature at the feeler in one corner of the stove. The actual temperatures at various sections of the oven may vary considerably and the effect of heat radiated from the oven walls or from other articles to be dried has also to be considered. Automatic air circulation in the stove equalizes the temperatures of a convection type oven practically at all points, provided that the damper has been properly adjusted, prevents overheating and reduces the fuel cost and the drying periods of coatings (however, not on brass). Thick-walled metal parts such as, for instance, castings, delay the cooling and may therefore lead to underestimation of the drying period. It is always advantageous to try out the best drying conditions by actual experiment. For instance, where black enamel on an oleoresinous basis is to be engraved, each film should be dried at 120°C (250°F) for one to one and half hours. Excessive drying may cause chipping of the film during the engraving process.

(c) In the case of p.v.c. plastisols and organosols a prerequisite for good adhesion to metals, e.g. iron and steel, aluminium and aluminium or magnesium alloys or brass, is usually the application of an 'adhesive' priming coat. This is generally stoved. If the metal can be surface-roughened by one of the known methods the better will be the adhesion of the 'adhesive' coat and indeed the whole film system.

All solvents must first be removed from the adhesive primer before applying organosol or plastisol, as over areas where this has not been the case the adhesion can suffer.

P.v.c. dispersions, which fuse at approx. 177–182°C (350–360°F) in around 10 to 20 minutes to an irreversible hard film must be stabilized, otherwise they will, if overstoved, not only discolour and sweat plasticizer out, but also suffer embrittlement.

Where objects are completely enveloped by a plastisol, etc., film as in wire or tubular work, and the film thickness is high so that a certain amount of shock is buffered, the effects of cohesion and shrinkage suffice and no primer coat is required. Roughening of the metal surface is for such use beneficial.

One coat organosols of good adhesion are also available.

(d) Excessively heavy films which (after thorough drying) are nearly always less elastic. The top layer is not well anchored to the base. Heavy coatings of this type usually require more and longer heat to effect thorough drying, leading possibly to local overbaking. Differences of stress and strain within such films can lead to checking. Such films may at first deceive by false adhesion if air dried or insufficiently stoved.

In one case, a stoving alkyd finish containing aromatic solvents had been dipped after a coat of primer, drying by evaporation of solvents. This dipping operation lasted for 20 minutes. During the long dipping of the finish its solvents partly dissolved the primer again. When the two-coat system passed through the oven, the excessively wet layer of films dried only superficially in the given time and the differences of tension between the dried rigid upper system and the softer primer caused cracking (J.L. Francis). In other cases a dipping coat may lift a soft film by wrinkling it.

Apply a thinner coat or, where a certain thickness is required, apply a number of them instead of a heavy one. Hot spraying can be considered.

A reduction of the dipping time led in the case mentioned above to faultless results.

A particular problem can arise with thermoset powder coatings applied at excessively high film thicknesses of 400–500 μm (15–20 thou) for chemical resistance or pipe coating applications. To achieve this high film build it is necessary to preheat the article to temperatures which will cure the coating at the substrate by residual heat. The temperature gradient between the inner and outer film surfaces falls off so rapidly that uneven cure will occur resulting in severe internal stresses. When the film relaxes upon ageing, loss of adhesion and cracking becomes evident.

Highly reactive curing agents tend to aggravate this condition and quench cooling of the coated article should be avoided. A reduction in the preheat temperature followed by a period of post-curing should largely overcome this problem.

Depending on the type of binder used, say, medium oil length alkyd resin varnishes, two or more coats may be applied, each film being dried, e.g. for one hour at 120°C (250°F), if colour does not matter up to 160°C (320°F). Other synthetic enamels may dry at lower temperatures and in shorter time. It is advisable not to bake the first one or two coats too hard, i.e. not entirely scratch-hard, although drying must have proceeded far enough to avoid lifting or cracking of the films when the next coat is applied.

This method of treatment ensures a good anchorage of the upper film in the lower and eliminates unnecessarily long heating and embrittling of priming coats, if there are any. Thus, also intercoat adhesion difficulties of films of thermosetting lacquers are reduced.

(e) Engraving or cutting is effected too long after lacquering.

Most nitrocellulose lacquers harden through comparatively quickly. Engraving should be effected on the first day, the latest two days after lacquering, but attention is called to the fact that not all nitrocellulose lacquers are suited for engraving. Articles which are to be cut can be dipped into hot water before this operation to soften the films.

(f) Excessively heavy application may cause peeling of a normally perfectly adherent film if general conditions are unfavourable.

(g) False elasticity can be caused by solvent retention due to the content of certain resins, in cellulose lacquers, e.g. due to shellac, rubber, and in particular polymerized products such as polystyrene.

See observations by H. Staudinger [11] and O. Jordan [12]. Solvent retention of some film formers can be extremely long lasting. 'Measurable amounts of specific solvents were found to be present in, for example, an acrylic lacquer film which had been exposed outdoors in New Jersey for two and half years' [13].

A nitrocellulose lacquer film, which appears to be dry may under normal atmospheric conditions still retain 10 to 20% volatiles for a considerable time. These include solvents of low volatility. Solvent retention is typical for high polymers. Thinners, etc., of the more quickly evaporating type should be preferred.

(h) Lack of adhesion also occurs if a too short oil primer or filler for an oil-base coating system has been applied directly on to the metallic surface.

The adhesion of fillers directly on to smooth metal is nearly always unsatisfactory. It is, as a rule, well worth while to apply an air drying or stoving elastic primer first.

(i) Overpigmented films, for instance those of some oil-base primers, in which a proper orientation of the vehicle molecules is prohibited—especially when the film sets rapidly—tend to exhibit poor adhesion [14]. This fault arises from the poor wetting qualities of heavily pigmented films. Hot application can help to overcome this defect.

(j) The base has not been sufficiently cleaned from grinding dust, fluff, mill scale or rust or the traces of a dried-on rust remover previous to varnishing [4]. If hydrochloric or sulphuric acid has been used for cleaning purposes (which are only completely removed with difficulty), the iron may even contract a certain degree of brittleness (the so-called 'etching brittleness') induced by the formation of hydrogen during the etching process which reacts with the iron.

(k) Insufficient degreasing and cleaning, particularly of castings, e.g. of residues of mineral oils applied as corrosion protective or of cutting oils [4]. Most sensitive in this respect are transparent nitrocellulose lacquers (zapon lacquers), especially on highly polished brass and still more so on silver, from which the lacquer may peel off. They are less sensitive over slightly roughened surfaces, such as obtained by polishing fine parallel lines (strichpolieren) and brushing [5].
Iridescence may also be caused by insufficient degreasing and cleansing previous to lacquering.

Cellulose lacquers and enamels form soft films, at least in patches, above greasy or waxy surfaces, e.g. caused by paint remover, frequently resulting in peeling from highly polished metals such as steel, tin, brass.

Careful degreasing has to precede every metal finishing. The failure is usually visible at once.

(l) Careless degreasing of aluminium and light metal surfaces by means of substances containing alkalis or acids. Special care is required with castings.

L. Valentine reported that in systematic experiments at the Paint Research Station, Teddington, regarding the effects of degrees of cleanliness of steel surfaces on adhesion of paint films, stearic acid was found to be one of the worst contaminants [4], [15].

Aluminium parts can be degreased by a short exposure to a 10 per cent sodium hydroxide solution at about 40–60°C (100–140°F), provided one neutralizes immediately afterwards with weak nitric acid and rinses thoroughly in water (which may contain a small amount of sodium chloride). Alkali carbonates and cyanides are also in use; they have a slight roughening effect on the surface. Mixtures of xylol and alcohol and various other organic solvents have also been recommended.

Chemical pretreatments e.g.: Alocrom is the preferred surface conditioner for aluminium. Light weight anodizing is a suitable but more costly alternative.

Atmospheric corrosion of aluminium alloys under a paint film can manifest itself as filiform corrosion (see Figs. 7.15, 716).

(m) Coated aluminium and light metals can corrode and lift the film subsequently through the influence of acids, such as tannic acid contained in leather cases, etc., e.g. in the case of field glasses which are stored for too long a time.

(n) Continuous influence of elevated temperatures may also cause embrittlement of the lacquer or varnish films (for instance, in the case of cookers). It is clear, of course, that very thin films embrittle more rapidly under these conditions than heavier ones, while very thick coatings may develop cracks on cooling (due to contraction) even if heat resistance is satisfactory in normal film thickness.

Select special coatings of particularly strong resistance to ageing. Do not apply too heavy coats, but a number of thinner films if a certain degree of thickness must be attained. Thermosetting material is likely to withstand fairly well. Where exceptional resistance to elevated temperatures forms a chief requirement, aluminium paste should be used in a suitable medium, e.g. silicone resin solution.

(o) Inferior humidity resistance of stovings combined with premature film softening can be observed if the exposure is carried out before films are completely hardened through or before a thermosetting process has been completed. Heavy film layers of plasticized urea-formaldehyde-resin lacquers, for instance, can show this tendency quite distinctly. The possibility of such an effect is increased when two layers have been applied wet on wet, or at any rate before the first has dried through, and are then stoved in one operation.

Though it is not intended to condemn this method of working as such, it must be mentioned that comparison of urea lacquer films showed a better humidity resistance if each film had been stoved separately only half of the time of the total of the 'in one operation stoving' of two films.

The addition of a catalyst before use can be beneficial. If the humidity resistance of the urea formaldehyde resin based lacquer cannot be improved by use of a catalyst i.e. *p*-toluene sulphonamide, then better results will be given by melamine formaldehyde resin lacquers.

Notes

[1] Burns, R.M., and Schuh, A.E., *Protective Coatings for Metals,* p. 313: 'If direct contact of a paint medium with the base metals is made possible (e.g. by complete removal of grease, dirt and oxide, etc., layers) the electrical forces residing in the atoms and molecules of the metal surface can readily interact with the electrical forces residing in the polar molecules of the finishing medium. The distance over which these attractive forces can act is extremely short, being at most of the order of several molecular layers of the constituents

in the coating medium. Their sphere of influence drops to a negligible quantity at distances of the order of a few hundred Ångström units from the substrate surface, and such a distance comprises only about one-thousandth of the thickness of an average single-coat paint film.'

[2] Broockmann, K. (1954), *Farbe Lack,* **60**, 445.

[3] Bullet, T.R. and Rudram, A.T.S. (1959), *JOCCA,* **42**, 789. 'Measurements made at the Paint Research Station, Teddington, and elsewhere suggest that the maximum attainable adhesion force level (approximately) for non-polar materials such as polystyrene is probably of the order to 2000 lb in^2. For polar materials such as polyvinyl acetate, it is at least of the order of 5000 lb in^2. In practice with films cast from solution at room temperature the adhesion measured for the fully dried film is about 500 lb in^2 for polystyrene and 2000 lb in^2 for polyvinyl acetate.' 'Chemical bonds between coatings and substrates would be expected to give forces many times greater (than van der Waal's) and this expectation may well have been realised in the etch primer film.' 'Adhesion figures up to 12 000 lb in^2 have been measured for these films.'

[4] Selden, G. and Prutton, C.F. (1940), *Techn. Proc. Ann. Convention Fed. of P. and V. Prod. Clubs,* October 1940, p. 231, and *ibid.,* 1941, p. 149, point out that oil and grease inhibit adhesion through absorption forces (residual valencies), because they are so strongly absorbed on to metal surfaces owing to their polar character that they prevent the polar groups of the coating material from making intimate contact with the base. On the example of the 'greasy' stearic acid these workers demonstrate that long straight hydrocarbon chains, because they are 'slippery ones, without turns or twists', cannot anchor themselves in the organic coating film.

[5] Selden, G. and Prutton, C.F. *ibid,* 1940, pp. 252–4. distinguish between two types of interstitial adhesion: The first is that produced by the penetration of the coating into the interstices of the metal surface, and the second by the penetration through a densely adhering coat of metallic oxide, sulphide, phosphate, etc, to the base metal below.

[6] Hess, M. *loc. cit.* [9]. The acid value of such media, however, gives only an incomplete indication of the probable extent of the reaction. It can easily be demonstrated on tinned steel that oil-base media with a higher acid value may show less reaction, e.g. as visible by the crystallization of the surface layer of tin, than others with lower acid values. That these reactions take place even down to a considerable depth in the solid surface can be shown by a similar experiment using tinnings of various thickness. With the same acid varnish the appearance of the surface crystallization ('chemical frosting spangle effect') varies with the thickness of the tin layer. This proves that the character of the free acid is decisive. A slight acidity of a coating for metal surfaces is generally regarded as useful for attaining good adhesion. The beneficial influence of slight acidity on adhesion may, however, well be mainly due to polarity and not to a chemical reaction where the primary valencies are involved. This may account for the fact that the strong basic metals of the first and second groups

of the Periodic System, which would be expected to react easily with the carboxyl groups of the free acid, are the most obstinate with regard to adhesion of coatings. No doubt these metals influence structure and gelation when the film is being formed. Should, however, the chemical reaction go so far as to create a layer of reaction products, the thickness of which is greater than the reach of the secondary valencies between base and coating, bad adhesion is to be expected as in the case of oil and grease layers. As 'the effect of chemical reaction thus far reported has been a deleterious one to adhesion in each case'. Selden and Prutton, *loc. cit.,* [4] 1941, p. 155, believe that there may be 'an upper limit of acidity which can be present, beyond which corrosion of the metal surface may occur with a resultant loss of adhesion'.

Metallic soaps formed by chemical reaction between base and coating are likely to ease the wetting. In addition they act also, and so do many metals, as catalysts which further the oxidation and in the long run the possible deterioration of oil-type films.

[7] Brittleness due to drying in infra-red plants, see Section 4.12.

[8] Sutton, H. (1943), *J. of Sc. Instr.,* No. 61, 89.

[9] Hess, M., (1944), *Paint Tech.,* **9**, No. 98, p. 29.

[10] Marks, S. (1940), *Paint Man.,* p. 154.

[11] Staudinger, H. (1952), *Z.f. Angew. Ch.,* **64**, p. 152. How tenaciously parts of solvents can be occluded purely mechanically by macromolecules may be judged from the observation that cyclohexane or carbon tetrachloride held in such a manner by cellulose could not be removed in spite of applying high vacuum for several days and in addition temperatures at 80–100°C.

[12] Jordan, O. *Technology of Solvents,* p. 83: 'Solvent retention is the result of the molecular size and form of the binder. Even resins whose molecules are small compared with those of (the micellar) cellulose esters, such as ester gum, cyclohexanone resins, and simple polymers of poyvinyl acetate, possess strong powers of solvent retention. The large molecule of the binder can exert a considerable influence on the 'geometrical retention' (due to its frequent three-dimensional structure).'

[13] Scherzinger, R.A. (1960), *Off. Dig.,* 1198.

[14] It is assumed that the polar groups of a coating tend to turn to the surface of the base 'in order to satisfy their affinity by some form of polar orientation on that surface, producing a more or less permanent interface between the solid and liquid either through strong secondary forces operating at the surface or through chemical reactivity.' (W.E. Wornum). Overpigmented films tend to exhibit poor adhesion owing to insufficient numbers of polar heads of the medium confronting the metal surface. Polar compounds must, however, be 'firmly retained in the coating material', according to G. Selden and C.F. Prutton, *loc. cit.,* [4] 1941, p. 155. The above workers mention as consequence of the heat also the displacement of air and the maintaining of the coating in a fluid condition until solidification by oxidation and polymerization.

It is probable that by stoving, more forces, presumably mainly of secondary valency, are set free and more chance contacts are made possible owing to the heat, promoting excitement and rotation (M. Hess, *loc. cit.* [9]).

[15] Valentine, L.A., (1963), *JOCCA*, **46**, 674.

6.3 Insufficient adhesion to zinc, zinc alloys and galvanized sheets

A detailed discussion of adhesion disturbances generally is contained in the introductory remarks to Section 6.2.

The tendency of zinc to change its physical properties suddenly (high expansion relative to the paint films) and its smooth surface offering little possibility for keying necessitates the use of paint materials of an exceptionally high degree of elasticity. In addition a zinc surface is fairly reactive. The pronounced crystalline structure of cast zinc is changed by rolling, and other forms of compression, at room temperatures. Heating the rolled zinc sheets to temperatures between 75° and 350°C induces recrystallization. The optimum drying temperature of coatings should be round about 120°C, the highest for rolled sheets up to about 150°C, with a maximum drying time of 2 hours. Alloyed zinc may be heated at rather high temperatures–soft-rolled to 250°C (480°F) for 2 hours, hard-rolled to 150°C (300°F), but soft-rolled zinc should be baked for the schedules suggested for hard-rolled if they have been subjected to severe deformation during fabrication. For the zinc die casting alloy to B.S. 1004:1955 Alloy A, copper-free (corresponds, e.g. to Mazak 3, 4·1% aluminium, 0·05% magnesium, balance almost all zinc), the following stoving treatments are recommended (maximum):

½ h	1 h	2 h	3 h
218°C (425°F)	190°C (375°F)	163°C (325°F)	150°C (300°F)

The above alloy is the only die casting alloy 'that retains its dimensions and mechanical properties virtually unchanged after higher temperature stoving treatments'. Only zinc die casting alloys which are copper free can fulfil these requirements. For Alloy B, which corresponds to, e.g. Mazak 5, which is of similar composition, but contains 1% copper, low stoving temperatures of up to 93°C must be applied.

If the zinc sheet is plated by brass (brass zinc), the maximum temperature must be lower, because the thin brass layer may otherwise partly become matt or may disappear at 100–120°C (200–250°F) [1]. Although chromium-plated zinc (chrome-zinc), may not show any visible alteration at temperatures up to 180°C, or aluminium plated at still higher temperatures, plated zinc sheets should, if possible, not be exposed to higher temperatures than unplated zinc.

While a zinc sheet can easily be bent previous to the baking of a coating, the change of structure induced by the heat is indicated by the cracking sound emitted by the sheet on bending. The sound is the louder the higher the baking temperatures and already very pronounced at 160°C (320°F) [2]. Extremely small amounts of foreign metals dissolved in the zinc (tin!) may cause intercrystalline corrosion followed by sudden changes of the physical characteristics. This is greatly accelerated by somewhat elevated temperatures, a warm humid atmosphere in service or in storage.

That a chemical reaction between zinc and a coating may also take place has been shown by H.J. Wing. Since brass contains from 20 to 80% of zinc, besides copper, which it hardens, it is feasible that the behaviour of brass in relation to paint materials is somewhat similar to that exhibited by zinc.

As a rule, paints do not adhere well to the smooth, freshly galvanized iron or to rolled zinc or zinc die casting alloys, although they do so to zinc coatings applied by the metal-spraying process. Zinc sprayed steel can be painted immediately after spraying. There appears to be no significant advantage in weathering sprayed zinc for 6 months [3] nor has any influence of the structure (roughness) of the coating on performance of the paint system been observed.

Under normal conditions of atmospheric exposure the rate of corrosion of zinc is approximately one fifth to one tenth that of untreated steel; where it is exposed in enclosed and polluted conditions such as railway tunnels (steam), zinc may be as corrodible as steel. Various estimates for the durability of unpainted galvanized steel have been given by several authors. For instance, a zinc coating applied at 1 oz/ft^2 (approximately 40 μm) could last up to 25 years in a rural atmosphere, 15–20 years in a marine environment, but only 5 years in industrial surroundings.

(a) Use of paint materials unsuited for the purpose (see Fig. 6.1). There are coatings which adhere well to zinc and zinc alloys which have been cleaned and degreased only.

Use only highly elastic coatings for air-drying, decorative purposes. Tar paints adhere even to zinc sheets which were not subjected to a preparatory treatment. Bituminous paints without priming coats often show good results, as also do linseed oil–iron oxide paints.

In particular, priming paints based on calcium plumbate adhere excellently even to new galvanized surfaces. Etch primers are also effective, but should be regarded rather as a form of pretreatment than as a coating in themselves.

Special enamels, e.g. containing medium oil alkyds should be used for stoving. Media giving acidic breakdown products should be avoided. Suitable lacquers include those based on polystyrene, styrenated alkyds and nitrocellulose.

Metallic pigments – with the exception of zinc dust – and graphite tend to contribute to the corrosion of zinc and give rise to adhesion troubles. Zinc-rich primers have been used satisfactorily on new galvanized surfaces, provided these have been cleaned thoroughly [4].

In many tropical developing countries corrugated galvanized sheeting is frequently used as a roofing material. Whiteley [5] has suggested that priming with calcium plumbate should be adopted, but as rainwater is often collected from the roofs for drinking purposes, it is desirable to use non-toxic paints free from lead compounds. Acrylic emulsion paints [6] have proved eminently suitable.

(b) The paint was applied to new and therefore smooth metal. (See Fig. 6.1.)

The difficult adhesion of coatings on new galvanized iron sheets is believed to be due to the formation of a very fine skin of basic zinc chloride, remaining from the use of ammonium chloride, etc., in the galvanizing process, and/or the presence of a thin layer of oil. Such impurities impair adhesion.

It is a frequent custom in the decorator's trade not to paint new zinc or galvanized sheets, but to leave them exposed to weather for, say, three months until the bright surface has become dull and somewhat rough, followed by scouring with a wire brush. This is believed to be too short a period. Six to twelve months, depending on conditions is more realistic but it is difficult to assess when the surface is fit for painting [7]. Firmly adhering weathering products protect the metal beneath against further corrosion. If on account of extraordinary chemical or electrolytic influences, e.g. by contact with nobler metals, a strong corrosion of the zinc has taken place, the layer of white zinc salts formed is sometimes called 'white rust'. It consists mainly of basic zinc carbonate.

Unweathered zinc surfaces require chemical or mechanical pretreatment. Careful consideration must also be given to effective degreasing [4]. This can consist of a solvent wash with white spirit or if small components are involved, a vapour degreasing process with trichlorethylene or other organic solvents. Large structures can be degreased with a steam jet containing a mildly alkaline solution. Mechanical pretreatment with wire brushes or light blasting with rounded shot is effective but great care must be taken to avoid removal of the thin layer of zinc. Abrasive blasted zinc is very reactive and must be painted immediately.

Treatment with mordant solutions, such as copper sulphate or hydrochloric acid – once much favoured – is not advised, but a phosphoric acid wash is useful and promotes good adhesion of paint. Thorough rinsing with clean water should follow the phosphoric acid wash. A short dip of zinc or galvanized articles into a solution of acid sodium or potassium dichromate (Cronak process) [8] is a simple and effective pretreatment before application of stoving primers and finishes.

Notes

[1] Hooper, W.H.L. and Inglis, N.P. (1954), *Chem. Ind.*, **44**, 1344.
[2] Hess, M. (1944), *Paint Tech.*, **9**, No. 98, 29.
[3] Wallington, R.H. (1970), RS/T/21/70, Paint Research Association.
[4] Zinc Development Association (1966), *Painting of hot dip galvanized steel.*
[5] Whiteley, P. (1962), *Tropical Building Studies*, No. 4, 25, HMSO.
[6] Low, C. and Vorster, O.C. (1972), *JOCCA*, **55**, 901.
[7] Reeve, J.S. (1966), *Review of current literature on Paint and Allied Industries*, **39**, 1.
[8] U.S. Patent No. 2035 380.

6.4 Insufficient adhesion to smooth, frequently polished metal surfaces

An explanation of the physical and chemical processes in relationship to adhesion failures has been attempted in the introductions to Sections 6.1 and 6.2

Insufficient adhesion may be particularly evident in chromium-nickel steel, nickel- and silver-plated surfaces, aluminized surfaces, aluminium, the light metals, Electron and Dow metals, brass, zinc, chromium and cadmium plating, leaded surfaces, and plastics, e.g. bakelite.

(a) The difficulties of adhesion of a film to the surfaces of the above materials increase somewhat in the order named. Of course, no definite rule can be given in this respect, especially in the case of alloys which show different degrees of paint adhesion even if composed of the same basic metal. The chief difficulty is that films of coatings may be unable to find sufficient key on smooth metal surfaces (especially on light metals) to overcome the pulling stresses exerted by the expanding and contracting metal base. A strongly cohesive and elastic film such as, for instance, a thin nitrocellulose lacquer film is fairly satisfactory, although it can often be peeled off when damaged.

The coefficients of expansion of the various substrates play their part. That of aluminium, e.g. is twice that of steel, and it consequently requires more elastic films.

(i) A smooth base may, in many cases, be roughened by sand or grit blasting, rubbing down, pickling and anodic pretreatment of metals. On surfaces so treated nearly every paint will adhere.

(ii) In many instances it will be impossible to roughen the smooth metallic surface in this manner, either because the smooth surface or colour of the metal is required for artistic or technical reasons or because the metal would be weakened too much by the treatment. This is also important in the case of plated metals. It is necessary in these cases to employ suitable special lacquers of sufficient elasticity, etc., which have been developed for practically all metals, lead glasses, etc., although stoving may become necessary where films of high resistance to rubbing and scratching are required.

(b) Adhesion of coatings to chromium will be poor if after plating:

(i) Thorough rinsing has not been carried out with distilled or de-ionized water.

(ii) Lacquering did not take place within 2–4 hours after exposure to the plating-room atmosphere.

(iii) The hardness of the water in the final rinse was over 2 p.p.m. [1].

Recommendations for rinsing are to use 1·5 to 3 parts by volume of chromic acid in de-ionized or distilled water at 60–82°C (140–180°F) for 10 to 120 s followed by de-mineralized water of pH 7.

Air drying water dispersible systems have been tried with success on some difficult metal surfaces. Where stoving is permissible an epoxy based coating will give best adhesion to smooth surfaces.

Note

[1] Scott, F.L. (1959), *Plating* (U.S.A.), **46**, No. 3, 261.

6.5 Poor adhesion of stoving finishes on brass surfaces [1]

In some respects adhesion disturbances on brasses are reminiscent of those on zinc, of which metal they contain varying proportions, the rest being copper [2]. Brasses are, however, not nearly so sensitive to higher temperatures as zinc. There are also brasses which contain further additions of a few per cent lead and less of the other metals.

So-called 'brassed zinc', a zinc base coated with a very thin layer of brass, shows similar adhesion troubles as brass, e.g. towards some nitrocellulose lacquers. Otherwise highly adhesive finishes may fail on clean annealed brass [3].

Vapour degreasing is usually considered satifactory on copper and copper alloys, provided it is 'followed by a hot alkaline cleaner or electrolytic cleaning in a hot alkaline bath. Cyanide dips are commonly used for stain removal' [4]. Fairly bright [3] wrought material possessing good surfaces may be completely cleaned by a 'bright acid dip of the sulphuric-nitric type' (H. Sutton) [4].

(a) The surface of many alloys, including especially those of copper [2] and also of aluminium, is hardened by cold working, i.e. hammering, rolling, drawing, etc., at room temperature. Internal stresses are introduced thereby which may lead to *season cracking* of the metal, especially in the case of deep-drawn cups and drawn tubes. This form of cracking can happen after articles have been stored or in service for varying periods of time in an atmosphere containing only small amounts of certain corrosive substances, particularly ammonia and its derivatives. In the less extreme cases which occur frequently and are hardly visible to the naked eye, the change in surface structure shows itself by a delayed embrittling of the superimposed paint films.

A somewhat similar behaviour has been claimed for some plastics containing such synthetic resins as polystyrene.

In order to avoid season cracking of cold-worked brass, it is necessary to relieve internal stresses in the metal by heating to 250–280°C (482–540°F), or by annealing at higher temperatures. Heating to the low temperatures first mentioned does not remove the rolling hardness. If this heat treatment of the metal is not carried out satisfactorily before coating, adhesion disturbances may occur sooner or later after baking, even if the varnish or enamel itself has satisfactory properties.

Brasses require, as a rule, longer baking periods than, e.g. iron.

(b) Residues of chemicals, especially of copper compounds as used for blackening for the purpose of increasing adhesion (kishy metals), can have a detrimental effect.

Such treatments should be followed by sufficient washing to remove even the slightest traces of salts.

Notes

[1] This section has been improved by the incorporation of suggestions by Mr P.J. Macken, Metallurgical Department, Copper Development Association, London, whose assistance is gratefully acknowledged.
[2] The compositions of the most commonly used types of brass are red brass, 80% or more copper, remainder zinc; yellow brass, copper with 20 to 50% zinc plus additions of tin and iron; white brass, copper with 50 to 80% zinc.
[3] The highly glossy brass, i.e. a product annealed in the brightest condition possible in practice, is close annealed in an airtight oven. The matt-finished brass is oxidized in the air and then acid cleaned.
[4] Sutton, H. (1943), *J. of Sc. Instr.,* **20**, 90.

6.6 Bad adhesion to lead or leaded surfaces

It is quite possible that bad wetting of coatings sometimes plays a part in the case of bad adhesion on lead. Many oil varnishes and paints exhibit insufficient adhesion to lead and leaded surfaces, particularly cold- and hot-water pipes, leading to flaking because of the catalytic action of the lead and lead oxides (intentionally made use of in driers) on the drying of oils. This favours the drying of the films, but as the action continues after the drying, the ageing of the film is accelerated and the films can embrittle within a comparatively short time [1].

Lead shows excellent resistance to atmospheric exposure but may be susceptible to corrosion in contact with some acidic timbers like certain species of oak and others [2].

If lead has to be painted, careful degreasing is necessary and a solution containing equal parts of trisodium phosphate and sodium carbonate is recommended. The surface must be completely dry before painting. Condensed water can be removed with methylated spirit. Paint usually adheres better to old lead than to new lead. This may be due to an adherent layer of lead oxide.

Abrading with sandpaper is helpful in promoting adhesion but is best avoided owing to the risk of lead poisoning.

An etch primer is the most satisfactory first coat as the normally smooth surface provides only a poor mechanical key. Conventional undercoats and finishes can then be used to follow on.

It has been claimed that paints free from resins show a reduced tendency to flake from lead surfaces.

Suitable coatings are those which contain as media:

(i) Alkyd [1] varnishes, also in combination with nitrocellulose.
(ii) Appropriately formulated chlorinated rubber and asphalt varnishes, which are not catalysed and embrittled by lead and also are resistant to chemicals.
(iii) Thermosetting or *in situ* setting heat resisting acrylic lacquers.

Notes

[1] Hess, M. (1944), *Paint Tech.,* **9**, No. 98, 27. The bad adhesion of films of an oil-base medium on lead and lead glass, the surfaces of which are very smooth, commonly supposed to be due to the strong catalytic action of lead and lead oxides on the oil of the medium, must also have other reasons. Media consisting of highly polymerized oil-modified alkyds show much better adhesion on such surfaces.

[2] Shrier, L.L. (1976), *Corrosion,* Vol. 2, Chapter 19, p. 87, Newnes-Butterworths, London.

6.7 Delayed loss of adhesion of paint films over some metals

There are cases where adhesion which appeared to be perfect on inspection immediately or shortly after finishing, became unsatisfactory after some service.

(a) On aluminium, normal boiled linseed oil, alkyd resin varnishes, and nitrocellulose alkyd resin combination lacquers show signs of embrittlement after four to five months outside exposure due to oxidation, if not protected by pigmented top coats (K. Broockmann) [1].

Copolymers of, e.g. vinyl chloride–vinyl acetate–maleic acid, or acrylates and thin films of nitrocellulose, maintain their adhesion on aluminium over a long period, whether applied as transparent or pigmented coatings. In contrast lacquers containing chlorinated rubber, polyvinyl chloride, copolymers of vinyl chloride–vinyl acetate and/or vinyl ether as the only film formers, possess unsatisfactory adhesion from the start [1].

(b) It is claimed [2] that mechanical roughening and chemical mordants only help initially with keying of paint films to galvanized iron, and that in spite of such pretreatments, after a few months the usual cracking and flaking is possible just the same. The breakdown is explained by the permeation of water vapour through the film which allows zinc to react with constituents of the paint binders, forming zinc soaps. These loosen the film.

Satisfactory paints for galvanized iron are suggested in Section 6.3 (b). Zinc phosphate, produced by phosphating, prevents the formation of undesirable soaps [2].

(c) The consequences of season cracking of brass on a paint film become noticeable after varying periods of time.

Notes

[1] Broockmann, K. (1954), *Farbe Lack,* **60**, 445.

[2] I.C.I. *Pretreatment News* No. 14, July 1958.

7

Defects of coatings developing in service

7.1 Pigment volume concentration (PVC) in relation to film properties

The binder content of any paint is dictated by many factors, which include not only the physical and chemical characteristics of the pigment or pigments used but also the gloss and performance properties desired in the paint.

In early investigations on oil paint, H. Wolff [1] showed that there is a 'critical oil content' below which there is marked falling off in performance. Asbeck and van Loo [2] studied alkyd type paints and showed that film properties and performance were related to the volume rather than the weight of pigment present. This is a more logical approach to comparisons of paints since specific gravities of pigments vary so widely. They introduced the term 'pigment volume concentration' (abbreviated to PVC) which is the percentage of pigment by volume in the dry film. It is calculated from the expression:

$$\text{PVC} = \frac{\text{Volume of pigment} \times 100}{\text{Volume of pigment} + \text{Volume of non-volatile binder}}$$

Pigment volumes are readily calculated from specific gravities.

The 'critical oil content' of Wolff has a modern counterpart in the 'critical pigment volume concentration' (CPVC) which can be taken as the value of the PVC of a film in which there is just sufficient binder present to coat the pigment particles and to fill the voids between them. The value of the CPVC is specific to a given system and will vary from one type of paint to another. At the CPVC a paint will be practically devoid of gloss but the gloss will increase rapidly on addition of binder. On the other hand an increase in pigmentation will give an underbound porous film. Asbeck and van Loo studied the effect of varying PVC on several paint properties and found these to change rapidly at a particular value. By plotting the properties against PVC a graph of the form shown in Fig. 7.1 is obtained. The point of rapid change is the CPVC.

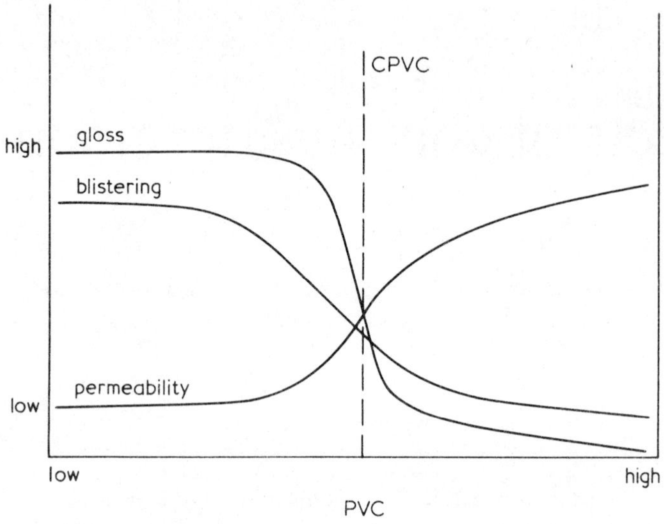

Fig. 7.1 Graph showing relationship between pigment volume concentration (PVC) and various paint properties.

The value of the CPVC depends on a number of factors, viz. oil absorption, size and shape of the pigment particles, the type of medium i.e. the percentage of the high molecular weight species and the packing characteristics of the pigment particles. The latter will be controlled by the size and nature of the adsorbed polymer envelope and also by the degree of flocculation of the coated particles.

Low sheen paints, such as semi-gloss and eggshell finishes used for interior work, as well as undercoats are formulated with a PVC approaching the critical point. The very limited amount of 'free' binder in these paints renders them unsuitable for exterior exposure. Undercoats are, of course, always protected by a finishing coat.

High gloss paints contain a high ratio of binder to pigment and possess a PVC considerably lower than the critical value. From Fig. 7.1 it will be seen that it is possible to formulate a gloss finish over a range of PVC. At low values the opacity may be poor (depending on the type of pigment) but at the higher values the amount of 'free' binder, although sufficient to produce a gloss, will be insufficient to give a durable film. Poor gloss retention and the early onset of chalking are likely on exterior exposure. For each paint there is an optimum PVC consistent with opacity. For finishes on a long oil alkyd pigmented with rutile titanium dioxide the PVC is usually 16 to 18%. It is not possible to quote figures for coloured finishes since they depend on the nature of the coloured pigment used. The lower limit is probably represented by carbon black. The low pigmentation of a black gloss paint (as a consequence of the high opacity and fine particle size of the pigment) results in a PVC of about 2%.

Determination of CPVC

A great deal of work has been done on the correlation of PVC with the properties of liquid paints as well as paint films. The determination of CPVC values has, of course, been an integral part of these studies. A review of much of the work has been published [3].

A simple practical method for CPVC determination was employed by G. Guillaume [4] who measured the gloss of a system over a PVC range using a goniophotometer. By this means small changes in gloss could be measured accurately and rapidly. The initial paint is prepared above the CPVC and measurements made as the PVC is reduced by successive additions of medium.

The calculation of CPVC from pigment properties has been described by G.P. Bierwagen [5]. The pigment properties involved are oil absorption and a 'densest packing factor' for the pigment with the adsorbed layer of resin. Use is made of an algorithm for pigment packing suggested by D.I. Lee [6].

Notes

[1] *F.Z.*, 1931, **36**, 374; 1932, **37**, 1962, 1723; 1933, **38**, 158, 1271. H. Wolff determined (and it was confirmed by the results of investigations carried out by F. Wachholtz) that the water resistance of oil paints decreases with increasing oil content from a certain point onward. In some cases this decline starts in connection with some pigments very suddenly and abruptly (increased tendency to whiten and correspondingly reduced weather resistance). Since the oil content also expresses itself in the degree of viscosity, a critical point can be determined from the viscosities as measured by a turbo-viscometer (H. Wolff and G. Zeidler, *F.Z.*, 1935, **40**, 26, p. 667). The characteristic properties above indicated commence to appear within a certain critical range above or below this point, and the values thus determined for various pigments were called 'Critical Oil Contents'.

[2] Asbeck, W.K. and van Loo, M. (1949), *Ind. Eng. Chem.*, **41**, No. 7, 1470.
[3] Suryanarayana, N.P., Yaseen, M. and Aggarwal, J.S. (1966), *Paint Tech.*, **30**, No. 6, 28.
[4] Guillaume, G. (1969), *Paint Tech.*, **33**, 16.
[5] Bierwagen, G.P. (1972), *J. Paint Tech.*, **44**, 45.
[6] Lee, D.I. (1970), *J. Paint Tech.*, **42**, 579.

7.2 The porosity of surface coatings

Paint films are permeable to water, water vapour, oxygen and other liquids and gases. This implies that there are channels or capillaries present in the film. Yet very careful examination of films with the electron microscope has not shown any evidence of such channels [1]. However, the permeability of paint films is selective and not all types of molecules can pass through. Paint films are therefore usually classified as 'semi-permeable'.

The precise mechanism of permeability has not been fully explained. Kooisman [2] has introduced a concept of inter- and intramolecular porosity and found water vapour permeability to be inversely proportional to film weights for the lower weights. At greater weights permeability decreased to a lesser extent. Funke [3] studied permeability for water and water vapour in relation to film thickness and pigment volume concentration (CPVC–see Section 7.1). He concluded that permeability was governed by the swelling properties of the films. Strong swelling gave rise to increase in permeability, but permeability and water absorption did not run in parallel in films showing low swelling. The effect of pigmentation was to reduce permeability but in most cases permeability increases above the CPVC. In general, permeability was inversely proportional to the film thickness.

Permeability is governed by two processes:

(i) Diffusion as determined by the diffusion coefficient D in Fick's law.
(ii) Absorption as determined by the solubility coefficient S.

The permeability coefficient P is represented by the product $D \times S$ [4]. Fick's law simply states that diffusion is directly proportional to the difference in concentration of the water vapour on either side of the film, to the time of exposure and to the area of the film. It is difficult to determine the coefficient of diffusion by direct measurement, but several methods are available for determining permeability, the best known one being the Payne cup. The solubility coefficient can be determined by sorption methods e.g. with a quartz spiral balance [4].

An elegant technique for the measurement of permeability, based on the use of radio isotope tracers has been described by Matsui [5]. The transport of chloride ions through surface coatings has been investigated by Murray [6].

In the context of paint film defects we are more concerned with channels which lead through the film to the substrate and which are more aptly described as pores, holes or '*holidays*'. The main differentiation in this nomenclature is by size [7]. They are often not noticeable in freshly applied films but form in many cases after drying from occluded air bubbles or centred on dust particles. Mayne [1] conjectured that paint films may be free from capillaries when dry and only develop them owing to the up-take of water.

It is important for the paint user to know whether a coating is free from pores and holidays which could be detrimental to the performance of the coating. There are several instruments available for measuring porosity [8]. They are based on a high voltage breakdown test and can only be used if the coating is on an electrically conducting substrate. An electrode, either a metallic brush or flat rubber sponge type, is passed over the film. The substrate is connected to the instrument and a high voltage (500 to 1500 V) is applied. Where there are pores in the film the voltage will break through from the electrode to the substrate and this will register on the meter. Naturally the voltage must be carefully selected to match the film thickness as too high a voltage would break through a thin film even if no pores were present [9]. An instrument measuring the flow of current from an electrode to the coated test panel has also been described [10].

The degree of porosity has also been measured by immersing a coated metallic test panel into a solution of copper sulphate and observing the deposition of metallic copper in the pores. More recently a method has been described by which radioactive tracers (^{60}Co or ^{64}Cu) are deposited by electrolysis in the pores [11]. To avoid detrimental effects from film porosity it is most desirable not to rely on a single coat but to use at least two coat systems.

The causes of pore formation are:

(a) The type of medium. Chlorinated rubber lacquers, for instance, are particularly prone in contrast to nitrocellulose lacquers. Also, the higher the content of medium-solids, the fewer the number of pores, and the smaller their diameter. Adequate plasticization minimizes porosity.

(b) Overpigmentation. This may be an advantage in special cases, as in antifoulings, but in most cases, e.g. with screen pastes, it is a grave set-back.

(c) Drying conditions. Slower drying helps towards the formation of a sound film structure. Solvent retention is, however, disadvantageous. The type of solvent affects the number of pores, but their evaporation rate or water solubility were found to be no guide. The older a film, e.g. of chlorinated rubber paint, the fewer pores can be found.

(d) Faulty application, e.g. by dry spraying.

(e) Condition of substrate. A moist base, for instance, can create blisters and pores by rising moisture.

(f) Dust from the atmosphere, pigments, and contamination of coatings with solid impurities are pore promoters. Observe cleanliness during application and initial drying or flash-off period.

(g) Condensation of moisture, e.g. on drying of cellulose lacquer films where clouding indicates throw-outs. This is produced by rapid evaporation of solvents as a consequence of which the surface is cooled down.

(h) Aeration in paints applied by dip, hot spray, brush and hand roller methods:

(i) Aqueous-based paints.

Water paints and emulsion paints often form highly porous films due to excessive aeration caused by surface-active constituents of the material. This becomes accentuated on overbrushing or too vigorous application of a paint roller. Certain rollers are more prone to cause this effect than others. Retained air bubbles giving this phenomenon of porosity are more permanent in paints based on high proportions of thickeners and colloids which limit the escape of the air during the drying period.

Very often this defect may be eradicated by an adjustment of the viscosity of the water paint or emulsion paint. Where permissible, thinning with water may be sufficient to ensure the liberation of the air bubbles and the subsequent satisfactory flow out of the paint. Generally, however, a more careful application will minimize or remove the defect. Brush application with subsequent 'laying off' is generally satisfactory.

Careful stirring with an eccentrically placed stirrer in the container is advised for de-aeration. The addition of proprietary anti-foaming agents, such as silicone fluids, often assists in reducing aeration.

(ii) 'Oil-based' paints.

Many 'oil-based' paints can deposit porous films when applied by hand roller. The effect is usually created by vigorous roller application, which is responsible for surface foaming. Alkyd-based paints, flat, eggshell or gloss can be affected in this way, the air bubbles being trapped in the early setting stage of the film and being observed under the miscroscope or hand lens as minute craters in the film.

The remedy in this case is the change of application method to one of brushing. The brush application is usually responsible for bursting the air bubbles and assisting flow during the 'laying off' stage.

(i) Aeration in epoxy powder coatings and entrapment of volatile condensation products released during the curing cycle of polyester power coatings.

Bubbles of air or vapours are trapped beneath the surface of the film (see Fig. 7.2). Small additions i.e. 0.1% of benzoin during manufacture of polyester powder coatings, have been found to be beneficial.

Notes

[1] Mayne, J.O.E. (1957), *JOCCA,* **3**, 183.
[2] Kooistra, M.F. (1957), *Fatipec Congress Book.,* p. 95.
[3] Funke, W. (1967), *Farbe Lack,* **73**, 707.
[4] Selier, P. (1970), *Verfkroniek,* **43**, 391.
[5] Matsui, E.S. (1969), *J. Paint Tech.,* **41**, 455.
[6] Murray, J.D. (1973), *JOCCA,* **56**, 210.
[7] Anon. (1965), *POCJ,* **148**, 742.
[8] Latter, T.D.T. (1971), *Polym. Paint Col. J.,* **160**, 549.
[9] Elser, W. (1966), *Ind. Lack. Betrieb,* **34**, 64.
[10] BASF (1965), *Ind. Lack. Betrieb,* **33**, 253.
[11] Luczak-Teichert, M. *et al.* (1966), *Przem. Chem.,* **45**, 250. See *Review of current literature,* PRA, No. 295, 1257, 1967.

7.3 Some general reasons for the premature destruction of paint films applied to building structures

The term 'weather resistance' as applied to paint and varnish materials is meant to imply maximum protective capacity especially against corrosion of the materials covered, accompanied in some cases by chalking resistance, fastness of colour, stability of gloss, resistance to disintegration (life) and cracking. Specially suited as weather-resistant materials are paints based on long-oil varnishes, or stand oil or certain long-oil synthetic resin varnishes whose high weather resistance is due chiefly to the large molecules formed during the manufacturing process. The retention of gloss, ability to shed dirt, and general integrity of gloss paints based on alkyds far surpasses that of the best products based on linseed stand oil or long-oil varnish; e.g. for dark shades durability has been extended 50%, and for light shades up to 100%. Their faster set-up renders the paintwork less vulnerable than the older types to marring by sudden showers or dust coming immediately after application.

The principal factors causing the breakdown of paint systems are ultraviolet radiation from the sun, moisture, atmospheric pollution and excessive fluctuation of temperature [1].

A life of at least five years can be expected of a good exterior paint system under normal conditions of exposure. In testing films for weather resistance, they are usually exposed at an angle of 45° facing south. This is commonly considered to accelerate the breakdown by a factor of 2 compared with the vertical exposure on most structures. In order to assess the full potential of an exterior paint it should be exposed in rural, marine and industrial environments. With many 'long life' coatings the testing of weather resistance described above, would be too slow, and the process can be speeded up by utilizing the intensive sunshine and the high humidity prevailing in Florida, U.S.A., where a number of exposure sites are available to paint manufacturers and users. Another interesting development in this field has taken place in Arizona, U.S.A., one of the three sunniest parts of the world [2]. It enjoys 4000 hours of sunshine a year. An equatorial mount was designed for exposing panels to the sun all day at an angle of close to 90°. These samples receive about one third more radiation than stationary samples facing 45° south. This system was further developed by using highly polished aluminium mirrors to reflect 70–80% of the u.v. radiation on to the exposed paint film in the equatorial mount. A water spray compensates for the absence of rain in Arizona and an air blower keeps the surface temperature at about the same level obtaining when exposing panels by the conventional method at 45° facing south. Whilst this method allows rapid assessment of ultimate durability and type of breakdown to be expected it does not take into account the atmospheric pollution one finds in industrial surroundings.

Many laboratory devices are available for accelerated weathering of paints. In most of them the painted panels are exposed to ultraviolet radiation from an arc lamp and to intermittent water spray for 1000 to 2000 hours. The method of test described in British Standard 3900, part F3 gives generally accurate predictions of the behaviour to be expected of airdrying alkyds, oleoresinous and many other paints on natural weathering. There are some limitations with respect to colour change and corrosion of substrate [3].

Causes of failure:

(a) Moisture. This can originate from the construction water (new concrete), water gaining access through defects in the structure, or condensation, and can lead to early loss of adhesion, blistering and flaking. It is imperative to allow walls to dry before decorating and to take measures to prevent rain penetration and rising damp. Where it is impossible to achieve a dry wall, alkali-resistant, permeable (porous) paints can be used.

The moisture content in walls can be determined by several methods, most accurately by drying a sample, obtained by drilling, in an oven and calculating the moisture content from the weight loss. There are also conductivity and capacitance meters available. The former measure the resistance between two needle probes forced into the wall, but soluble salts present in the wall can affect the result. Capacitance meters measure the moisture content in the uppermost 1 to 2 mm layer of the wall and do not work well on rough surfaces. A fairly reliable test is based on the determination of the equilibrium humidity in an air space in contact with the wall. Such an air space can be created by a sealed and insulated box mounted on the wall or by fixing a sheet of polyethylene to the wall with a hygrometer inside. If the relative humidity in equilibrium with the surface is below 75%, painting can proceed without problems. If the reading is between 75 and 90% certain limitations are imposed, but flat emulsion paints can be used without too much trouble, unless the humidity is close to the upper limit. If the surface can be seen to be wet (equilibrium humidity 100%) it is preferable to postpone decoration until conditions improve. When dealing with internal walls, ventilation and heat should be applied to speed up the drying process. On external surfaces not much can be done, but if at all possible plastic sheeting should be arranged in such a way as to create ventilation and to protect the surface from further wetting by rain or condensation, which could cause rain spotting and early loss of gloss.

(b) Application during unfavourable weather conditions. Very high and very low temperature and high humidity should be avoided. Dust and dirt will produce a 'bitty' finish. See above and Section 3.5.

(c) Insufficient pretreatment of the base, be it steel, concrete, masonry or brick. This includes:

- (i) Imperfect removal of loose, old paint which should be removed mechanically by wirebrushing, carborundum or glass paper, pumice stone or by paint remover, in which case the surface must be thoroughly washed and allowed to dry before painting.
- (ii) Insufficient preparation of old porous cement paint. All loose and spongy material must be removed by scraping or wirebrushing and a suitable sealer applied. This can be an oil-based type or a solution of a low molecular weight p.v.a. resin in a suitable solvent such as methylated spirit. Direct application of emulsion paint without binding the porous surface may cause flaking, due to contraction on drying.
- (iii) Imperfect removal of mill scale, rust or grease. See Section 7.9.
- (iv) Friable surfaces arising from poor hydration of cement or plaster; they provide little adhesion for paint films and should be left unpainted. The surface can be tested by applying a piece of self-adhesive tape. If this shows little adhesion

when pulled off and material adheres to it there is not much prospect of paint adhering. Unless the friable material can be removed until a sound base is obtained, replastering is necessary. If the surface is not too weak, a thin coat of emulsion paint might prove satisfactory.

(d) Wrong choice of paint. The paint may not be suitable for a particular substrate or for the climatic conditions, or for atmospheric contamination with acidic or alkaline dust or gases such as sulphur dioxide.

In any paint system it is also necessary to match primer, undercoat and finish. If, for instance, a short-oil coating is applied over a long-oil paint or vice versa, stresses will develop in the system, leading eventually to cracking and flaking.

(e) Salts and alkalinity. Many types of bricks contain soluble salts. Unwashed sand in mortars and plasters, or cement additives, are other sources of such salts. These salts are dissolved by water in the structure, rise to the surface where the water gradually evaporates, depositing salt crystals either in a fluffy form or as a hard layer of crystals. This is known as '*efflorescence*' and can force an impermeable paint film of the surface, causing cracking and flaking (Figs. 7.10–7.12). Emulsion paints may allow the passage of salts without disruption of the film, but adhesion may be weakened. Deposits of salts can be brushed off when they are formed and this process repeated until the structure has dried out.

The soluble salts in cement and lime plasters are associated with alkalinity and in this case permeable alkali resistant primers should be used. Oil and alkyd paints which are subject to attack by alkali should be avoided, otherwise '*saponification*' will take place. This implies splitting of oils into fatty acids and glycerol and leads to sticky, soft, brownish films, water blisters and oily brown droplets and runs.

Masonry paints based on synthetic rubbers or certain acrylic polymers have good resistance to alkali and are also reasonably permeable. Cement surfaces can be treated with dilute phosphoric acid to neutralize lime.

(f) Patchiness of gloss or colour. The cause of this phenomenon is uneven density and suction of the substrate (cement and plaster), permitting varying penetration by the paint binder. Excessive suction calls for the application of an 'anti-suction' primer consisting of a first coat of very dilute emulsion paint or a well-thinned alkali resistant primer. A high degree of suction can remove nearly all the binder from a paint, leaving a powdery deposit of pigment on the surface.

(g) Mould growth. The appearance of coloured spots or stains usually signals the growth of moulds (Fig. 7.3). The most frequently encountered colours are grey, black, purple, green, red and pink. The problem occurs in persistently damp, badly ventilated areas, both on the inside and outside of buildings. To achieve a permanent cure these conditions must be remedied. In suspect locations a fungicidal wash before painting is recommended. In the early stages of growth, scrubbing will remove the organisms. If dampness cannot be cured use frequent fungicidal washes on the paint-work and special fungicidal paints. Fungus is less prone to grow on hard-drying and chemically cured paints such as epoxies.

(h) Special factors influencing the films:

(i) Exposure to excessive ultra-violet radiation on high mountains, on the ocean or in the tropics, particularly if the exposure is in one direction (weatherside, etc.) only, can lead to discoloration, premature loss of gloss, chalking, embrittlement, cracking, peeling or several of these defects.

It is absolutely necessary to use paint materials, e.g. alkyd based, adapted to these specially severe requirements and to build up the paint coat by a sufficient number of layers.

(ii) Accumulation of exceptionally unfavourable weather conditions (hail, hoar frost, freezing, etc.). Sides of buildings exposed to rain showers, wooden window frames, etc., must be painted with special care, and without omitting any area in order to prevent the absorption of water through unprotected parts (Fig. 7.4) leading to attack of the films from beneath (especially in the case of wood).

(i) Unsuitable or excessive thinning of paints.

(j) Insufficiently long periods of drying between consecutive applications.

Conversely, excessive periods of drying between coats can also lead to flaking if the underlying coating is too hard and presents an inadequate key for the next coat. This applies especially to two-pack epoxy- and polyurethane paints but also to alkyds pigmented with calcium plumbate.

(k) Excessively high drier content leads to embrittlement and subsequent cracking.

(l) Selection of unsuitable pigments. Pigments should be resistant to water, acids and alkalis and non-fading when exposed to ultraviolet rays. Organic dyestuff pigments are more prone to fading and must be used with particular care when blending with white pigments to obtain pastel shades. An example is toluidine red (helio red) which is quite light-fast at full strength, but pink shades produced by mixing with a titanium dioxide white base are very fugitive.

In atmospheres containing sulphur dioxide, paints containing yellow lead chromate pigments tend to fade rapidly, owing to the reduction of hexavalent chromium to trivalent chromium.

(m) Excessive pigmentation causes a paint to behave as one being too short in oil length. In the case of emulsion paints early chalking takes place. Too high a percentage of binder, however, creates softer films with excessive dirt retention.

A proportion of one to one is generally recommended in the case of emulsion paints based on oil-modified alkyd resins.

(n) Excessively thick paint films unable to dry through entirely or which may even form wrinkles and folds.

Amongst other effects, solvent retention, to which quite a number of binders tend, is likely to increase susceptibility to moisture.

If possible, dry thoroughly before exposure. In some instances this may be achieved with the aid of portable infra-red drying units or by a force-drying method which would greatly benefit the protective value of the paint films.

(o) Re-softened films show usually a much reduced life.

Notes

[1] *Building Research Establishment Digest* No. 198, 1977. An excellent review on which parts of this section have been based.
[2] Desert Sunshine Exposure Tests Inc. The machines described here are known as 'EMMA' and 'EMMAQUA'.
Scott, J.L. and Anderson, T.E. (1976), *JOCCA*, **59**, 383.
[3] Hipwood, H.A. and Weaver, W.R. (1970), *JOCCA*, **53**, 341.
Hoey, C.E. and Hipwood, H.A. (1974), *JOCCA*, **57**, 151.

7.4 Swelling and poor resistance of paint and varnish films exposed to water and moist atmospheres

Swelling is the continuous absorption of moisture by a solid mass accompanied by an increase in volume. It can be caused by liquid water or water vapour [1].

The most common effects are whitening or milkiness and softening. Old films are less affected than new films. As long a drying period as possible should be allowed before the film is brought into contact with water.

Some films increase considerably in volume in contact with water. Among these increases of 20 to 50% are common.

Poor resistance on continuous immersion in water or prolonged exposure to high humidity is shown by linseed oil, linseed oil varnishes and long oil alkyds.

(a) Alkyd films frequently permit the passage of water under very humid conditions, and this may collect at the interface between finishing coat and under coat, resulting in loss of adhesion of the former. Adhesion may be restored on drying, but frequent alternations of wet and dry conditions will eventually produce an irreversible loss of adhesion.

Loss of adhesion of alkyd finishing coats may be reduced or prevented by selecting a suitable undercoat, since the nature of the undercoat contributes substantially to the incidence of the defect. There appears to be no guiding rule, however, for the selection, and empirical methods are necessary to decide which type of undercoating is most suitable.

According to P. Whiteley [2] *et al.* paint films based on alkyds or polyurethane alkyds are weakened appreciably by immersion in water. The probable cause is thought to be hydrolysis rather than leaching out of plasticizing ingredients.

(b) The environment within the film when immersed in water or exposed to water vapour seems to be of crucial importance. It was found that a high pH furthers swelling, and many pigments in common use, like those mentioned below, tend to be alkaline and promote swelling. Clear alkyd films from resins having a significant acid value swell little. The high weight gain found when testing alkyd films pigmented with alkaline pigments could presumably be due either to OH ions from these pigments or to soaps formed by the reaction with free acidic groups in the alkyd [3].

(c) Certain pigments promote the swelling and whitening of films depending on the character of the medium. Zinc oxide, titanium dioxide coated with zinc oxide or aluminium oxide, ochre, calcium carbonate, white lead and ultramarine belong to this group. Suitable pigments, carbon black and some chromates [4] and extenders, e.g. barium sulphate, should be selected for outdoor exposure.

(d) From their work on non-pigmented films, R. Katz and B.F. Munk [5] conclude that the polarity of the solvent, which affects the molecular orientation in the film, has a definite influence on the water vapour permeability.

(e) W. Funke [6] considers that the liquid water absorption of paint films is related to pigment/binder interaction and he points out the influence exerted by pigment-binder adhesion and by the Pigment Volume Concentration.

(f) High resistance to moisture is shown by tung oil-pure phenolic varnishes and these have long been used as boat and spar varnishes.

The following media also show high resistance to water and moisture –

Acrylics	Polyurethanes
Chlorinated rubber	Vinyls
Epoxies	

(g) The water resistance of boat varnishes applied at relatively low temperatures, about 14–15°C (57–60°F), may be much different from that exhibited by varnishes applied at more normal temperatures, 20–25°C (68–77°F) [1].

If, therefore, a varnish film must not turn white in the water, and if it must be applied at lower temperatures (for general climatic reasons) a more suitable varnish must be selected.

The widespread use of glass-reinforced polyester (GRP) in boat construction has considerably reduced the use of varnishes on hulls, particularly those of small craft. These are often polished only.

(h) Rain and condensation may attack a film from above, and if it has not dried through, cause spotting (‘*rain spotting*’) which may partly be due to whitening. If, however, the film had already deteriorated to a large extent by ageing or unsuitable composition, the general breakdown with the consequences of cracking, peeling and blistering, owing to humidity will ensue. This does, of course, also happen if the moisture attacks from below.

(i) Delay in the drying of oil-based paints and varnishes due to dampness of substrate.

Apply on dry substrates in dry weather.

(j) The absorption of water by latex films has been studied by J. Snuparek [7], who finds that the absorption is related to the stability of the latex particles. Emulsions containing fine particle polymers can be less stable and films deposited from such emulsions show higher water absorption. On increasing the stability of the emulsions by addition of a secondary emulsifier the films showed lower water absorption, probably through closer packing of the polymer particles in the film.

Notes

[1] Brunt, N.A. (1964), *JOCCA*, **47**, 31.

[2] Whiteley, P., Rothwell, G.W. and Kennedy, J. (1975), *JOCCA*, **58**, 200.
[3] Long, J.S. *et al.* (1965), *Off. Dig.*, **37**, 1050.
[4] Ledwith, R.J. *et al.* (1946), *JOCCA*, **29**, 67.
[5] Katz, R. and Munk, B.F. (1969), *JOCCA*, **52**, 418.
[6] Funke, W. (1967), *JOCCA*, **50**, 942.
[7] Snuparek, J. (1976), *JOCCA*, **59**, 19.

7.5 Blooming, hazing and bronzing of paints and varnishes

Hazing and blooming. There appear to exist three different main groups:

(i) Permanent blooming under the influence of moisture or high humidity. To this belong the phenomena 'Crater bloom', 'Syneresis bloom' and 'Wrinkle bloom'. A bluish-white, sometimes oily, exudation on parts or the whole of the surface of completely air-dry films of mainly oil base (including oil-modified synthetic resin) varnishes. It is accompanied by loss of lustre at the affected areas. This type of bloom develops in a humid atmosphere on certain films which are characterized by good water resistance and their tendency to swelling in contact with water.

(ii) Permanent hazing emerging from the film typical of air-drying bitumen coatings or pitch varnishes insufficiently dried by stoving ('Pitch bloom'), or from enamels containing dyestuffs or certain pigments due to sublimation on the film surface ('Sublimation bloom').

The types (i) and (ii) are characteristic in that the bloom will always reappear after a certain time even after repeated wiping.

(iii) Temporary hazing from the deposition of salts, 'Crystalline bloom'. To this group belongs also the haze formed on some kinds of oil base or alkyd stoving varnishes due to products of combustion ('Oven bloom'). It is removable by wiping or flushing with water and will not reappear.

The bluish-white deposit, as on grapes, makes the film wholly or partly lustreless and may develop a few days after application of the varnish, or later on, but always only after complete drying. There still remains a considerable divergence of opinion with regard to the basic reasons for this fault.

It seems to be definitely dependent on the surface condensation of moisture, which may cause changes in the structure of the facc of a varnish film. Changing conditions, particularly when a fall in temperature is involved, accelerate bloom formation. Other factors such as humidity at the time of setting and drying of a film are also instrumental. Delayed or defective drying due to frost or fog may cause blooming of newly painted films.

There are permanent and temporary forms of bloom. The latter are characterized in most instances by the presence of ammonium sulphate, and can be removed readily by rubbing with a cloth, dry or mostened with water. The different forms, however, are probably further subdivided into other varieties, and it should be emphasized that no single explanation or theory has been found completely satisfactory, but only operative in a majority of instances.

(a) Permanent bloom occurs if moisture condenses on the film before it is fully dry, the droplets leaving their imprints as minute craters ('Crater bloom'). It is visible as soon as the film is dry, and arises only under adverse conditions. Many schools of thought do not consider this a genuine bloom, but that this term should be reserved for those phenomena caused by a deposit or exudation on the surface.

(b) The cause of 'crystalline bloom' (removable) has interested chemists for a number of years. It was described by N.K. Adam and R.S. Morrell [1] who considered that a low contact angle between the varnish film and water (i.e. good wetting) reduced the tendency to bloom. The results of W.E. Wornum [2] were in line with these when he found that the more water resistant films tended to bloom more (poor wetting). Later workers in the field were J. Ames and S. Gourley [3] and G.L. Holbrow [4] who consider that the ammonium sulphate can be formed in condensed droplets of water from ammonia and sulphur dioxide absorbed from the atmosphere (Figs. 7.5 *a* and *b*). According to Holbrow the formation of ammonium sulphate is slow and it remains as a bloom when the droplets evaporate. The presence of calcium naphthenate in the paint film appears to induce moisture condensation as a film rather than as droplets. Evaporation is then more rapid and the time available for ammonium sulphate formation correspondingly reduced. Blooming is then reduced by improved wetting.

G.V.G. Hill [5] considers that the presence of droplets of condensed moisture is not essential for the formation of crystalline bloom. He has shown that it can be produced on painted panels by thermal precipitation of particles of ammonium sulphate. These particles are mainly below 2 μm in size and appear to be charged since they could also be precipitated by an electric field.

(c) All types of coating can be affected to differing degrees. Pentaerythritol-based alkyds are more prone to bloom than those based on glycerol, see below. Severity of blooming is affected by many of the ingredients used in formulating the coating, e.g. lead combined with lead naphthenates often increases blooming, whereas calcium naphthenate will reduce it.

(d) Blooming is more frequently observed in countries with prevailing humid weather, and under special local conditions (where sulphur oxides are present, or in presence of ammonia). Draughty locations are also more prone to the production of bloom. The fault is rarely observed in dry and warm regions. Thus varnishes which exhibit the failure in the British Isles may often be applied on the Continent without trouble.

Use if possible materials which have been found satifactory under the prevailing conditions. The time of the year when the finish is applied also has considerable effect. Dry and warm weather usually produces the most resistant coatings. Late spring and autumn are probably the worst times, since the fluctuations in temperature and humidity between night and day are greatest at these times.

Bloom of the temporary 'crystalline' type may often be removed by wiping off with wash leather and water, and will then not reappear for some time.

Since alkyds based on pentaerythritol are more prone to bloom than those based on glycerol, the latter should be selected for severe blooming conditions. The difference is, however, only one of degree, and under some conditions at the present state of our knowledge all alkyds will bloom.

(e) 'Pitch bloom' is a form of permanent haze which can appear on gloss finishes if the thinners contain a proportion of high boiling solvent in which the resin is virtually insoluble. It takes the form of slight precipitation.

(f) 'Sublimation bloom' is sometimes evident on the surface of stoved enamels containing certain pigment dyestuffs [6]. It appears when the articles cool and can be wiped from the surface. The pigment dyestuffs most prone to give sublimation bloom are the arylamide yellows and toluidine red which are very slightly soluble in the paint medium at the stoving temperature. (See Fig. 4.7.)

(g) '*Bronzing*' is the term used to describe the metallic lustre seen on the surface of films pigmented with certain pigments at full strength, e.g. some Prussian and phthalocyanine blues and some organic red pigments.

'Bronzing' is in some ways related to blooming, since it is a phenomenon apparent to the eye because a surface film has formed on the finish involved. It is now thought to be due to a raft of small pigment particles which have exuded to the surface of the coating. At some angles this film is seen as a metallic appearance. It may be evident as soon as the film is dry, or may develop on weathering. Sometimes bronzing is deliberately encouraged, e.g. in some printing inks.

(h) Bronzing is generally observed on freshly prepared films or after weathering with certain pigments only, e.g. Prussian blue, Reflex blue, some Fanal colours (phosphotungstomolybdic acid lakes), and phthalocyanine colours, organic reds, e.g. 4B toner, magentas, etc. The phenomenon seems to be correlated with the presence in the pigments of a substantial amount of material of particle size below about 0·1 μm [7]. (See Fig. 7.6.)

The use of alternative pigments in the original paint formulation can eliminate bronzing, but this is not always possible if a specific shade is required for which non-bronzing pigments are not available. Also a better wetting medium or addition of surface active agents can be tried.

'Bronze' in a film can be readily removed by superimposing a coat of clear varnish, which 'submerges' the fine particles giving rise to the appearance.

Notes

[1] Adam, N.K. and Morrell, R.S. (1934), *J. Chem. Ind.,* **53**, 255.

[2] Wornum, W.E. (1934), *JOCCA,* **17**, 132.

[3] Gourley, S. (1953), *JOCCA,* **36**, 8.

[4] Holbrow, G.L. (1962), *JOCCA,* **45**, 701.

[5] Hill, G.V.G. (1974), *JOCCA*, **57**, 342.

[6] Geissler, G. (1973), *Farben-Zeit.,* No. 10, 462.

[7] Crowl, V.T., Read, A.R. and Tilleard, D.L. (1961), Paint Res. Association, Research Memo. No. 293, April 1961.

7.6 Chalking and erosion

'*Chalking*' is the condition of a paint surface which, having lost most of its gloss, is coated with a white powder or 'chalk'. This can be removed on light rubbing with the finger or a cloth. Generally, chalking appears on whites or light tints and is the result of photochemical breakdown of the surface layer of binder with consequent release of pigment.

Chalking is associated with certain pigments which are activated by ultraviolet radiation of which the most active wavelengths lie in the 290 to 350 nanometre band [1]. The most well known chalking pigment is the anatase grade of titanium dioxide; a paint based on this pigment and exposed to the weather suffers rapid loss of gloss followed by heavy chalking. Other pigments which cause chalking are lithopone and white lead.

Pigments which are resistant to chalking include rutile titanium dioxide, red oxide of iron, antimony oxide and carbon black. These are thought to absorb ultraviolet radiation [2], the energy being converted to heat. In this way the medium is protected and paints made on these pigments show good durability. Tropical exposure may lead to chalking of red lead, oxide of iron and chromium oxide paints. By mixing the two grades of titanium dioxide, paints can be produced with predetermined rates of chalking. In white paints slow chalking can show certain advantages. The surface retains its whiteness since dirt is shed and slow erosion of the film is not accompanied by checking or cracking. With tints chalking is objectionable since it tends to mask the colour and the paint appears to have faded.

The main causes of chalking and erosion are as follows:

(a) The use of unsuitable pigments (as mentioned above). Examples are anatase titanium dioxide and lithopone which give heavy chalking (in lithopone the photochemical action is accompanied by slow oxidation of zinc sulphide to soluble zinc sulphate which is leached out of the film). White lead also causes chalking but for various reasons is now little used.

(b) Certain types of highly pigmented paints containing rutile titanium dioxide will chalk after 2–3 years exterior exposure. Examples are emulsion paints and some types of masonry finish in which the binder content is low. In these cases chalking probably results from causes other than the straight photochemical breakdown. For example, emulsion paint films contain water-soluble colloids and surfactants which are slowly leached from the film. The chalking and general durability of emulsion paints varies with the type of polymer used. Vinyl co-polymer emulsions are superior to the externally plasticized homopolymer in which plasticizer migration can lead to early embrittlement. Greater durability is shown by the '100%' acrylic emulsions.

(c) The solvent type masonry finishes contain high proportions of extenders and these can vary in their reaction to ultraviolet radiation. Many masonry paints are white or very pale tints and are designed to chalk slowly, thereby shedding any dirt and presenting a permanently clean appearance. Paint films which erode gradually are generally free from tendency to check or crack. Preparation for repainting consists in washing off the chalk and allowing to dry.

(d) Claims have been made that alkyd modified exterior emulsion paints adhere well to chalking substrates.

(e) Chalking is a characteristic of some highly cross-linked polymers, even when pigmented with rutile titanium dioxide or other pigments which are not usually inclined to chalk. The epoxy resins are characterized by this tendency. In polyurethanes the tendency is lower but the types cured by aromatic isocyanates yellow on exposure.

(f) The onset of chalking is accelerated by repeated condensation on the film followed by drying out. The surface layers of the film are ultimately disrupted by the repeated swelling and contraction.

(g) Chalking has been observed in undercoats, pigmented with anatase titanium dioxide, under finishing coats containing rutile titanium dioxide and leading to a breakdown of the system.

(h) The mechanism of chalking in paints containing titanium dioxide is based on the formation of OH-radicals, formed by the interaction between the pigment particles, u.v. light and water [4]. The OH-radicals break down the binder by oxidation. Whilst loss of gloss is usually a significant indication of the onset of chalking, the progress is best followed by measuring the loss of weight which occurs before chalking or loss of gloss becomes apparent; it is claimed that the rate of erosion of paints in their later life can be predicted from the measurement of early weight losses [5].

A loss of material amounting to 5 μm of film thickness per year is not detrimental to the qualitative properties of the film [6].

A simple assessment of chalking can be made by applying a piece of self adhesive tape to the film, pulling it off and measuring the reflectance of the pigment layer adhering to the tape [7]. The intensity of the white spot obtained by wrapping a piece of black velvet around a finger and moving it across the chalking surface under medium pressure over a distance of 10 cm is another simple qualitative means of assessing the degree of chalking. Photographic standards are available for comparison of the brightness of images obtained by pressing the wet, exposed gelatin film of a photographic paper onto the chalking surface. (See Fig. 7.7, and ref. [8]; for standards of erosion, see Fig. 7.8.)

(i) The exceptional resistance to chalking associated with most modern grades of rutile titanium dioxide is due to incorporation of a small amount of zinc oxide into the pigment particle surface and following this with a coating of silica and aluminium oxide. Such rutile grades are nowadays used in most exterior coatings, often as the sole pigment. In latex house paints which are widely used in the United States, the addition of zinc oxide to the pigment is also claimed to bring about an improvement of chalk resistance, which is coupled with fungistatic properties and a synergistic effect on other bactericides and fungicides present in the formulation [2].

(j) Among extenders, natural barium sulphate and natural calcium carbonate give better chalk resistance in combination with rutile titanium dioxide, than their precipitated counterparts.

(k) Too thinly applied films may show early chalking.

On tarry surfaces, for instance on wood previously impregnated with creosote, most coatings do not adhere well, so that even if their original appearance is satisfactory, they may be washed off by heavy rains.

The question as to the extent to which the insulation with special paint materials, emulsion paints, shellac varnishes, aluminium paints, etc., for outside work and with hard-drying wood-oil type varnishes for interior applications is recommendable, must be carefully examined in every instance. Absolutely reliable results can only be obtained if another tar oil type coat is applied.

(l) Under conditions of very high relative humidity i.e. 80% and over, the drying and film formation of emulsion paints can be seriously delayed. In such circumstances the coating is in danger of being washed off if rain occurs.

(m) Lack of oil or binder, e.g. of linseed oil in bituminous paints, may cause decomposition commencing with chalking under unsuitable weather conditions. This, in the case of bituminous paints is termed 'black chalking'.

(n) Toluidine red mixtures with extenders may exhibit a behaviour corresponding to chalking.

Casein paints if applied to old plaster may remain water soluble. They must also not contain excessive amounts of pigments, otherwise chalking and dusting off may possibly ensue.

Faultless results are to be expected only if the casein reacts with the reactive calcium oxide of the still new plaster, thus forming the insoluble calcium caseinate. It should, however, be presumed that the casein paint had been made up correctly by the manufacturer in such a manner that it becomes insoluble after application. Alkali casein paints should never be applied on to old plaster outdoors. Faulty casein paint films, which may sometimes exhibit peeling, have to be removed.

(o) Insufficient additions of media of all kinds to pigments and, vice versa, excessive additions of pigments to media, make their films susceptible to wiping off or powdering. This phenomenon is somewhat similar to chalking. Similarly, excessive absorption of binder by a 'thirsty' porous surface can cause this effect.

Notes

[1] Hoffman, E. (1971), *J. Paint Tech.,* **43**, No. 563, 97.

[2] Larson, L. (1972), *Am. Paint J.,* **56**, No. 43, 47.

[3] Merkel, K.W. (1966), *J. Paint Tech.,* **38**, 543.

[4] Volz, H.G., Kampf, G. and Fitzky, H.G. (1970), *Fatipec Congress Book,* p. 107.

[5] Ritter, H.S. (1970), *Am. Paint J.,* **54**, No. 34, 74.

[6] Gabriel, G. (1970), *Farbe Lack,* **76**, 563.

[7] Bohmann, T. and Papenroth, W. (1968), *Farbe Lack,* **74**, 347.

[8] Kempf, R. (1929), *Farbenzeitung,* **35**, 1, 650.

7.7 Breakdown of water-based and solvent-based paints on new or damp brickwork, concrete, cement-rendered surfaces, asbestos cement, lime plaster

The principal cause of failure of paint systems of most types on new or moist brickwork, cement rendering, lime plasters, etc., is moisture. Secondary causes, arising out of the presence of moisture, occur due to the presence of alkali and salts causing chemical attack and efflorescence.

Efflorescence is the development of a crystalline deposit on the surface of brick, cement, etc., due to water, containing soluble salts, coming to the surface and evaporating so that the salts are deposited. In somc cases the deposit may be formed on the top of any paint film present, but usually the paint film is pushed up and broken by the efflorescence under the coat.

New brickwork and concrete can contain large quantities of water – up to 150–170 litres per cubic metre. Troubles can arise due to painting when the surface appears dry, and water is subsequently brought to the surface from the backing materials. This can occur especially when heat is applied internally in the building and the passage of moisture inwards is accelerated. Impervious films will most likely be affected, loss of adhesion and blistering may result.

Chemical attack by alkalis, giving saponification, and mechanical rupture of the paint film or blistering by efflorescence are largely due to the presence of moisture which activates the compounds responsible and transmits them to the site of the breakdown.

Calcium sulphate plasters may be only mildly alkaline, but where lime has been added precautions should always be taken. Mortar, concrete cement renderings, lime plasters and asbestos cement, especially when new, should be considered as alkaline surfaces. A rough guide may be obtained by applying moistened red litmus paper to the surface. (If paper turns blue there is a risk.) After prolonged external weathering the above surfaces will probably have lost their alkalinity due to the leaching out of soluble material, but it is preferable to experiment on a trial area as a precaution before painting. If time permits, an area can be coated with a flat wall paint (alkyd type) tinted with Prussian blue. The blue colour is destroyed by alkali but the test may require several days to give a result.

Cement sets essentially by a hydration and crytallization process, whereas lime sand mixes, as used in plasters, carbonate and dehydrate to form calcium carbonate and silicates. Unwashed sea sands may be responsible for the presence of salts, and sands used should always conform to specified standards [1]. Building limes [2] may be subdivided into:

(i) Non-hydraulic limes suitable for plasters.
(ii) Semi-hydraulic limes which contain some clays and may give rise to higher alkalinity due to the liberation of caustic alkali from salts of potassium and sodium. Both these types are used in lime–and calcium sulphate–plasters, and their effects may therefore be variable.
(iii) Hydraulic limes contain higher proportions of clays, and they are used

principally in mortars and lime/sand renderings due to their good setting properties. Alkalinity may be high with this type.

(a) The presence of moisture in the substrate results in paint adhesion failures (Fig. 7.9) and is also responsible for the activation of alkalinity and efflorescence.

Peeling and blistering occur when paint films, applied to damp surfaces, or to surfaces which later become damp, have high cohesive strength and are pushed away from the surface by the water pressure from behind. Oil and synthetic media paints have generally poor adhesion to damp wall surfaces, e.g. cement renderings, brick-work, plaster, etc., especially when the surface is smooth, giving relatively poor mechanical key. Damp, although not present at the time of application, may be brought to the wall/film interface from:

(i) The substrate, e.g. from brickwork, plaster backing, etc., in new structures (attempts to speed up the drying-out process by the use of radiators, central heating etc., will often accelerate the breakdown);
(ii) The seepage of rainwater from faulty gutters, down-spouts, window sills, radiators, etc., and rising damp where damp courses are faulty or non-existent.
(iii) The presence of damp due to faulty cavity walling on external walls or to porosity of non-cavity construction such as found in old buildings.

Non-permeable films, e.g. of gloss paints, applied over a water-resistant primer, may blister due to water pressure from behind, or, peel if the surface is ruptured, e.g. by a glancing blow from a hard object. More brittle films such as old flat oil paints may break down by cracking and flaking rather than blister. Oil-based and alkyd films tend to swell under prolonged influence of moisture resulting in blistering, more especially when the back of the film is damp than when heavy condensation occurs on the surface. The presence of distemper or a similar water sensitive layer at the wall surface beneath a gloss or oil-based system can aggravate these defects.

The presence of mica in sand has also been found to be a potential source of bad paint adhesion.

It is advisable that new walls and plaster surfaces should be allowed to dry out at least six months before the application of a non-porous system. Masonry, stucco or concrete require three to twelve months for drying. If early decoration is necessary a porous or permeable paint should be used such as flat emulsion paint, which will permit gradual evaporation of residual moisture.

When damp areas are present on old surfaces, the source of the trouble (e.g. structural defects as given above) should be sought, and where possible remedied before redecoration is attempted. It is unsatisfactory to attempt to seal off damp from these causes, as blistering, peeling, etc., will generally follow, or the damp will be transferred to another area. It is important to note that walls which are damp should not be sealed on both sides, e.g. in the case of partition walls, by impervious paints, tiles, etc.

Old films of soft distempers, etc., should be removed, and the surface allowed to dry thoroughly before an oil-paint system is applied. A good water-resistant primer should then first be applied, preferably giving two coats to obtain an effective seal.

(b) Emulsion paints and oil-bound distempers can lose adhesion in a similar manner to that described above, i.e. relatively non-porous films fail by blistering and peeling, more permeable films tend to fail by cracking and flaking after prolonged saturation.

Soft distempers (glue or size bound), ceiling whites, etc., do not blister, but become softened under damp conditions if the film has originally dried out. If applied to a persistently wet surface the film remains wet, runs down the wall and in extreme cases may be washed down completely by the flow of moisture from the substrate. Excess moisture at the time of application can be due to:

(i) Insufficient drying out of the surface, e.g. new plaster.
(ii) High humidity conditions, poor air circulation, prolonged excessive condensation.

Oil-bound distempers and emulsion paints may also be washed down the walls in extremely wet conditions, giving sagging, curtaining or runs which may show as channels in the film, terminating in a hard blob of dried-out paint.

Patchiness may be exhibited by water-thinned paints where uneven porosity of the wall surface has resulted in some areas having dried rapidly and rewetting having occurred in others. Highly trowelled areas hold moisture longer than the more porous areas of plaster walls, resulting in a thinner application of water-based paints on these parts. Uneven porosity can also cause variation in gloss and apparent fading in patches without the presence of moisture and with all types of paint.

Preferably apply only porous paints to new walls in order to permit moisture to dry out. If prolonged wet conditions have caused failure, wash off as much of the paint or distemper as possible, allow to dry out and then redecorate.

Satisfy uneven porosity of wall surfaces by the application of a wash (or 'mist') coat of well-thinned emulsion paint.

(c) Emulsion paints can blister and peel on external walls especially if the surface on which they are applied has a non-porous nature, e.g. smooth-faced brickwork, old glossy painted surfaces, tiles, etc.

Showers of rain or water from an overflowing gutter can permeate the paint film, especially during its early life, causing swelling and blistering due to poor surface key. If conditions change and the film dries back satisfactorily, the effect may not occur again, but in many cases, once adhesion to the surface has been impaired the high film strength of the paint will cause peeling or stripping.

Good mechanical key should be obtained before many p.v.a. emulsion paints are applied to external surfaces. Apply water-based exterior paints in dry weather.

(d) The alkalinity of lime plasters, cement-rendered surfaces, concrete, etc. results in the breakdown of oil-based paints by saponification. The degree of breakdown depends on the strength of alkali present, the sensitivity of the paint film in contact, and the duration of wet conditions to maintain the activity of the alkali.

Table 7.1 Pigments which may be affected by caustic alkali or lime

Inorganic

Chrome yellows
Chrome and molybdate chrome oranges and reds (there may be some grades unaffected)
Chrome (Brunswick) greens
Prussian (Milori) blues
Co-precipitated greens containing chrome yellows

Organic

Lakes of acid dyestuffs
Lakes of basic dyestuffs
Azoic toners, e.g. manganese, calcium and barium toners of azo Reds B, 2B and 4B
Lithol Reds
Lake Red C
Manganese calcium and barium B.O.N. Toners
Arylamide (Hansa) yellow 5G
Maroon toners (e.g. manganese)

Pigments which are generally fast to lime and caustic alkalis at the concentrations encountered in cement, plaster and new brickwork, etc.

Inorganic

Chromium oxide
Hydrated chromium oxide (Guignets green)
Zinc yellow (zinc chromate)
Ultramarine blue and violet
Carbon blacks, iron oxide blacks and graphite
Iron oxides (reds, browns, yellows), earth colours (ochres, umbers, sienna, etc.)
Cadmium yellows, oranges and reds (cadmium/selenium colours)
Antimony, titanium and zinc whites, lithopones
Extenders: Asbestine, barytes and blanc fixe, kieselguhr, silica

Organic

Phthalocyanine blues and greens
Indanthrone blue
Vat dyes (anthraquinonoid, thioindigoid, etc) (yellows, oranges, reds, violets)
Arylamide (Hansa) yellows G and 10G
Certain benzidine yellows and oranges
Nickel azo yellow
Pigment green B
Madder Lakes (includes alizarine lakes)
Some grades of toluidine red (helio)
Para red (paranitraniline red)
Permanent red R (ortho chlor. para red)
Permanent red 2G (dinitraniline orange)
Certain B.O.N. arylamide reds and maroons, such as Permanent Red 4RH, Bordeaux F2R

Paint films attacked by alkali first become soft and sticky and partially water soluble. Under severe attack, and where moist conditions prevail, e.g. externally from rain and internally from condensation or the substrate, sticky runs may develop of a yellow or yellowish-brown coloured liquid, or watery sticky blisters may be present. Upon drying out the stickiness may disappear, but peeling, cracking and flaking may then occur, as the film will have lost binder and have become brittle. Gloss paints lose gloss and as with flat oil paints, etc., show brownish discoloration when saponified by alkali. Heavily attacked areas may readily be washed away with water, showing complete breakdown of the oil or medium.

In lime plasters, etc., the lime may be unevenly distributed, resulting in areas of higher alkalinity. Areas of high lime content harden slower and hold moisture longer than areas of low lime content, hence alkali attack may be prolonged and randomly distributed. Lime by itself rarely causes trouble, but when salts of sodium or potassium are also present caustic alkalis may be formed. Under perfectly dry conditions saponification does not occur, but it must be remembered that the presence of alkali in a surface is a potential source of trouble. Re-activation may readily be observed when structural faults cause dampness, or when a water-based paint is applied.

Where saponification has been severe, the paint film (sticky runs, discoloration, etc.) should be removed preferably by scraping and only if necessary, washing down with water, as the use of water may result in further reactivation of alkali in adjacent areas. Allow to dry out thoroughly and then seal the surface by application of (preferably two coats) an alkali-resisting plaster or cement primer.

Standard types of finishes may then be applied. The use of a system which is unsaponifiable may be considered after removal of the defect. Chlorinated rubber paints applied without the necessity for an alkali-resistant primer are suitable, although these may be subject to blistering or peeling if moisture persists in the substrate. Polyvinyl acetate copolymers and acrylic emulsion paints may also be applied after suitable preparation. The possibility of reactivating the alkali present should be carefully considered when using water-based paints.

(e) Efflorescence (Fig. 7.10) often wrongly referred to as 'saltpetre', the crystallization of salts, is often evident as a white fluffy deposit of sodium sulphate crystals on the surface of a wall. Paint faults occur when efflorescence develops through a film, causing its disintegration and powdery deposits, or when crystal growth forms beneath the film resulting in blistering, peeling or flaking. Although plasters are often thought to be responsible for efflorescence, it is usually found that the salts are brought to the surface by water from brickwork at the back during the drying-out process. Mortar and plaster undercoats may also be responsible, especially when poor-quality sand had been used. Relatively small concentrations of water-soluble salts, such as sodium sulphate, present in the substrate layer of a wall or ceiling may result in bulky deposits of efflorescence, which continue to develop as long as moisture is present. Stripping, flaking, etc., of paint films may be due to very fine efflorescence, not readily seen with the naked eye. The examination of the back of

Table 7.2 Materials suitable for the painting of alkaline surfaces, such as lime plasters, concrete, new brickwork, cement renderings, asbestos cement

Type of paint	Medium	Remarks
1 Emulsion paint	Vinyl, acrylic or styrene/acrylic copolymers	Give porous films. Can be textured for exterior use
2 Solvent type paints	Styrene/acrylic copolymers	Used in exterior masonry finishes. Give porous films with good durability
	Plasticized chlorinated rubber	As above
	Alkyd	Usually for interior work, can be textured for exterior work. Requires primer for alkaline surfaces
	2-pack epoxy	Impermeable films. Good durability but chalks
3 Cement paints	Portland cement	Used on outside walls. Moderate durability

paint films with the aid of a good lens will usually confirm that the adhesion to the surface has been broken down by the growth of minute crystals at the wall/film interface. The paint defects may vary between general peeling or flaking and small nodules or eruptions.

Efflorescence on brick walls (Figs. 7.11 and 7.12) may originate in the bricks or the mortar. In the former case this may not necessarily be the customary white crystals but they may be brownish due to the presence of iron salts in the interior of the brick. Isolated bricks may show efflorescence which, on plastered partition walls may develop in areas of approximately brick size and shape. When emanating from mortar in the brickwork, efflorescence may appear in narrow bands corresponding to the mortar joints.

Other forms of crystal growth, having the appearance and causing breakdown typical of normal sodium sulphate efflorescence, may be due to the presence of contaminating salts in wall surfaces. The use of unwashed sea sand may be responsible giving mainly chlorides. The crystal growth of photographic chemicals spilt on the walls of a room which had previously been a dark room caused blistering and flaking of both flat oil paint and emulsion paint, due to the reactivation of the soluble carbonates and thiosulphates present in the plaster when heavy condensation occurs.

Similar troubles are also possible on walls which have been contaminated by residues from the use of fire extinguishers, etc.

Alkaline salts, e.g. sodium carbonate, present in some household cleaners, may be absorbed in like manner into the plaster substrate, and may subsequently be responsible not only for efflorescence but also softening of the actual paint film.

The use of alkaline cleaners and detergents on permeable paint films may be responsible for subsequent efflorescence if concentrated solutions had been used. The soluble salts which could have been absorbed into the plaster substrate, e.g. by passage as solutions through porous films of distempers, emulsion paints, flat paints, etc., especially sodium sulphate used as a bulking agent in certain detergents, would ultimately recrystallize at the back of the paint film on evaporation of the water from the surface.

The formation of zinc sulphate crystals below the paint film may be responsible for breakdown. The use of zinc sulphate or zinc silicofluoride 'petrifying' solutions to neutralize the alkalinity of plaster is an old practice, but there is some doubt of its efficacy owing to the possibility of efflorescence.

It is essential that all salts and sources of moisture should be removed before painting is commenced. Since efflorescence depends upon the presence of salts and moisture, the growth of crystals can continue so long as both are present, i.e. salts from within brickwork and plaster, and moisture present either from new plaster or from some building defect such as a faulty damp course, leaking gutter, etc., or plant growth near the wall.

Painting should be delayed if efflorescence continues, salts being brushed off with a dry cloth or brush whenever deposits are formed. On no account should salts be washed off, since the use of water will aggravate the defect. Check all possible sources of moisture, allow walls to dry out thoroughly in the case of new structures and rectify any building defects.

The use of water-based paints should not be contemplated on any surface where efflorescence has been heavy. Application of an alkali-resisting oil-type primer, e.g. of the type formulated for plaster and cement surfaces, is satisfactory when the surface is thoroughly dry and all crystal growth has ceased. It is unsatisfactory to attempt to 'seal in' an active efflorescence with such a primer, as blistering and loss of adhesion will invariably be the result. Further decoration should not be carried out until the drying out is complete.

Where efflorescence has caused partial stripping of an existing decoration, all loose material should be removed by scraping and dry brushing and then proceed as above.

Oil-paint systems are generally more satisfactory on surfaces which have dried out after heavy efflorescence. Emulsion-paint films of high binder content are relatively impermeable to salts although permeable to water. Paints of this class therefore having eggshell or semi-gloss finishes may be pushed off by efflorescence where flat finishes may allow the solutions of salts to pass to the surface, giving a growth which can be removed without damage to the film.

Various solutions have been recommended to reduce the persistent growth of efflorescence, e.g. dilute acids, vinegar, etc., but none have given conclusive results on heavy deposits. The use of soluble calcium salts such as 1–2% calcium chloride [3] and bromide are claimed to inactivate efflorescence by converting the sodium sulphate to calcium sulphate with resultant formation of sodium chloride or bromide. Calcium chloride and calcium bromide are claimed not to 'cause any efflorescence even at very high concentrations, and sodium chloride and sodium bromide are much less troublesome than sodium sulphate' [3]. An analytical method for the determination of concentrations of salts in a wall surface is described by E. Hoffmann [4] and it is claimed that 5 g/m^2 (0·5 g/ft^2 approx.) of sodium sulphate in the plaster does not cause paint failure. If a higher concentration is found, then the wall should be allowed to dry out or be treated with calcium chloride or calcium bromide solution as given above.

Where a heavy growth of efflorescence takes place, the plaster may be disintegrated at the surface, or in some cases the upper plaster layer (skimming or finishing coat) may be pushed off. Re-plastering must be carried out before painting is attempted, all defective plaster being removed, and all salt deposits brushed away. The application of new plaster may reactivate salts present in brickwork and the surface should therefore be allowed to dry out thoroughly before being painted. After removal of defective plaster, success is sometimes achieved by sealing the brickwork before replastering. The sealing of plaster surfaces can result in further peeling or flaking where efflorescence develops behind the paint film.

(f) Discoloration due to the influence of a wall substrate can be caused by the alkaline attack of lime and caustic alkalis on sensitive pigments together with saponification of the paint binder. Causes other than saponification can not be excluded.

Certain pigments are attacked by alkali and may change colour—see Table 7.1 'Limefast colours'. This may not necessarily result in a complete bleaching out, but more probably the loss of the colour of one component of the pigment blend, e.g. a sensitive red with a fast iron oxide yellow, could change from a salmon pink to a cream. This type of change may be patchy because of uneven distribution of lime in plaster. Under heavy condensation or in the early stages of drying out, coloured soluble salts, e.g. of iron, may be leached out of plaster or especially from brickwork behind the plaster, causing brownish runs which may easily be mistaken for saponification. These discolorations, however, are not responsible for film breakdown, and can usually be removed with a damp cloth without affecting the paint film itself. Permeable films such as distempers, water paints, emulsion paints, etc., are most likely to be affected.

Emulsion paint films have also been stained by brownish runs under severe condensation conditions. These stains are usually slightly sticky and water soluble, and can therefore be removed by washing, although sometimes demarcation lines following the track of the runs remain faintly visible. C. Bondy [5] attributes these exudations to water-soluble constituents of the paint leached out in the 'early stages, especially at low ambient temperatures before film integration is complete'. R.J. Woodbridge [6] attributes the formation of sticky brown blobs to soluble materials in the system, particularly the emulsion, and describes a laboratory test to reproduce the conditions necessary for formation.

Apparent 'fading' and loss of gloss on interior coatings of plasters is sometimes caused by uneven suction due to porous areas in an otherwise very dense surface. It frequently happens on old plasters due to repair patches.

Remedy without repainting is rarely possible.

It is advisable to allow the surface to dry out, and if the paint film is sound, to repaint with a paint containing a more alkali/lime resistant colour. It may be advisable to use an emulsion paint for repainting if the original failure was with a distemper. If this is so it is preferable to strip the latter before repainting. An alkali resistant primer applied after the wall surface has thoroughly dried will give a satisfactory surface for subsequent coats.

Discoloration due to water-soluble material from the substrate coming through an emulsion paint film cannot be arrested by application of further coats of the same material. Any matter on the surface should be wiped off with a damp cloth and the surface allowed to dry. It should then be coated with a solvent-based material such as a flat wall paint.

(g) Loss of adhesion of paint films causing peeling and flaking can be due to a surface condition of the plaster surface: the paint film strips from the surface, not necessarily damp at the time, and a dry powdery deposit of plaster is found adhering to the back of the film. The plaster surface is often friable or may have completely disintegrated beneath the paint. This is, of course, a plaster fault and may be due to 'dry out' 'delayed expansion', etc., which may have become aggravated either by water from water-based paints, or by the contracting forces of oil or synthetic resin based paints such as alkali-resisting primers, etc., during the drying process. This peeling or flaking may also be associated with premature decoration with an impervious oil paint which prevents effective evaporation of water.

Isolated 'pockets' of breakdown, apparent as surface eruptions and originating from the plaster, may appear as small craters after the paint film has been removed. Small sections of plaster remain adhering to the paint film. Such 'popping', 'pitting' or 'blowing' of plaster is due to the rapid expansion of small pockets of dehydrated material, usually lime, when water-based paints cause hydration.

The breakdown of a plaster surface may also be due to excessive efflorescence, which may develop and expand within the plaster, causing either the flaking of the plaster finishing ('skimming') coat or its disintegration.

Uneven trowelling, bad mixing or retempering of plaster (i.e. remixing with water) after having partially set can cause surface cracking. This would show as the cracking or crazing of the final paint system with ultimate possible flaking.

When plaster surfaces have become friable the only satisfactory remedy is to strip the paint and plaster to a sound surface, to replaster as necessary and then to paint as for a newly plastered surface. Isolated areas may be patched or filled with a plaster mix, but there is danger that the water absorbed into the surrounding areas may cause further breakdown. The isolated craters caused by 'popping', etc., may be filled with a hard-drying filler, and the decoration touched up with paint to match surrounding areas.

Cracks or crazes in plasters can often be filled with a quick-setting plaster mix, provided the plaster surface is not crumbling. The whole surface should then be rubbed down with fine abrasive paper and redecorated. The application of a plaster primer to seal the newly filled cracks is desirable before further decoration. If the crazing is very fine, i.e. the fissures are merely 'hair' cracks, the filling properties of a flat paint may be satisfactory to cover the defect. On the other hand, if cracking is extensive although the plaster surface is otherwise sound and adhering well, the use of a lining paper has much to commend it, followed either by painting e.g. with a flat paint, emulsion paint, etc., or by wallpaper.

(h) Asbestos cement is encountered in several forms in buildings, notably in sheets used in wall partitions, imitation marble, external cladding, prefabricated wall units, roofing sheets, plain and corrugated, etc., for the exterior of buildings. Sheds, garages, greenhouse sections, coal bunkers, etc., may be constructed wholly or in part in asbestos cement. As this material is very absorbent, often highly alkaline and may contain relatively high concentrations of salts such as sodium sulphate, various paint breakdowns may occur.

Stripping, peeling and blistering may be caused by moisture (as above), which may persist due to entry from the back of the structure (e.g. when gutters, down pipes, etc., are not internally painted, and the continued flow of rainwater maintains damp conditions within the asbestos cement). Discoloration, patchiness, loss of gloss, stickiness, etc., may also be present as a result of saponification. A white powdery or crystalline deposit, either developing at the back of the paint film or growing on the surface, may be due to efflorescence. On asbestos sheets breakdown may be concentrated towards the centre of the individual sheets due to sheets being stacked before fixing and drying out of residual moisture. There is thus an increasing moisture gradient towards the centre of the sheet with resultant greater alkaline and efflorescent effect and consequently greater paint breakdown, i.e. saponification, efflorescence, discoloration.

The behaviour of paints on asbestos cement surfaces outside differs according to the degree of compression of the sheets.

Asbestos cement should be dry before painting. Whenever possible where there is any chance of it being re-wetted, e.g. by damp, rainwater, etc., from behind, both faces should be painted (this may only be possible before fixing). An alkali-resisting paint should be used, e.g. plaster and cement type primers, preferably two coats, or a complete system of alkali-resistant, e.g. chlorinated rubber, or bituminous paint. Primers based on tung oil/phenolic varnish are also used. Emulsion paints based on polyvinyl acetate and its copolymers, acrylic copolymers, etc., are often very satisfactory provided good adhesion is obtained with the first coat. The high porosity of very dry asbestos cement may require a sealer coat of well-diluted emulsion paint or of a clear emulsion sealer. Cement paints are usually satisfactory on internal or external surfaces.

Asbestos cement roofing which is often left unpainted externally may give rise to blistering and peeling of impervious paint films inside the building. Sheeting of this kind remains wet for considerable periods especially in winter months.

When thoroughly dry, external surfaces should be painted or sealed, e.g. with a bituminous roof paint. Satisfactory internal painting is then possible. Emulsion paints have been used successfully on interior roof surfaces, even when the exterior has not been painted.

Pipes and gutters of asbestos cement which may be internally wet due to rainwater, etc., can rarely be successfully painted with an impermeable system such as an alkali-resistant primer and coats of gloss paint. An emulsion-paint system may be reasonably satisfactory, due to its more permeable nature. If possible the insides of pipes and gutters should be painted before fixing, e.g. with chlorinated rubber or bituminous paint.

When paint failures have occurred repaint as for similar surfaces described above, removing old paint and repriming as necessary.

In view of the hazards associated with asbestos, the use of this material is now diminishing.

(i) On brickwork, paint failures may be isolated to individual bricks and may be due to great variations in porosity, resulting in the flaking of very thick coats. The presence of salts causing efflorescence, and its attendant flaking, and the presence of coloured substances such as iron salts, causing severe staining especially of water-based paints, may also be responsible. Certain types of bricks have highly glazed surfaces and offer poor key for paint systems.

Dirt and grime may be firmly adherent, giving a greasy film which may cause subsequent flaking or peeling. It is recommended that brickwork should be left unpainted whenever possible.

Old brickwork, especially on exteriors of buildings, may have a loose powdery surface and pointing may also be defective. To paint this type of surface will invariably lead to subsequent failure.

Paint failures may often be rectified by the removal of the isolated damaged area, priming with an alkali-resistant primer and then redecorating. The painting of brickwork and stonework, however, is often difficult, due to the age and condition of the surface.

Where the mortar joints in brickwork have become badly weathered, but the brickwork is otherwise sound, repointing should be carried out before painting is attempted. The surface should then be treated as alkaline and either left to weather or painted with an alkali-resistant paint or primer. In the north of England external brickwork joints are often pointed with a mastic (e.g. red lead and sand) bound with boiled linseed oil, and the whole of the brickwork subsequently treated with two coats of boiled linseed oil. The exterior of factory chimneys and other areas of buildings exposed to severe weather can be treated successfully in this way, giving a glossy, dark weatherproof brickwork.

(j) Masonry finishes: this term is generally used to refer to paints used on the exterior walls of buildings but it is sometimes used for coatings for interior walls and floors. The increasing diversity of materials used in building construction has created many problems for the paint technologist. Composite materials, synthetic building boards, impregnated wood and primed metals all require specific treatment.

Furthermore, coatings often require renewal for aesthetic reasons rather than breakdown; dirt collection, colour change and mould growth are among the common defects [7].

The most common exterior surfaces are concrete, cement render, roughcast, stucco, brickwork and stone blocks. Paint for such surfaces, will, ideally, possess the following properties:

(i) Ease of application.
(ii) Flow out to a uniform coating, especially on roughcast.
(iii) Fill (and not bridge) fine cracks in the surface.
(iv) Dry to a uniform sheen.
(v) Resist the growth of moulds and algae.
(vi) Show good durability.

The following types of paint are in use:

(i) Solution types based on styrene-acrylic copolymers. 2-pack epoxies and polyurethanes have also been recommended [8]. Alkyd types, often textured with grit or short fibre, can be applied to exterior surfaces but if these are alkaline a suitable primer must be used. Plasticized chlorinated rubber paints are widely used.
(ii) Emulsion paints, usually of the acrylic or vinyl types. Used as flat finishes or textured. The latter give thicker coatings with considerably greater durability.
(iii) Proprietary cement paints. These are supplied as powders to be mixed with water before use.

Types of failure experienced with exterior masonry finishes are:

(i) Chalking. This leads to gradual erosion of the surface and to apparent change of colour. The severity of chalking is often more severe on south and west aspects than on north and east. A chalking surface sheds dirt and is the least objectionable form of failure. In fact a controlled slow rate of chalking in white finishes is often regarded as an asset.
(ii) Flaking (Fig. 7.13). This is usually the result of poor surface preparation and applies to all types of masonry surface [7]. Some of these are difficult to clean but must, nevertheless, be free from dirt, powdery surface layers and chemical contamination.

New concrete surfaces can present difficulties due to high alkalinity and to the presence (particularly on horizontal surfaces such as floors – see below) of a loosely bound cement layer on the surface (known as 'laitance'). This must be removed if good adhesion of paint coatings is desired. Methods of removal can be either grit blasting or the more usual treatment with dilute (about 10%) hydrochloric acid followed by thorough washing with water. The surface is then allowed to dry out. If there is any risk of moisture in the concrete the paint films should be porous to allow the water to escape and to prevent blistering.

Concrete floors often require anti-dust treatment. This is applied after removal of laitance and must be abrasion resistant. The types of coating used include 2-pack epoxies, 2-pack polyurethanes and moisture-cured polyurethanes, the latter as unpigmented finishes.

(k) Multicolour paints are used as interior wall finishes in public buildings, schools, etc. They are poly-phase systems consisting of one or more pigmented organic binders dispersed in a pigmented or non-pigmented aqueous phase. The consistency of the latter is important and is controlled by the use of thickeners such as cellulose ethers or acrylates. The commonest type of multicolour paint contains pigmented cellulose nitrate as the dispersed phase. It is made by stirring the pigmented lacquer into the aqueous phase to give a dispersion of lacquer droplets, from 20 to 200 μm in diameter. The size depends on the speed of stirring. If more than one colour is required, separate dispersions are made and then mixed. Application is by spray to give a pattern of coloured droplets of varying size and shape, the latter being influenced by the viscosity of the aqueous phase which forms the continuous matrix.

The can stability of multicolour paints requires close watching. The specific gravities of the organic and aqueous phases should be as close as possible for dispersion stability and, in order to preserve the droplet sizes and shapes, the organic phase(s) should contain no water-miscible solvents.

Multicolour paints are not weather resistant. Films on interior walls are hard wearing and are not disfigured as easily as plain surfaces.

(l) Coatings of emulsion paints and other inorganic and semi-organic materials are often nearly impermeable to air. This can delay the carbonization of lime containing plaster and rough cast. Early weathering failure of the substrate occurs, accompanied by cracking and flaking of the coating. The defect can be avoided by allowing thorough hardening of the lime plaster before painting [9].

Notes

[1] British Standards Specification 1198, 1199, 1200: 1976, *Building Sands from Natural Sources.*
[2] British Standards Specification 890: 1972, *Building Limes.*
[3] Hoffman, E. (1961), *JOCCA,* **44**, 40; (1962), **45**, 266.
[4] Hoffman, E. *ibid.*
[5] Bondy, C. (1955), *JOCCA,* **38**, 512.
[6] Woodbridge, R.J. (1955), *JOCCA,* **38**, 291.
[7] Whiteley, P. and Rothwell, G.W. (1971), *JOCCA*, **54**, 855.
[8] *Paint Oil and Col. J.,* **151**, No. 3578, 927, (1967).
[9] Rossmann, E. (1975), *Farbe Lack,* **81**, 107.

7.8 Failures of paint systems on calcium sulphate plasters

While the scope of this book does not allow for a full treatise on calcium sulphate plasters, it is necessary to give some details of the various types used on internal building surfaces.

Four main classes of gypsum plasters are listed [1] in order of increasing hardness and varying in surface character, and also in setting time. Class A, the plaster of Paris group and Class B, the retarded hemihydrate gypsum plaster group, are chemically similar [2], the latter having retarding agents added to delay the setting time. Classes C and D are anhydrous calcium sulphate plasters having slow setting times, generally with accelerating agents added.

All types set by chemical action with water to form gypsum, excess water evaporates on drying and a porous mass of gypsum remains; the greater the proportion of excess water the higher the porosity.

The setting time of these plasters varies. Plaster of Paris sets extremely rapidly and retarded hemihydrate gypsum plaster somewhat more slowly, depending on the particular grade used (e.g. for an hour or for several hours depending on the proportion of retarder added. The anhydrous calcium sulphate plasters set slowly and defects can arise if drying is premature – see below.

Since the amount of water required in the plaster mix varies according to type, the hardness of the surfaces also varies. As set plaster consisting of crystalline gypsum is partially soluble in water (0·2% at 20°C), deterioration occurs under persistently wet conditions, hence such plasters are not used externally in most countries.

Lime and calcium sulphate plasters are commonly used together in the same mix with or without sand, but it should not be assumed that all types can satisfactorily be used in this way [3].

Classes A and B plasters may be used with lime, although most manufacturers of the retarded type B plasters recommend a 10% maximum addition since setting times are shortened with increasing lime content.

Class C and some of Class D of the Keene's or Parian cement type may be used with lime, but setting times may be lengthened, and the finishing reduced in hardness. Manufacturers' instructions should be carefully checked. Gypsum plasters should never be used in a mix containing Portland cement, and gypsum plasters of different classes and proprietary brands should not be mixed.

Faults in plasters may lead to painting defects, and under certain circumstances the application of paint may precipitate a plaster failure which may otherwise have remained unnoticed. The hardening times, porosity and chemical nature of a plaster surface may be greatly affected by the nature of the backing material, and this should be remembered when analysing any defect.

(a) Gypsum plasters vary greatly in porosity according to the type of plaster, the substrate, surface finish and degree of drying out reached. When surfaces have very high suction, often referred to as 'hot' surfaces, water-based paints may show patchiness and paint 'build up' due to the creation of thick films notably on overlapping edges. The thick build-up of material may result in subsequent flaking, most prevalent in distempers due to the underbound nature, water-soluble binder having been rapidly absorbed into the highly porous substrate.

When porosity is uneven, often caused by irregular trowelling or patchy distribution of lime early decorations of distempers or emulsion paints may exhibit variations in sheen and colour, paint build up in the more porous areas and lack of opacity due to an excessive spreading rate on polished areas; possibly, also, flaking due to loss of binder. Uneven porosity caused by brickwork behind the plaster can result in uneven sheen, colour, spreading etc. of water-based paints shown as a 'pattern' staining effect corresponding to the shape of the brickwork. This may be particularly noticeable where moisture absorption from the plaster has been much greater into the bricks than into the mortar joints.

It is advisable to try out the first coat application of water-based paints such as emulsions on plaster surfaces in order to estimate the degree of porosity and then to adjust the paint accordingly. For very porous ('hot') surfaces it may be necessary to dilute the paint with an appreciable amount of water. Even so, two coats may be necessary to seal the surface completely. For the application of alkyd-based wall paints the plaster should be allowed to dry out completely and should then be sealed by application of a plaster primer.

When, on drying out, unsightly shade variations, irregular sheen, etc., are noticed on a well-adhering paint film, it is better to apply a sealing coat such as a plaster primer followed by normal redecoration rather than to risk a repetition of the defect by just giving one more coat.

'Pattern staining' caused by uneven porosity may disappear when the surface becomes thoroughly dry; if not then apply a sealing coat as above.

(b) Peeling and blistering can occur on paint films applied to damp plaster surfaces. These defects can occur in large areas with emulsion paints having high binder contents (often those with eggshell to gloss finishes). Since it is essential that plasters should dry out before the surface is sealed, it is dangerous to apply oil and synthetic media paints prematurely.

Paint films which are permeable to moisture (e.g. distempers and many emulsion paints of the p.v.a. or similar type) may 'sweat' due to the passage of moisture to their surface from the substrate. This may also result in the leaching to the surface of water-soluble materials causing staining, unsightly runs, etc.

Blisters on non-porous films on new plaster may sometimes dry back satisfactorily, but often it is necessary to strip blistering and peeling films with paint remover, wash, clean and allow to dry out thoroughly before repainting.

Surface staining, runs, etc., may often be removed by washing with warm water after the surface moisture causing the defect has dried out. Repainting after thorough drying out may be necessary.

(c) Gypsum plasters of the Keene's type, having a slower set and a harder surface than hemihydrate plasters, are frequently trowelled to a very smooth, hard, glossy surface. Distempers (oil bound) and soft distempers (size bound), often give poor adhesion to these surfaces and subsequently peel and flake. Condensation giving cycles of repeated wetting and drying out can contribute to early breakdown. Roughening the surface by means of abrasive paper, etc., may lead to build-up of distempers, emulsion paints, etc., on the abrasion marks, showing as more opaque, slightly ridged, score marks.

When paint has peeled or flaked from a highly trowelled, polished plaster, or when such a plaster has fully dried out before priming, a hard drying plaster primer should be used (after suitable preparation if being redecorated). The smooth regular surfaces obtained with the anhydrite and Keene's types of plaster are very suitable for decoration with oil paints, alkyd enamels, etc. It is important, however, that the plaster should be completely dry before being sealed by plaster primers to avoid other plaster defects – delayed expansion, dry-out, etc.

(d) The alkaline nature of many plaster mixes used can lead to saponification of oil paints and discoloration, colour change of non-resistant pigments, etc. In general alkalinity of calcium sulphate plasters should not be very high unless high proportions of lime are used in the mix. Lime is frequently used in quite high proportions with the retarded hemihydrate plasters, giving a softer and more porous plaster. Bad mixing of the lime plaster mix, and uneven trowelling can give rise to areas of higher lime content. These therefore coincide with areas of higher alkalinity and porosity.

Saponification of oil paints, and, in extreme cases of high moisture and lime content, of oil-bound distempers also on plaster surfaces will result in the discoloration, sticky, water soluble runs, etc. The alkalinity may, of course, be brought to the surface from structures beneath the plaster, concrete, brickwork, mortar, etc. Cement should never be used in calcium sulphate plaster mixes, because of interaction which can result in its disintegration. Cement is, however, often used in the plaster ground coats (backing coat) in admixture with sand and lime, and may therefore contribute to the alkalinity brought to the plaster surface during the drying-out process.

Patchiness of colour can occur when non-lime-fast pigments are attacked, sometimes in heavily limed areas as given above. Some alkali- and lime-fast pigments may show apparent colour loss in highly limed areas due to flocculation in aqueous paints. This fault is less common with modern emulsion paints than it was with the older oil-bound distempers.

Patchiness of sheen with some emulsion paints can occur on heavily limed areas due to the precipitation of insoluble calcium salts of certain thickeners.

Patchiness does also occur due to the uneven porosity created when lime is badly distributed in plaster surfaces.

(e) When plasters are incompletely hydrated due to a rapid removal of water, the chemical change necessary for the formation of crystalline calcium sulphate will not have taken place, and the plaster will not have attained maximum hardness. In some cases this will result in 'dry out'. Subsequent application of water, however, either from a structural source or more usually from aqueous-based paints, distempers, emulsion paints, etc., can result in 'delayed expansion'. This effect results in a surface rippling, softening or expansion causing peeling or blistering of any paint film present, due to the delayed hydration of the incompletely formed calcium sulphate.

The premature drying out of the anhydrous plasters causing either 'dry out' or

Table 7.3 Recommended paints for early decoration of new calcium sulphate plaster

Material	Uses and properties	Limitations
1 Distempers (soft, size bound, ceiling white, etc.)	Used only on ceilings	Water sensitive, Flake easily on highly trowelled plasters. Must be removed before redecoration
2 Distempers, casein and oil bound (now little used)	More permanent decoration May withstand light washing when hardened. Porous. Can usually be redecorated with oil paints	Could flake on highly trowelled plasters especially if repeatedly wetted and dried out. Sensitive to acids. Inclined to be softened in extremely alkaline conditions
3 Emuslion paints (synthetic polymers, e.g. p.v.a. etc.)	Flat or eggshell paints, permeable to water, and permit drying of new plasters. Non-saponifiable by alkalis in plasters. Very good washability and durability. Permanent decoration	May blister or peel under wet plaster conditions, or under heavy condensation depending on type
4 Emulsion paints (alkyd based)	Comparable to flat oil or alkyd paints. Not so porous as synthetic polymers, e.g. p.v.a. types. Use only on dry surfaces. Durable permanent form of decoration	Unsuitable for damp and alkaline surfaces. Could blister under damp conditions and saponify with alkali Require non-saponifiable primer on alkaline surfaces
5 Flat oil or alkyd paints	May be sufficiently porous for application to reasonably dry plaster, allowing completion of drying out process. Durable and a permanent decoration	Not suitable for alkaline or damp plasters. Require non-saponifiable primers on alkaline surfaces

'delayed expansion' may also be caused by excessive suction of the background instead of by too rapid drying conditions.

In severe cases it is necessary not only to remove the affected paint film, but also to hack off the plaster down to a sound foundation and replaster. The anhydrous plasters such as Keene's cement may suffer from this defect, and the use of a 'sharp coat' primer following the trowel is recommended to minimize surface evaporation in the early drying stages as well as giving a suitable surface for subsequent painting.

(f) 'Popping', 'blowing' or 'pitting' are plaster defects resulting in small crater-like eruptions appearing through a paint film, removing with the film a section of the plaster surface. The defect arises from expansion of particles of extraneous material, often calcium oxide partly hydrated or anhydrous which have become pocketed within the plaster, and which subsequently expand, usually due to hydration. It is stated [4] that 'two things are necessary before blowing can take place, an expansive particle and moisture'. The moisture may penetrate the structure from outside or it may occur as condensation. The coincidence of the decoration and the appearance of blowing is probably related either to the introduction of water by the decoration or to the reduction in the rate of loss of water occasioned by the decorative film. The appearance of blowing may be delayed for years, and may then appear after the building has been left unoccupied in damp weather.

The holes may be quite small when emanating from faulty material in the surface plaster skimming, or much larger when the trouble arises from the backing, giving areas of breakdown up to approximately 7·5 cm (3 in) diameter.

It is often impossible to decide whether all areas of breakdown have occurred, and it is therefore unwise to carry out expensive redecoration until a period of some months has elapsed, provided no further blowing has taken place. Wetting the wall surface thoroughly may bring out any further defects, although one cannot be certain of thus having completely eliminated the risk of any future breakdown. In very severe cases replastering should be carried out. In isolated areas the craters formed may be filled either with a hard knife filler or a quick-drying plaster mix, and repainted with an oil or alkyd system. It is preferable to avoid aqueous-based paints unless the whole wall area is previously sealed with a plaster primer. The use of a lining paper to conceal the patches after repainting is recommended, followed either by painting, distempering or the application of wallpaper.

(g) The uneven distribution of alkalinity particularly due to lime is essentially a plaster fault. This often becomes evident only after painting with faulty results. Lime when permitted in a plaster mix should be used only to the plaster manufacturer's directions, and should be well mixed, preferably by the correct pre-soaking of hydrated lime, preparation of lime putty, etc., rather than the use of dry hydrated lime in the plaster mixing.

Where patchiness of colour or sheen has occurred, redecoration when the plaster is fully dried out may sometimes be satisfactory. It is usually better, however, to seal the surface with an alkali-resistant primer before further painting is carried out.

Saponified areas require to be completely stripped and allowed to dry out before the use of a primer.

(h) Peeling or flaking may be due to the presence of efflorescence or salts beneath the paint film. The efflorescence may occur in the plaster mix due to the presence of water-soluble impurities, but more often is brought to the surface from the backing materials, brickwork, mortar, etc. Efflorescence may:

(i) Push off an impervious paint film, causing peeling or blistering.
(ii) Break through the paint film in areas where concentration is greatest, causing eruptions.
(iii) Grow through a porous paint such as a flat paint, emulsion paint, etc., giving a white powdery surface 'deposit'.

Cement paints fail on calcium sulphate plasters due to interaction of cement and calcium sulphate often with resultant efflorescence.

The affected areas should not be washed, as water can re-activate further salts which may be present, causing further breakdown. For the same reason, distempers, emulsion paints, etc., should not be applied without first sealing the area with an alkali-resisting plaster primer (the wall must be thoroughly dry before this is attempted). All loose material, powdery deposits, disintegrated paint films, etc., should be removed by dry brushing or scraping before repainting.

Certain types of very porous plasters are claimed to give less trouble with efflorescence. Those containing vermiculite in its exfoliated form (i.e. a highly porous packing filler) have lower specific gravity and hence a bulkier, more porous structure, allowing the absorption of a certain amount of soluble material without transmission to the plaster surface.

E. Hoffman [5] has stated that: 'Addition of 1 to 2% of calcium chloride to a mix of gypsum plaster largely eliminates efflorescence. This makes it possible to paint surfaces that have not completely dried out, without the risk that the paint film will be disfigured or disrupted by efflorescence'.

Efflorescence from gypsum glass board (glass fibre reinforcement) can be lessened by the incorporation of sodium methyl siliconate to reduce the tendency to absorb moisture. The amount used is critical (about 0·1% on the calcined gypsum) [6].

(i) 'Dry-out' takes place when the moisture is abstracted from the plaster surface too rapidly to allow complete hydration of the plaster mix. This can occur due to excessive currents of air, or too early use of central heating. The result is a weak and powdery plaster surface which in paint terminology appears 'underbound'. Plasters having a slow set should not be allowed to dry too quickly. Hemihydrate plasters may also show this defect under very rapid drying conditions. Oil paints applied to this type of surface may not key well, and may subsequently peel off.

If the plaster surface is in the above condition before painting, the surface should be sealed by applying a penetrating primer, providing the surface is not alkaline, otherwise saponification may result. A hard drying plaster primer could be unsuitable, as this type does not penetrate sufficiently to give adequate binding of the surface. Water-based paints should not be used unless also preceded by the primer suggested above. It may be necessary in cases where the plaster is severely disintegrated to remove and replaster. Attempts to dry plaster out rapidly by the use of heat and/or excessive ventilation are largely responsible for this defect.

Notes

[1] B.S. 1191. *Gypsum building plasters.* Part 1: 1973. Describes Grades A, B, C and D but excludes premixed lightweight plasters. Part 2: 1973. Premixed lightweight plasters. Specification includes methods for determination of soluble salt content, dry block density, etc.
[2] Both Plaster of Paris and the retarded hemihydrate are partly hydrated and have the formula $2CaSO_4\ H_2O$. (This is often erroneously set out as $CaSO_4$ $\frac{1}{2}H_2O$). The fully hydrated gypsum is $CaSO_4\ 2H_2O$. Anhydrous plasters contain no water of crystallization. The setting of Plaster of Paris, the hemihydrate and anhydrous types, involves combination with water to give crystalline gypsum.
[3] See B.S. 890: 1972, *Building limes*; also B.S. 1198: 1976, *Building sands from natural sources.*
[4] Building Research Station Digest No. 26, *Blowing, popping or pitting of internal plaster.*
[5] Hoffman, E. (1962), *JOCCA*, **45**, 266.
[6] King, G.A., Ridge, M.J. and Walker, G.S. (1974), *JOCCA*, **57**, 127.

7.9 Corrosion of painted iron and steel

The formation of rust would be impossible to any appreciable extent in the absence of moisture and oxygen. Chemically, rust consists of several oxides and hydroxides of iron, together with small amounts of iron carbonate and sulphate, and other salts in some environments. The initial product in normal corrosion is ferrous hydroxide, which in presence of oxygen turns to hydrated ferric oxide. Magnetite (Fe_3O_4) is also present. The colour of rust varies with age, through yellowish-red and dark brown to black, as the water content decreases, but may vary on different irons and steels.

The resistance of bare steel to corrosion can be increased by alloying with suitable metals. High quality stainless steel (18% Cr, 8% Ni) is several thousand times more resistant than ordinary structural steel, but its high price restricts its use to specialized purposes. Even small additions of alloying elements can double the corrosion resistance; the 'low alloy' steels currently in use have been principally designed to improve the mechanical properties of the steel. For instance, a steel containing 1% Cr and 0·5% Cu rusts three times less rapidly and has a tensile strength 50 times greater than common mild steel. Nevertheless, such steels still need protection by painting, but generally speaking 'low alloy' steels benefit the performance of paint systems only to a very limited extent [1].

Steel that has rusted during exposure to the atmosphere nearly always carries deposits of ferrous sulphate, formed in the presence of moisture and oxygen from sulphur dioxide in the air and iron. Ferrous sulphate crystals are very difficult to remove by scraping or wire brushing a rusty surface and if left behind are an important source of the breakdown of painted steel surfaces. The hygroscopic crystals promote rust under the paint film, give rise to filiform corrosion and since one volume of ferrous sulphate expands to approximately three volumes of rust, blistering of the paint film is inevitable.

Other soluble salts present may be derived from atmospheres contaminated with sulphur dioxide or hydrochloric acid, or from contact with sea water. Clean ferric hydroxide may be innocuous beneath a paint film, but soluble contaminants are probably always present.

The total cost of corrosion to the national economy is quite large. Many estimates have been made and they vary considerably. The figures arrived at in 1971 by the Committee on Corrosion and Protection of the Department of Trade and Industry [2] show that the cost of corrosion in the United Kingdom was then £1365 million per annum. The committee expressed the belief that approximately £310 million could be saved by better use of current knowledge and technique. More recently the total cost of corrosion in Australia has been put at $ A 470 million and the avoidable costs at $ A 250 million [3].

Corrosion is best explained as an electrochemical phenomenon. The main components of a corrosion cell are the anode which sustains the process and from which electrons pass through a conducting path to the cathode which accepts the electrons and thus keeps the anodic attack going. The aim of corrosion prevention is therefore to impede the functioning of corrosion cells by eliminating cathodic sites (millscale or contact with other metals), by electrical separation of the anode from the cathode, which is usually difficult to accomplish, and/or by eliminating the electrolyte [4]. The latter objective can only be achieved by means of an impervious coating keeping out moisture and oxygen. Coatings based on chlorinated rubber, vinyl resins, epoxide resins, especially coal tar epoxides, polyurethanes and bitumen possess a low degree of permeability to moisture and oxygen, but no coating based on an organic material can be claimed to be completely impervious. It is therefore necessary to assist inhibition of corrosion by additional means. It can be made difficult for ferrous ions to leave the base metal by supplying electrons from metallic coatings or pigments, less noble than iron, such as zinc. This makes the steel play the part of a cathode and is called cathodic protection. Zinc corrodes and protects the iron. Anodic passivation depends on creating a stable oxide film on the surface; certain metallic chromates can provide this protection, but must be present in sufficient quantity to be effective. Whilst the coatings mentioned earlier can greatly reduce the access of moisture to the steel substrate it is far more difficult to keep out oxygen which plays an essential part in forming ferrous hydroxide and in converting it into ferric hydroxide which, in its dehydrated form, is called rust. Under immersed conditions, oxygen can be removed from water by oxygen scavengers like sodium sulphite or hydrazine.

Passivation of steel, when painting is not practicable, is also possible by forming thin absorbed films of inhibitors, like glue or octadecylamine, on the surface.

Since rusting is an electrochemical process, steel structures buried in damp soil or immersed in sea water may be protected by impressing an electric current to render the whole structure cathodic. Alternatively this result may be achieved by connecting magnesium anodes to the structure. If the surrounding material is conductive (e.g. damp soil, sea water), the steel remains the cathode in the cell so set up, and corrosion is transferred to the magnesium anodes, which can be readily replaced from time to time. Such protection (known as 'Cathodic protection') should be reinforced by a protective coating applied to the steel structure, so that in effect it will come into play only to protect 'holidays' or weak points in the coating.

The areas where the under-rusting (rusting beneath a paint film) is beginning are first indicated by the yellowing of the film, then the rust gradually breaks through, as the volume of the rust is larger than that of the iron from which it is derived and due to its hygroscopicity.

The degree of rusting is now usually classified in European countries, by a system based on photographs supplied in 1961 by the Corrosion Committee of the Royal Swedish Academy of Engineering Sciences. The paint protection efficiency is measured on a scale from 10 to 1. A score of 10 represents a completely rustfree surface and 1 denotes a completely rusty surface (Fig. 7.14).

The rate of corrosion varies greatly with climatic conditions, such as temperature and degree of humidity, and with atmospheric contamination, e.g. sulphur oxides in urban districts and adjacent to heavy chemical industries, and salt in coastal districts. For instance, in Sheffield, England (heavy industry), the rate is one hundred times as high as in Khartum (Sudan). In central U.S.A., where humidity is low, much less thorough painting than that required in Europe will provide adequate protection against corrosion. If the relative humidity is below 60% no corrosion will take place.

Prevention of corrosion starts at the design stage. Since the presence of water is essential for the corrosion of iron and steel all structures should be so designed as to avoid trapping of moisture and dirt. This means elimination of crevices e.g. between riveted plates by sealing with mastics or welding and coating of plates before assembly. Joints and fastenings should have clear uninterrupted lines. Welds are to be preferred to bolted joints and butt welds to lap welds. Sharp edges and corners are best avoided. Good circulation of air round the structure is most desirable to accelerate evaporation of condensed moisture. Holes should be drilled at the lowest points of recesses to allow water to drain away and horizontal surfaces should be given a slight slope to ease drainage of surface water. Easy access for painting is of great importance, otherwise the likelihood of skimping the job increases. Contact between steel and non-ferrous metals should be avoided to prevent the setting up of galvanic corrosion cells.

It has been shown that one way of protecting iron and steel against corrosion is to exclude moisture and oxygen by means of impervious coatings to prevent the setting up of galvanic cells. However, there are very few coatings which are completely impervious. Metallic coatings, such nickel/chromium, zinc, applied by galvanizing or stainless steel cladding are virtually impervious. Organic coatings rarely provide a complete barrier, although thick coatings of bitumen or pitch are relatively cheap and give good service within the limits of their durability when exposed to the weather. Epoxide and polyurethane coatings are also effective, but rather expensive.

The method usually adopted is to use paints formulated with so-called 'inhibitive' pigments. Some are alkaline compounds, capable of reacting with fatty acids to form metal soaps. The fatty acids stem from the linseed oil or other vegetable oils in the paint vehicle. The soaps are adsorbed on the metal surface and are believed to isolate it from water and oxygen. This mode of action is not followed by all anti-corrosive pigments.

Red lead, Pb_3O_4

A mixture of lead oxides, $PbO_2.2PbO$. This is the most important pigment for the protection of structural steel work and is most effective when used in a linseed oil medium; on well prepared surfaces outstanding performance can be obtained as is shown in Table 7.4.

Table 7.4 (From Fancutt, F. and Hudson, J.C. (1968), *The Protective Painting of Structural Steel,* Chapman and Hall, London.

	Duration of protection, years †			
Method of surface preparation	Paint system A 2 coats of red lead in linseed oil paint 2 coats of red iron oxide in linseed oil paint Average total dry film thickness, c. 100 μm (4 thou)		Paint system B 2 coats of red iron oxide in linseed oil paint Average total dry film thickness, c. 50 μm (2 thou)	
	Calshot (marine)	Sheffield (industrial)	Calshot (marine)	Sheffield (industrial)
Weathering and wire-brushing	6·2	2·3	Not tested	1·2
Degreasing only; paint applied over intact millscale	6·5	8·2*	3·6	3·0
Pickling	15·3*	9·6*	5·3	4·6
Blast-cleaning ‡	16·1*	10·5*	5·2	6·4*

† The figures are averages for duplicate specimens of eight structural steels at Sheffield and of three steels at Calshot.
The criterion of failure was rather more severe than that (0·2–0·5% of rust) laid down later in the BISRA scale and corresponded roughly to 0·1% rust.
An asterisk (*) denotes that the paint had not failed on all the surfaces when the test ended.

‡ Leighton Buzzard sand was used as the abrasive.

Such paints have excellent wetting and penetrating properties and are more tolerant than many other paints to less than surface preparation. They are always applied by brush or roller because of their toxicity.

Mayne and his co-workers [5] have shown that lead and calcium salts of several mono- and dibasic organic acids inhibit the corrosion of iron, and lead azelate in particular was found to be highly effective. This compound is formed during the degradation of the red lead soaps formed in paints. More recently [6] it has been shown that linseed oil yields more azelaic acid than oiticica oil, an alkyd resin, or tung oil (in that order) and the protective properties of red lead paints based on these media similarly give less protection than linseed oil/red lead paints.

Metallic lead [7]

Because of its great reactivity, finely divided metallic lead particles are supplied in form of a paste in linseed oil. It acts in much the same way as red lead and is used for similar purposes. It confers better levelling properties on the paint and makes it dry and harden faster.

Calcium plumbate, $PbO_2.2CaO$

A cream coloured effective alkaline pigment, best used in linseed oil, but is quite often associated with varnish media. It tends to form very hard films and if left to harden for too long before overcoating it can impair the adhesion of the topcoat. Calcium plumbate primers are often recommended for new galvanized steel and they have been successfully used for priming hard woods.

Basic lead sulphate, $PbO.PbSO_4$

A white pigment which has been successfully used in combination with aluminium powder especially in an alkyd medium under marine conditions.

Lead silico-chromate (Oncor M50)

A bright orange pigment of low tinting strength. The particles consist of a core of silica coated with lead chromate. The development of cored pigments in general has been motivated by the desire to reduce the amounts of expensive 'active' ingredients in the pigment to a minimum. Lead silico-chromate is often not sufficiently inhibitive in a priming paint, but is frequently used in undercoats and topcoats to strengthen the anti-corrosive function of the primer.

Lead cyanamide, $PbCN_2$

A lemon-yellow pigment sometimes used in priming paints, claimed to reduce blister formation under immersed conditions.

Zinc dust

Zinc is a less noble metal than iron and can confer cathodic protection on iron. The pigmentary grades of zinc dust are used in zinc-rich paints giving films containing 92–95% of the metal, thus providing electrical contact between the particles themselves and the substrate. Among the vehicles suitable for the formulation of zinc-rich paints, solutions of polystyrene, chlorinated rubber or epoxide resins are the most common ones. Inorganic vehicles have also gained importance and sodium silicate, lithium silicate, quaternary ammonium silicate and partially hydrolysed ethyl silicate are pre-eminent in this field.

Initially zinc provides protection by sacrificially corroding in preference to the iron substrate. The white corrosion products formed ('zinc rust') progressively seal the pores of the coating creating a virtually impermeable barrier. If corrosion products are allowed to form before overcoating, blistering and flaking of subsequently applied coatings can take place. It is essential to pay great attention to the surface before overcoating, to wash it thoroughly, or better still apply a sealer or tie-coat, like an etch-primer to the cleaned surface. Non-saponifiable paints should be used for overcoating, e.g. chlorinated rubber.

Inorganic zinc-rich paints are two-pack products and are very suitable where resistance to high temperature (500°C) is required.

Zinc phosphate

A relatively new, white pigment [8], giving very good performance in many types of vehicles. Its activity takes some time to develop and for this reason accelerated tests, started after a short drying period, give unfavourable results when compared with natural exposure [9]. The anti-corrosive mechanism by which this pigment functions has not been fully clarified, but it appears to form protective layers in anodic areas. The chemical composition is $Zn_3(PO_4)_2.2H_2O$. British Standard 5193 (1975) prescribes a zinc content of 50·5 to 52%, a loss on ignition of 8·5–10% and stipulates the pH of the aqueous extract as 6–8.

With the decline of red lead, due to its toxicity, as the most important anti-corrosive pigment, zinc phosphate has in recent years become one of the most widely used pigments in anti-corrosive primers and has proved its value during the ten years or so since its commercial introduction. It is frequently used in chlorinated rubber and two-pack epoxy primers, but also performs well in alkyds, phenolic/tung oil varnishes and styrenated alkyds. It has also been recommended as a pigment in undercoats, thus providing 'defence in depth' [10].

Potassium-zinc chromate, $4ZnO.2CrO_3.K_2O.3H_2O$

Pale, greenish yellow pigment, which is soluble in water to the extent of 2·6 g/l. It is mildly basic and inhibits the corrosion of ferrous and non-ferrous metals by anodic passivation. For this reason, and because of its good compatibility with many vehicles, especially fast drying types yielding low permeability films it is much used in the formulation of metal primers. It is often combined with red oxide of iron, but must be present in a sufficiently large quantity (at least 40% of total pigment) to give the beneficial passivating effect. It is often used as a colour pigment in finishing paints.

Zinc-tetroxy-chromate (Basic zinc chromate), $4Zn(OH)_2.ZnCrO_4$

A pale yellow pigment of poor tinctorial properties, much less soluble in water and widely used in polyvinylbutyral resin solutions for the formulation of etch primers. Because of its low solubility it performs much better than potassium-zinc-chromate under immersion conditions in water.

Strontium chromate, $SrCrO_4$
Pale greenish-yellow pigment of very low solubility in water used in some etch primer formulations.

Chromic phosphate $CrPO_4 . 3H_2O$
Bluish-green pigment. Its main use is in one-pack etch primers, often in combination with strontium chromate or basic zinc chromate. Chromium phosphate is derived from trivalent chromium which is considered non-toxic.

Zinc-phospho-oxide complex (Nalzin SC-1)
Proprietary anti-corrosive pigment. It is recommended for electro-coating primers and because of its high zinc content (61%) excercises mildewicidal effects, an advantage in aqueous electrocoating paints [11].

Zinc salts of organic nitro compound (Sicorin RZ)
A corrosion inhibitor recommended as an additive (0·5%) to zinc phosphate for reinforcement of anti-corrosive properties [12].

Organic dichromates (Corinac)
Yellowish guanidin derivatives developed as pigmentary grades and as anti-corrosive additives.

Tannin
An acidic extract from the bark of trees. Many varieties are available. They are composed of polyhydric phenols with central carbohydrate groups. Aqueous solutions of certain tannins have been found effective in reacting with iron surfaces, forming blue-black iron tannate film. Claims have been made that they are particularly suitable for treating slightly rusted surfaces after removal of all loose rust and scale. Tannate films are resistant to temperatures of up to 250°C and give a good key for subsequent paint films. Without protective overcoating, tannate films will resist exposure to weathering for up to 3 weeks before showing signs of rusting [13].

Rust stabilizers
Several are on the market.'*Colorol* Rustbinder' based on modified esterified oils, is claimed to penetrate deeply into rust forming a hard, water-repellent film over which any standard paint system can be applied.

Corroless, described as a rust stabilizer, is incorporated in priming paints and is said to convert hydrated ferric oxide into magnetite over a period of 9–12 months.

All loose rust and scale must, of course, be removed before application. Standard painting procedures are then followed. The use of 'Corroless' is especially recommended where thorough surface preparation is impossible and in locations such as breweries, food processing plants etc. where its non-toxic nature is a distinct advantage [14].

Corrosion inhibitors

Many organic compounds have been investigated and some have been found promising as additives to primers for aluminium and its alloys to replace the frequently used toxic chromates [15]. Sodium nitrite and sodium benzoate are often included in the formulation of emulsion paints and have been found effective in preventing can corrosion. N-acyl sarcosinates are claimed to protect steel when incorporated in paints or mineral oils. They are used in low concentrations [16].

Zinc molybdate (Moly-white 101)

Basic zinc molybdate, a white anti-corrosive pigment for use in solvent vehicles of many types. It is non-toxic and enjoys fairly widespread use in the U.S.A.

Calcium-zinc-molybdate (Moly-white 212)

Also a white basic compound designed for water-thinned and latex systems like electrocoating paints. Molybdate pigments are believed to form a film of ferric molybdate on steel substrates.

Barium metaborate

A white, modified barium metaborate, marketed as Busan 11-M1. It is said to combat corrosion, increase fire resistance and reduce chalking. This product is perhaps better known for the controlling effect on moulds and bacteria. (See Section 7.18).

Non 'inhibitive' pigments and extenders are those that do not form metallic soaps or passivating ions. Nevertheless, they play an important part in the formulation of anti-corrosive primers and finishes. Quite recently it has been suggested that a number of factors, like particle size and shape, the tendency to flocculation, the presence of aggregates and the wettability by paint media, influence the capability of the pigment to prevent the adsorption of water on the pigment/vehicle interface which is crucial to the permeation of water through the film, a major source of corrosion [17].

Red oxide of iron

Because of its capability to provide effective barriers against water permeation in many vehicles it is frequently used, often in conjunction with basic zinc chromate.

Micaceous iron oxide

A natural, chemically inert ore, containing 85–93% Fe_2O_3, possessing a lamellar structure. The platelets, having a length-to-thickness ratio of about 50:1, overlap in the film forming an extremely impermeable barrier. Finishing paints containing a high percentage of micaceous iron oxide are being used on bridge structures, transmission towers, oil production platforms in the North Sea and in many other fields. The dark grey colour of this pigment imposes certain limitations from the point of view of appearance.

Talc
A white magnesium silicate also has a platelet structure and is available in many specially treated forms of high purity and good wettability. These properties contribute to the absence of bubbles and pinholes and to the impermeability of the paint film. At the same time they confer desirable viscosity characteristics and freedom from settlement, together with good application properties on the paint.

Calcium carbonate, $CaCO_3$
A white extender pigment, available in many specially prepared grades in a range of particle sizes, and sometimes coated with stearates or other compounds. It is resistant to water and its mildly alkaline nature is of assistance in the formulation of anti-corrosive primers. Calcium carbonate is also found as the mineral calcite which is very hard and in its micronized form finds use in the paint industry.

Dolomite, $MgCa(CO_3)_2$
A white, very hard mineral, normally used in micronized form. Other useful hard and chemically inert extender pigments of some importance are diatomaceous silica and wollastonite.

Green chromium oxide (Cr_2O_3) and rutile titanium dioxide (TiO_2)
Very inert prime pigments, resistant to atmospheric contaminants and prolonged weathering, often used in anti-corrosive finishes.

During recent years the application of sprayed coats of metallic aluminium or zinc has become widespread. The surface to be sprayed is first cleaned by shot blasting and the metallic coat applied to a thickness of 50–100 μm by either the powder or the wire pistol [18]. The coating so produced is uneven and rather porous, since the metallic droplets are rapidly cooled on contact with the cold surface, and cannot flow or fuse together perfectly. For this reason it is preferable to apply one or more paint coats to seal the surface, and also to protect the metal of the coating. Priming coats for this purpose should be chosen carefully–good quality zinc chromate paints or etch primers being satisfactory for both zinc and aluminium. There are no adhesion problems as with galvanizing. Protection is electrochemical, so that if the coatings are scratched to expose the bare steel, the scratch mark will remain protected. Even in London, steel given one coat of sprayed aluminium plus one coat of paint was 'almost perfect' after seven years' exposure. It must be remembered, however, that shot blasting will greatly enhance the life of conventional paint coatings when compared with normal commercial surface preparation, and comparisons of sprayed metal against e.g. red lead priming should be carried out using shot-blasted steel in each case.

Metal spraying is recommended where steel is exposed to high temperatures and to atmospheric corrosion. In such circumstances aluminium and certain aluminium alloys are recommended (British Standard 2569). For the best protection of structural steel under tropical marine conditions a coat of sprayed aluminium under the paint system is considered essential [19].

The British Standards Institution has issued a Code of Practice, BS 5493 (1977), formerly CP2008, and this provides a useful guide to the best practice of painting structural steel.

Paint failures on steel can be attributed to the following causes:

(a) Insufficient cleaning of old paint, rust or millscale from iron or steel surfaces previous to application or belated painting.

Hot rolled structural steel always carries a layer of millscale which, if completely intact and unbroken, would provide good protection to the steel. Millscale has a laminated structure with a layer of FeO nearest to the surface of the steel and accounting for 40–95% of the total. The next layer consists of 5–60% Fe_3O_4 and the top layer of 0–10% Fe_2O_3. The thickness of the millscale varies tremendously; it can be anything from 5–70 μm. Millscale is nearly always broken or cracked and for satisfactory protection of steel by painting, must be removed. This cannot readily be done by wirebrushing, mechanical tools or flame-cleaning. A common method of preparing new hot rolled steel for painting used to be exposure to the weather for up to one year. It was assumed that during the weathering period all the millscale would have been converted to rust which was considered to be amenable to removal by wire brushing by hand or to mechanical cleaning. Table 7.5 shows how unsatisfactory this procedure is.

Table 7.5

Surface preparation for painting	Duration of exposure, years			
	To 0·1% of rust	To 0·2% of rust	To 0·5% of rust	To 1% of rust
Weathering and hand-cleaning	0·8	3·7	4·7	6·7
Weathering and flame-cleaning	3·3	4·0	6·3	7·4
Blast-cleaning*	7·4	–†	–	–

* Leighton Buzzard sand was used as the abrasive.

† The paint on the blast-cleaned plate had not reached the grade 0·2% of rust after 14 years, when regular inspections were discontinued; by then there was 2–5% of rust on the other two plates.

Since millscale is cathodic to steel, such conditions are ideal for further rusting to ensue. Even after painting, moisture penetrating to the steel surface will cause electrochemical action, the millscale will be undermined by rust formation and it will eventually flake from the surface over large areas, bringing the paint film with it.

This is the cause of innumerable spectacular failures on steel, and can only be prevented by complete removal of millscale.

Less severe forms of rusting can also take place beneath a paint film, their degree depending upon the amount of rust trapped beneath the paint. Usually small pin-head blisters are seen, which eventually burst through the paint film. Even if no rust is originally present, soluble salts not removed from the steel may cause 'filiform corrosion'. (See Figs. 7.15 and 7.16). This is seen as threads of loosened paint weaving over the surface, and can only occur if the relative humidity of the atmosphere in which the articles are stored is over 82%. Flux residues from soldering or welding are a common cause, but when imperfectly washed, phosphate coatings can also exhibit the defect, particularly if the rinsing tank contains soluble chlorides (derived from underground water supplies). These will form small crystals on the steel, from which the threads of corrosion originate.

A properly applied paint system should under rural conditions, remain intact up to five years. Frequently structural members have been completely painted before delivery. 'The improved resistance to corrosion thus obtained more than compensated for any damage to the painted surface that may occur during transit or erection [20]'.

The iron and steel surfaces must be carefully cleaned from rust, in the field by mechanical means, such as by hammering, scraping, grit and shotblasting, flame cleaning, wire brushing (including mechanical wire brushing) or rubbing down with sandpaper. Machine de-rusting is considered more thorough and is, obviously, more rapid than rust removing by hand, e.g. wire brushing.

Abrasive blast cleaning is now accepted as being the best means of achieving a clean surface free from millscale, rust and dirt. Although this has been known for some time, the high cost of the process has deterred potential users. Only in recent years has it been fully recognized that the durability of paint systems can be dramatically improved by applying them to really clean surfaces and that the cost of blast cleaning combined with the use of high quality, albeit more expensive coating materials, is markedly outweighed by the prolongation of durability and the substantial reduction in maintenance costs. (See Fig. 7.17.)

There are two main types of blastcleaning machines:

(i) The compressed air type.
(ii) The impeller or wheelabrator type.

The former projects abrasive particles through a nozzle in a high velocity airstream against the surface, whereas in the latter the abrasive is directed at the surface at high speed through centrifugal force from a rotating wheel. The wheelabrator is very suited for automatic plants which are used for the blasting of large components like ships plates. Many types of abrasives are in use. Sand, one of the first to be used, has now virtually disappeared in the United Kingdom owing to the health risk for the operators. Wet sandblasting is practised in some countries and has the advantage of assisting in the removal of the water soluble ferrous-sulphate associated with rust. Chilled iron grit and shot, hardened steel shot, chopped steel wire, aluminium oxide and slag from metal refining plants are in use [21].

Slags are cheap and effective and are favoured where blasting is carried out in the open and recovery is difficult.

Although careful blast cleaning with abrasive grit will produce an absolutely clean surface, the impact of grit or shot inevitably damages the steel surface and produces a degree of roughness that varies with the size of the grit, and leaves peaks which must be covered and troughs which must subsequently be fully filled with paint if protection is to be gained. To the practical engineer and the painting contractor it is important to know how rough the surface is so that the amount of paint needed can be calculated and adequate covering assured. The roughness is often characterized by the CLA-value (centre line average) which gives the deviation from a mean line of profile. A 12.5 μm CLA would mean a peak to bottom of the valley distance of between 34 and 40 μm [9]. There are instruments available to measure the surface profile; one of them, the Talysurf, uses a finely pointed stylus which is drawn in a straight line across the surface and records the peaks and valleys. However, this method is time consuming and more suited to laboratory conditions. In the laboratories of the Paint R.A. a roughness meter has been developed which can be used on site [22]. A portable cylinder supplies compressed air reduced to a standard pressure of 50 in w.g. This enters a pressure head fitted with a special gasket. In use, the head is pressed on to the test surface by a spring forming an escapement orifice between the surface contour of the metal and the surface of the gasket. On a flat surface the pressure is that of the air supplied by the reducing head through a needle valve, (P_1). With an open orifice the pressure falls almost to zero. On the rough surface the pressure drops to P_2. The difference $P_1 - P_2$ is a measure of the roughness.

The cleanliness of blast cleaned surfaces, another important factor, can be checked by measuring the diffuse reflection of light from the surface. As blasting proceeds the reflection increases until a limiting value is attained. The Paint R.A. has devised a portable Reflectometer for use on site and it has been found that this value can be established in each case on a sample area after a certain blasting time. Thereafter a practical acceptance value, somewhat below the limiting value, can be agreed between contractor and client. Blasting is continued until the agreed level of cleanliness is reached and unnecessary blasting time is thus avoided. A suitable instrument is available from Elcometer Ltd.

The standard of surface preparation by abrasive blasting is often assessed by a pictorial scale devised by the Swedish Academy and in Table 7.6 is compared with U.K. and U.S.A. specifications.

Table 7.6

Sweden	United Kingdom (BS 4232/1967)	United States
Sa 3	First quality	White metal
Sa 2.5	Second quality	Near white
Sa 2	Third quality	Commercial
Sa 1	–	Brush-off

Sa 2.5 is usually accepted as being the highest commercially attainable standard of surface preparation and is often specified by the user.

Coarse rust and millscale, which on account of their often rigid adherence to the iron are frequently mistaken for metallic iron, should be knocked off with a hammer, etc., even if the general de-rusting operation is effected mechanically. However, seams, angles and edges need particular care. Mechanical descaling with pneumatically driven bundles of chopped wire (Jason Pistol) is often practised. On small-scale work the use of a blow-lamp for drying and warming the surface before painting is recommended [20]. Attention is also called to the fact that the heating accompanying the riveting operation strongly induces 'oxide' (millscale) formation.

A surface which has been previously rusty will, in spite of all precautions, corrode more easily than a new smooth surface. Millscale has a tendency to rise in places beneath the paint film, causing blistering and other failures (Fig. 7.18). Large areas of paint flake from substrate within a few months of painting. This failure is most likely if the hot rolled steel has been weathered for a short period in an aggressive atmosphere. Millscale will be found to adhere to the back of the flakes. A similar failure occurs if a millscale-free, but rusted surface is painted (Figs. 7.19–7.21). However, in this case the flaking is unlikely to happen so soon after painting [23].

Flame cleaning is a more recent device for removing millscale, rust and old paint, and consists of burning the surface with a wide multi-jet oxy-acetylene burner at an angle of 45 degrees at a rate of 0–20 feet per minute. Hudson thinks 'that for badly rusted and inaccessible structures, the flame process will be the only practicable one' [24]. Fire risk must, however, be considered, and the process is obviously unsuitable where inflammable liquids and gases are stored. There is also danger of lead poisoning to the operatives if the previous paints contain lead pigments. Properly used, flame cleaning results in cracking-off of the millscale, loosening of heavy rust and dehydration of any rust remaining. After loose particles have been removed, priming can take place while the surface is still warm and dry. The American Institute for Steel Construction suggests that acetylene and oxygen should be used in approximately equal parts, and that the size of the inner cone should be one-eighth of the total flame length. For removing old paint and moisture only, flame cleaning is regarded as expensive, but on a badly failed previously painted surface the advantages of removing paint, rust and scale simultaneously are obvious. Another advantage of the method is its mobility as the equipment can easily be taken to existing structures, but complete removal of scale and rust cannot be guaranteed. Good results have been achieved by flame cleaning weathered steel and if painting follows flame cleaning after only a short interval, the warm dry surface is beneficial because moisture has been removed and the drying of the paint is assisted by the warmth.

It is understood, of course, that it is practically impossible in the case of large iron or steel surfaces exposed to weathering conditions to free them entirely from rust, but it is absolutely necessary nevertheless to remove as much of the rust as possible. The smoother the surfaces, the less exposed are they to the formation of rust. The rougher and more porous the surfaces, the more they retain moisture, rust and other deleterious substances.

An inhibitive primer should be applied directly after the rust and scale have been removed, since a fresh layer of rust begins to form at once (often this is hardly visible at first). Application of shop coats of red lead increased the average life of the coatings by 50%. A red lead priming will only need touching up at damaged points after transport and erection of the components.

The preliminary treatment can also be effected chemically. When the articles can be immersed, this is known as 'pickling'. Acids such as hydrochloric, sulphuric or phosphoric are used, usually hot, and often with the addition of inhibitors to prevent attack on the metal itself, e.g. the tolylthioureas or N-ethylquinoline. Rust and scale are thereby dissolved, and a slightly roughened surface is left, which assists anchorage of subsequent paints. (Thorough washing off of acid residues is, of course, essential, although traces of phosphoric acid are not deleterious, see below, Footners' Duplex process.) Phosphoric acid will remove rust, but not millscale. Highly alloyed steels, e.g. stainless steel, are descaled by immersion in a bath of nitric acid or a mixture of nitric and hydrofluoric acid. Scale removers in jelly form are now available for use on site. These can contain hydrochloric acid or phosphoric acid or nitric acid and a water soluble gelling agent. They can be used on vertical surfaces and produce a valuable saving in acid consumption. After contact with the surface for several hours rust and scale are dissolved, the jelly and the corrosion products are washed off by rinsing with water. Because of its relative cheapness, hydrochloric acid has, in many instances taken the place of sulphuric acid in the descaling of steel strip in the 'coil-coating' process before the application of organic coatings.

Inhibited sulphamic acid is marketed as a powder which dissolves in water to yield a strong acid solution which effectively removes rust and scale at 60 to 70°C. Exhaustion of the acid bath is indicated by a colour change of the bath. Such products obviate the dangers associated with the transport of liquid acids.

To achieve the necessary removal of grease, oil and dirt, a degreasing operation is essential in preparing steel for painting, particularly before the application of industrial finishes on articles which cannot be abrasive blasted. It can be carried out by immersion in hot alkali solutions, by dipping in organic solvents or by spraying on or dipping in solvent/emulsion cleaners. Another method is vapour degreasing, usually with chlorinated solvents. Alkaline degreasing necessitates subsequent rinsing with hot water, and solvent emulsion cleaners must also be rinsed off. It is absolutely essential in all cases of rust removal or degreasing to apply the inhibitive primer or paint without delay.

Chemical pre-treatment of iron by which phosphate or other layers are produced (Bonderite, Atrament, Parker, Coslett, Granodine, Negradizing, Walterizing, etc., treatments) which protect the treated iron surfaces against rusting for a time and improve the adhesion of coatings. Iron subjected to such treatments must be clean and free from copper (which is frequently incorporated to improve its corrosion resistance).

The thin layer of metallic phosphates which is formed on the metal surface by the application of certain phosphate solutions, is non-conducting and therefore, stops eddy currents.

H.B. Footners' Duplex process combines pickling and surface pretreatment. It descales with sulphuric acid, and after rinsing with water the steel is dipped in phosphoric acid containing iron phosphate, leaving a thin protective film on the surface of the steel.

In modern practice fast drying pretreatment primers are often used for application immediately after blastcleaning or pickling of structural steel. The most popular types are:

(i) Single pack or two-pack etch primers pigmented with zinc chromate.
(ii) Two-pack epoxy/polyamide primers pigmented with zinc dust.

Such primers can give good protection for up to 9 months, depending on film thickness, which can vary from 10 to 25 μm [25, 26]. Overcoating at a later date requires only a minimum of further preparation.

In some cases it may be desirable to protect cleaned metal surfaces by coatings which can be removed at a later date if no further protection is needed or a permanent coating is to be applied. These temporary protectives are materials used to protect clean metallic surfaces during fabrication, transport and storage [27]. They must be readily removable by cold solvents, by mechanical stripping or by mildly alkaline strippers, when it is desired to apply a permanent protective coating. According to conditions one expects a temporary protective to prevent corrosion for 3 to 12 months. There are some circumstances where a permanent protection cannot be given and a temporary protective must be relied upon for long term protection.

There are many types of temporary protectives from which to choose. The choice will depend on environmental conditions and the likely length of exposure, on the geometry of the article and the presence of other materials such as plastics which may be sensitive to solvents.

Some protectives are based on hard resins or waxes dissolved in white spirit or distillate, others are soft coatings based on lanolin or its derivatives or on blown bitumens. Very soft, greasy coatings are formulated with petroleum jelly. Petroleum sulphonates are also in use and water displacing temporary coatings are obtained by adding suitable cationic surfactants. The latter type is much used for articles after degreasing with aqueous solutions and rinsing with water, and also for electrical components. There is also a good market for hot dipped strippable coatings (often containing ethyl cellulose and mineral oil), and solvent deposited strippable coatings.

Volatile corrosion inhibitors, like those based on cyclohexylamine, also known as vapour phase inhibitors, must be mentioned here. Di-cyclohexyl-ammonium nitrite is mostly used, both in strippable coatings and packaging materials. Uncoated steel vessels (boilers, for example) are sometimes protected for limited periods by inserting trays containing silica gel which absorbs moisture avidly.

(b) Touching bright iron or steel surfaces with sweaty fingers, residues from soldering solution, etc. The impressions of the fingers frequently shine through comparatively thin films and also tend to form rusting centres. Fingerprints on iron that are not removed chemically may cut the life of the coating as much as 60–70%, as has been proved many times [28]. Some persons' fingerprints are more corrosive than others and may show through several coats of enamel [20].

Operators suffering from sweaty hands should either wear cotton or leather gloves or better be employed elsewhere. Bright metal surfaces should only be handled with gloves, if possible.

(c) Preparatory work and painting should, if possible, be carried out under cover. In painting shops, parts must be at room temperature and be entirely free from moisture of condensation before starting with the first coat. In the case of conveyor work, a special preliminary drying process should be inserted (for instance, in the case of enamelling of mudguards, etc).

Painting operations, and even preparatory work, should be interrupted during fog or rain or in cold, damp weather. It is necessary in any case under conditions of this type not to commence painting early in the morning. It is obvious that this may lead quickly to strong surface under-rusting and a corresponding loss of adhesion of the paint to the base. Favourable weather conditions for painting (May to September and not before 9 a.m., recommended by the Protective Coatings Sub-Committee for the Protection of Iron and Steel against Atmospheric Corrosion by Means of Paint), exert an important influence on the protective effect of paint films. The relative humidity should be below 90%.

H. Wolff and G. Zeidler [29] observed that superficial moisture on smooth surfaces is emulsified with the oil contained in paints by the movement of the brushes and can be considered as fairly harmless. More serious, however, is the moisture located in the pores of the surfaces to be painted, which is out of reach of the bristles of the brush.

Paints have been developed which, it is claimed, can be applied to dew-moist steel. These contain an emulsifying or water-displacing agent, such as ammonium linoleate, or cationic surfactants. J.E.O. Mayne and U.R. Evans [30] experimented with linseed oil and emulsion paints on moist surfaces and suggest that only water in crevices caused serious corrosion, and that the use of emulsions shows no advantage.

Bitumen emulsion paints have been recommended for corrosion-resisting coatings over humid steel surfaces. Emulsion paints are sometimes used, but they must contain inhibitive pigments.

A phenomenon known as 'flash rusting' often appears as soon as blast cleaned steel has been primed with a water-based primer. It takes the form of brownish rust stains. The Paint R.A. has shown that the defect can be avoided by removing contaminants remaining after blast cleaning of rusty steel by scrubbing the surface with clean hot water or by a chemical treatment before application of the primer. (P.R.A. Newsletter No. 7, 1978.)

(d) Lacquers, the solvents of which contain a high percentage of water, e.g. spirit varnishes made in low-grade methylated spirit, or which cause water condensation owing to the rapid evaporation of some of their thinners (cold of evaporation) may promote rust formation on an iron base.

The use of anti-chill thinners (which contain solvents with a comparatively slow evaporation rate) or air conditioning are to be recommended as remedies.

(e) Storing in the open of unpainted or inappropriately painted steel parts previous to completion of the job. The tendency of any freshly descaled iron or steel surfaces is to rust immediately; this tendency is particularly pronounced in the humid and polluted conditions that are common in the industrial areas of Great Britain [31] and many other countries. In the words of J.C. Hudson: 'The chief source of this pollution is the smoke from factories and chimneys which contains large quantities of noxious solid particles and highly corrosive sulphur compounds; in fact, in some localities, the rain water is virtually a dilute solution of sulphuric acid [32]'. (The rain water of one city has an average pH value of 3·0 [33]).

Unless only temporary protection is desired the best protection under such conditions is obtained with red lead, zinc phosphate or zinc-rich primers. Zinc chromate primers are less suitable if they are likely to be exposed to weathering before repainting, as soluble chromate may be leached from the film. Zinc-rich primer has also the advantage that it does not interfere with subsequent welding. Other primers have been advised, but are inferior. Red oxide of iron in boiled linseed oil is often used as a shop primer, but has no inhibitive properties.

(f) Degreasing or pickling with acids, especially with substances containing free chlorine, or pickling fumes, may cause rusting. Under some conditions chlorinated solvents may be hydrolysed and form small amounts of hydrochloric acid if no inhibitors are present. Due attention should therefore be paid to these points.

Proximity of galvanic baths (plating) to the varnishing shops may cause condensation of acid vapours on the articles to be coated. Steel articles in a laboratory are also prone to rust formation.

(g) Location of a soldering or welding plant near the varnishing shops.

If degreasing is not carried out with neutral, organic substances, it is absolutely necessary to wash, neutralize and dry the degreased surfaces carefully. Parts contaminated with acid should be brushed over with milk of lime or sodium carbonate solution and washed thoroughly with hot water. Degreasing and welding plants, as well as plating baths, should be well separated from the varnishing shops.

(h) Influence of sea water (sea salts), marine conditions and also of water vapour, air containing salt (salt pits, salt-grading works, etc.) or tropical climate. Corrosion is most pronounced in industrial areas (see Introduction and sub-section (e) above). Rusting in rural atmospheres proceeds rapidly during the first 18 months of exposure, but a thin layer of rust formed provides protection and the rate drops to 1/9 of its former level [34].

Smooth surfaces of steel, which usually present no great obstacle to the adhesion of a good coating, prove to be among the most difficult objects when it is desired to provide a lasting protective film under humid or sea mist conditions. Much, of course, depends on the other metals present in the alloy.

Resistance of a film to humidity is in general proportional to its thickness, provided that it is thoroughly cured or hardened through. Imperfectly cured thermosetting media are more vulnerable in the thicker layers, although these may appear sufficiently hard for use (M. Hess) [35]. Bituminous paints exhibit in a pronounced way the beneficial effect of film thickness, and are often reinforced with asbestos fibre to enable films of 0·5 mm (20 thou) thickness or more to be applied even to vertical surfaces.

V.G. Jolly [36] found by experiments that there is 'only a minor difference between the effects of a marine and concentrated industrial environment'. J.C. Hudson, however, considers the former not so pronounced as the latter. 'Corrosion in the soil bears a general resemblance to corrosion in sea water'. It leads to solid cakes of rust so that 'in some cases at least, the attack should stifle itself' [32].

In some areas, particularly in certain tropical climates, the atmosphere is sufficiently pure and dry to enable rust-protective measures to be dispensed with. Paint films, however, break down more quickly in these areas owing to high solar radiation [32]. During, and since, the 1939–45 war, 'mothballing' of components and even complete warships and planes was successfully used. The articles are enclosed in a sprayed 'cobweb' of vinyl resin, which is built up with bitumen, etc., to form a completely impenetrable coating, inside which is located a desiccating agent. Provided that the humidity within the 'cocoon' was maintained below 30%, storage could continue indefinitely without corrosion.

Oil-modified alkyds have proved their value as finishes for marine and industrial conditions. In the latter case low dirt retention of the alkyds (compared, e.g. with ready-mixed oil paints) contributes to a clean-looking surface. Softer paints permit the industrial grime to penetrate the film surface, so that it is not removed by rain. Rust-inhibiting primers, however, are still essential. Bitumen and refined tar coatings are also useful if applied sufficiently thickly (not less than 100 μm (4 thou) preferably).

If chemical attack is so severe as to predominate over normal weathering, it is advisable to use, e.g. chlorinated rubber, epoxy or pure phenolic resin-tung oil based paints, all of which possess excellent chemical resistance, but somewhat inferior weather resistance compared with, e.g. alkyds.

Baked primers, often containing zinc chromate, are used for bicycles very commonly, particularly in northern Europe. With alkyd-based stoving finishes, or high-temperature bituminous blacks, excellent performance is achieved.

(i) If rust formation is of the dotted type, it can be assumed that the peaks created by abrasive blasting have not been covered by paint. If pigments contain alkali or alkaline earth metals (e.g. sodium, calcium ions), free alkali can be formed at cathodic areas on the metal surface, with the result that saponification of the medium and loss of adhesion can follow at those areas. This can also occur if a film absorbs alkali salts from the environment, e.g. sea water [37]. The phenomenon is known as 'alkaline peeling' [38].

The choice of the most suitable pigments is of considerable importance besides that of the vehicle. If an inhibitive (passivating) primer is used, 'the differences between finishing paints of good quality and suitable colour are relatively insignificant' [32].

Passivated surfaces not only do not rust, but prevent the spreading of rust if the thin passivated layer of 3–12 μm thickness is damaged. Good protecting top-coat paints usually contain a number of pigments. Certain pigments, if used alone, may even assist the formation of rust. Anti-corrosive pigments must not swell.

Coal-tar coats appear at first to give satisfactory paint film characteristics, but often lead to dangerous under-rusting phenomena if the tar was not refined and combined with suitable drying oils, tallow, etc. See also above, sub-section (c).

(j) In chemical works contact with acidic solutions or gases (e.g. hydrochloric acid, sulphuretted hydrogen, sulphur oxides) may induce severe corrosion.

The choice of anti-corrosive primer is important. Chromate primers may lead to accelerated corrosion (often as 'pitting') under acid conditions, as the chromate may depolarize cathodic areas where action would otherwise have been stifled. Primers based on metallic lead are particularly suitable, but those based on red lead and zinc-rich also give good results, although the latter should be avoided if hydrochloric acid is present. Use acid-resisting top coats, e.g. pure phenolic tung oil media. Thick coats of chlorinated rubber or bitumen may also be used, without special primer if the thickness is sufficient. Vinyl acetate-chloride copolymer paints are also effective. Epoxy paints, particularly with a primer containing metallic lead, are excellent. Urea, because of its hygroscopicity, is very corrosive to steel. An epoxide/zinc chromate primer followed by 2 coats of a phenolic/epoxide finishing coat, the system providing a total film thickness of 100–110 μm, has given good protection in a fertilizer plant [39].

(k) Steel and cast iron are heterogeneous in structure, and hence numerous cells for electrolytic action are always present, which tend to destroy the protective action of paint films. Mechanical stresses within the metal may contribute to the action.

Such corrosion can only occur in presence of moisture and can therefore be prevented by applying completely impervious coatings, e.g. a heavy bitumen coating of 250 μm (10 thou) or more thickness. Zinc-rich paint is ideal for such cases, since the film is electrically conducting (by containing 92–95% of zinc dust in a chlorinated rubber or polystyrene medium) [40]. Sacrificial protection is thus conferred, as with galvanizing or Sherardizing, if moisture penetrates to the metal-paint interface. Such paints are more tolerant of slightly rusty steel substrates than are other paints [40], but it must be known how much rust and scale a particular paint can tolerate [41]. The age of the rust and the season when painting are also critical. Mayne [41] states that rust should preferably be not more than six months old, and that better results are obtained by painting in summer.

Because of the distributed graphite contents in cast iron, local elements (graphite-iron) cause rust to be uniformly spread over the entire surface in the presence of moisture. In Martin-ingot iron, which is comparatively poor in carbon, local elements formed are a greater distance apart, so that deep local rust zones (e.g. on the steel hulls of ships) are formed. The best preventive is the selection of paints containing corrosion-inhibitive pigments.

Stress corrosion cracking is produced when either external or internal tensional stresses are present where corrosive conditions prevail. Alloyed materials are more sensitive in this respect than mild steel and austenitic stainless steels are particularly prone to this form of failure. The presence of even very small amounts of chloride ions accelerates the corrosion of these stainless steels dramatically and it is often thought necessary to prevent the access of chloride ions to such substrates by providing a chloride-free barrier coating. Successes are claimed for products based on graphite pigmented silicone resin varnishes. To promote adhesion to the stainless steel, shot blasting or a phosphoric acid treatment precedes the application of the coating.

(l) Stray currents of electric plants on board ships or emitted by electric power works, etc., and conducted along by water or in the ground, may also cause rusting (electrolytic corrosion). This may also affect pipe lines. These may also be corroded by long line currents [32].

Pay attention to satisfactory insulation of all electric machinery and avoid all bare electric conduits. It appears that no really efficient protective coatings, etc., have yet been found to withstand wandering currents, and the only reliable remedy is to detect and eliminate them. Stray-current corrosion has been noted [42] near plastic/metal joints where the non-conductive plastic causes de-earthing, leaving the metal inadequately earthed.

(m) Contact of iron with a more noble metal is harmful due to the formation of electrolytic elements. For instance, contact of iron with copper in sea water increases rust formation by 47%. These potential differences make themselves felt even in contact of ingot steel with cast iron [43], or in the contact of chemically homogeneous metals of different physical conditions.

Grinding wheels having been in contact with other metals, especially mild and low-alloy steels or with iron oxide (possibly contained in a polishing compound), must not be used for stainless steel. Such particles could impregnate the latter's surface and reduce corrosion resistance.

Points of contact between different metals must be avoided wherever possible, or at least insulated or completely surrounded by a suitable coat of paint or jointing compound.

(n) Rust formation may be caused by roughly or badly ground pigments. This does not apply to the scaly-shaped pigments such as aluminium bronze powders (not unanimously considered as rust protecting if used without other pigments), graphite and mica (which should be stirred thoroughly). Undispersed rough pigment particles remain partly unwetted, permitting moisture to penetrate through the lumps to the iron surfaces.

The shape of the ground-in pigments may also exert a certain influence, i.e. spherical-shaped pigments such as zinc oxide and splinter-shaped particles of barium sulphate if mixed together yield a much denser and more protective texture than if used unmixed [43]. The same applies to needle-shaped structures due to their felting tendencies as long as they are not ground too finely or if they are very hard.

These troubles are sometimes caused by the fact that the price needed for a sufficiently fine grinding of the paint is considered excessive or that the pigments are stirred into the medium by the consumers themselves. Rust formation is most rapid where badly wetted pigment lumps are located.

(o) If conditions are in favour of under-rusting or if steel surfaces are painted which are not entirely free from rust, the electrolytic formation of rust is increased with the number of layers of oil paint applied, irrespective of the kind of pigment used in the linseed oil. This results in the formation of entire crusts of rust and remainders of paint [43]. It has specially to be faced in the case of under-water paints.

Structures immersed in water, particularly low-conductivity water, usually show rapid blistering of most coatings. An inorganic zinc-rich coating followed by a phenolic modified epoxide top coat often gives the best result. A polyurethane coating, pigmented with red oxide of iron has shown exceptional promise in hot distilled water. The tendency to blister is much less in sea water than in fresh water. Strong pitting on an underwater structure coated with a zinc chromate primer and a vinyl resin topcoat was ascribed to insufficient passivation of the substrate [44].

Claxton and Carter [45] have reported the failure after a few months of the paint system on a zinc sprayed bridge structure primed with a red oxide/zinc chromate primer. It was found that the primer was very low in its zinc chromate content. Repainting with a primer rich in zinc chromate provided adequate passivation and prevented repetition of the failure. Blistering and loss of adhesion on zinc-sprayed steel coated with an etch primer and an epoxide system was observed in one case and believed to be due to hydrogen formed from excess phosphoric acid in the etch primer. Similarly, a loss of bond between an epoxide/phenolic topcoat and an inorganic zinc coating has been reported and was ascribed to hydrogen formation by the zinc/steam reaction [46].

J.A.W. van Laar [47] has observed the influence on the formation of blisters on a painted steel panel of the presence or absence of a paint coat on the opposite side of the panel. For instance, he found no blistering on a steel plate immersed in water at areas from which the coating on the reverse side had been removed. He calls this the 'blister dia-phenomenon', and attributes it to the action of atomic hydrogen as a consequence of corrosion stimulation. J. van Loon [48] made somewhat similar observations on painted building timber under humid conditions.

Coatings which are continuously exposed to condensation, for instance in mines, or to contact with moisture, for instance in the ground below the water level, have to meet similar conditions to those exposed to water immersion. In such cases chemical and electrochemical attack can take place. Corrosion due to bacteria can also come into play.

Under anaerobic conditions, especially in clay soils containing sulphates, e.g. gypsum, the bacterium *Desulfovibrio vulgaris* is a widespread cause of corrosion of buried iron and steel. Bacterial corrosion has been found in underground pipes, oil well installations, aircraft fuel tanks, cooling towers, central heating systems, sewage works, ships' plates in polluted estuarine conditions, at the bottom of gasholders and even in pipes carrying water, if local conditions of stagnation prevail. With cast iron pipes 'graphitization' may take place, the pipe retaining its shape, but becoming soft at affected places. The mechanism of this corrosion process is believed to be based on cathodic depolarization caused by removal of the hydrogen from the surface, thereby reducing SO_4^{2-} to S^{2-}, forming iron sulphide (FeS) which in itself has been identified as highly corrosive. Under aerobic conditions it can be oxidized to sulphate. Thick coatings of bituminous materials or polyamide cured coal tar/epoxide systems seem to offer the best means of protection. Laboratory experiments indicate that certain biocides, like phenol derivatives, or some quaternary ammonium compounds could with advantage be incorporated in the coatings to minimize bacterial corrosion [49].

In very wet conditions, i.e. under water or in damp soil, it is necessary to avoid paints with a tendency to swelling, like those based on linseed oil or alkyd resins. It is essential to select non-saponifiable vehicles like bitumen, chlorinated rubber, isomerized rubber, synthetic rubbers (chloroprene) or certain phenolic resins. Such paints exhibit only minimal swelling and are unaffected by alkali formed at cathodic areas.

In the absence of inhibitive pigments protection can be ascribed to the high electrolytic resistance of the paint films [50] hampering the movement of ions.

(p) Like metals, most plastics are subject to chemical attack and therefore to corrosion. The degradation products of plastics and plasticizers can in turn cause corrosion of metals. The most noted problems occur with p.v.c., particularly in fires when hydrogen chloride is generated in large quantities with severe corrosive effect on steel and other metals. There are cases known when the damage caused by HCl in a fire far exceeded the actual fire damage. This is particularly so if expensive and very delicate electronic equipment is close to the source of the fire. Postchlorinated p.v.c. is liable to split off HCl under less strenuous conditions and cause corrosion of the substrate which may not be apparent to the observer as the coated surface appears to be intact.

Mild steel exposed to hydrochloric acid vapours from the thermal decomposition of chlorinated polymers and subsequently to condensation, has been shown to develop black, horn shaped (corniform) corrosion products (Fig. 7.22), which have been identified as magnetite [51]. Corniform corrosion has also been found in emulsion paint containers in the moist acid environment above the level of the paint.

An unusual case of corrosion has been reported [42] on the colder end of the steamside of a stainless steel heat exchanger, where a corrosive deposit of iron and polystyrene was found. The cause of this was a carry-over of the polystyrene from the ion exchange resin in the feedwater de-ionizing plant. It was discovered that hot instead of cold sodium chloride solution had been used for the resin regeneration process.

(q) E. Meier has been able to show that paint films disintegrate more easily in the grooves drawn by the paint brushes. On account of the smoother films obtained he recommends spraying. M. Ragg, on the other hand, assumes the opposite position with regard to ships' paints, as these are used in regions where the relative humidity is high, which may lead to the formation of smeary emulsions, causing considerable delays of the drying processes and under-rusting.

Whether the application of paint materials by brushing or spraying is more advisable depends also on local conditions.

(r) Covering steel parts of skeleton constructions with concrete or with cement mixtures (torkretizing) has yielded satisfactory results if the protective layers of concrete or cement are sufficiently adhesive, although for reasons not yet explained this method has led in some cases, e.g. on a bridge, to strong rusting on unpainted inlets, accompanied, of course, by corresponding weakening of the structures, so that the protection afforded by these methods is not entirely reliable (K. Würth) [52].

This defect does not come in where cement coatings are used for the protection of pipes against soil corrosion. 'They have given good results in cases where pipes had to be safeguarded against attack by stray electric currents' [32].

Aqueous cementiferous paints for use on rusty steel have been developed at Cambridge University. These take up any moisture present on the surface, can be made to emulsify, and thus render harmless small amounts of oil or grease.

It should be insisted that in cases as described the iron parts must be painted, if possible, with a bituminous paint (compare with (c) above).

(s) Vehicles containing acids, as in the case of catalytically hardening synthetic resin varnishes, or containing compounds which release acid, may induce under-rusting if the base consists of steel, or promote other forms of corrosion in case of other metals.

Acid-containing varnishes should be preceded by an insulating coat free from acid, so that there is no direct contact with the metal. If the acid concerned is phosphoric acid, such precautions can be dispensed with, however.

In a machine shop corrosion was noted on steel parts after milling. The coolant used was an alkaline soluble oil emulsion containing an amino alcohol. The coolant was stored in a concrete tank coated with a two-pack epoxide filler. The curing agent was of an acid nature and had been used in an excessive quantity. This acidity had neutralized a large part of the amino alcohol, thus rendering the coolant corrosive [44].

Care is required in the use of two-pack coatings. It is desirable to allow the mixed product to rest for half an hour so that air bubbles can escape before use. This presupposes a reasonably long pot life, which is also desirable to prevent early cross-linking after application and gives enough time for air to rise to the surface and leave the film. The importance of this is obvious when coating concrete from which the solvent will displace large quantities of air which could cause formation of pores in the cured film.

(t) Bituminous coatings of substantial thickness may be split, exposing the underlying steel, if repainted with certain types of coating. V.G. Jolly describes such an effect when protein-bound water paints are applied to bituminized steel [36]. Oil paints can also produce such splitting, which may be followed by severe corrosion.

Use bituminous paints (solution or emulsion types) in these circumstances. The splitting is caused by shrinkage of the finishing paint tearing the bituminous coating, and can only be prevented with certainty by ensuring that the finish will retain the thermoplastic characteristics of the bitumen beneath.

(u) Acidic components of wood can cause corrosion of steel, and lead to blistering of paint films on prolonged contact. Examples of rapid failure of enamelled steel articles packed with oak shavings in moist conditions have been often observed. Rain water dripping from wooden shingles can cause corrosion of lead and other metallic flashings to the point of perforation. Severity of attack varies with the type of wood and degree of seasoning.

Alternative packing materials should be used, or the cases kept perfectly dry.

Notes

[1] Fancutt, F., Hudson, J.C., Rudram, A.T.S. and Stanners, J.F. (1968), *Protective Painting of Structural Steel,* Chapman and Hall, London.
[2] Report of the Committee on Corrosion and Protection, H.M.S.O., 1971.
[3] Potter, E. (1975), *Australasian Corr. Eng.,* **19**, No. 3, 5.
[4] Neufeld, P. (1975), *Elementary Aspects of Corrosion,* Portcullis Press, Redhill.
[5] Appleby, A.J. and Mayne, J.E.O. (1967), *JOCCA,* **50**, 897.
[6] Appleby, A.J. and Mayne, J.E.O. (1976), *JOCCA,* **59**, 69.
[7] Wild, G.L.E. (1965), *JOCCA,* **48**, 43.
[8] Barraclough, J. and Harrison, J.B. (1965), *JOCCA,* **48**, 341.
[9] Biddle, G.J. (1974), *Australasian Corr. Eng.,* **18**, No. 5, 5.
[10] Harrison, J.B. (1970), *Trans. Inst. Metal Fin.,* **48**, 83.
[11] Davidson, S.L. (1975), *JOCCA,* **58**, 435.
[12] Pantzer, R. (1975), *Defazet,* **29**, 13.
[13] Knowles, E. and White, T. (1958), *JOCCA,* **41**, 10.
[14] Maass, W.B. (1975), *Defazet,* **29**, 334.
[15] Alwis, K.W. and Wallington, R.H. (1970), P.R.A. RS/T/50/70.
[16] Anon., (1976), PPCJ, **166**, 153.
[17] Kresse, P., *Farbe Lack,* **83**, 85, 1977 and Defazet, **28**, 459, 1974.
[18] Old, A.R. (1952), *JOCCA,* **35**, 20.
[19] Whiteley, P. and Sperling, R., *Paint Manuf.,* **33**, 221, 1963 and **34**, 43, 1964.
[20] Committee convened by the Paint Research Association on *The Painting of Buildings,* par. 100, 119.
[21] British Standard Specification No. 4232, 1967.
[22] Bullett, T.R. and Dasgupta, D. (1969), *JOCCA*, **52**, 1099.
[23] Chandler, K.A. and Reeve, J. (1966), *JOCCA,* **49**, 464.
[24] Hudson, J.C. (1945), *JOCCA,* **28**, 27.
[25] Bayliss, D.A. and Wall, D.C. (1968), *JOCCA,* **51**, 792.
[26] Grey, C.A., Hill, L.A. and Marson, F. (1976), *JOCCA,* **59**, 22.
[27] Elgar, D.J. (1973), *Anticorrosion,* **20**, No. 10, 17.
[28] Wampler, R.H. (1946), *Modern Organic Finishes,* p. 67, New York.
[29] Wolff, H. and Zeidler, G. (1934), *Deutsche Farben Zeitschrift,* **39**, 1071.
[30] Mayne, J.E.O. and Evans, U. (1949), *J. Soc. Chem. Ind.,* **68**, 212.
[31] British Iron and Steel Institute Corrosion Committee (1941), *Protective Painting of Structural Steel,* p. 11.
[32] Hudson, J.C. (1940), *The Corrosion of Iron and Steel,* London.

[33] Jordan, L.A. and Whitby, L. (1936), *The Protection of Iron and Steel by Means of Paint,* p. 119, 16th Bulletin, P.R.A., Teddington.
[34] Spedding, P.L. (1971), *Australasian Corr. Eng.,* **15**, No. 8, 27.
[35] Hess, M. (1944), *Paint Tech.,* **2**, 29.
[36] Jolly, V.G. (1944), *JOCCA,* **27**, 103.
[37] Gay, P.J. (1949), *JOCCA,* **32**, 488.
[38] Blom, A.V. (1939), *JOCCA,* **22**, 104.
[39] Naram, S. and Mitra, S.K. (1973), *Paintindia,* **23**, No. 10, 23.
[40] Evans, U.R. (1946), *J. Chem. Soc.,* **24**, 234 and 242.
[41] Mayne, J.E.O. (1954), *J. Iron and Steel Inst.,* 140 and 143.
[42] Kennett, A.C. (1972), *Australasian Corr. Eng.,* **16**, No. 10, 13.
[43] Ragg, M. (1954), *Schiffsboden and Rostschutzfarben,* Berlin.
[44] v. Oeteren, K.A. (1974), *Metalloberfläche,* **28**, 229.
[45] Claxton, A.E. and Carter, E.V. (1969), *Brit. Corrosion J.,* **4**, 187.
[46] Singleton, W.T. (1973), *Mat. Prot.,* **12**, No. 11, 41.
[47] van Laar, J.A.W. (1959), *Fatipec Congress Book,* p. 263.
[48] van Loon, J., *Verfkroniek,* **32**, 35 (1959) and **35**, 191, 291 and 331.
[48] van Loon, J. (1959), *Verfkroniek,* **32**, 35 and (1962) **35**, 191, 291 and 331.
[50] Mayne, J.E.O., (1956), *POCJ,* **129**, 484.
[51] Hill, G.V.G. (1972), P.R.A. Technical Report TR/37/72.
[52] DMV 38, *Anstrich im Stahlskelettbau,* p. 3 (1936).

Proprietary materials

Busan 11-M1	Buckman Laboratories Inc., U.S.A.
Colorol	Croxton and Garry Ltd.
Corroless	Lawrence Industries, Michigan, U.S.A.
Moly-White 101 and 212	Sherwin Williams Chemical Inc., U.S.A.
Nalzin SC–1	National Lead Industries Inc., U.S.A.
Oncor M50	National Lead Industries Inc., U.S.A.
Sicorin RZ	BASF, Bereich Siegle, Stuttgart.

7.10 Lack of effectiveness of chemical pretreatment of steel

The following materials and methods are used for pretreating iron and steel.

Group 1

Dilute hydrochloric or sulphuric acid, sulphamic acid, formic, hydroxyacetic, hydrofluoric, citric acid and alkaline gluconate solutions are all used for the removal of rust and/or millscale. Acids must contain inhibitors to restrain attack on the base metal.

Group 2

Phosphoric acid or phosphate solutions, with or without accelerators which deposit firmly adherent, chemically bonded layers of iron, zinc, manganese phosphate or mixed phosphates of controlled thickness. The thickness can be varied to suit requirements and normal limits are from 2–10 μm. Several commercial processes are available and known by their trade names like Granodine, Parkerizing, Bonderizing and Walterization. These treatments can be applied by brushing, spraying or dipping. Consistent quality and uniform thickness can be achieved by automated monitoring and control of the process.

Group 2 treatments normally precede the painting of manufactured articles like motor car bodies, domestic appliances, etc. and are usually incorporated in conveyorized finishing lines.

Clean steel surfaces, especially those which have been treated by pickling or abrasive blasting, are very susceptible to rusting. Surfaces which have never been contaminated by rust formation usually behave better than those from which rust had been removed previously. It is always advisable to apply one of the surface-conversion treatments and to paint pretreated articles with as little delay as possible or to apply at least a temporary preservative. All deposits of grease, oil, millscale and non-ferrous metals should be removed before the treatments. The prescribed temperatures, exposure times and after-washes should be watched. The protection against corrosion afforded by the phosphate coatings alone is only of a moderate order.

(a) Chemicals of the Group 2 mentioned above are sometimes without effect, if even traces of other metals are present. Copper, zinc, even remains of old paint are mentioned as detrimental impurities. Simple washes, e.g. those containing dilute phosphoric acid will only remove slight surface rusting, not heavy deposits.

Modern rust removers are available which dissolve slight grease and rust in one operation.

For some alloys or mixed metal combinations, such as steel and zinc, special processes are available.

(b) Where chemicals of Group 1 are used, the steel surfaces must be carefully cleaned afterwards to eliminate all danger of failures in the paint films.

Acids scarcely remove any oil or grease. Films of the latter, if sufficiently substantial, may even prevent the action of the acids in places.

Apart from the usual alkaline, etc., washes after pickling, pre-cleaning, e.g. by means of emulsion washes, to remove oil, grease and dirt is advisable.

(c) Ring blistering of motor body finishes has been observed, caused by condensation water droplets falling on phosphated surfaces before painting, if they carry calcium and chlorides in solution.

Use of deionized water and a chromic rinse has been found effective in such cases.

(d) 'Snail track' blistering has been described by Nicholson and Jenkins [1]. This takes the form of lines of minute blisters found on phosphated and painted steel during prolonged transit under tropical conditions. These were thought to be associated with small amounts of water-soluble material at run marks or dry-off marks from the final rinse. The defect could be reproduced even on glass when rinses which had been deliberately contaminated were used.

The defect was eliminated in this case by running the final chromate rinse cold, to give maximum drainage, and to dry the articles in an upright position without rotation so that draining took place away from important areas.

(e) The agent in the solution may have become too weak.

Pickling or phosphating baths require frequent renewing. Periodic determinations of free acidity will indicate whether the strength has fallen, and the required concentration must be maintained by addition of acid or acid phosphate. Automatic control equipment is now available. The bath must also be kept free from high concentrations of metal salts, e.g. iron chloride in hydrochloric acid pickling. In large plants this may be effected by cooling the liquor to induce crystallization and returning the liquid to the bath.

(f) Local ineffectiveness in recessed areas can be due to the formation of gas pockets.

Keep the bath in movement.

(g) Hands or cloth are attacked.

Wear rubber or p.v.c. gloves, boots and aprons when using the chemicals, or select a milder type of derusting agent. Protect eyes by goggles.

Note

[1] Nicholson, A. and Jenkins, H.A.H. (1955), *Trans. Inst. Metal Fin.*, **33**, 2.

7.11 Corrosion and efflorescence followed by blistering and lifting of paint and lacquer films on magnesium and aluminium alloys as well as on refined aluminium

The main difficulties in coating light metal alloys [1] are due to their 'strong electro-positive nature, their high coefficient of expansion and their susceptibility to corrosion'. The attack on these metals manifests itself quite differently from that on iron by deep, uneven, sometimes filiform, corrosion which, of course, reduces their resistance. Aluminium alloys in particular show, apart from surface pitting, intergranular crystallization with a loss of tensile strength and flexibility. Magnesium alloys are particularly prone to surface attack due to hydrolysis under saline conditions forming bulky corrosion products with development of hydrogen.

Different alloys, however, differ greatly in their resistance to corrosion, and cladding (coating with a resistant alloy, e.g. 99·5% aluminium, on both sides) lowers the corrosion tendency considerably. Paint life on the less corrodible surfaces is obviously much longer than on more corrodible alloys.

Generally, very satisfactory results for the protection of light metals from corrosion are obtained by the use of primers based on zinc chromate in suitable media, e.g. phenol formaldehyde-containing dispersion resins. Very successfully applied can also be one or two pack 'etch' or 'wash' primers mostly based on a vinyl butyral resin, chromate pigment and phosphoric acid. (See, however, below, sub-section (j)). It should be emphasized that etch primers do not afford adequate protection to magnesium alloys unless these are pretreated. Etch primers have been referred to before in connection with other metals, especially iron and steel.

Magnesium alloys

The chief constituent of the magnesium alloys represented by the Dow and Electron metals is magnesium in amounts of above about 90%, the other metals making up these alloys being aluminium (in amounts of up to about 10%) and various others, mainly zinc, manganese and tin. Because of the low specific gravity of magnesium, which at a temperature of 20°C (about 70°F) is 1·74 as against a specific gravity of 2·70 for aluminium at the same temperature (sp. gr. of iron 7·84, of titanium 4·5), magnesium alloys are the lightest. They are not attacked by dry air, whilst under the influence of moist air they become covered with a protective layer of oxide. They corrode in marine air [2] and acid-contaminated atmosphere, as well as by the continued action of water, acids and salt solutions [2], the only efficient protection against these actions being pretreatment of the metal plus the selection and application of suitable paint materials (see sub-section (d) and (e) below).

Magnesium alloys are practically insensitive toward the action of hydrofluoric acid and alkaline solutions within certain ranges of concentration, and at temperatures below 120°C (250°F) in contrast to aluminium. Indeed, magnesium parts have been made more corrosion resistant by treatment with hydrofluoric acid, producing a protective layer of magnesium fluoride exhibiting properties similar to those achieved by the oxidation of aluminium as discussed below.

Tropical conditions with their damp, warm climate and atmospheres rich in carbon dioxide and subject to great temperature fluctuations require coatings of particularly high quality.

If the alloy is to be coated, attention must be paid to the fact that the oxide layers already present on the metallic surface may be a source of failure for the film, because magnesium oxide reacts alkaline in presence of moisture. This may break down the coating, although the metal itself is passive towards this reaction.

Before coating or protective treatment, careful degreasing is a most important operation. Washing with alkaline detergents or with trichlorethylene or 1, 1, 1–tri-chlorethane is frequently used to advantage. Good results are frequently obtained by rubbing down with pumice powder or whiting and baths of petrol, methylated spirit and varnish thinners.

The corrosion resistance of magnesium metals, as well as the keying of films, is noticeably improved, and a highly protective layer only obtained if, previous to painting, one of the following surface treatments is applied.

Castings are frequently pickled by immersion in a solution of nitric and/or sulphuric acid, followed by a dip in nitric acid-sodium bichromate solution, which produces a brass-yellow coloration. This may be followed by an additional boiling with sodium bichromate. The Dow No. 3 treatment consists of boiling in an ammoniacal solution of ammonium sulphate and ammonium and potassium dichromate. The hydrofluoric acid-dichromate process is also regarded as very effective.

Electrolytic processes may also be used. An anodizing process using alternating current, which removes heavy corrosion products and is claimed to confer resistance to salt-spray and to heat has been described [3]. Electrolytic oxidation of magnesium alloys [3] in presence of oxalic acid (Elomag process) has been recommended. In other similar processes phosphate-chromate solutions are used instead. One of these treatments is imperative if the metal is to be exposed to acid, saline or similar conditions. They alone do not provide, however, any very effective protection without suitable painting. There should not be any great interval of time between the chromate treatment and the application of the primer.

In view of the fact that the various magnesium alloys react differently to various conditions, it is advisable to select the most suitable alloy for each job [4].

The wrought and cast aluminium alloys

These [5] consist chiefly of aluminium and of varying percentages of magnesium (up to 13%), copper (up to 13%), silicon (up to 21%), manganese (up to 2%), and/or iron, nickel, antimony, chromium, zinc, titanium (and other grain refining elements), lead, tin, bismuth, etc., see the various specifications [5]. They are characterized by high strength and easy workability.

The strength of many wrought aluminium alloys corresponds to that of ingot iron. Chemical resistance and electrical conductivity of the purest grades of refined aluminium appears to surpass that of copper.

Many of the pretreatments recommended for steel are suitable for aluminium, particularly in respect of degreasing. Aluminium when rolled, extruded or cast, gives relatively poor paint adhesion. This is mainly due to the natural skin of oxide which is present on aluminium surfaces and, to a lesser extent any oil and grease which it may contain.

A thorough degreasing of aluminium metals in order to remove oil, grease and cutting compounds prior to painting or at least some preparatory treatment is just as necessary as that of the magnesium alloys above discussed (see also sub-section (a) below). Suitable processes are: trichlorethylene vapour bath, weak alkaline baths [6], chromic and sulphuric acid and phosphoric acid baths with additions of solvents and application of emulsion cleaners.

Cast aluminium alloys have a lower degree of shrinkage than pure aluminium [5]. If the grey nitride skin of the aged aluminium alloys, or the very thin 'natural' oxide skin of all other aluminium alloys, are sand blasted either for removal or for surface

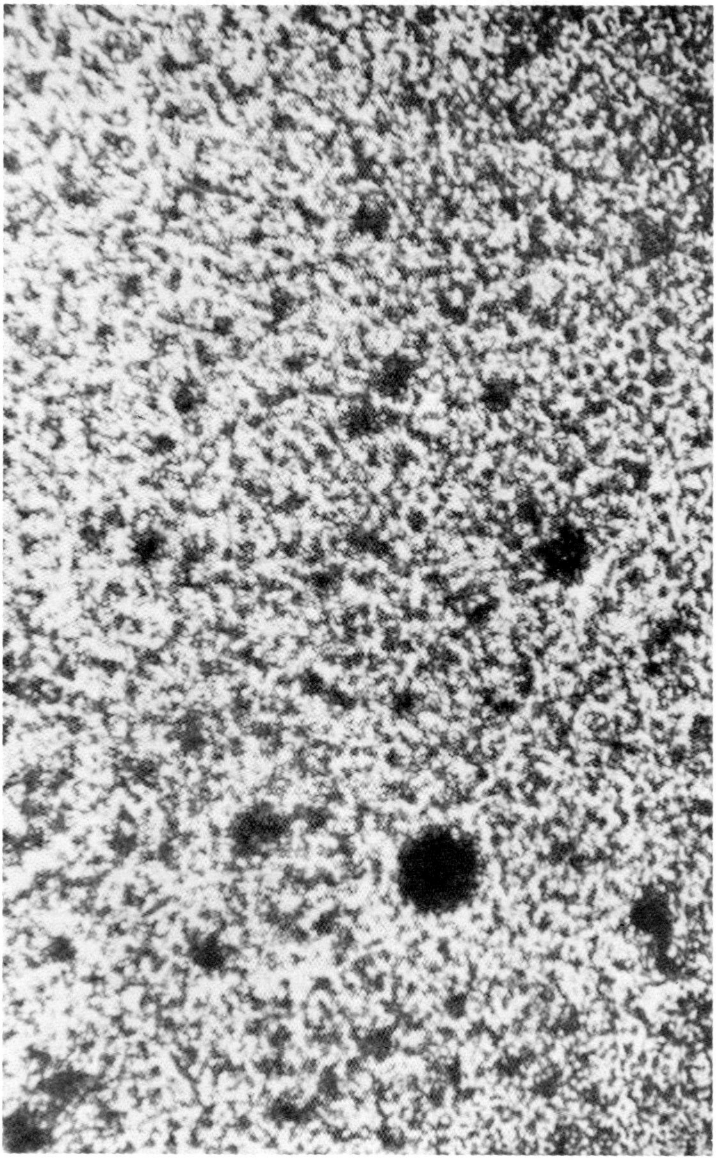

Fig. 1.2. Flocculation of blue pigment. (*By courtesy of Laporte Industries Ltd.*)

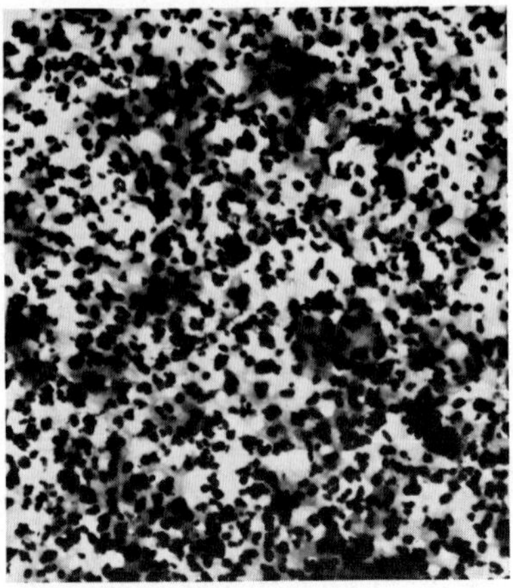

Fig. 1.3. Flocculation of titanium dioxide. (*By courtesy of the Paint Research Association, Teddington.*)

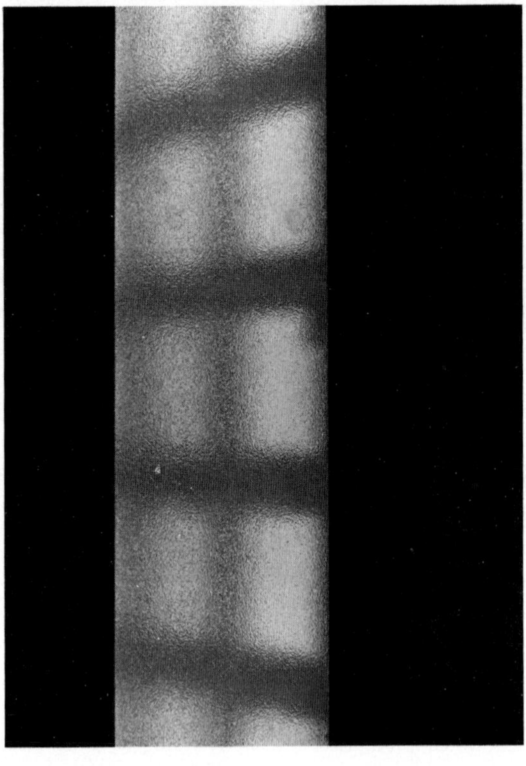

Fig. 1.4. Powder cured partially during storage. (*By courtesy of BTP Tioxide Ltd.*)

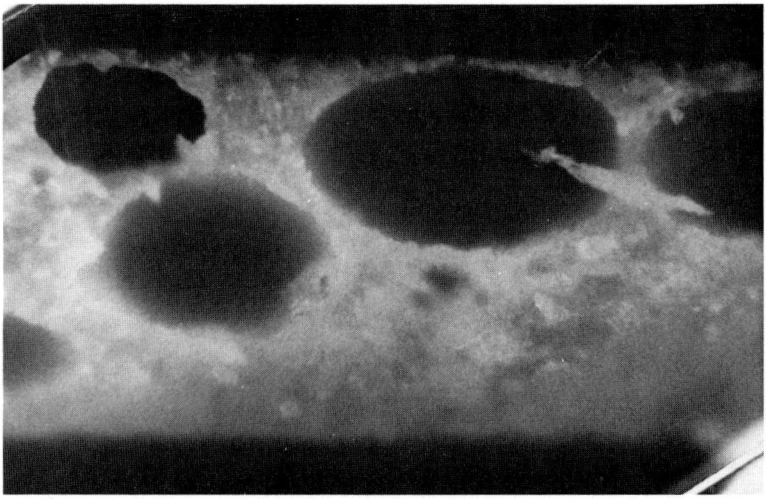

Fig. 2.1. White chlorinated rubber paint film showing vacuoles ($\times$ 45). (*By courtesy of the Paint Research Association, Teddington.*)

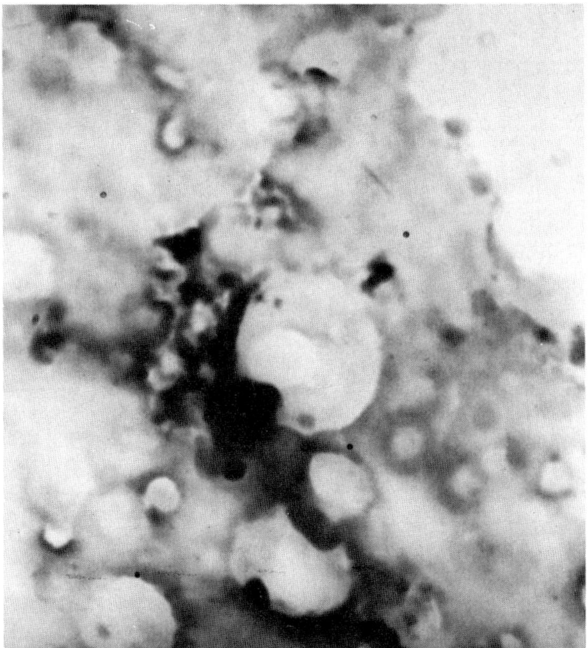

Fig. 2.2. Ion-beam etched film showing incomplete fusion of polyvinylidene fluoride granules. (*By courtesy of the Paint Research Association, Teddington.*)

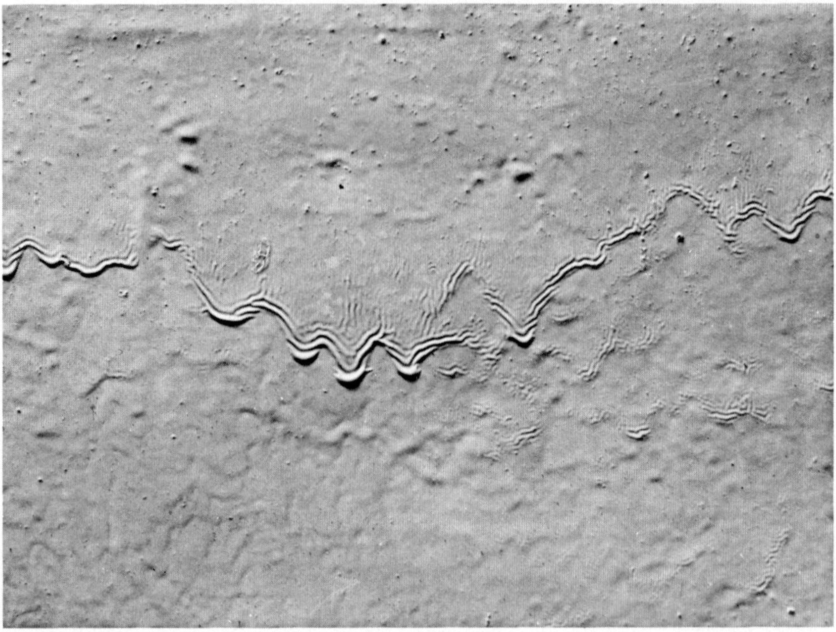

Fig. 3.1. Curtaining due to incorrect application, also showing wrinkling of thick film. (*By permission of the British Standards Institution.*)

Fig. 3.2. A run out of moulding. (*By permission of the British Standards Institution.*)

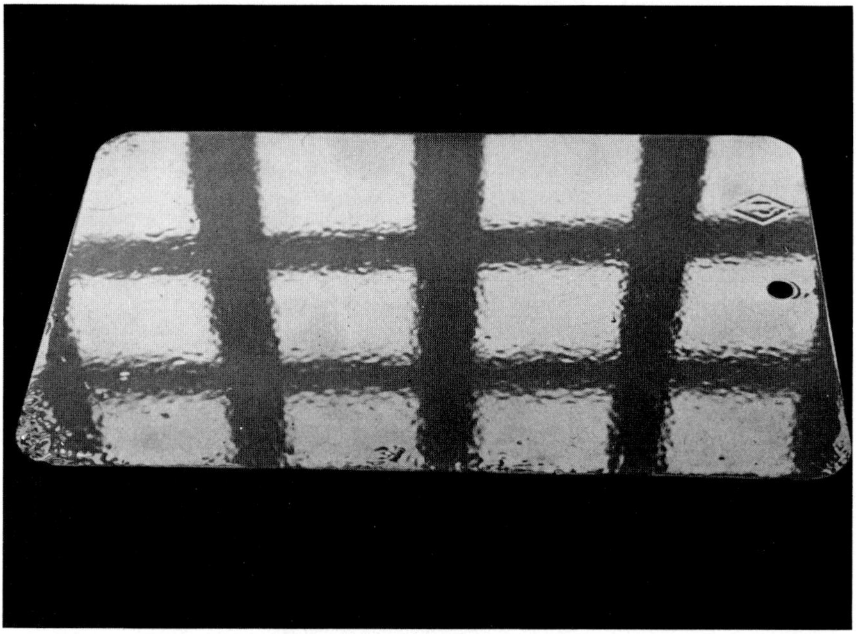

Fig. 3.3. Orange peel on powder coating. (*By courtesy of BTP Tioxide Ltd.*)

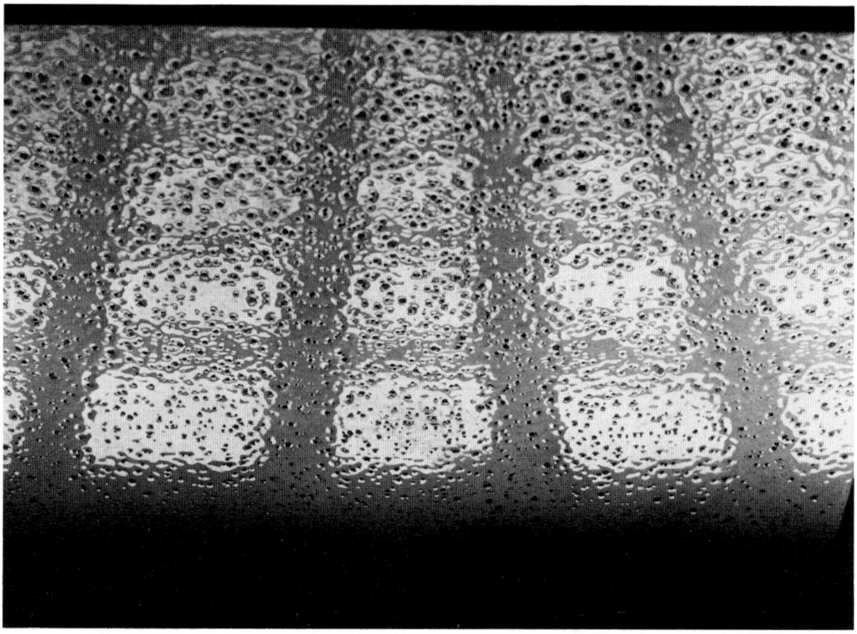

Fig. 3.4. No flow control agent in powder. (*By courtesy of BTP Tioxide Ltd.*)

Fig. 3.5. Different forms of pinholes in stoving enamel films. (Highlights are due to illumination.) (*Photo: M. Hess.*)

Fig. 3.6. Formation of blisters due to the contamination of the paint with condensed water from the pipeline of the spraying plant. (Gives the impression of pinholes in the film.) (*Photo: M. Hess.*)

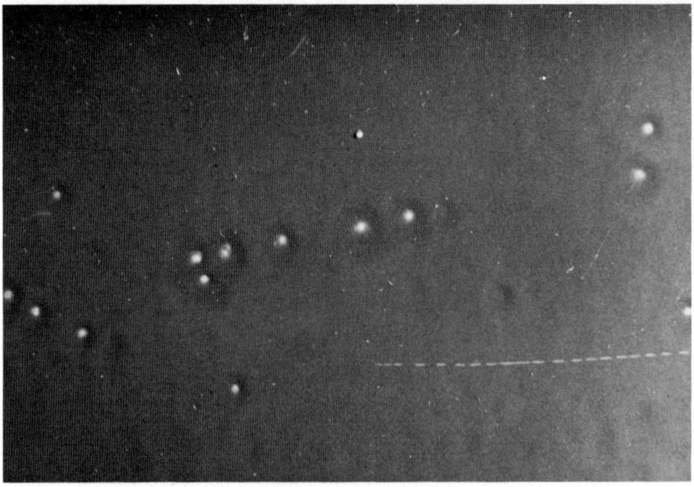

Fig. 3.7. Cissing. (*By courtesy of the Paint Research Association, Teddington.*)

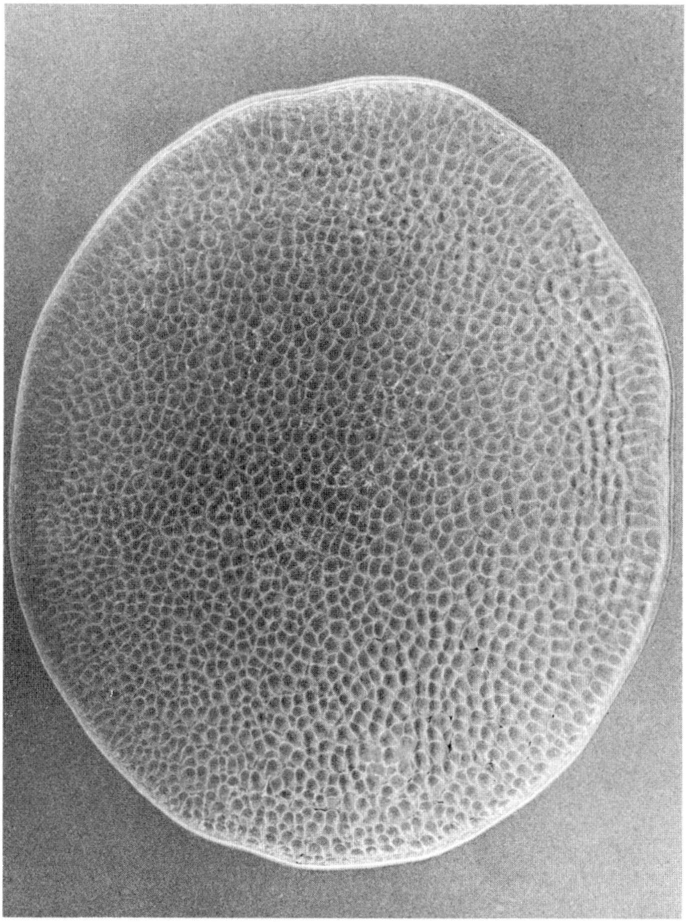

Fig. 4.1. Bénard cell formation. (*By courtesy of Laporte Industries Ltd.*)

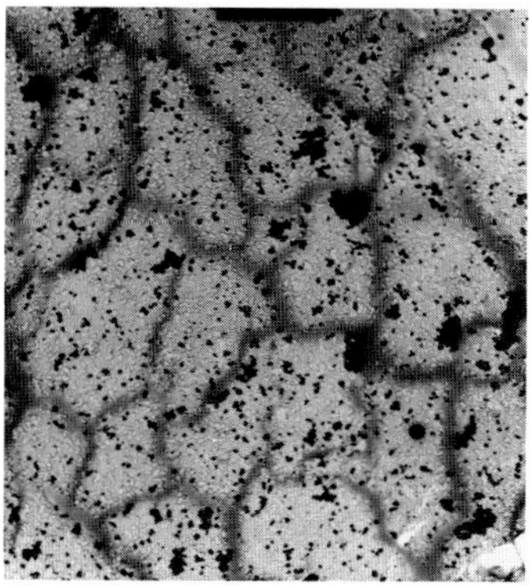

Fig. 4.2. Titanium dioxide in soya alkyd medium showing Bénard cell formation and flocculation. (*By courtesy of the Paint Research Association, Teddington.*)

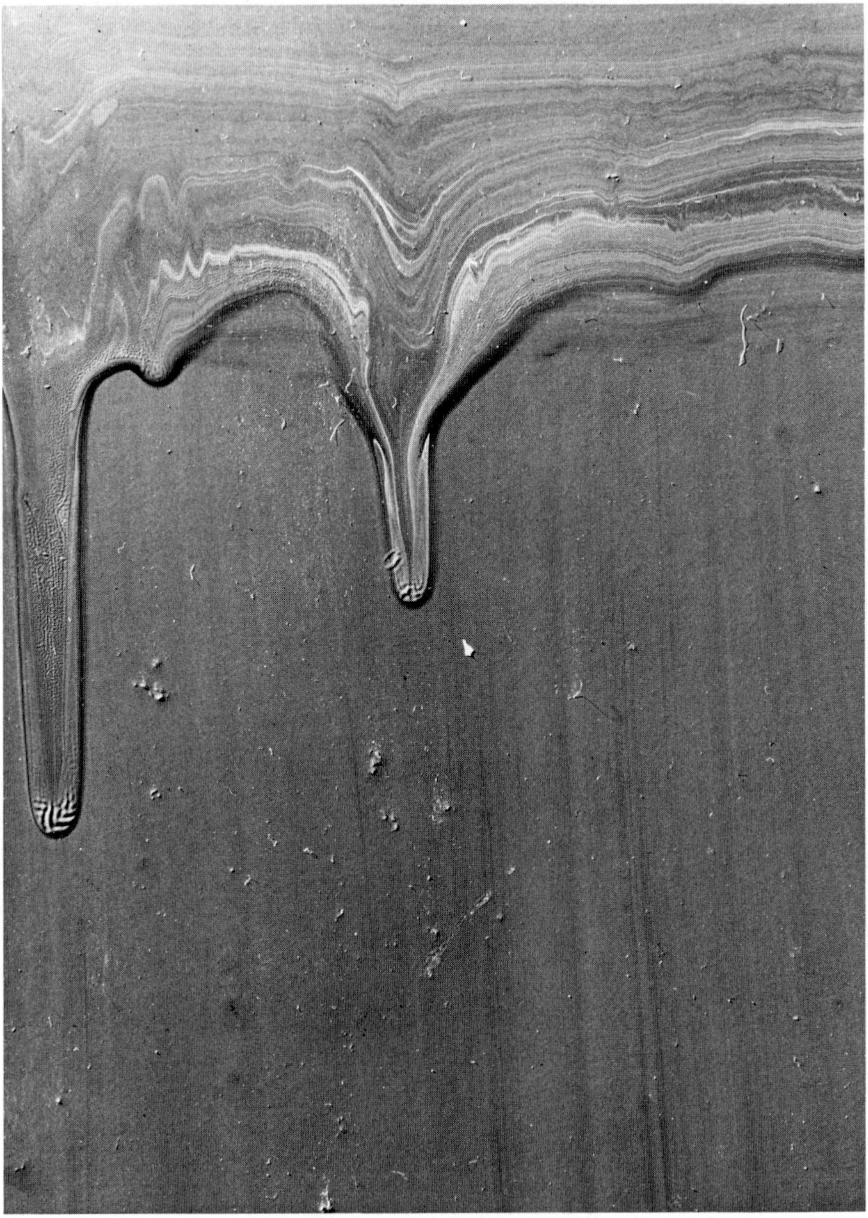

Fig. 4.3. Curtaining and running accompanied by flotation. (*By permission of the British Standards Institution.*)

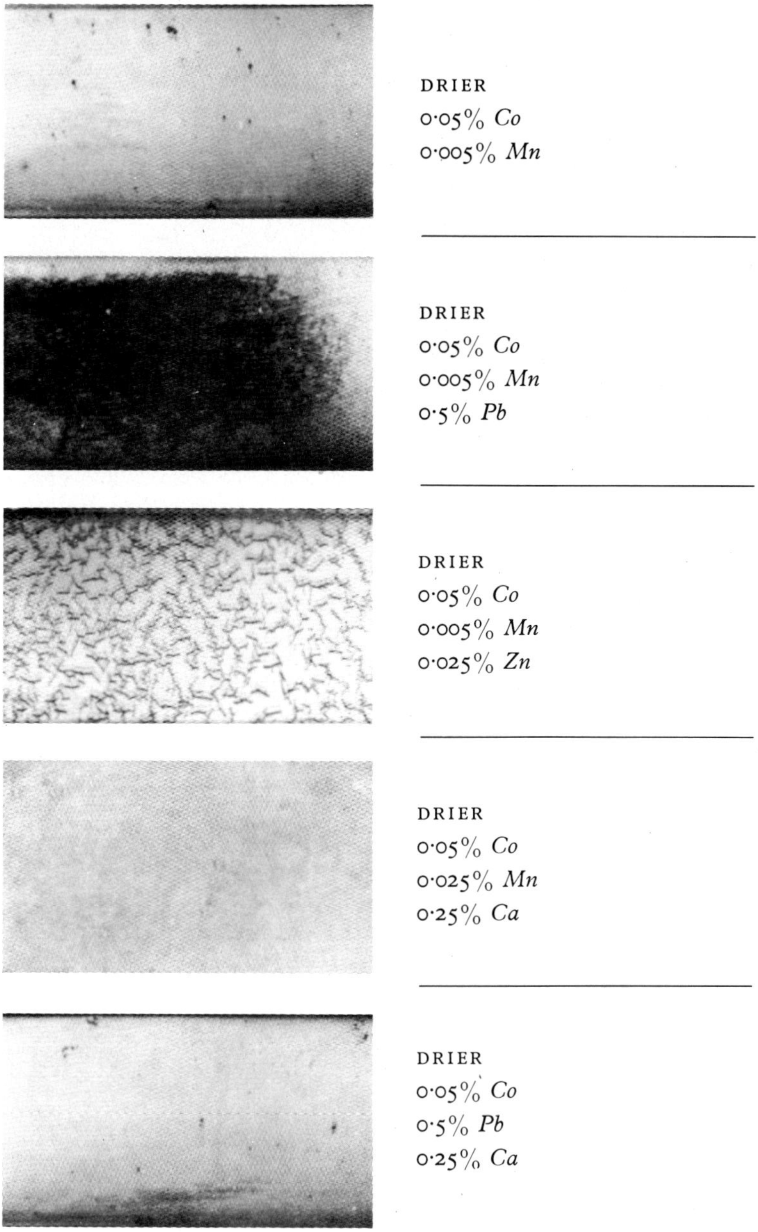

Fig. 4.4. Gas checking of air-dried wood oil alkyd medium, showing effect of various drier additions. (*By courtesy of the Paints Division, ICI Ltd.*)

Fig. 4.5. Lifting due to solvent attack (on a lighting column). (*By courtesy of BIE Anti-corrosion Ltd.*)

Fig. 4.6. Bleeding tendency of red pigment dyestuffs at various concentrations, showing superior performance of Pigment Red 112. (*From Geissler, G.* (*1973*), Defazet, *No. 10*, *p. 462.*)

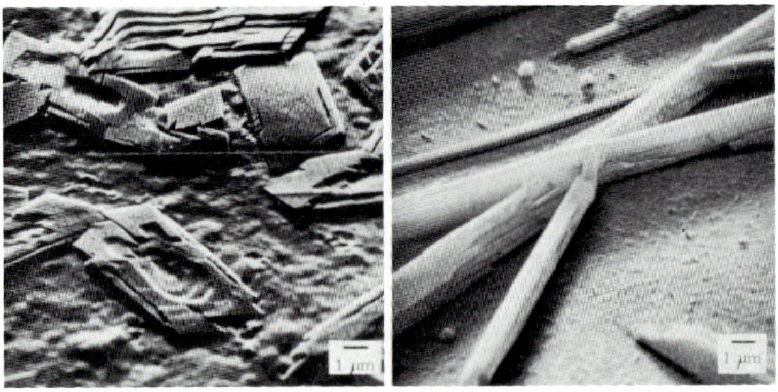

Fig. 4.7. Electronmicrographs of blooming of Hansa Yellow G in stoving finish, showing migration of pigment to surface. (*From Geissler, G.* (*1973*), Defazet, *No. 10*, *p. 462.*)

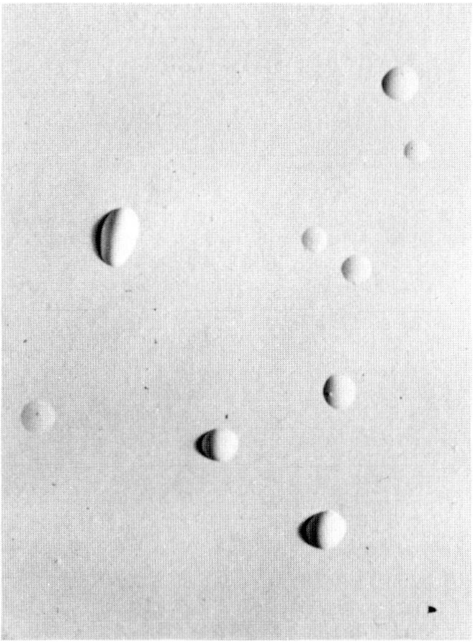

FEW

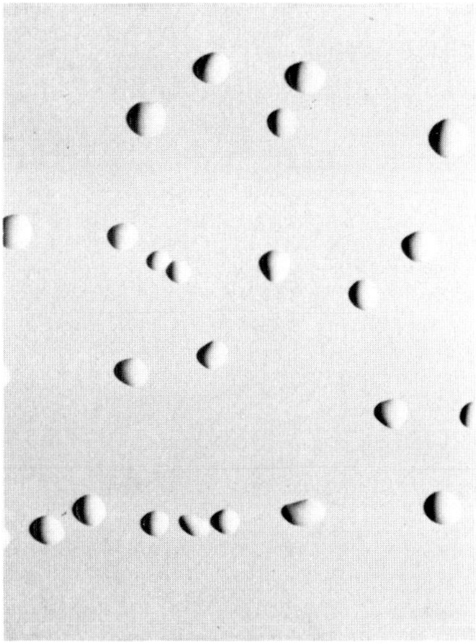

MEDIUM

BLISTER SIZE NO. 2

MEDIUM DENSE

DENSE

BLISTER SIZE NO. 2

Fig. 5.1. Blistering – standards for comparative evaluation. A.S.T.M. Designation = D.714–56. (*By courtesy of the American Society for Testing Materials.*)

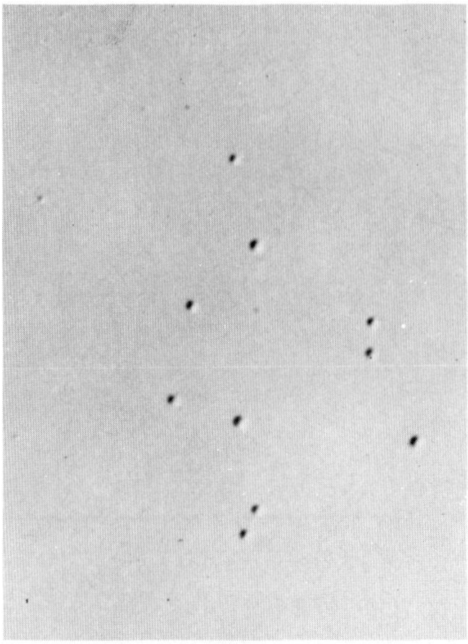

FEW

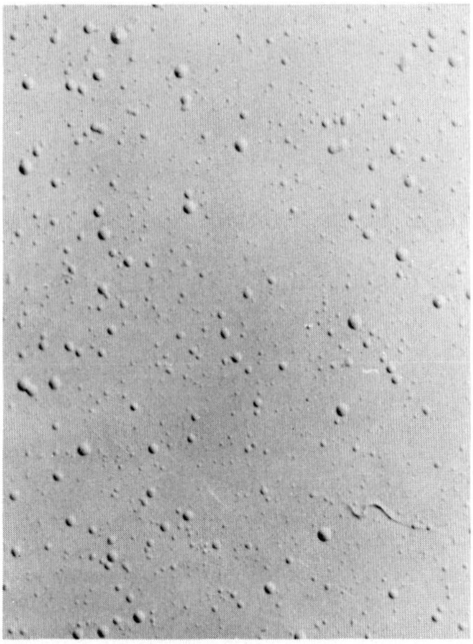

MEDIUM

BLISTER SIZE NO. 4

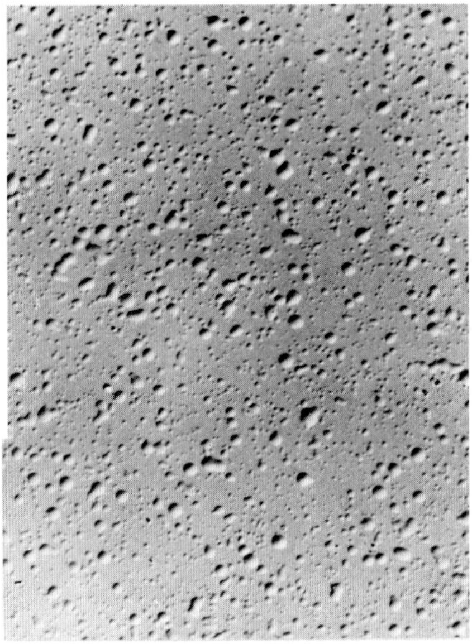

MEDIUM DENSE

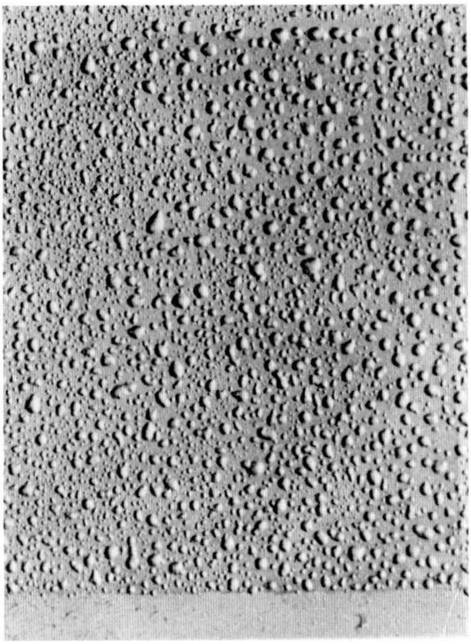

DENSE

BLISTER SIZE NO. 4

Fig. 5.1. A.S.T.M. Designation = D.714–56. (contd.) (*By courtesy of the American Society for Testing Materials.*)

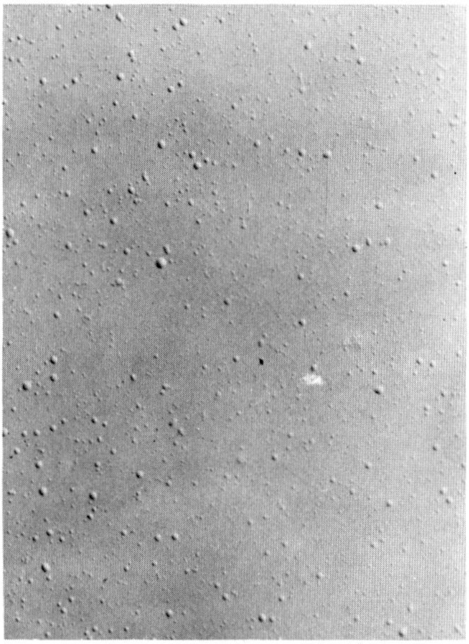

FEW

MEDIUM

BLISTER SIZE NO. 6

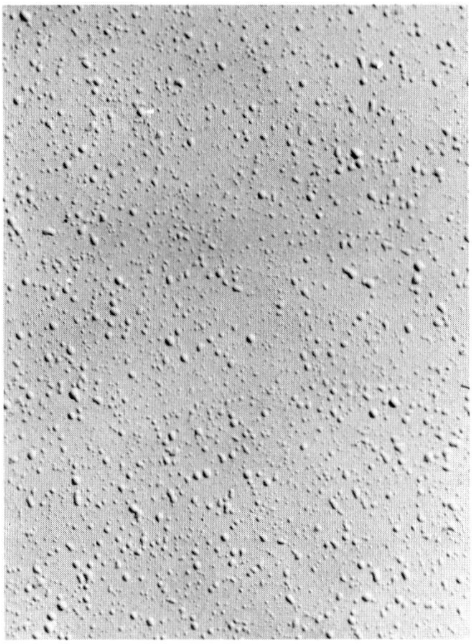

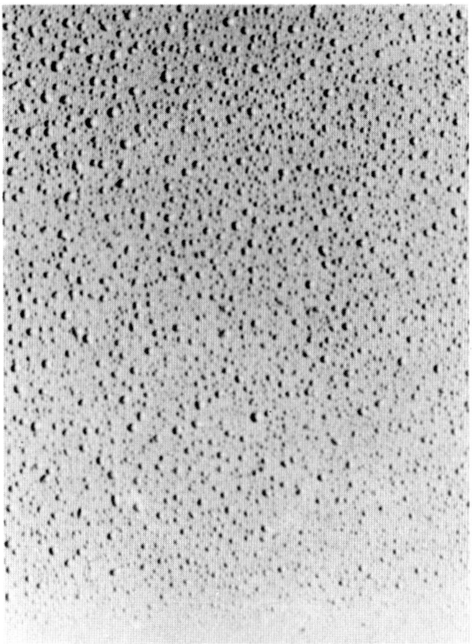

MEDIUM DENSE

DENSE

BLISTER SIZE NO. 6

Fig. 5.1. A.S.T.M. Designation = D.714–56. (contd.) (*By courtesy of the American Society for Testing Materials.*)

FEW

MEDIUM

BLISTER SIZE NO. 8

MEDIUM DENSE

DENSE

BLISTER SIZE NO. 8

Fig. 5.1. A.S.T.M. Designation = D.714–56. (contd.) (*By courtesy of the American Society for Testing Materials.*)

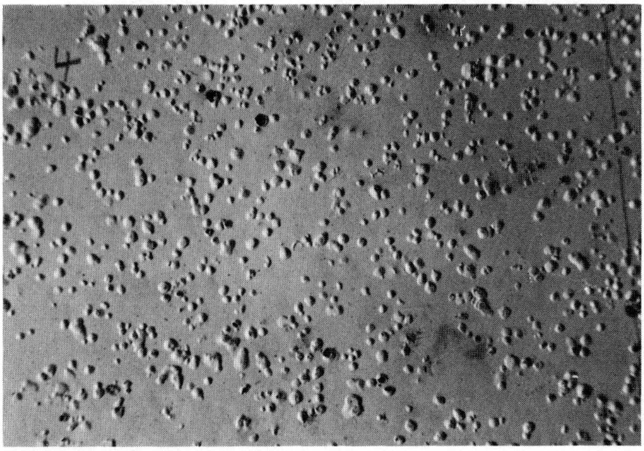

Fig. 5.2. Blistering (general). (*By courtesy of the Paint Research Association, Teddington.*)

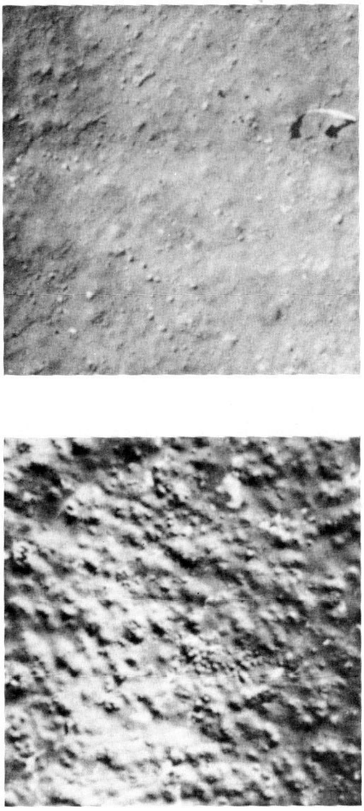

(*a*)

(*b*)

Fig. 5.3. (*a*) Electronmicrograph ($\times$ 4500) of a replica of the surface of a good gloss film. (*b*) Electronmicrograph ($\times$ 4500) of a replica of the surface of a poor gloss film. (*By courtesy of the Paint Research Association, Teddington.*)

(a)

(b)

Fig. 5.4. (a) Electronmicrograph (× 1150) of a section through a good gloss film. (b) Electronmicrograph (× 1150) of a section through a poor gloss film. (*By courtesy of the Paint Research Association, Teddington.*)

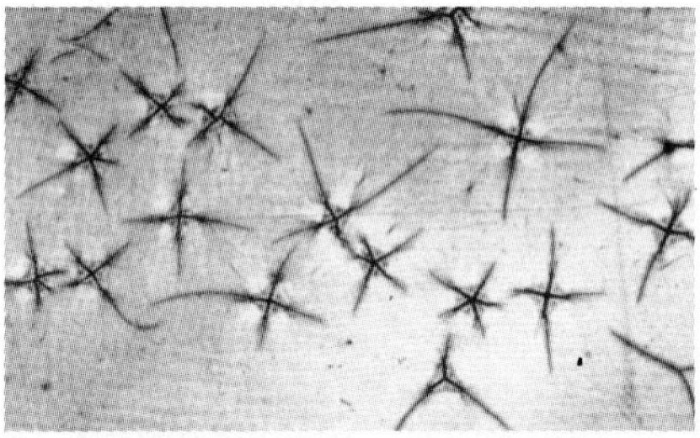

Fig. 5.5. Effects of insufficiently curing thermosetting base lacquer (× 65). Vacuumized aluminium film disrupted due to lacquer-absorbing solvents which penetrate through pinholes in the metal coating when the top lacquer is applied. (*By courtesy of L. Holland.*)

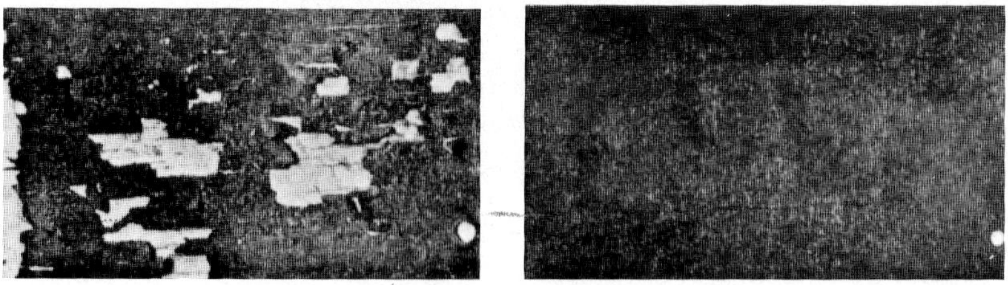

Fig. 6.1. Failure of paint on new galvanized iron. In the left-hand picture the paint flaked badly, owing to application without previous treatment. In the right-hand panel proper adhesion has been ensured by treatment of the surface before painting. In particular calcium plumbate priming paints adhere extremely well even on new galvanized iron. (*By courtesy of the Paint Research Association, Teddington.*)

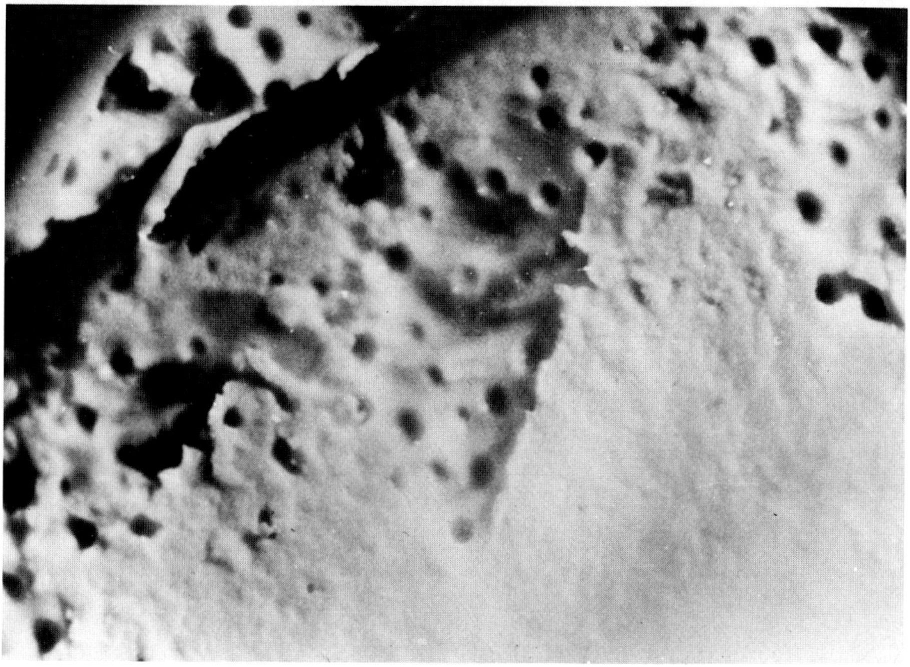

Fig. 7.2. Entrapment of air beneath the surface of a powder coating film. (*By courtesy of Scado N.V.*)

Fig. 7.3. Mould growth on fibre-filled exterior masonry paint. (*Crown Copyright: published by permission of the Building Research Establishment.*)

Fig. 7.4. Water entry through crack in corner of window resulting in disruption of paint on adjacent panel by fungus growth. (*By courtesy of the Paint Research Association, Teddington.*)

Fig. 7.5(*a*). Electron micrographs (× 5000) showing deposit of ammonium sulphate crystals on surface of varnish film. *Left*, pentaerythritol alkyd after three days in laboratory. *Right*, pentaerythritol alkyd after two months in laboratory. (*By courtesy of the Paint Research Association, Teddington.*)

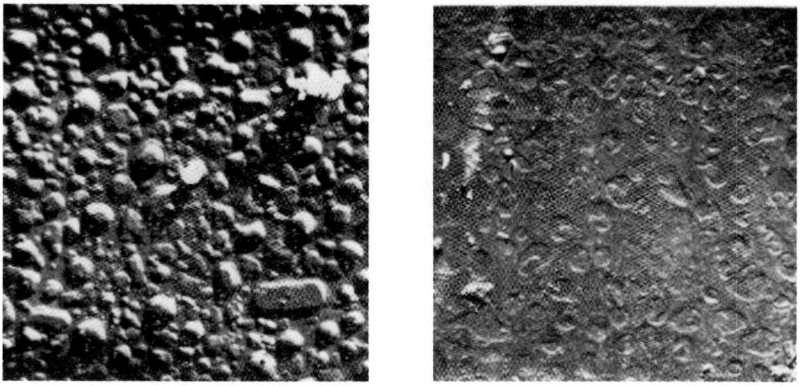

Fig. 7.5(*b*). Electron micrographs ($\times$ 5000) showing damage by crystals to surface of pentaerythritol alkyd film after two months in laboratory. *Left*, showing ammonium sulphate crystals; *right*, after removal. (*By courtesy of the Paint Research Association, Teddington.*)

Fig. 7.6. Shadowed replica ($\times$ 11 250) of bronzing film, containing Calcium 4B Toner. (*By courtesy of the Paint Research Association, Teddington.*)

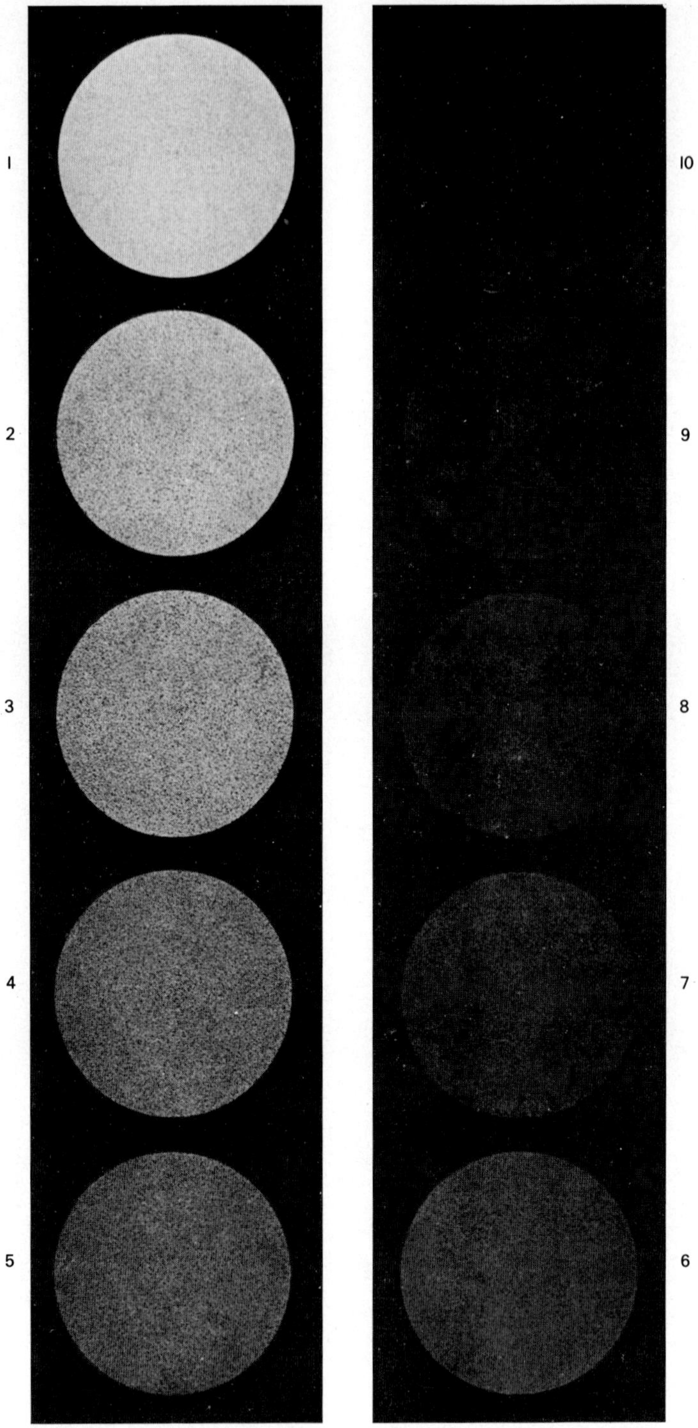

Fig. 7.7. Degrees of chalking (Kempf scale). (*By courtesy of the Paint Research Association, Teddington.*)

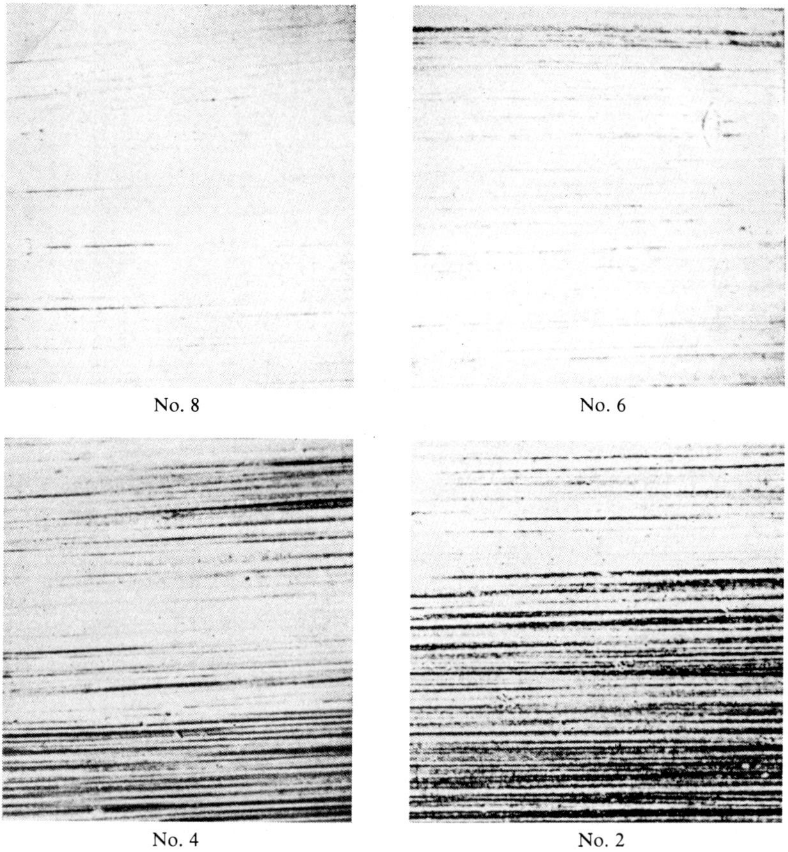

Fig. 7.8. Degrees of erosion. These standards are intended for comparative evaluation. (*By courtesy of the American Society for Testing Materials.*)

Fig. 7.9. Flaking of emulsion paint on plaster due to dampness. (*Crown Copyright: published by permission of the Building Research Establishment.*)

Fig. 7.10. Efflorescence. (*By permission of the British Standards Institution.*)

Fig. 7.11. Efflorescence on brickwork. (*Crown Copyright: published by permission of the Building Research Establishment.*)

Fig. 7.12. Flaking of emulsion paint caused by efflorescence on bricks. (*Crown Copyright: published by permission of the Building Research Establishment.*)

Fig. 7.13. Flaking of various masonry paints on brick walls. (*Crown Copyright: published by permission of the Building Research Establishment.*)

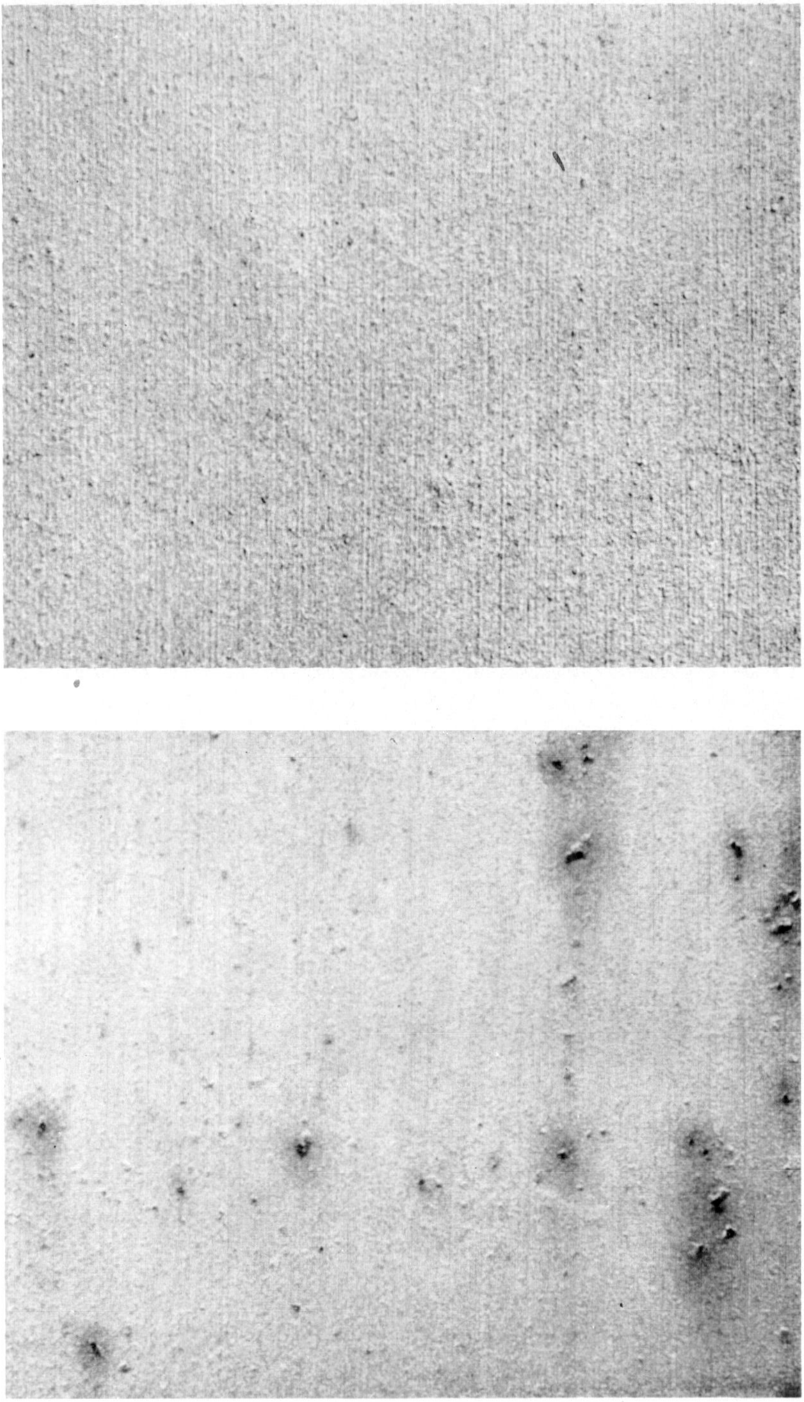

Re 0

⑩

Re 3

⑦

Fig. 7.14. Degrees of rusting: ReO = rust-free; Re9 = whole surface rusty. Degrees of paint protection efficiency: 10 = rust-free; 1 = whole surface rusty. (*By permission of the Standardiseringskommissionen i Sverige, Stockholm.*)

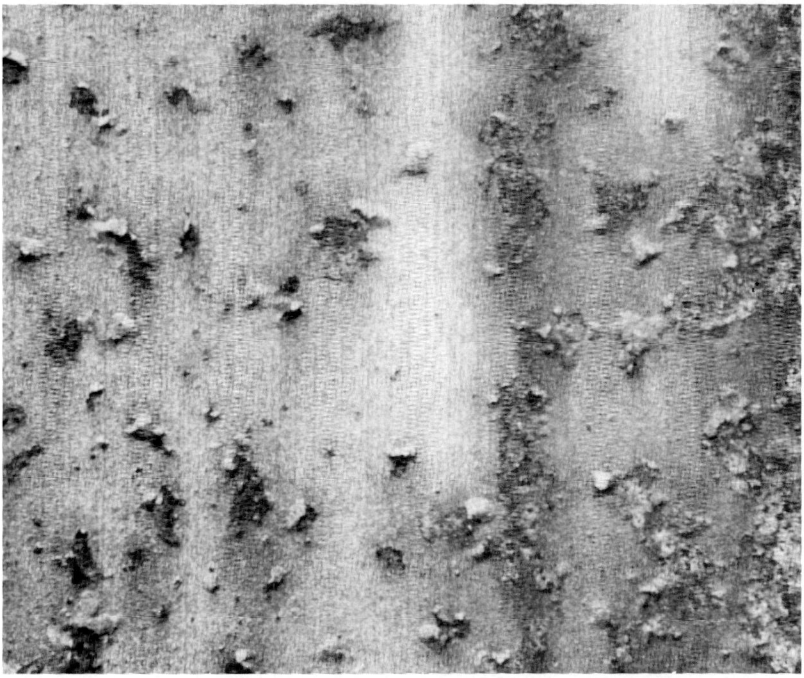

Re 6

④

Re 9

①

Fig. 7.14. (*contd.*)

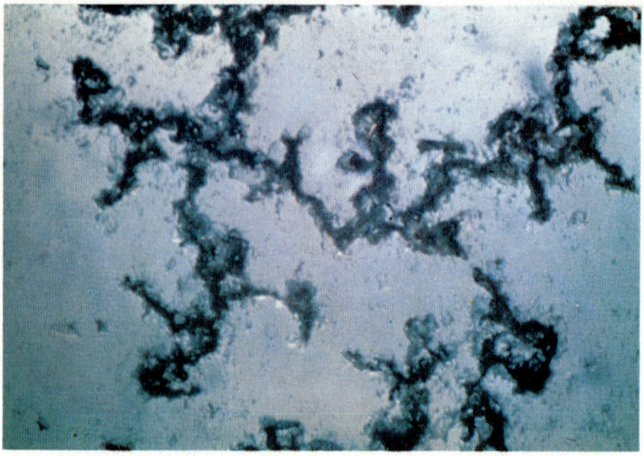

Fig. 7.15. Filiform corrosion. (*By courtesy of the Paint Research Association, Teddington.*)

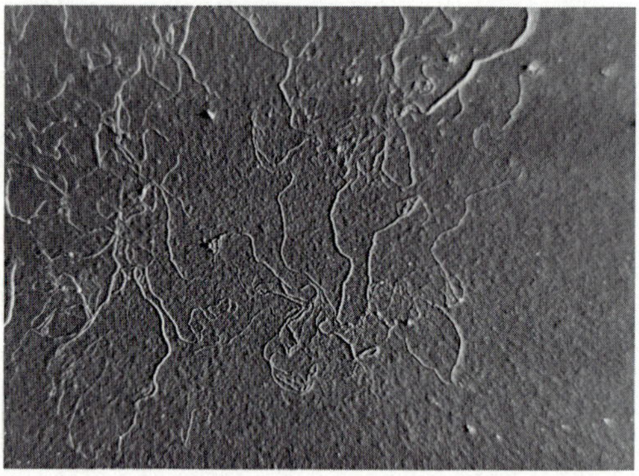

Fig. 7.16. Filiform corrosion. (*By courtesy of the Paint Research Association, Teddington.*)

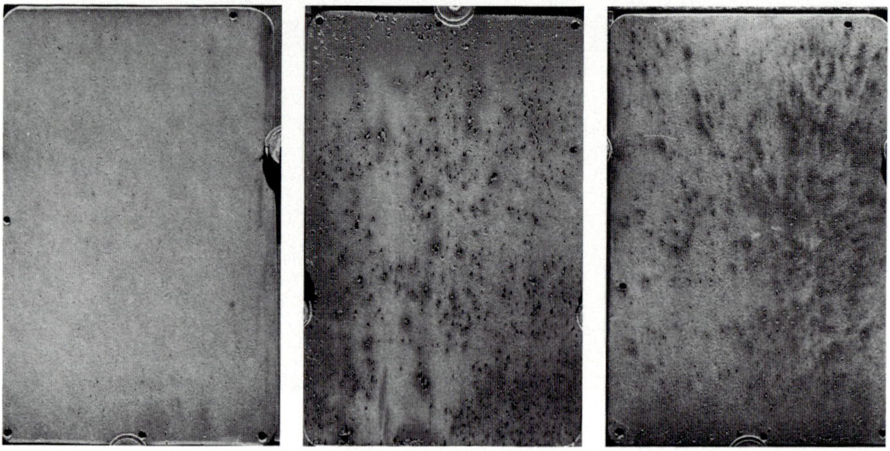

Fig. 7.17. Effect on a two-coat painting scheme of surface preparation. Grit-blasted (*left*), weathered and flame-cleaned (*centre*) and weathered and wire-brushed (*right*). Exposed at Birmingham and Battersea for 8.6 years. (*By courtesy of the British Iron and Steel Research Association.*)

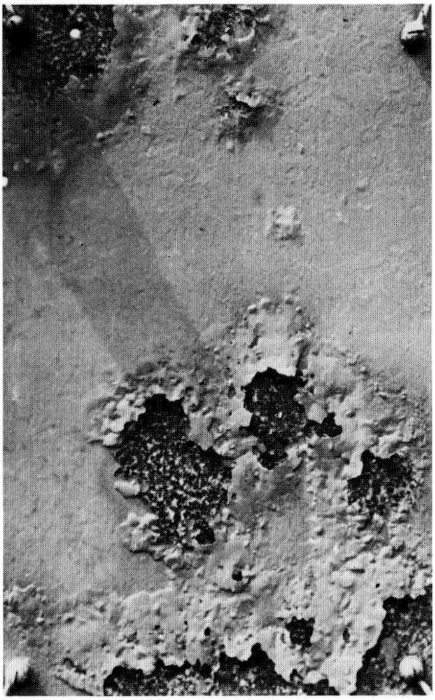

Fig. 7.18. Failure of paint over millscale. (*By courtesy of the Corrosion Advice Bureau, British Steel Corporation.*)

Fig. 7.19. Detail of toll bridge. It is difficult to paint a complicated assembly such as this effectively. (*From Fancutt, F., Hudson, J. C., Rudram, A. T. S. and Stanners, J. F.* (1968), Protective Painting of Structural Steel, *2nd Ed. Chapman and Hall, London.*)

Fig. 7.20. Corrosion due to painting bare metal. (*By courtesy of the Corrosion Advice Bureau, British Steel Corporation.*)

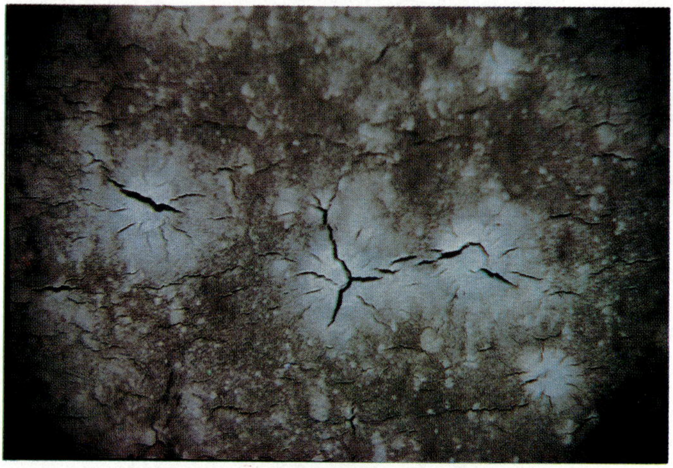

Fig. 7.21. Checking enhanced by corrosion. (*By courtesy of Sub-Tropical Testing Service Inc., Miami, Florida.*)

Fig. 7.22. Corniform corrosion. (*By courtesy of the Paint Research Association, Teddington.*)

No. 8

No. 6

No. 4

No. 2

Fig. 7.23. Degrees of flaking. A.S.T.M. Designation = D.772–47. These standards are intended for comparative evaluation. (*By courtesy of the American Society for Testing Materials.*)

Fig. 7.24. Paint peeling from window frame. (*By courtesy of the Paint Research Association, Teddington.*)

Fig. 7.25. Paint peeling from window frame. (*Crown Copyright: published by permission of the Building Research Establishment's Princes Risborough Laboratory.*)

Fig. 7.26. Blistering on plywood door due to water entry through unprotected edge. (*By courtesy of the Paint Research Association, Teddington.*)

Fig. 7.27. Blistering of oil-based paint on woodwork. (*Crown Copyright: published by permission of the Building Research Establishment.*)

Fig. 7.28. Blistering over knots in timber. (*Crown Copyright: published by permission of the Building Research Establishment's Princes Risborough Laboratory.*)

Fig. 7.29. Flaking following the grain of the timber. (*By courtesy of the Paint Research Association, Teddington.*)

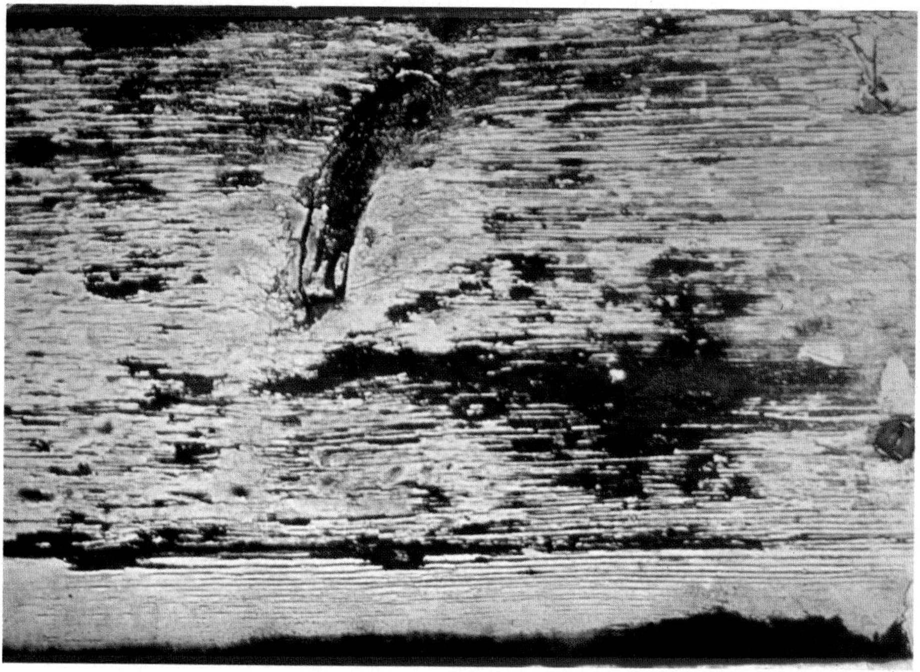

Fig. 7.30. Breakdown of badly weathered paint film following pattern of wood grain. (*By courtesy of the Paint Research Association, Teddington.*)

Fig. 7.31. Flaking of brittle paint. (*Crown Copyright: published by permission of the Building Research Establishment.*)

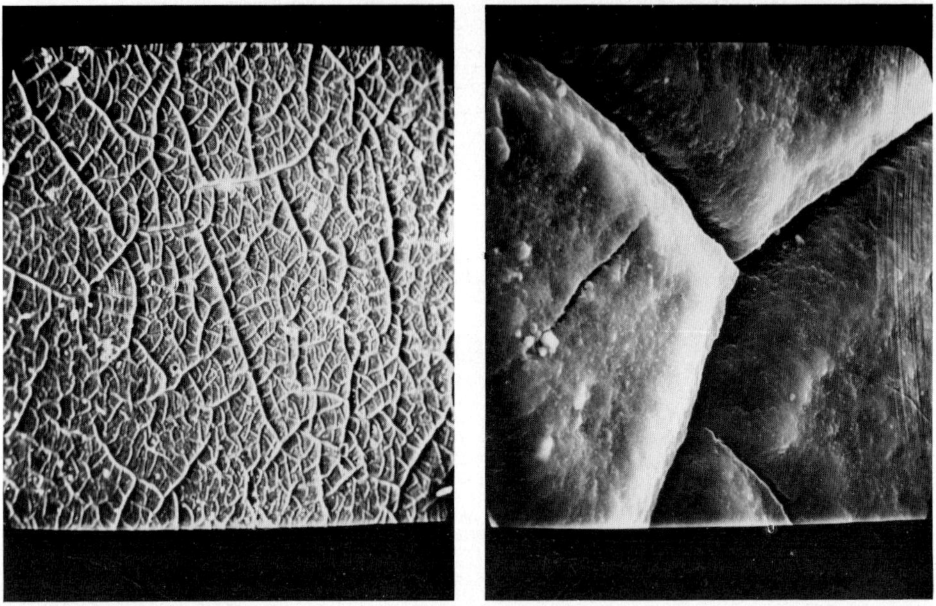

Fig. 7.32. Stereoscan photograph of crazing caused by weathering of inelastic gloss paint. (*Crown Copyright: published by permission of the Building Research Establishment.*)

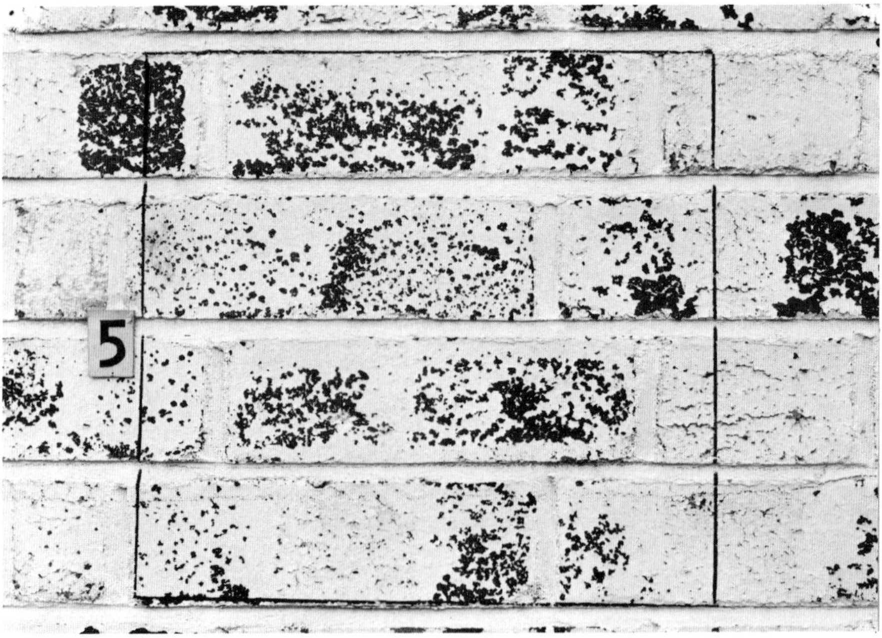

Fig. 7.33. Efflorescence and flaking on Fletton bricks. (*Crown Copyright: published by permission of the Building Research Establishment.*)

No. 8

No. 6

No. 4

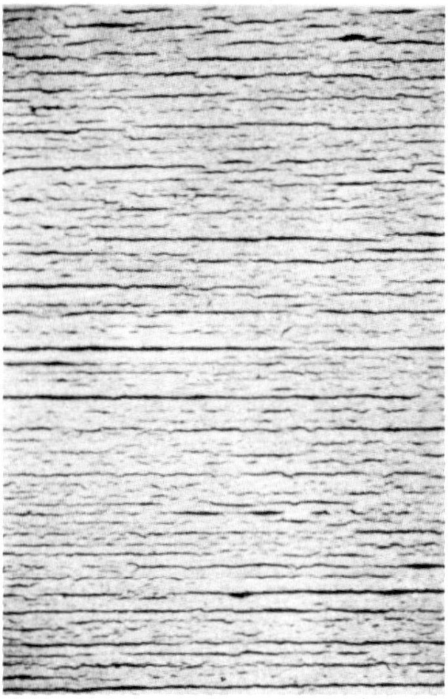

No. 2

Fig. 7.34. Degrees of cracking. These standards are intended for comparative evaluation. (*By courtesy of the American Society for Testing Materials.*)

Fig. 7.35. Cracking. (*By courtesy of Sub-Tropical Testing Service Inc., Miami, Florida.*)

Fig. 7.36. Cracking caused by insufficient elasticity of top coat. (*Crown Copyright: published by permission of the Building Research Establishment.*)

Fig. 7.37. Severe cracking and flaking. (*By courtesy of Sub-Tropical Testing Service Inc., Miami, Florida.*)

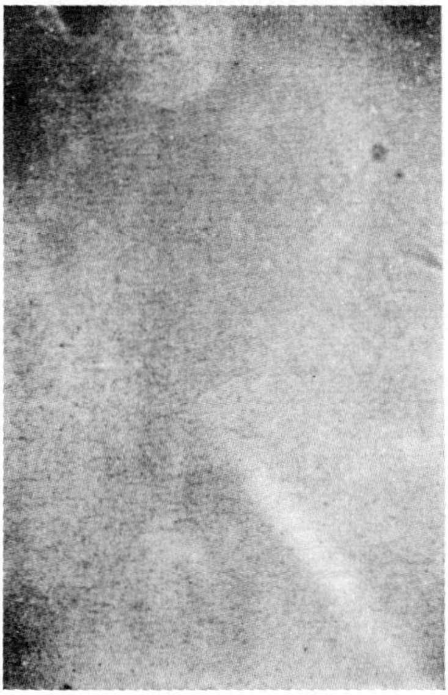

No. 8

No. 6

No. 4

No. 2

Fig. 7.38. Degrees of checking. These standards are intended for comparative evaluation. (*By courtesy of the American Society for Testing Materials.*)

Fig. 7.39. Crowsfoot checking. (*By courtesy of Sub-Tropical Testing Service Inc., Miami, Florida.*)

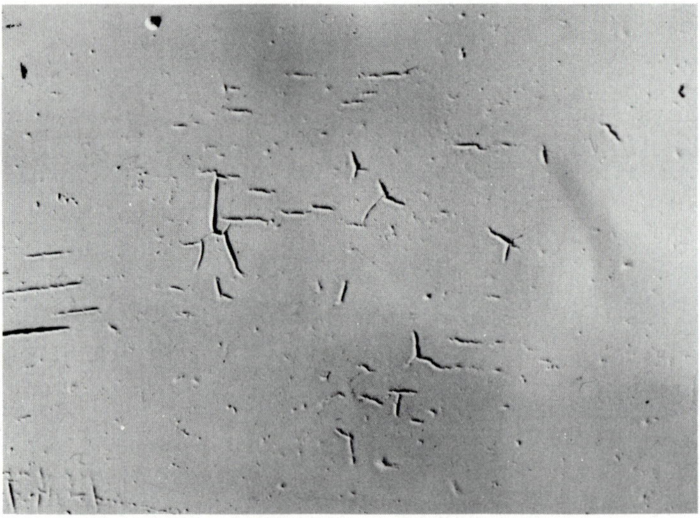

Fig. 7.40. Crowsfooting. (*By permission of the British Standards Institution.*)

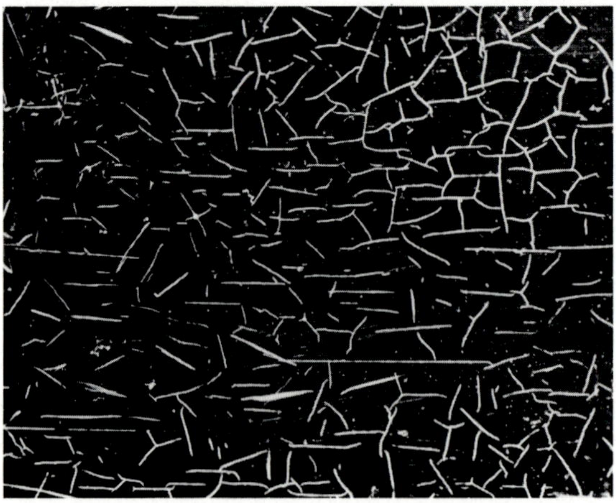

Fig. 7.41. Small crowsfoot checking. (*By courtesy of the Paints Division, ICI Ltd.*)

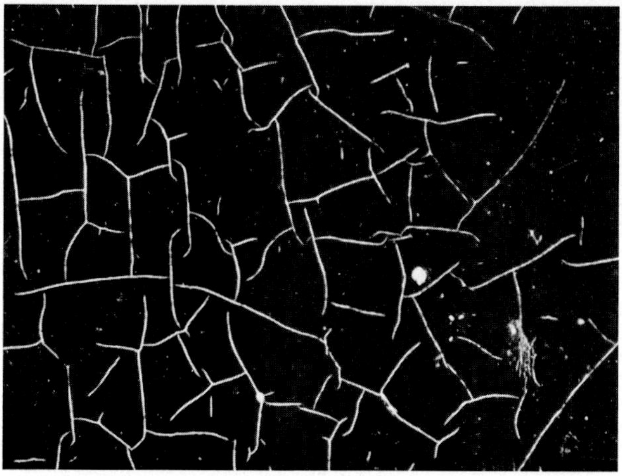

Fig. 7.42. Enlarged crowsfoot checking. (*By courtesy of the Paints Division, ICI Ltd.*)

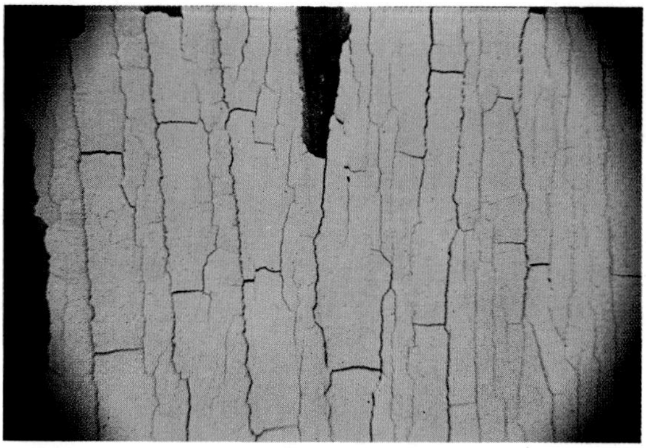

Fig. 7.43. Severe checking and flaking. (*By courtesy of Sub-Tropical Testing Service Inc., Miami, Florida.*)

Fig. 7.44. Alligatoring of short oil top coat over long oil primer. (*Photo: M. Hess.*)

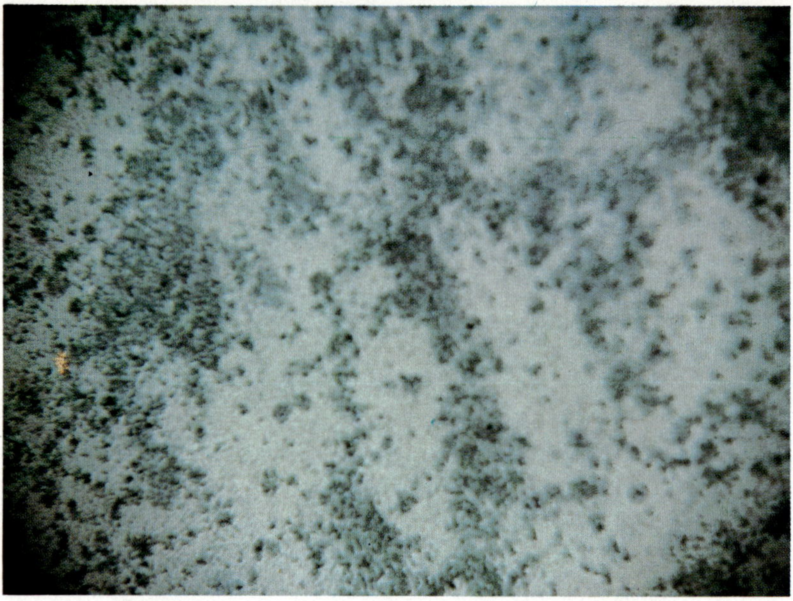

Fig. 7.45. Mildew growth. (*By courtesy of Sub-Tropical Testing Service Inc., Miami, Florida.*)

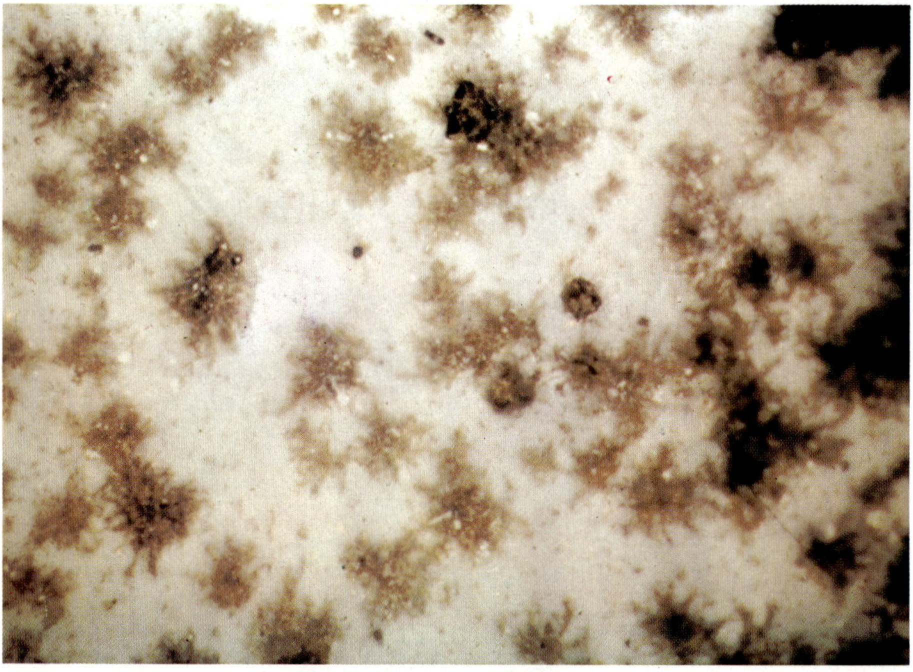

Fig. 7.46. Fungus growth on paint flake from brewery. (*By courtesy of the Paint Research Association, Teddington.*)

Fig. 7.47. Mould growth on internal wall (West Indies). (*Crown Copyright: published by permission of the Building Research Establishment.*)

Fig. 7.48. Mould growth on external painted timber (West Indies). (*Crown Copyright: published by permission of the Building Research Establishment.*)

Fig. 7.49. 'Blue' stain breaking through paint film. (*Crown Copyright: published by permission of the Building Research Establishment's Princes Risborough Laboratory.*)

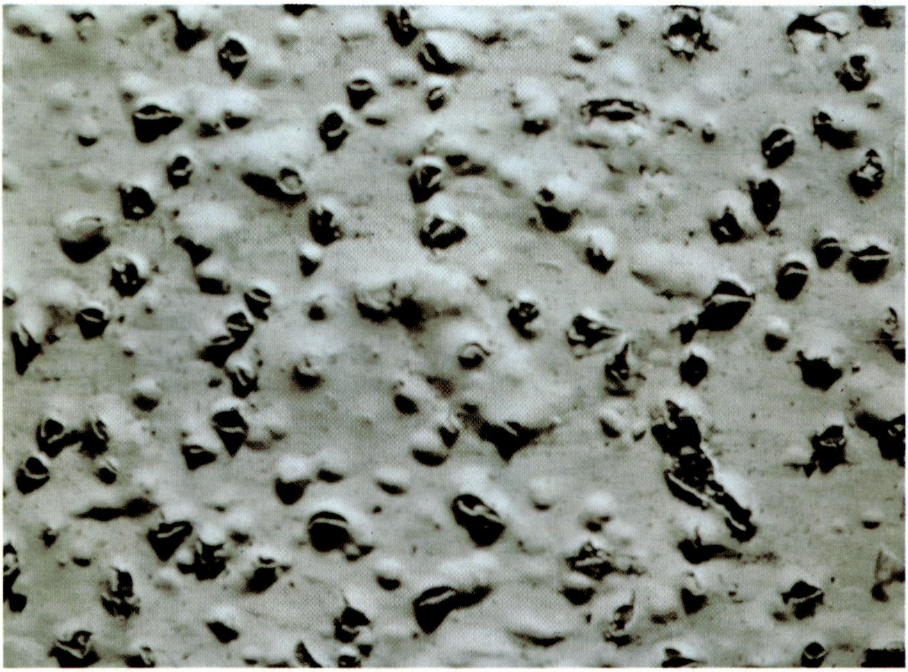

Fig. 7.50. Mould breaking through paint film. (*Crown Copyright: published by permission of the Building Research Establishment's Princes Risborough Laboratory.*)

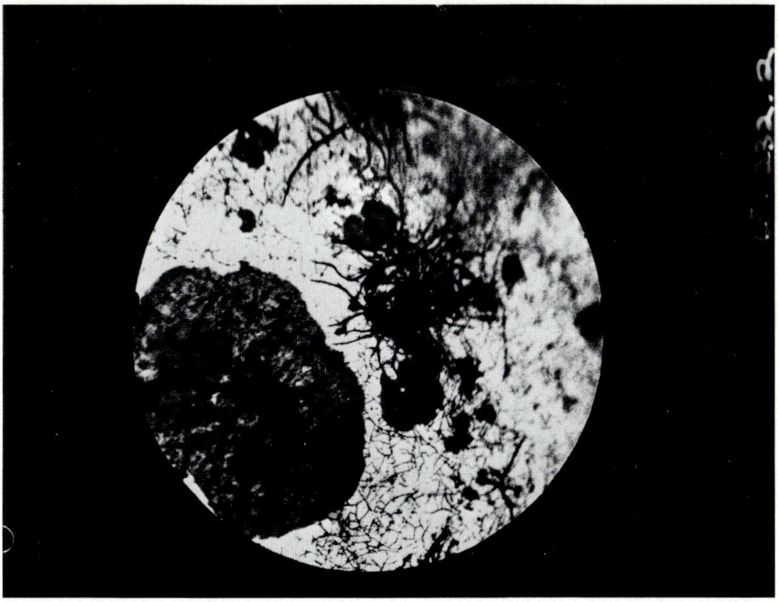

Fig. 7.51. Lichen, algae and mould mycelium (*Cladosporium herbarum*) on paint in the tropics. (*Crown Copyright: published by permission of the Building Research Establishment.*)

Fig. 7.52. Algal growth on a horizontal ledge on a house in Surrey. (*By courtesy of the Paint Research Association, Teddington.*)

Fig. 7.53. Rear surface of paint flake from joinery showing contamination by dirt and fungal material, formed on the primer during bad site storage. (*By courtesy of the Paint Research Association, Teddington.*)

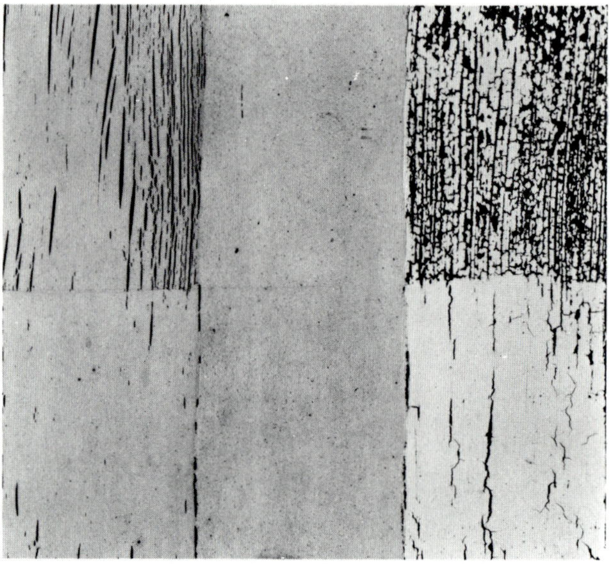

Fig. 7.54. Cracking and splitting of unsuitable paints over bitumen. (*Crown Copyright: published by permission of the Building Research Establishment.*)

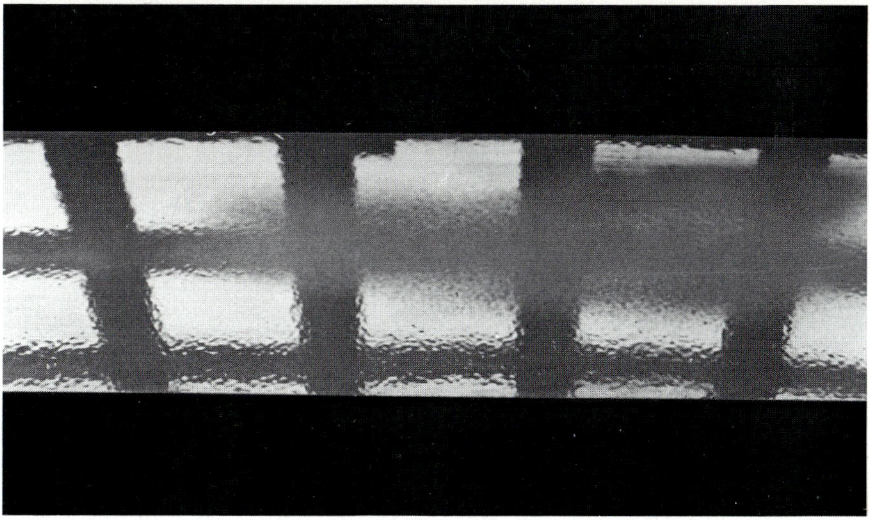

Fig. 7.55. Powder coating film attacked by solvent. (*By courtesy of BTP Tioxide Ltd.*)

Fig. 7.56. Breakdown of blast-primed steel after six months' exposure; the horizontal lines show rusting along areas damaged by contact with rollers as plates emerged from the blast-priming machine; staining on the upper parts is the result of uneven spraying due to incorrect adjustment of spray guns. (*By courtesy of Torpedo Paints Ltd.*)

Fig. 7.57. Damage to paintwork on the exterior of a ship caused by internal welding. (*By courtesy of Torpedo Paints Ltd.*)

Fig. 7.58. Barnacle fouling on raft test panels after ten months' immersion. The top left-hand panels are still effective against barnacles, but the others, painted with antifouling compositions with reduced cuprous oxide contents, have fouled to varying degrees. (*By courtesy of Torpedo Paints Ltd.*)

Fig. 7.59. Algal fouling on raft test panels after eight months' immersion. Tufts of the green weed *Enteromorpha* are attached mainly to edges of panels. (*By courtesy of Torpedo Paints Ltd.*)

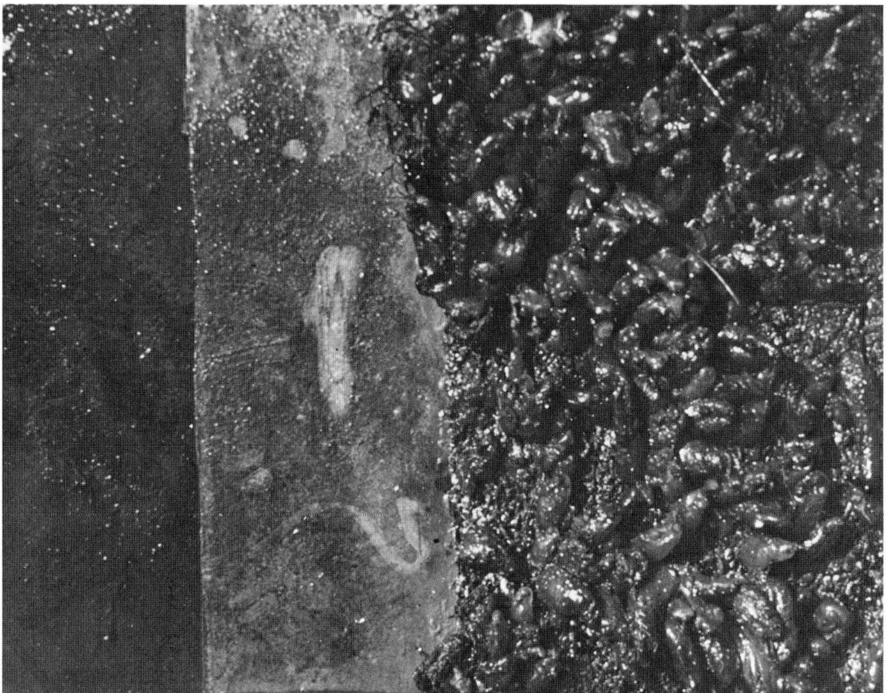

Fig. 7.60. Heavy fouling with a variety of organisms on a raft test panel after fourteen months' immersion. The left sections are still effective but those on the right, with low poison contents, have lost all antifouling properties. (*By courtesy of Torpedo Paints Ltd.*)

Fig. 7.61. Baltic redwood attacked by *Teredo* after only a few years of service. (*Crown Copyright: published by permission of the Building Research Establishment's Princes Risborough Laboratory.*)

roughening, the sand used for this purpose must be of a specially small and uniformly grained texture, and sandblasting should be effected only under low pressure. The resistance of unpretreated aluminium to corrosion is due to the presence of the above-mentioned fine layer of oxide, which is, however, dissolved by strong alkalis.

However, the oxide film of aluminium/magnesium alloys contains magnesia, particularly nearest to the surface, and the higher is the magnesium content, the more it contains. Such oxide films are hard, brittle and slightly alkaline. The higher the magnesia content, the more precautions are necessary to avoid defects of the coating on deformation and subsequent exposure to warm and humid atmospheres (R.D. Guminski and F.M.P. Meredith [7]).

As to the nitride skin, it is not absolutely necessary to remove it. Indeed, its presence can be of advantage under conditions of exposure to humidity or weathering (see the remarks on p. 239 as to tropics), but only if it is not covered with a coating. However, the removal of this skin is advisable where considerable mechanical stresses are to be expected.

The removal of the oxide skin by sandblasting, brushing, grinding (with emery) or etching [6] offers a very satisfactory anchorage to varnish and paint films and a corresponding improvement of quality and life of the films (some polished light metal plates are much smoother than corresponding iron or steel plates). Roughening of the metallic surfaces, on the other hand, always influences, to some extent, the mechanical strength and the resistance of the metal to fatigue, so that thin-walled light metal sections, and such as are designed to resist considerable stresses, should not be surface-roughened solely to improve the adhesion of paint films. Do not use steel grit in such cases, since particles of steel embedded in the metal surface can set up local electrolytic cells, giving rise to rapid corrosion. 'Wash' or 'etching' primers, which react chemically with the metal surface, promote improved adhesion and corrosion resistance.

Many chemical treatments have been proposed for aluminium, and probably the oldest of these is the alkaline chromate treatment known as M.B.V. (Modified Bauer-Vogel). This does not give as protective a film as anodizing but is quite useful as a base for paint providing conditions of exposure are not too severe. Later developments with acid fluoride/chromate processes have given a very much higher degree of paint adhesion and corrosion resistance, painted and unpainted. The last-mentioned processes are extremely rapid in action and therefore lend themselves to mass production use. In a spraying plant, for example, a treatment time of fifteen seconds is quite sufficient to form a good key for paint. Some dipping installations are also worked with this short processing time and are proving very successful indeed. If it were not for speedy processes of this nature it would be impossible to process strip continuously without an excessive length of plant; as it is, strip is regularly treated nowadays at speeds varying from 60–300 ft/min (0.3–1.5 m/s) [8].

The E.W. process is reported to form more dense, glassy layers and to afford a superior degree of protection than that provided by the M.B.V. process, 'especially in regard to coatings on copper-bearing-aluminium alloys. Owing to the density of the layers obtained by the E.W. process, they are less efficient as a basis for paint or varnish coatings than the M.B.V. layers'.

The E.W. process is, however, not applicable to aluminium-magnesium alloys which contain more than 5% of magnesium.

The Alrock process (Aluminium Corporation of America) which uses a hot solution of sodium carbonate and potassium dichromate, followed by sealing in a hot 5% solution of potassium dichromate is described as 'almost as good as the anodic process in protective qualities'.

There is, finally, the simple immersion in a boiling dilute solution of chromic plus phosphoric acid.

The anodizing processes for treating aluminium are, of course, well-known particularly as a base for dyes and for providing a protective finish. Anodized coatings can form an excellent base for paint but a word of warning is necessary as anodized films readily pick up oil and grease, and under these circumstances paint adhesion is very bad [8].

Three main processes for anodic oxidation of aluminium and its alloys are in use [9]. The chromic acid (as electrolyte) process was developed in Great Britain by Bengough and Stuart and modified in the U.S.A. The layers produced, although less hard, are more elastic and mechanically resistant than in other anodic processes. Alloys containing more than 5% copper, however, may develop pitting and final breakdown of the oxide layer. On aluminium and aluminium/magnesium alloys an opaque light grey film is produced, while alloys containing copper or silicon give a dark film, more or less intense in colour [9].

The Sheppard or the American Alumilite (Aluminium Company of America) process uses an aqueous solution of sulphuric acid and an alcohol as electrolyte. The layers produced by this type of process reproduce on aluminium and magnesium alloys the quality and colour of the original surface. The copper and silicon containing alloys yield a darker layer, but not so dark as that given by a chromic acid process, their hardness is only exceeded by the layers obtained by the oxalic acid process. Under extreme conditions the corrosion resistance of the chromic acid process layer is superior [10].

The Eloxal process, developed in Germany, mainly employs oxalic acid in its electrolyte. It is of some advantage that either D.C. or A.C. may be used. Its layers have a buff colour on pure aluminium, but may even turn brown on alloys rich in copper. They are the hardest of all three anodic processes, and as to their general properties they are similar to those produced by the two other anodic methods.

All three anodic processes can be applied to pure aluminium and to most of its alloys containing a maximum of 5% of alloying ingredients with the exceptions mentioned above. Silicon, however, may be contained up to 13%, zinc up to 8% [10] and magnesium up to 10%. To avoid corrosion, heavy metals must be absent.

Other special anodic oxidations are, e.g. the Alzak and Brytal processes, which produce reflecting surfaces and require metal polishing, and the Irotka process. By these various processes protective oxide layers of considerably greater thickness are formed than would develop under ordinary atmospheric conditions, being up to ten times the thickness of the natural oxide film. Anodizing provides good corrosion resistance by itself, and also affords an excellent base for painting [11].

For aluminium/magnesium alloys in particular, anodic pretreatment in sulphuric acid has been found to be efficient for producing lacquered strip. The anodic film of this type affords good key for lacquers and protection for the metal on storage [7].

Only short reference can be made here to some processes and it is recommended that the literature be consulted in full.

The oxide films produced by these processes must be sealed in order to give best results. The newly treated metals are in a very receptive state for the absorption of impurities, such as grease from fingerprints, or for grease-bearing dust which falls into unprotected washing vats, etc. Therefore a freshly formed film must be kept away from any likely source of contamination until it has been finally stabilized by some form of sealing. The absorptive power of the anodic film decreases with time, and hence it is highly desirable that the finishing processes of dyeing and sealing, etc., should take place as soon after anodizing as is possible [10].

The sealing is effected as follows:

(i) By organic methods, using paraffin or other waxes of high melting point, special resin solutions, lanoline solutions [10] varnishes or boiled linseed oil. Waxes or lanoline are not advisable if subsequent painting is to take place, as drying troubles, bleeding or blistering may then occur.

(ii) By inorganic methods, using water glass or heavy metal acetates (Siemens and Halske). Potassium dichromate solution is also recommended. Immersion in boiling water or a hot solution of a salt of a weak acid is suggested by R.H. Warring [12].

(iii) If one of the special paint primers is applied to an anodized surface the above sealing is omitted. Such paint forms a very tenacious bond with the newly oxidized metal.

Corrosion of the various painted light metals may be due to one of the following causes:

(a) By some organic degreasing (see above) and paint-removing process or by chemical surface treatment with alkalies [6] (as to magnesium metals see above) or acids, or substances containing or producing such alkalis or acids by decomposition (see below (d)). If, for instance, certain chlorinated hydrocarbons are used, aluminium or magnesium chloride may be formed which tends to expand beneath the varnish film. This kind of corrosion starts with the formation of tiny blisters which become more and more numerous, gradually grow in size and finally break open showing white salt crystals inside. Magnesium alloys corrode very easily [2] in the presence of chlorinated aliphatic solvents, e.g. methylene chloride or by free hydrochloric acid formed by decomposition by traces of water present as an impurity in the solvents applied Acid may also be formed by the hydrolysis of esters containing traces of water, e.g. in the case of ethyl acetate, free acetic acid is produced, and this has a very serious effect on magnesium.

A somewhat similar corrosion effect is possible if articles have been degreased are again touched with sweaty fingers before they are protected by pretreatment or coating.

It is necessary to wash alkaline or acid-treated material thoroughly by repeated rinsing with fresh water. If possible, some neutral cleaner should be employed. Under no circumstances must the parts still moist with chlorinated hydrocarbons, such as trichlorethylene, e.g. after a vapour bath, be exposed to light, especially to the rays of the sun (formation of aluminium chloride). The presence of moisture and oxygen accelerates the decomposition.

The time of exposure in such baths must be restricted to the minimum if the reaction of the organic medium is not completely neutral. It can be observed that baths which were neutral to begin with show an acid reaction after being in use some time.

(b) Greases and oils in contact with the light metals must be acid-free as well. Acid residues remaining on the metal, when the latter is being lacquered, may lower the adhesion of paint films as may do sweaty fingers prints (see above (a)).

(c) The coatings applied are not free of acid, e.g. those containing phenol or cold curing amino resin; alkyd lacquers, which contain acids or compounds which give off acids, as curing catalysts; or, as is occasionally possible, cellulose ester lacquers which contain acid, e.g. as a consequence of decomposition, see above (a). Thus metal salts are formed which may cause similar troubles to those described above, (a).

(d) Continuous aggressive action penetrating from the outside through the lacquer films to the magnesium or aluminium alloy bases, such as is caused by keeping articles in leather cases (tannic acid), or by wrapping them in traditional packing materials, i.e. felt, cork, etc., for a considerable time, especially in damp, tropical, marine air or saline conditions, depending on the alloy.

Aluminium alloys are more sensitive towards alkalis than towards acids. In the case of magnesium alloys it is just the reverse.

The suppliers of the paint should be informed exactly as to the special purpose for which it is intended in order to enable them to supply the correct material.

Where the varnished metal is in constant contact with leather, the latter should be as free from chlorine and acid as possible. It is necessary that such films should show a great resistance to the penetration of moisture in order that the formation of alkaline liquids at the interface may be avoided. Though the metal itself may not be affected by such conditions, the adhesion of a coating is naturally impaired.

Light metals intended for contact with sea water should not be painted with the usual ship's paints, etc., containing metallic copper or lead, or lead pigments. Very good results have, however, been obtained by the German Navy with synthetic resin varnishes containing zinc oxide.

Chromate pigmented-primers, temporary preservatives or jointing compounds are recommended.

An effective primer recommended for magnesium alloys is based on an alkyd varnish containing wood-oil type oils and zinc chromate amongst the pigments. Up to three additional finishing coats are given to ensure exceptional performance. It is interesting to note that experiments by R.L. Wray proved that a fourth coat extended the life of a film system which was exposed in an American coastal climate by a further two years.

Oiled paper is a suitable packing material.

(e) The behaviour of the different alloys under, say, a paint, varnish or lacquer film in presence of humidity, etc., is not the same by any means.

Magnesium alloys which are to be exposed to dampness or action of acids particularly require correct pretreatment and a paint system of pronounced impermeability to moisture, owing to the reactivity of magnesium [2]. The adhesion of films may be considerably impaired (formation of blisters and peeling), owing to the formation of a whitish efflorescence consisting of alkaline magnesium hydroxide and possibly other salts.

Metal-clean surfaces of magnesium alloys must be roughened by a chromate treatment or electrolytic oxidation which renders the metal more resistant to corrosion. These treatments must be effected after cutting, because corrosion first starts along the freshly cut edges. It is necessary after cutting to remove all burrs and to smooth the edges so as to remove all possible points of corrosive attack. It is recommended to repair slightly damaged chromate films and cut edges of chromated sheets with a 10% aqueous solution of selenious acid. (Beware of contact with skin.) Chromate treatment is also required after sandblasting or welding. Castings and sheets are usually supplied in a chromate-treated state, and this process needs to be repeated after mechanical working. Thus also a good key for the paint is obtainable. A second degreasing is, however, not required after the chromate treatment if the painting follows at once, unless the parts to be varnished are actually greasy or oily. (See sub-section (b) above.)

Most of the aluminium alloys which are free of copper and zinc are as corrosion resistant as pure aluminium. However, if one of the two above metals is present in the alloys these must be chemically or anodically oxidized whenever they are likely to be subjected to strong corrosive influence.

The number of coatings required for aluminium alloys may be up to four (above (d)). However, where suitable baking varnishes are employed, two coats frequently suffice. Stove-drying has been found very beneficial for numerous metals, with the exception of those alloys sensitive to elevated temperatures which should be coated with suitable alkyd or special varnishes or with aluminium bronze paint, especially used as a primer and dried either at comparatively low baking temperatures or in the air. It appears that red and white lead are not suitable particularly if continuously exposed to the action of sea water. If the above-mentioned coatings are used as primers, finishing can be carried out with similar varnishes or nitrocellulose lacquers, or one of the old-fashioned weather-resisting oil-base paints. In view of the fact that primers and top coats must be adapted to one another, it is necessary to procure them from the same manufacturer, who should be accurately informed as to the purpose for which these paint materials are intended.

For instance, oil-paint primers containing graphite, lead compounds and some other pigments can form local elements with the metal base, although quite suitable when used with ferrous metals. Aluminium plates marked with crayon containing lead chromate have been observed to develop deep pitting along the crayon marks when immersed in sea water, the remaining metal surface remaining bright.

The painting of aluminium is discussed in detail in British Standard Code of Practice C.P. 231, 1966.

Aluminium alloys which exhibit sufficient corrosion resistance even in the unvarnished state, should be provided with a suitable coat of paint if they are exposed to precipitates of ash and coal, etc. [13].

Apply a primer of good adhesion, and impermeability to water, and a weather-resisting last coat (see above). Joints should always be painted on both sides in order to prevent the effect of moisture. It is claimed that corroded aluminium and aluminium alloys can be prepared for re-painting if treated with a chromate solution.

(f) In order to avoid corrosion, structures of mixed metals, including magnesium and aluminium alloys, and pure aluminium, require special insulation against the possibility of parts of the assembly acting as cathode or anode whenever 'moisture and saline matter, accessibility of air, relative position of the mixed parts, temperature, etc.,' have to be considered. 'The solution potential of metals and alloys depends upon the nature of the solution to which they are exposed. Differences of solution potential are probably the greatest sources of corrosion troubles, e.g. in the case of instrument work.' Regarding electro-deposited coatings proposed to act as insulators, thus avoiding contact with light metals, reference must be made to the relevant literature [14].

Insulation is required in co-structure with:

(i) Other metals, especially heavy metals such as steel, iron, copper [4] brass [4] and other copper alloys, and even with other light metals. In the case of the materials containing magnesium this insulation can be dispensed with only if jointed to pure aluminium and aluminium-magnesium alloys. In the case of other light metals this is permissible only in contact with metals containing manganese, zinc, chromium or cadmium or with platings of these metals (cadmium or zinc plated iron, stainless steel, etc.). Ordinary, untreated iron should not be used in direct contact with these light metals on account of the unfavourable influence of rust on aluminium.

Where the use of dissimilar metals is unavoidable the contacting surfaces should be properly insulated.

This is most readily achieved by the use of a chromate pigmented jointing compound (e.g. barium chromate and extender in a suitable medium).

(ii) Wood (exuding acid substances):

Insulation as described in (i).

(iii) Concrete, mortar, cement (forming alkaline setting substances).

Insulation as described in (i). Other insulations suggested are neutral roofing felt free from coal tar (without sand), ship's felt impregnated with bitumen, oiled paper, laminated phenol resin in sheets, and leather free from chlorine and acids.

(g) If stray currents have to be considered, insulation is also required.

(h) Baking of age-hardening copper-bearing light alloys can reduce their corrosion resistance.

(i) K. Broockmann has observed premature embrittlement of shellac, oil-base or alkyd (and other) resin media, particularly when unpigmented, on aluminium alloys. This is considered to be due to the strong reflective action of the metal surfaces, and to a catalytic effect of the aluminium oxide layer.

The addition of an ultra-violet absorber to a clear coating can reduce the risk of premature breakdown.

(j) The importance of using on light metals suitable primers, especially such containing zinc chromate pigments in appropriate media, e.g. alkyd resin or polyvinyl butyral resin solutions (etch primers), etc., together with chemical pretreatments, has been stressed in the Introduction.

A suitable pretreatment is of particular importance where the metal will only be protected by a clear lacquer and be subject to the effects of weather and strong light [13]. Severe filiform corrosion could make such metal parts soon unsightly. Therefore at least two coats of a non-discolouring lacquer are essential.

Where etch primers are applied under opaque paints certain precautions must be taken. If the primer consists of two components, they should be mixed immediately before use in the proportions indicated by the manufacturers, but not applied until about 15 min after mixing and according to the make, not after about 6 hours. They may otherwise be ineffective. The application to the precleaned metal should be as a thin coating. Heavy coats should be avoided. Adequate protection is afforded by a film barely sufficient to hide the surface of the metal. The dried film (air dried or stoved) is sensitive to moisture, and should be protected as quickly as possible by the next coat of paint. Magnesium alloys are only given adequate protection by etch primers if previously chromated by one of the recognized methods.

Acrylate or cellulose acetobutyrate-based clear lacquers are amongst those mostly chosen for the protection of bright metal.

See also above (i) Ultra-violet absorbers.

Notes

[1] Light metals have a specific gravity below 5, heavy metals above 5.

[2] Hanawalt, J.D., Nelson, C.E., and Peloubet, A. (1941), *Trans. Am. Inst. Mining Engrs.,* Inst. Metals Div., Tech. Pub. No. 1353; 26. Magnesium and its important alloys are resistant to salt water corrosion, but are sensitive to certain elements and combinations of elements. The existence of 'tolerance limits' or discontinuities in the rate of corrosion as a function of the percentage present of Al, Mn, Zn, Fe, Ni, Cu, Si and Pb singly and in all combinations, has been demonstrated.

The corrosion resistance against salt water of magnesium alloys that meet the purity requirements is 10–100 times greater than that of the common qualities.

[3] Ryan Aeronautical Corp. San Diego, 1958, **1**, p. 40.

[4] Tables are published by the manufacturers. See also the literature, e.g. *The A.D.A. Information Bulletins* of the Aluminium Development Association, London, or *Aluminium Taschenbuch.*

[5] As to the manifold compositions of the numerous light alloys, reference is made to the various specifications, e.g. D.T.D. and B.S. specifications for Great Britain, A.S.T.M.–*Service Characteristics of Light Metals and Alloys,* for the United States of America, and DIN, for Germany, and also to the appropriate literature, e.g. R. Irmann's, *Aluminiumguss in Sand und Kokile,* which deals particularly with cast aluminium alloys, and A.V. Zeerleder, *Technology of Light Metals.*

[6] The most common solution used for this purpose is a 10 to 20% caustic soda solution (of about 16 to 20° Baumé) applied at a temperature of about 50–80°C (120–175°F) for 1 to 2 min. (*Aluminium Taschenbuch.*) Solutions containing di- and tri-sodium phosphate, sodium metasilicate, sodium carbonate or soap are also in use.

[7] Guminski, R.D. and Meredith, F.M.P. (1961), *JOCCA*, **44**, 93.

[8] *Preparation of Metal for Painting,* I.C.I. Paints Division booklet.

[9] Burns, R.M. and Schuh, A.E., *Protective Coatings for Metals,* p. 320. Electrolytically produced oxide coatings on aluminium and aluminium and magnesium alloys are usually about 100 μm (a few ten-thousandths of an inch) in thickness.

[10] Bovey, E. (1943), *J. of Sc. Instr.,* **20**, 92.

[11] Harper, W.H. and Inglis, N.P. (1954), *Chem. and Ind.,* **44**, 1344.

[12] Warring, R.H. (1956), *Ind. Fin.,* **9**, 91, 323.

[13] A good or high degree of corrosion resistance to the action of atmosphere and sea water spray is exhibited by the following copper-free alloys:

Al-Si	aluminium-silicon wrought alloy, high silicon content.
Al-Si-Cast	aluminium-silicon cast alloy, high silicon content, no additions.
Al-Si-Mg-Cast	aluminium-silicon magnesium cast alloy, higher silicon, lower magnesium content.
Al-Mg-Si	aluminium-magnesium-silicon wrought alloy, low silicon and magnesium content, if heat treated.
Al-Mg-Si-Cast	aluminium-magnesium-silicon cast alloy, medium silicon and low magnesium content.
Al-Mn	aluminium-manganese wrought alloy, low manganese content.
Al-Si-Co-Mg-Cast	aluminium-silicon-cobalt-magnesium alloy.

Satisfactory corrosion resistance of magnesium alloys and of aluminium-copper alloys can only be obtained by applying suitable coatings.

Very high sea water-proofness is exhibited by the following groups of alloys:

Al-Mg	aluminium-magnesium wrought alloy.
Al-Mg-Cast	aluminium-magnesium-cast alloy.
Al-Mg-Mn	aluminium-magnesium-manganese wrought alloy.
Al-Mg-Mn-Cast	aluminium-magnesium-manganese cast alloy.

[14] The journal *Light Metals, Transactions,* Institute of Metal Finishing and the *Monthly Review of the American Electroplaters' Society* contain the most important contributions.

7.12 Flaking and peeling of air-drying coatings on wood

Flaking or *scaling* is defined as the lifting of paint from the underlying surface in the form of flakes or scales. Flaking is generally preceded by cracking or checking.

Peeling is the spontaneous removal in ribbons or sheets of a paint, varnish or lacquer film from a surface due to loss of adhesion.

The American Society for Testing Materials [1] has issued reference standards of 'Degrees of Flaking (Scaling)' for the above cases.

See Fig. 7.23 for the only type of flaking recognized as defined.

The factors significantly affecting the performance of exterior paints on wood are:

(i) The kind and quality of the wood, in particular its water content and its paint holding properties.
(ii) The design and construction of the building or structure which should be such as to avoid unnecessary exposure of the timber to the penetration of water.
(iii) The quality of the paint and its suitability for the environmental conditions at the site.
(iv) An adequate painting specification.
(v) Supervision to ensure proper application of the required number of coats by skilled operators.
(vi) The severity of the climate and the local conditions of exposure.

'A wide variation will be found in the paint-holding properties of woods. The woods which hold paint longest and suffer least when repainting is neglected are:

(i) Cedar, redwood and cypress.
(ii) Next in order are northern white pine, western white pine and sugar-pine.
(iii) The third group in order is Ponderosa pine, spruce and hemlock.
(iv) The last group consists of woods which have the poorest paint-holding properties: Douglas fir, western larch and southern yellow pine' [2].

In Great Britain the most commonly used wood species for joinery work are: Baltic redwood, including Swedish fir, Scots fir, red and yellow deal (pine). This is soft and resin-bearing. There is a considerable variation of grain throughout any one timber; a considerable number of small knots and also some resin pockets or pitch streaks exist.

British Columbian pine (Oregon and Californian pine or Douglas fir) varies according to district. This species is widely distributed over the western portions of Canada and the United States. 'It may be relatively free from knots, but bad pitch streaks are commonly found. Resin exudation is a most serious problem when this wood is used in joinery work. It has poor paint holding qualities. Special care is necessary with the surface treatment and the application of the priming coats' [3].

Other contaminants in woods from a lacquering point of view are hydroquinone, vanillin, catechol, phenolic compounds and some which contain aromatic amine groups [4]. Some of these occur in valuable woods and have a detrimental effect on the drying and the durability of drying oil-based varnishes and of polyester lacquers. These 'accessorial substances' occur mostly in quantities of 1 to 6%, but frequently up to 25% and more [5].

'Origin of wood, structural features of the timber as cut and extent of seasoning', etc., affect the success of painting. 'Poor quality wood may warp, shrink and change shape to such an extent as to be beyond control of any paint film, and scaling or flaking is almost inevitable' [3].

The varnish holding properties of many timbers have been investigated by K. Jones [6].

According to van Loon [7], the penetration of primer into the wood does not aid the performance of the paint and on softwoods quicker drying thixotropic primers are to be preferred; otherwise, owing to the variability of porosity, uneven penetration will occur, resulting in a coating of uneven thickness. On hardwoods having very wide pores a suitable filler must be used to prevent excessive penetration.

The Paint Research Association has developed a Porosimeter, operating on an air leakage principle, and originally designed for measuring the profile of blast cleaned steel (see Section 7.9). This could be a useful tool in the joinery industry for measuring porosity [8].

(a) The main damage to painted timber (Figs. 7.24 and 7.25) is due to water [9]. For economic reasons the most frequently used timber is the sapwood, which is the most porous part of wood and is capable of holding a lot of water. To achieve satisfactory performance of a paint system on timber, the moisture content should be less than 15 to 17%, and if chemically cured coatings are to be used, less than 12% [10]. If the water content exceeds 20 to 25%, fungal decay can be expected. The crucial point at which blistering of paint films followed by flaking takes place is reached when the fibre saturation point, which can vary according to the timber between 22 and 30%, is exceeded and liquid water accumulates below the film.

Because of the vulnerability to fungal attack it is now universally accepted that a treatment with a preservative is essential before painting. Care must be taken in selecting and applying the preservative to minimize its interference with the paint system which can manifest itself in poor drying, loss of gloss, staining and rivelling. Brush application of the preservative is usually unsatisfactory and vacuum or pressure application or dipping is to be preferred. Pentachlorophenol and tributyltin oxide solutions or a combination of both are currently favoured as least objectionable, but high-boiling solvents should be avoided. In any case sufficient time must be allowed before painting so that the solvent can evaporate. Troubles can occur if the wood has zones of especially high porosity where an excess of preservative can accumulate. Solutions of pentachlorophenol laurate in white spirit are reported to be relatively free from effects on paints [11]. Water-repellent preservatives can also be used provided they are free from wax which could cause cissing and inhibit drying.

The design factor cannot be overestimated. Areas of condensation should be avoided by providing good ventilation and reducing cold surfaces. Condensation moisture and rainwater should be directed away from timber, moisture from the ground must be prevented from rising into timber structures. It has been pointed out by van Loon [12] that high moisture content does not only arise if rain penetrates into wood, but also from migration of water vapour from the inside of a structure or building into painted timber exposed to the weather. The water vapour will tend to condense at the cold substrate surface below the film and lead to blistering (Figs. 7.26 and 7.27) and flaking [13].

Timber surfaces in contact with masonry, concrete or steel should be well coated. Iron compounds can create severe staining of timber; the colour of the stains can vary from orange-brown in permeable softwood to blue-black in oak or other hardwoods. All fastenings, fittings and nails should be made from non-ferrous metals. Primed joinery timber is often stored on site under very bad conditions for long periods. Covering of such components with waterproof sheeting is a minimum precaution and part of good site management. In painting of wood, film thickness plays an important role. It is suggested that two coats of paint, a total thickness of at least 50 μm, are necessary to control the uptake of moisture [14]. Three coats of oil-based paint provide an effective barrier.

Coatings should be applied only to perfectly clean dry wood, while the side opposite to the weathering side should also be impregnated, e.g. with boiled linseed oil in order to prevent the penetration of moisture. Timber structures should be primed as early as possible. Priming paints should be well worked into end grains, joints and other places where decay is most likely to occur or where moisture tends to penetrate. Joints should be made with waterproof glue and properly sealed. A viscous primer is helpful and also protects sharp edges better. Where priming paints have chalked or flaked, the surface should be rubbed down with glass paper and reprimed before the application of undercoating. Good results have been obtained with several layers of nitrocellulose or polyurethane lacquers, but they must be applied to perfectly dry wood. The moisture content can be checked with handy instruments. New wood should be painted as soon as possible to avoid the absorption of water.

(b) Peeling off takes place several months after painting with an oil-base coating over an old and smooth oil varnish or paint film, without previously roughening the old film (smooth on smooth gives no chance for keying).

Bad wetting is the reason here just as it is for cissing.

Remove old glossy paint films by the use of a paint remover; or by burning off, if still adherent enough, roughen with pumice stone or emery paper.

(c) Flaking or peeling of coatings on the wooden boarded undersides of eaves, soffits, etc., often occurs after repainting when the paint film is relatively new or in otherwise sound condition. These areas, protected from direct weathering, are generally subject to deposition of moisture, e.g. by dew formation which may remain on the surface for long periods due to the still air pocket created in these situations and the absence of direct sunlight to assist in drying off the film of moisture. The presence of the moisture on repainting or the absorption of moisture into the woodwork giving swelling of the wood, causes poor intercoat adhesion with ultimate peeling or flaking.

Ensure the surface is thoroughly dry when repainting surfaces of the kind as described. Give good mechanical key by rubbing down such surfaces with glass or emery paper.

The cracking and flaking of exterior oil-based house paint of the type widely used in the U.S.A., has been studied in the laboratory by Yamasaki [15]. Repeated wetting and radiation caused the breakdown, but the durability depended also on the condition of the surface before repainting; this should always be done before the original system has lost its integrity.

(d) Unusual circular cracking of exterior wood finishes was observed by Underhaug [16]. He was able to show that this was caused by wind-driven hailstones.

(e) Relatively low oil content in oil-base varnishes or paints (for instance, containing much ochre), or if the films were applied on to unoiled wood.

(f) Paste filler is applied directly on to wood and therefore does not adhere well.

Filling pastes, which do not contain much oil, should be applied only to primed wood.

(g) Faulty adhesion may be due to the application of a nitrocellulose lacquer of insufficient elasticity on wood.

Select another type of lacquer, or interpose a suitable intermediate layer possessing good adhesion, but which is not lifted by the solvents of the lacquer. Apply, e.g. a coat of polyurethane sealer (diluted) or of a sealer consisting of a strongly diluted phenolic resin or medium oil length alkyd varnish previous to, if necessary, the application of a pore filler.

(h) Paint material of insufficient weather resistance with low oil or high rosin content, has been used for exteriors by mistake.

(i) The old film consisted of paint or varnish containing some wax, rendering keying impossible. May also be due to incomplete removal of wax containing paint stripper.

The old waxy film should be removed.

(j) Wood which is comparatively rich in rosin may also induce peeling of paint and varnish films, especially at the sides of buildings facing the sun where cracks may precede the peeling. The areas above knots are obviously the most exposed (Fig. 7.28).

Interaction of resin acids with certain pigments may reduce the mechanical properties of a film [3]. As to the paint-holding properties of various wood species, see Introduction above.

Insulate or cut out the resinous wood. Visible exudation must be rubbed off and insulated, e.g. with knotting, before painting. An effective insulation is the application of aluminium primer on such wood on which the finishing coat should be applied only after it has completely dried.

Prolonged air seasoning or kiln drying of the wood hardens the resinous components and therefore minimizes the possibility of troubles.

Some other contaminants of woods cause stickiness in places, non-drying, darkening, cissing and reduced durability outdoors of oil-base and polyester coatings. Pigmented lacquers and varnishes suffer more than clears [5]. Unsuitable or excess stains can cause bad adhesion of polyester lacquers and prevent their hardening.

Sealing is essential. Some woods can be successfully sealed with nitrocellulose or polyurethane lacquers, which, however, detract somewhat from the natural wood colouring. But even this does not in all cases guarantee trouble-free results with polyester finishing lacquers.

Reddening of some woods occurs under acid hardening lacquers.

(k) Peeling often happens in cases where the pores of very coarsely porous material had been incompletely filled. Jordan and Gay draw attention to the fact that coarse grades of British Columbian pine (usually highly resinous) may induce scaling and flaking of paints [3]. The penetration and therefore also anchorage of a paint coating as shown on the example of pinewood, is greater on the softer spring wood than on the harder and denser summerwood, thus the junction of both woods forms critical areas for coatings. The differential swelling and shrinking of the cells of the two growths also results in the raising of the grain.

Different swelling and shrinking and modifications of the wood cells can also be caused by the pressure of the planing blade. It burnishes the relatively dense summerwood portions more markedly. As burnishing is due to the fact that broken cell walls partially seal the surface, the movement of moisture and the adhesion of paint may be affected [3].

Painting of the coarse timber with primers containing aluminium powder is recommended to minimize possible trouble.

(l) Priming films are sometimes deficient in adhesion, when sprayed.

Work primer in with a brush if spraying gives bad results. It is preferable to brush-apply priming paints on wood.

(m) A principal requirement in joinery primers is a high degree of elasticity to accommodate the dimensional changes caused by water absorption and sunshine. These can be quite considerable and impose a severe strain on the paint system. Long-oil paints usually meet this need. Short-oil paints do not and often crack and flake badly within months of application (Figs. 7.29–7.32).

In recent years extensive investigations into the painting of joinery timber have been carried out by the Paint Research Association and the Building Research Station. Valuable guide lines to correct procedures and the selection of the most suitable primers have emerged [11]:

(i) The traditional white lead/red lead oil-based primer, conforming to British Standard No. 2521, fulfils all requirements, but must now be excluded from many uses because of the toxicity hazard.
(ii) Thixotropic alkyd and thixotropic single-pack polyurethane primers with a standard low-lead pigmentation.
(iii) Polyamide cured two-pack epoxy primer pigmented to 50% with a standard pigment mixture and 1% colloidal silica.
(iv) Long oil oleoresinous primer pigmented with aluminium powder.
(v) 'High quality' acrylic emulsion primers. These must have a low pigment volume concentration and it has also been suggested that a low titanium dioxide content is desirable [17]. Naturally the opacity of such products is not good.

None of these primers meets the joinery manufacturers demand for a fast drying, hard primer, except possibly (v), but whilst drying properties of acrylics are acceptable, the films tend to be soft. This precludes stacking of mass-produced components soon after priming.

The performance of some commercially available fast drying primers can be substantially upgraded by prior treatment of the wood with a preservative composition of the type containing a resin and possibly a water repellent.

'Preservative' primers which are claimed combine a preservative action with the priming function have given promising results. Although there are many good quality 'lead-free' or 'low-lead' primers on the market only a few were found to match the performance of the 'pink' primer described in British Standard No. 2521 [18]. The investigation also showed that poor quality primers had severe adverse effects on good quality gloss finishes, while most commercial undercoats and gloss finishes performed well over a good primer [19]. The Princes Risborough Laboratory of the Building Research Establishment found that whilst poor weather conditions lengthened the drying time of most paints, the ultimate performance of paint systems on timber was little influenced by the climatic conditions during application [20].

Water repellent, wood preserving coloured stains have received much attention in recent years. It is claimed that they not only protect wood against wood-destroying organisms, but also cannot peel, flake or blister and do not obliterate the grain of the wood. Although they provide a watershedding finish they breathe and allow moisture regulation in the wood. However, it would seem that such stains cannot prevent dimensional changes of timber. They also need renewal at fairly frequent intervals (see also Section 7.19).

Notes

[1] A.S.T.M., Philadelphia, D772-44T, 1944.
[2] Walker, P.H. and Hickson, E.F. (1945), *Paint Manual,* U.S. Department of Commerce, Washington.
[3] Jordan, L.A. and Gay, P.J. (1938), *Priming Joinery Timber,* Bulletin No. 26, Paint Research Association, Teddington.

[4] Meyer, K. and Mebes, K. (1952), *Farbe Lack,* **58**, 215.
[5] Sandermann, W. and Schwarz, E. (1956), *Farbe Lack,* **62**, 134.
[6] Jones, K.L. (1966), *JOCCA,* **49**, 314.
[7] van Loon, J. (1966), *JOCCA,* **49**, 844.
[8] Holbrow, G.L. *et al.* (1972), *JOCCA,* **55**, 35.
[9] Mc Kelvie, A.N. and Sherwood, A.F. (1975), *JOCCA,* **58**, 345.
[10] Langrock, (1969), *Ind. Lack. Betrieb,* **37**, 255.
[11] Anon. (1973), *Pigment Resin Tech.,* No. 9, 19. A full summary of BRE work on priming of joinery timber.
[12] van Loon, J. (1966), *Fatipec Congress Book,* p. 58.
[13] Dooper, R. (1970), *JOCCA,* **53**, 653.
[14] Dooper, R. and Veenendaal, D.A.H. (1969), *Verfkroniek,* **42**, No. 6, 178.
[15] Yamasaki, R.S. (1972), *J. Paint Tech.,* **44**, No. 568, 68.
[16] Underhaug, A. (1973), *Färg och Lack,* **19**, No. 4, 65.
[17] Hopkins, R.P. *et al.* (1972), *J. Paint Tech.,* **44**, No. 568, 85.
[18] Whiteley, P. and Rothwell, G.W. (1975), *JOCCA,* **58**, 90.
[19] *BRE News,* No. 29, HMSO, 1974.
[20] *BRE News* No. 38, HMSO, 1976.

7.13 Flaking and peeling involving water-based coatings

Definitions of flaking and peeling have been given in Section 7.12. Water-based coatings used for decorative work are usually of the polymer emulsion type and recoating such surfaces generally presents little difficulty. With older properties, however, oil-bound distempers can be encountered and special care is necessary in view of the variable quality of these materials. Glue-bound ceiling whites can also be met. Before any redecoration is carried out the water sensitivity of the old surface should be ascertained by wetting and then examining a test area.

The faults which can arise and suggestions for their avoidance are set out below.

(a) Ceiling white is bound with glue only and before any form of redecoration is commenced the ceiling white must be washed off completely. When a film of emulsion paint dries it develops considerable tension and will adhere only if the substrate is hard and compact. Over water-softened materials such as ceiling white , emulsion paint will crack and peel soon after application. Similar results are obtained with solvent-based coatings applied over ceiling white. In recent years many ceilings have been coated with an emulsion paint of very high pigment volume concentration in order to obtain 'dry hiding'. 'Dry hiding' is the term used to describe the high obliteration or opacity shown by underbound films. It results from scattering of the incident light by the surfaces of the pigment or pigment and extender particles. Such coatings are porous and should be overcoated with similar material or treated with a suitable sealer before painting.

(b) Old distempered surfaces. The water sensitivity test should be carried out – a small area is wetted with water and the ease of removal ascertained:

(i) Readily removed. The material must be completely removed as for ceiling white.

(ii) Softened to some extent by water but not readily removed. Provided the dry film is firmly adherent it can be given a coat of a sealing solution. Such solutions are marked as 'distemper sealer' or 'sealer for distempered surfaces' and often consist of a lightly pigmented polymer emulsion. After sealing the surfaces in this way either polymer emulsion paint or flat wall paints can be applied.

(iii) Completely resistant to water. No difficulty should be encountered. After washing down, the surface can be coated with either emulsion paint or solvent-type flat wall paint.

(c) Highly porous surfaces absorb much of the binder from any type of paint leaving an overpigmented and possibly friable film on the surface. This will possess very poor adhesion and is likely to peel off, particularly if subjected to the stresses of further coats. Such surfaces must be sealed effectively with one or two coats of suitable sealer.

Moisture in such a surface will check the penetration but the film of moisture will prevent adequate wetting of the surface and adhesion will be poor. If painting cannot be postponed until the substrate is completely dry then a slightly porous film should be applied through which the moisture can evaporate.

(d) Great carelessness is sometimes shown in painting very dusty (e.g. caused by building material), crazed, loose or weathered-out bases, e.g. plaster or any other surface including old paint. Cracks in the base are immediately transferred to the film and set up conditions for flaking. A white-washed plaster is an impossible base for an emulsion paint, however firmly it may seem to adhere. Flakes of emulsion paint which would be coming off in ever-increasing numbers within a few months would show layers of the distemper or whitewash on their undersides. Undercoats consisting of old distempers or emulsion paint, or if underbound, can behave similarly.

The base must always be well cleaned and all loose plaster or old films removed mechanically or by hand before painting. Emulsion and other paint films can be removed by chemical paint removers or can be burned off with a blow lamp, etc. Old distemper can be removed effectively by means of a steam jet.

(e) Highly impervious or smooth surfaces. Plasters which have been trowelled up to a hard glossy surface and old gloss paints provide little key for emulsion paints. Subsequent condensation on the surface followed by moisture absorption will cause the film to become detached. Such surfaces must be rubbed down with a suitable grade of glass paper and under these conditions adhesion is usually satisfactory. Nevertheless, the use of emulsion paint over old gloss paint where heavy condensation can occur, as in kitchens or bathrooms, is not recommended. Not should they be used over old gloss paints outside.

(f) *Efflorescence* usually takes the form of a crystalline deposit on the surface of plaster, brickwork, etc. (Fig. 7.33). It is caused by water, containing soluble salts, coming to the surface and evaporating so that the salts are deposited. A porous paint film on the surface will allow the solution to pass through and the deposit forms on the surface. More often the paint is not sufficiently porous and the deposit forms beneath the film. The latter is then pushed up and, ultimately, broken. Cases have been known when the crystalline deposit (usually sodium sulphate) formed behind the plaster resulted in flaking of the plaster with paint film attached.

When efflorescence results from the drying out of a wall the latter should be brushed clean when thoroughly dry and it can then be coated with polymer emulsion paint. If the efflorescence persists, moisture is probably permeating through the wall and an attempt should be made to seal the wall from the outside. If this is not practicable any interior decorative coats should allow the salts to crystallize on the surface from which they can be brushed off. This is best achieved by using a highly pigmented emulsion paint with a porous film. An impervious coating will be disrupted by efflorescence forming beneath it.

(g) Exterior surfaces can be coated satisfactorily with emulsion paints (some exterior types contain texturing agents) provided the substrate is suitable. Surfaces such as cement render and brickwork can be coated but great care must be taken with surfaces coated with limewash or old cement paints. Limewash must be removed completely. Old cement paint is porous and should be removed as far as possible with a stiff brush. Complete removal is impossible on coated rough-cast, pebble dash, etc. and so any residues in pits and crevices should be bound back to the solid surface with a penetrating sealer. Suitable materials are marketed by manufacturers of exterior emulsions.

(h) Buildings exposed to salt-laden air on the coast tend to acquire a thin deposit of sodium chloride crystals on the surface. The surface of the wall might appear to be in good condition for painting but the salt will impair the adhesion of any applied coating and early failure by flaking will result. Before painting, the walls should be scrubbed down with fresh water and allowed to dry.

Note

[1] Whiteley, P., Russman, H.D. and Bishop, J.D. (1977), *JOCCA*, **60**, 142.

7.14 Cracking; alligatoring; checking; embrittlement

These phenomena do not necessarily indicate that anything was or is wrong if they are the natural consequences of the normal deterioration of the film as judged and expected from the purpose and type of the coating applied. The elasticity of oil-base films diminishes as oxidation, etc., proceeds, and is connected with the ageing and weathering.

Cracking is the type of breakdown shown in paint films as a break extending through to the substrate (Figs. 7.34 and 7.35). Where this is difficult to determine, the break should be called a crack if the underlying substrate is visible. Three types of cracking are recognized:

(i) *Irregular pattern type.* Cracking in which breaks in the film are of no definite pattern.
(ii) *Line type.* Breaks in the film are generally in parallel lines, either horizontal or vertical. They often follow the line of brush marks.
(iii) *Sigmoid type.* Breaks in the film form a pattern consisting of curves meeting and intersecting, usually on a relatively large scale.

Cracking is often the result of inadequate elasticity of the paint system on a dimensionally unstable substrate (Fig. 7.35). It differs from alligatoring or checking in that the cracks extend through the entire paint system to the substrate. Subsequently the coating may become detached from the edges of the cracks leading to flaking (Fig. 7.37).

The most common causes of cracking are the following:

(a) The use of short oil quick drying systems on steelwork exposed to the sun. Considerable expansion and contraction of the metal takes place and the paint should be sufficiently elastic to accommodate this movement. The cracking follows an irregular pattern.

(b) The use of short oil quick drying systems on exterior joinery which has not been properly sealed (particularly at the ends) against the ingress of moisture. Apart from the overall changes in volume of the timber on moisture absorption, differential expansion and contraction takes place at the junction of spring and summer cells (see Figs. 7.29 and 7.30). Line cracking can take place along these seams. It is the commonest form of cracking on wooden window sills and the lower frame members. The cause is failure to seal the ends of sills and also the ends of joinery before assembly.

Joinery is frequently primed at works with a quick drying primer which gives a non-elastic film. For interior work this may be adequate but it should not be relied upon for exterior exposure; in this case a further coat of a more elastic primer may be applied but, even so, the durability of the system is always suspect if the most suitable primer has not been used. Cracking of a brittle primer may not be evident immediately but it will be transferred to the following coats as these oxidize and become more brittle.

Painting of wood must not be started before it is seasoned.

(c) The ideal painting condition is for the wood to have a moisture content in equilibrium (depending on humidity and temperature of air) with the average condition likely to be met in service. Paint coats exercise a controlling influence and reduce the amplitude of distortion and movement of wood due to moisture. The maximum amount of moisture before painting should not surpass 15 to 17%. Well-seasoned wood will stand ordinary changes in weather conditions without warping but not prolonged abnormal conditions.

In the case of fissures in wood coatings owing to swelling and shrinking of the softer cellular tissues interspaced between the hard grain, the pattern of the latter shows up first.

One of the most frequent causes of cracking of paints on wood is the fitting of uncoated wooden doors and windows into new buildings. The new walls always retain relatively large quantities of excess moisture for a considerable time.

(d) The application of an insufficiently weather-resistant paint material for exterior purposes soon (months or years after application) leads to checking and finally cracking, followed by loosening of the films in form of flakes which sometimes include the layer of filler or putty.

Scale-formation may also be connected with the presence of resinous occlusions in the wooden base.

The start of breakdown depends on the weather resistance of the paints or varnishes applied and on the severity of the variations in the atmospheric conditions and temperatures. After a period of exposure of only a few days the film develops checking which is visible to the naked eye at first as dullness, and confirmed only with the aid of a microscope. This is due to the destructive oxidation of the drying oil contained, especially when it is overtreated with driers. In linseed-oil varnishes this is far more noticeable than in those based on wood oil or linseed stand oil. Wood-oil varnish films may show cracks due to excessive content of rosin, without loss of lustre.

The weather resistance of quick air-drying coatings containing wood-oil or wood-oil-like oils, which include other conjugated oils, is frequently overestimated. They are very sensitive toward low temperatures and sun's rays. Gas-checking may be produced in films of faulty wood-oil type varnishes immediately on drying by the action of gases of combustion (nearby fires) or of draught.

Additions of short-oil varnishes, etc., to coatings which are to be used outdoors should only be made in small quantities, if at all.

(e) Good outdoor paints rarely crack due to working of the wooden base under normal conditions, because the wood maintains a more or less constant moisture content. This is, however, different in a very dry climate. There peeling, etc., of paint films due to intense shrinkage of the wood is quite frequent. The coefficient of expansion transverse to the wood fibre due to increased temperature is given by Hodgeman as 8 to 12 times of that parallel to it. If the finishing coats have been applied over an insufficiently elastic primer or filler which is unable to follow the movements of the wood, or if the application was made on to, perhaps, abnormally moist wood, then under such conditions, deterioration of outdoor coatings, due to the behaviour of the wood, will take place.

(f) Wood kept indoors and exposed to direct sunlight may cause the failure of the varnish by irregular cracking in a few years.

Cracks in the films of, e.g. oil-base films which are connected with the expansion and contraction of excessively dry or moist wood (which may gradually lead to chipping) proceed in the direction of the fibres. Such cracks possess the following characteristics:

(i) On wood which was too dry on painting and had subsequently stabilized its moisture content by absorption, they are smooth and flat.

Where the quality of the wood cannot be changed, it is advisable to provide for immediate and sufficient priming, especially where subsequent absorption of moisture is possible. The back of the wood should not be forgotten. Before painting, cracks and fissures in the wooden base have to be rubbed down and after priming to be filled with a hard-drying filler.

(ii) On insufficiently dry wood the edges of cracks are bent upward slightly, and can usually be felt to protrude.

It is, however, a fallacy to suppose that these protrusions can be satisfactorily fixed again by painting them over with a new coat. More often than not the films will lift at the edges even where these appeared to have good adhesion before the repainting. This is the same whether the surface is of wood or any other fibrous material.

It is impossible to obtain good protective films unless the wood is normally dry before being painted, although in the case of simple indoor articles the excess of moisture may possibly evaporate through the uncoated rear. Prepare surface for repainting by thorough rubbing down.

(iii) Many small fissures are produced in films of paint, etc., applied on insufficiently dry plywood, the edges of the cracks protrude upward and can partly be felt as burrs. While straight-cut veneers behave very much as normal wooden boards, the curled veneers tend more to fissuring and scaling, especially if the veneer layers had been glued before being sufficiently dry.

(g) If an old cracked paint layer is not removed, and the cracks are hidden by painting over, the irregularities can reappear at least in the form of grooves. This will happen only if the new film is, e.g. in the open, subjected to conditions where it can expand and contract. The form of the new cracks depends on the conditions which caused them on the old films.

Films applied on a smooth but checked plaster without the latter being pretreated may also show up the same checking, and it is no exception to find that the film lifts in many places, these originating from the fine cracks.

Old paint layers which are not absolutely safe must be removed before repainting. Defective plaster has first to be skilfully repaired.

(h) Application of paints to substrates containing depressions, cracks and fissures which have not been filled and faced up and which retain their sharp edges. The applied coating will be extremely thin on these sharp edges and when stresses develop as a result of oxidation and hardening, the film will fail by cracking at its weakest part i.e. along the sharp edges.

(i) Cracks can be caused by the presence of impurities in the atmosphere (coking works, chemical works, etc.), or indoors, products of combustion of gas-stoves or traces of mineral acids to which newly applied alkyd, modified urea or melamine coatings can be susceptible. The cracking caused by external effects is usually much more pronounced than that by 'internal' causes (i.e., due to properties of vehicles, pigments, driers, etc.), as the changes may commence while the films are still fairly wet, increasing with the degree of drying. Where cracking is induced by 'internal' causes the cracks are mainly of longitudinal type of small width, while 'external' influences, especially direct sunlight, alternating moist and dry weather, etc., cause the paint film to contract, to give in the worst cases the appearance of tree bark.

It is advisable to employ paint material specially suited to the conditions in question, e.g. chlorinated rubber, epoxy esters, acrylic resins, etc., and formulated with suitable inert pigments.

(j) Tropical exposure accelerates breakdown of films [1]. Surfaces may not only reach day-time temperatures up to 82°C (120°F), but a temperature range of up to 56°C (100°F) may be experienced in a 24-hour period. This in itself may lead to embrittlement, and heavy condensation also contributes. Paints which take up and lose water daily due to high permeability tend to chalk, craze and peel at a much higher rate than they do in temperate climates [2]. The effect of the greater occurrence of strong exposure of films to light rich in ultra-violet rays (high mountains, oceans or tropics), or exclusive exposure to the weather side only, as well as hoar frost and ice, results in bad cracking, which, on wood, may lead to the lifting of the film in long chips.

In the tropics other considerations such as high humidity, excessive and rapid changes of temperature and fungi contribute to the accelerated breakdown of films.

(k) Prior to peeling off as a consequence of lack of adhesion between two paint films, cracking is likely to occur. This may happen between an old and hard film and a fresh film of the same character or between two films of different properties, say, an oil-base film and a shellac-base film.

(l) Emulsion-based coatings (latex paints) may crack prior to peeling or flaking under exterior weather conditions or severe interior conditions. Highly pigmented paints crack under these circumstances due to the inability to accommodate the extremes of expansion and contraction, especially when applied on non-porous surfaces. Externally plasticized emulsion paints may lose plasticizer either by leaching out or by migration into substrates such as old oil paints or wood.

Where good durability is required specially formulated emulsion paints should be used. It is recommended that a primer or sealing coat should be used where the substrate is likely to give trouble. Copolymer emulsion paints give more satisfactory service due to the absence of migrating plasticizer.

(m) Deep, wide cracks, in severe cases curling up, termed *mud cracking*, of p.v.c. organosols can appear, when heavy films are built up by repeated applications without pre-setting (part-fusing) each coat at 150–165°C (302–329°F) in between, or by excessive film thickness in one-coat application. This temperature should also not be too low, nor should the thinners be too heavy, i.e. they must not be trapped (additional danger of bubbling). On the other hand, the finished parts must not be kept too long before the application of heat, so that the thinners have not been lost before the fusing of the discrete particles at, e.g. 165–180°C (329–356°F) for 10 minutes, or infra-red radiation.

Plastisols are free from this defect because they are practically free from volatile constituents.

Mud cracking can occur in films of zinc silicate paint due to the use of a zinc dust which is too fine or too uniform in particle size.

(n) If moisture is able to act on the paint or varnish films from beneath (e.g. on wooden buildings) or the rear of boards painted on one side (as in fascias) or if damp wood, etc. was used for painting, circular or spiral cracks of differing sizes are formed (concentric ring type).

(o) Cross-grain cracking (sometimes called 'alligatoring') is due to over-painting and an excessive number of coats on the surface. Tensions are set up in the hardened coating. Very heavy application, especially of fillers and primers, can result in cracking followed by peeling or breaking loose of the entire paint system. Heavy brush marks can result in 'line' cracking. Apart from the above, other troubles such as delayed hardening or tackiness may appear. The only remedy is removal of the complete system followed by repainting.

(p) *Checking* is the term used to describe shallow breaks in a paint film (Fig. 7.38); they do not penetrate to the substrate. (If the substrate is visible the break is called a crack). Three types of checking are recognized:

(i) *Irregular type.* The breaks in the film form no definite pattern. Irregular checking, particularly if severe, is frequently described as '*crazing*'.
(ii) *Line type.* The breaks are generally in parallel lines and often follow the lines of brushmarks.
(iii) *Crowsfoot type.* The breaks in the surface of the film form a definite 3-prong pattern with the breaks running from the centre and forming an angle of about 120° between prongs (Figs. 7.39–7.42). Crowsfoot and irregular checking often occur together.

Checking starts at the exposed surface, works progressively deeper into the coating, tends to take on a V-shaped cross-section with the open part at the exposed surface, but gives no sign of widening at the bottom by contraction of the coating though the checks may be widened slightly by erosion.

Slight checking is not a serious defect, as it indicates a relieving of the shrinkage stresses in a paint film. If a film does not check, because of its great tensile strength, it may crack with the expansion and contraction of the surface to which the paint is applied; flaking may then result (Fig. 7.43).

(q) *Alligatoring.* This term is used for pronounced wide cracks over the entire surface but the cracks do not reach the substrate; they may affect a single layer of film only. The failure resembles alligator skin (Fig. 7.44).

Alligatoring may begin as checking or cracking but the break tends to grow wider at the bottom as well as at the top. The top coating contracts, thus exposing portions of the undercoat. In the typical extreme cases of alligatoring, the 'islands' of coating between interlacing breaks have not only contracted in area but have increased in thickness and consequently often have become wrinkled.

The most common causes of alligatoring are the following:

(i) Appplication of a shorter oil type of finish over a longer oil type of undercoat or primer. The trouble is accentuated if the finish is applied before the undercoat or primer is completely hard.

Attention must always be paid to the chief requirement, i.e. the application of a longer oil coating over a shorter oil one. It is obvious, of course, that non-oiled or soft woods require comparatively long-oil primers, while the oil content of primers applied on hard woods should be correspondingly shorter.

Emulsion binders which are designed for use on plaster, brickwork and similar absorbent bases are exceptions to the above rule. On such a base, it may even be wise to apply as a first coat a paint of higher binder content than that of the finishing coat, e.g. by addition of emulsion.

(ii) The danger of alligatoring with subsequent flaking is increased if the layers applied prematurely are unduly thick.

The edges of alligatored areas usually show good adhesion.

It is possible to draw conclusions regarding the thickness of the defective films from the mesh size of the cracks and fissures. The narrower the meshes the thinner are the films, and vice versa.

Notes

[1] Footner, H.B. and Murray, G.R. (1947), *JOCCA*, **30**, 378. These authors group tropical climates in three general types:

(i) Hot humid. Humidity often up to 100%, temperatures of the order of 29·5–35°C (85–95°F). 'Day and night temperatures vary little, and conditions are worst during the monsoon or rainy period. Due to the large amount of water vapour in the atmosphere the ultra-violet content of the sunlight is low. Examples of areas in which these conditions occur are India (Indonesia), Malaya and Burma in the rainy seasons, West Africa, parts of Central America and the Amazon and Orinoco basins.'

(ii) Hot climates with average humidity. Shade temperatures as (i), notably 32–38°C (90–100°F) but humidity, while varying, is usually 60–80%, depending on the time of the day. 'The sunlight here contains a considerable amount of ultra-violet light. Such climates are experienced in Southern India, the East Indies, Malaya and Burma during the dry season, most African Tropics, and the coastal area of Central Southern America.'

(iii) Dry climates, but much hotter than (i) and (ii). 'Temperatures rise as high as 54·5°C (130°F) during the day with a sudden drop at night. The sunlight is high in ultra-violet light and articles exposed to the sun may attain temperatures as high as 82°C (180°F). Humidity may be high locally, for example, near large rivers such as the Euphrates. Climate here is not the only condition of which it is necessary to take cognisance, since abrasion by sand is often a serious factor. It is known that steel can be gradually eroded by this means, so the outlook for paint is poor. Iran, Iraq, the Arabian peninsula, Egypt and the Sahara are examples of this type.'

[2] Exposures of gloss paints based on long-oil alkyds put out in 1960 gave early

breakdown in Florida when containing antimony oxide or small additions of zinc oxide when compared with exposures from English sites over the same period. Rutile titanium dioxide pigmentations, although showing chalking to some degree in the eighteen-month exposure period showed no sign of checking or cracking as experienced on the other more chalk resistant formulations.

7.15 Cracking and flaking of lacquer films

(a) Unsuitable lacquer for a substrate. A nitrocellulose lacquer designed for good adhesion on a rigid metal base may not be suitable for the dimensional changes of wood surfaces. The lacquer manufacturer will advise the most suitable product for the surface to be coated.

(b) The binder used in a primer or undercoating can be unsuitable for a finishing coat.

The primer or undercoat are generally specified or applied at source to provide good anti-corrosive properties. If the long term performance of the finishing coat is in doubt then the maker's recommended undercoat should be applied to the existing primer before the application of the top coat.

(c) Cellulose lacquer has been applied over an oil-base coating or filler with a comparatively high oil content. This is especially risky if the former is applied too thickly. The strong, e.g. nitrocellulose solvents attack the undercoatings, etc. Alligatoring and cracking down to the base are the consequences and lifting is likely to occur. Lacquers based on cellulose ethers do not usually contain such strong solvents as esters, ethers and ketones, some of which always form part of a nitrocellulose lacquer.

(i) First spray a thin film of, e.g. a nitrocellulose lacquer over such an undercoating, so that rapid evaporation of the solvents is possible before they have sufficient time for softening, etc., provided that it is not considered more advisable to use another preparatory coating in the first place. Such primers, etc., as contain short-oil wood-oil type or cellulose lacquer in alcoholic solvents should be suitable.

(ii) Interpose an insulating layer, e.g. a shellac or polyvinyl butyral lacquer between oil-base filler, etc., and cellulose lacquer.

(d) Cellulose, polyvinyl acetate, etc., lacquers which air dry by evaporation, may crack, if applied over insufficiently dried undercoats which still contain high-boiling solvents. Such a system is most susceptible, especially if the fresh lacquer films are exposed to the sun.

Too heavily applied films, e.g. by hot spraying of nitrocellulose wood lacquers may be prone to '*cold checking*'.

Every lacquer film must be permitted to dry throughout before applying the next one. Do not apply excessively thick layers of lacquers. Some films may give up their low volatile solvents more easily than others, depending not only on the evaporation rate of the solvents (and thinners) but also on the solvent retention by the film-forming materials.

(e) The less volatile constituents of the lacquer films (plasticizer) are too quickly driven off by continuous exposure to heat.

Procure a special heat-resistant lacquer, if necessary. The plasticizers, resins, etc., must be suitable for the job.

(f) Exposure, particularly too early exposure of cellulose lacquer films to strong frost or continuous cold, may lead to crack formation. This may happen on transport or in unheated storage-rooms. Nitrocellulose lacquers containing the very low-viscosity grades of nitrocotton are not satisfactory at low temperatures.

Special cold-resisting cellulose lacquers must be employed, e.g. for various aeronautical purposes. Hot spraying of lacquers containing relatively high-viscosity nitrocellulose of normal film thickness is practised in order to obtain greater toughness and 'cold check' resistance [1] in particular on wood. Cellulose aceto-butyrate lacquers are claimed to resist 'cold checking' well, provided a lower-viscosity cellulose had been used.

(g) Application of clear (unpigmented) lacquers on parts or surfaces exposed to the influence of weather.

Some nitrocellulose lacquer films, for instance, require special protection, e.g. by car polish, as they are highly sensitive to ultra-violet rays. Well-formulated cellulose alkyd resin lacquers, however, are almost as weather resistant as, for instance, long-oil, drying-oil modified alkyd varnishes. The outdoor durability demanded of such lacquer systems, not necessarily unpigmented, is five to six years. This period can be considerably longer, if the films are treated with wax-pastes, etc. [2]. It has become the practice to add ultra-violet absorbers to clear films which are likely to be exposed. These additives are only effective for relatively short periods and colour deterioration can be pronounced especially when the film is also subjected to elevated temperatures.

(h) Excessive content of slowly volatile solvents in nitrocellulose lacquers.

(i) Chipping of laquers from metallic objects is likely to take place if:

(i) The metal has a brightly polished surface which renders the keying of a lacquer film difficult. Thorough mechanical abrasion or chemical pretreatment is necessary.
(ii) The lacquer contains insufficient plasticizer.
(iii) A high proportion of high boiling solvents is present and brittleness develops after weeks or even months, due to solvent retention.
(iv) On wood the use of too brittle fillers or sealers beneath the lacquer film should be avoided.
(v) Sharp edges on metallic and wooden objects are most prone to chipping.

Notes

[1] Anglo-American Council on Productivity (1951), *Metal Finishing,* London.
[2] Jordan, O. (1940), *Technology of Solvents,* p. 42.

7.16 Pronounced cracking, alligatoring or checking, staining or formation of brown weathered-out layers of bituminous or tar paints

The term 'bituminous' is now taken to refer to the range of products including natural asphalts (Gilsonite, Trinidad) and petroleum asphalts, together with blown bitumens, coal-tar pitch, and the special products derived from the distillation of fatty acids (stearine pitch, cotton-seed pitch).

As a binder for coatings, coal tar pitch is preferred to natural asphalt for several technical reasons, particularly with respect to resistance to petroleum crudes which will dissolve the natural asphalt. Coal-tar enamels are also unaffected by either bacteria or root growth and the polar nature of coal-tar confers properties of improved adhesion to the system. These coatings are widely used in the unmodified form for coating steel structures and pipe. The tar or pitch is blended with oil from the same source and often pulverized coal is dispersed into the heated tar oil mixture forming a colloidal suspension or 'structured' material.

Fillers are widely used to improve the rheology of bitumen materials [1]. The durability is considerably improved by blending bitumens with drying oils, chlorinated rubber [2], and even powder coatings have been prepared [3] by blending a molten mixture of bitumen, a urethane prepolymer and a polyisocyanate and heating until gas generation is completed. Films are claimed to have excellent corrosion resistance, chemical resistance and water resistance.

(a) Bituminous coatings are highly resistant to water and many corrosive aqueous solutions, but their resistance to ultra-violet radiation is generally poor. Nevertheless good results have been reported, e.g. twelve years' life on steel pylons in rural England [4]. In the case of coal-tar paint, sensitivity to light is aggravated by the presence of large amounts of free carbon and naphthalene [5].

Breakdown in light can be either by 'black' chalking or by alligatoring. The latter takes the form of wide fissures in the film, which may not for some time extend through to the substrate, and which are caused by oxidation of the surface layer while the layers below still remain thermoplastic. Results vary considerably. Water-gas pitch films are more resistant to chalking than petroleum bitumen, but generally fail eventually by checking and alligatoring. Natural bitumens modified with drying oils fail by chalking but much less rapidly than petroleum bitumen. The Joint Technical Panel J/P1 of the British Iron and Steel Research Association carried out a series of exposure tests on bituminous paints, and obtained outstanding results from Gilsonite-dehydrated castor oil combinations, and these pigmented with leafing aluminium [6].

Most bituminous paints are satisfactory if light is absent, but condensation present, e.g. on undersides of bridge structures. For exposed positions, select compositions which fail by chalking, if possible, and apply heavy coats, as with filled bitumen solutions (see Introduction, above). Alternatively use a paint containing aluminium pigment, which will protect the film beneath from ultra-violet radiation. Coal-tar paints must be suitably refined, and are improved by the addition of fatty oils, such as tallow, and also lime to neutralize excess acidity.

(b) Heavy coatings of bitumen or coal-tar pitch applied directly on to cold steel structures, e.g. on lock gates or to concrete as protective coats for underwater purposes, may not form a firm key owing to a film of moisture of condensation. The vapours of the latter blister the film and cause peeling and under-rusting.

Coal-tar films usually have poor adhesion to metal [5]. But they were found to be superior to bitumen on steel work completely immersed in sea water both as regards loss in weight and pitting.

Warming of such structures before painting, e.g. by the use of a mobile electric unit producing radiant heat (infra-red) may be useful. Hot refined coal tar (150–200°C = 302–392°F) on cold plates provides 'excellent protection against sub-aqueous corrosion' whilst if applied to hot plates the films become too thin. Adhesion of bitumen and tar paints can be improved by additions mentioned above in (a).

(c) Without priming coat:

Paints not designed for application in thick coats should not be applied too thickly or superficial drying will occur, leading to cracking and alligatoring. With filled bitumen solutions, or with bitumen emulsions, this risk is not present owing to the reinforcing effect on the film of the asbestos fibre present with the first or the clay compounds (e.g. Bentonite) with the second.

Under water or in water-logged soils cracking does not occur, and a minimum thickness of 1·58 mm (1/16 in, 62·5 thou) should be used. Even if thick bituminous films are built up by multi-coat systems, the tendency to alligatoring in sunlight is not removed. Single coats, however, often contain pinhole perforations (which may be located by electrical tests) and two or more coats will usually overcome this permeability [5].

(d) With priming coat:

In the case of natural asphalt enamels a primer is necessary and for shot blasted metal surfaces it has long been the tradition to specify red lead filled primers. It has been possible to overcome many of the problems of cracking in bituminous finishes by the use of synthetic primers based on chlorinated rubber, suitably plasticized and pigmented with carbon black.

Notes

[1] Hansmann, J. (1976), *Gummi, Asbest. Kunst.*, **29**, No. 5, 272.
[2] B. and C.G. Interbuild Conf. (1975), *Coatings and Preservatives on Concrete and Masonry: Coal Tar/Rubber coatings.* London.
[3] Fujita, M. (1976), *Jap. Pat. Rept.* **76**, No. 46, GpG4, Derwent, London.
[4] Hudson, J.C. (1941), *The Corrosion of Iron and Steel,* p. 141.
[5] Jordan, L.A. and Whitby, L. (1936), *The Preservation of Iron and Steel by Paint,* 16th Bulletin, PRA, Teddington.
[6] British Iron and Steel Research Association (1955), *Painting of Structural Steelwork,* 3rd Interim Report of Joint Technical Panel, J/P1, London.

7.17 Colour changes – fading, yellowing and darkening

Fading is the gradual decrease of colour intensity resulting from exposure of coloured paints (generally tints from white) to sunlight or to weather generally. It results from photochemical degradation of the tinting pigment by ultra-violet radiation and, for a given paint, the rate of fading will therefore depend on the conditions of exposure. In this country fading can be more rapid in rural or marine conditions than in towns and cities where much of the ultra-violet radiation is absorbed by atmospheric contaminants. Fading appears to be accelerated by the presence of moisture. Tropical sunshine causes fading to take place more rapidly than in temperature zones. Chalking of a paint surface can mask the colour and give the appearance of fading but this is less likely to occur if the chalk resistant rutile grades of titanium dioxide are used. The degree of fading can be assessed by means of the Geometric Grey Scale [1].

Some coloured pigments which are fast to light in full strength can show poor fastness when reduced with white pigment. A number of organic pigments show this defect and it is usual for manufacturers to give figures for lightfastness [2] of pigments both in the pure state and when reduced with various amounts of white (usually titanium dioxide). The lightfastness of an organic pigment can be affected by the type of white with which it is mixed.

Zinc oxide pigments, particularly the acicular grades and the co-fumed acicular zinc oxide/basic lead sulphate and lead silicate pigments, possess good tint retention properties, probably due to the high u.v. absorption. The inclusion of a small portion of zinc oxide in rutile titanium whites used for tinting purposes can give improved tint retention. This may be due to a combination of u.v. absorption and to the delay of breakdown by chalking. The use of high concentrations of zinc oxide, however, may lead to a hardening effect which causes breakdown by checking and cracking.

Fading of a pigment can take place as a result of admixture with other coloured pigment or pigments. For example, Arylamide Yellow G normally shows good fastness to light but it can fade under certain conditions when mixed with the more lightfast phthalocyanine blue.

The causes of fading can be summarized as follows:

(a) The use of pigment (usually organic) not sufficiently fast to light in the reduced state. Since figures for lightfastness are provided by the pigment manufacturer this type of failure can be avoided. Although organic pigments with high lightfastness in the reduced state are expensive this is usually offset by high tinting strength.

(b) Acidic contaminants in the atmosphere can affect pigments which are sensitive to acid. For example, traces of acid can destroy the colour of pale shades made with ultramarine blue and for this reason this pigment has acquired a poor reputation for lightfastness. Actually this is good but the pigment is readily destroyed by acid.

Yellow finishes based on lead chromes can become paler in colour (the effect is similar to fading) as a result of reduction of the chromate radical by sulphur dioxide in moist air. The defects can be prevented if the new paint film, when it is dry, is overcoated with a durable varnish.

(c) Not every pigment or dye is equally light fast in every type of medium. This applies particularly to lead chromes, indigo blue and bluish-red thio-indigo in oil-base media.

The fastness to light of pigment mixtures in paint media may be different from that of the individual pigments (see above).

(d) Cadmium pigments in admixture with rutile titanium dioxide are fast to light only if air and moisture can be excluded. Unfortunately this is not feasible on exterior exposure where severe discoloration, sometimes at a rather rapid rate and accompanied by chalking, has been observed, e.g. with stoved alkyd/amino finishes. Protection by a clear lacquer did not prevent the defect.

(e) Some coal-tar dyes and pigment dyestuffs used in connection with synthetic resin and cellulose lacquers, etc., also react differently with regard to their solubility, colour strength, shade and light-fastness, depending on the ingredients of the vehicles, especially on the solvents present.

Certain phthalocyanine greens are susceptible to formaldehyde released from synthetic resins on stoving and lose in light-fastness.

(f) Faults caused by weathering of the films. Loss of gloss of the film surface and change of the mixture ratio of the pigment constituents may cause fading even of paints containing light-fast pigments.

(g) Work in the U.S.A. and in Great Britain has shown the loss of colour and film integrity to be about twice as fast in country districts when exposed without protection and where the solar radiation was not so much reduced by the atmosphere as in industrial areas.

(h) Chalking, mainly in presence of certain pigments comes in here. Some paints chalk unevenly, resulting in a blotchy appearance, particularly in greys. Blotching or so-called fading of colours occurs in spots when the porous surface has not received sufficient coats of paint or a suitable primer. The final coat should be as weather resisting as possible and should not contain pigments embodying any water soluble component. Examples of these are zinc chrome and Venetian red. If the final coat is a transparent lacquer or varnish it should be of a non-yellowing type and should preferably contain some u.v. absorber.

Other colour changes:

(i) Many pigments discolour over alkaline surfaces such as new cement, plasters, etc. Prussian blues turn brown under alkaline conditions. When greens containing Prussian blues are subject to alkaline attack, e.g. from lime, cement, etc., the loss of blue colour can result in a change to a dirty yellow. This may be patchy according to the concentration of alkali present (e.g. mortar, lime plasters, asbestos cement).

In tropical exposure, Brunswick (chrome) green paints tend to fade by yellowing rather than by blueing in contrast to the effect observed in temperate climates due to the acid atmospheres.

Apart from colour changes and loss of gloss, green paints are reasonably durable (in the tropics) and like black and red oxide paints take much longer than whites to fail.

(j) For many years it was common for lead chromes to darken slowly on exterior exposure. This defect has now been largely overcome by improvements in methods of manufacture. Similarly the acid resistance has been improved.

(k) *Yellowing* becomes apparent in white and pale shades and takes place under the influence of light, heat or atmospheric contaminants e.g. ammonia. Yellowing caused by contaminants is most severe in the absence of light since the compounds responsible are more or less readily bleached by light. The yellowing can originate in the medium, the pigments or other ingredients (especially certain driers) or, as already mentioned, from outside influences such as contaminants in the air.

(l) Most drying oils tend to yellow to some degree, especially in the dark, and this tendency is transferred to alkyds containing fatty acids derived from the oils. High atmospheric humidity also accelerates yellowing but here light must also be present.

Yellowing under the influence of light is considered to be associated with the presence of linolenic acid in the alkyd or oil. Linseed type alkyds, which contain an appreciable quantity of linolenic acid, yellow most intensely. The 'semi-drying' oils such as soya bean, safflower and sunflower contain little or no linolenic acid and give mixtures of 'linoleic rich' fatty acids and these, in turn, yield the 'non-yellowing' alkyds. It is thought that the degree of yellowing increases with the degree of unsaturation of the fatty acids and this view is supported by L.A. Hill and W.H. Hall [3] in an interesting investigation.

Driers of certain metals can cause discoloration under certain circumstances. The use of manganese driers can result in yellowing in some media while cobalt can give greenish tints to whites if terpene type solvents such as dipentene or pine oil are present.

(m) Dyestuff toners can be used, not only to counteract the yellowing of white paints but also to boost their opacity. Commercial products are available for the purpose and well-known brands ('Tint-Ayds') are either blends of Mineral Violet and red shade Ultra Blue or mixtures of Carbazole Blue and Indanthrone Blue.

(n) Yellowing due to the pigment is uncommon since the discoloration is most noticeable in whites where the usual pigment is titanium dioxide.

Yellowing of green paints is generally associated with the use of chrome green pigments. Prussian blue tends to lose its colour when stored for some time in an oleoresinous type medium and the paint, when opened and examined, appears to have yellowed. On application and exposure to the air the colour of the blue returns and the paint dries to the original green shade.

A chrome green oleoresinous or alkyd type paint, if applied to an alkaline substrate, will turn yellow or, more often orange, as a result of decomposition of the Prussian blue by the alkali.

(o) The most well known agent for causing yellowing is ammonia. Oleoresinous or alkyd paints used on refrigeration plant or in hairdressing establishments yellow very rapidly. It has been shown [4] that the yellowing can arise from the interaction of ammonia with products of auto-oxidation in the paint film, e.g. 1, 4-diketones react with ammonia to give substituted pyrroles. Oil-free media such as chlorinated rubber and vinyls are resistant to yellowing by ammonia.

Yellowing of certain media, principally those derived from drying oils, can result from exposure to tobacco smoke (as in public houses and the top decks of buses) or from decomposition products of fats and oils (as in kitchens with no extractor fans).

Darkening of wood is caused by ultra-violet radiation and this can take place under clear varnish or when stained. Different degrees of darkening can appear if the same item contains both heartwood and sapwood. The use of u.v. absorbers in the clear varnish has reduced the trouble, particularly on hardwoods. u.v. radiation causes a degree of decomposition of the surface layers of wood. This is accelerated considerably in the presence of moisture. If moisture penetrates beneath a varnish layer, the decomposition will lead to breakdown of the adhesion as well as discoloration.

(p) Blackening and greying of exposed paint films was a common occurrence, particularly in industrial atmospheres, when white lead finishes were in general use and H_2S (hydrogen sulphide) was an atmospheric contaminant. The replacement of white lead by titanium dioxide together with the decrease in atmospheric contamination brought about by the Clean Air Act has virtually eliminated this type of paint defect.

Most frequently the darkening of films is due to contamination by dust, especially in urban and industrial atmospheres. The softer and newer the paint films at the time of deposition of the dust, the more rapid and intense is the discoloration. Where dust collection is a serious problem, the use of slow-chalking white finishes should be considered.

The darkening of walls above radiators and the circular dark patches on ceilings over lights are due to dust particles deposited from convection currents. These particles are often electrostatically charged.

(q) Mould growths develop on badly ventilated, damp and warm places devoid of sunshine, and are a cause of discoloration of nearly all sorts of coatings, especially of the softer oil-base paints and varnishes, emulsion, casein and distemper paints on wood, plaster and brickwork.

Not only may such a growth itself look like dirt, but mycelia will actually collect dirt and soot, which support further mould life until the paint presents a uniformly discoloured surface.

(r) Nitrocellulose ester lacquers are comparatively sensitive to u.v. rays. The nitrocellulose decomposes, giving rise to yellowish and brownish tints. In this reaction nitrogen peroxide is liberated, which simultaneously causes deterioration of the durability of the film. Some cellulose ether lacquers yellow also, but not as much as nitrocellulose ester lacquers. This defect is not only visible on transparent films, but also if light-coloured first coats are covered by transparent last coats, especially if the last coat film is thick. It should be mentioned in this connection that it is not necessarily the cellulose component that is solely responsible, but the presence of certain resins or plasticizers, and even certain solvents may also induce discoloration of cellulose lacquer films.

It is advisable to add, where possible, at least a slight amount of pigment to such transparent lacquer or a small addition of enamel containing suitable pigments. Particularly good results have been obtained, with a low percentage of aluminium bronze powders. Cellulose acetate and acetate butyrate lacquers possess excellent light stability. The use only of thinners as recommended by the suppliers of the lacquer must be emphasized.

Yellowing and rivelling of nitrocellulose/alkyd resin lacquers is prevented by small additions (0·05–4% by weight) of bis (hydroxyphenyl) alkanes [5].

Notes

[1] B.S. 2662: 1961, *Grey Scale for assessing change in colour.*
[2] Lightfastness figures are quoted with reference to the Blue Wool Scale described in B.S. 1006: 1971.
[3] Hill, L.A. and Hall, W.H. (1973), *JOCCA*, **56**, 259.
[4] O'Neill, L.A., Rybicka, S.M. and Robay, T. (1962), *Chem. Ind.*, **41**, 1796.
[5] Bayer, AG. German Patent 2435 511. 1976.

Proprietary product: Tint-Ayd from Daniel Product Co., Jersey City, N.J., U.S.A.

7.18 Mildew (fungi, moulds), algae and moss growth, bacterial and termite activity on paint films

Mildew (Fig. 7.45) or moulds belong to the family of fungi. The infection of a painted surface by moulds manifests itself in two ways. The growth may consist of mycelium which consists of fine filaments (mycelial type) or coloured spots with spongy structure (nodular type) or a combination of both. Frequently it is difficult to distinguish fungus growth from dirt collection and a microscopic examination may be necessary.

Fungi require nutrients, such as carbohydrates, nitrogen, sulphur, potassium and phosphorus for their development. Algae are green chlorophyll-containing organisms and meet their energy requirements by photosynthesis. Both algae and fungi will only flourish where there is an ample supply of moisture. Algae grow almost exclusively out of doors. Moulds are met out of doors, particularly in the tropics, but more often indoors and food factories, sugar refineries and breweries suffer most severely (Figs. 7.46–7.48). Many species of mould, called *Fungi imperfecti* have been isolated from paints. The most common are *Pullularia* (black), *Cladosporium* (black), *Phoma* (pink, purple or brown-red), *Penicillium* (green), *Alternaria, Trichoderma* and *Stemphylium* [1]. Usually their spores settle on the painted surface, often before the film has fully hardened. The nourishment is provided by vegetable oils present in many paint films or by plasticizers and cellulosic thickeners available in many emulsion paints. In some instances the spores are present in the substrate before painting. Mould growth has often been observed under wallpaper and lining paper when the paste provides a convenient source of food.

The appearance of mildew depends on various conditions:

(i) Continuously high atmospheric humidity. The critical point is at about 75% relative humidity. The higher, the more rapid and varied is the growth of the microflora. Similar conditions can be created if wood is in contact with moist ground or in a situation liable to high condensation.
(ii) High average temperatures; optimum temperature for most fungi is 25°C (77°F) [2]
(iii) Lack of light and sunshine.
(iv) Still air. Good ventilation reduces the danger of fungus attack.
(v) Character (smoothness or roughness) and composition of the base, whether a nutrient substratum or at least the cause of supplying conditions suitable for growth in the form of moisture.
(vi) Light colour of the film.

Mildew infection occurs in damp rooms and buildings such as lavatories, basements, laundries, baths and swimming halls, breweries, fermentation rooms and cellars or outdoors. In Europe it is most prevalent near the coastlines and in the British Isles, whilst in the moist-warm tropical regions, e.g. in Central America, it is very common, due to the moisture absorption of the 'dry' film, and frequently causes great damage. Damp wood, distempered or papered walls are the most common places for mould growth, but it may also be either building material or old paint or the new coating itself. Even highly pigmented stoved films, if, e.g. based on oil media, can be attacked by fungi under tropical conditions. Hundreds of species of fungi are known to exist. Mould growth can even be found in Antarctic zones [3].

The failures due to 'vegetation' on paint films are discoloration, which may look like staining or spotting, holing and flaking off, but may occasionally be mistaken for checking [3]. The defects may remain visible after the fungi have died.

While discolorations are more obvious on whitish films, mould growth can also be unsightly on red and brown oxide paints. However, as many coloured paints contain fungicidal materials, or these can be added, without impairing the colour, mould growth on them is less serious [4].

Wood can have been infected before painting with, e.g., sap-stain fungus ('blue stain', Fig. 7.49).

Pesticidal coating compositions consisting of plasticized amino resins with D.D.T., Aldrin, etc., are claimed to be effective against insects [5].

In general, harder paint films are less liable to be affected by fungus growth; similarly water repellent paint surfaces can act as barriers to the settlement of fungus spores, but such paints are rare. Unpainted porous substrates like brick and concrete acquire some resistance to algal and fungal infection by treatment with water repellent silicone solutions [6].

(a) The appearance of mould may be first noticeable either by a dotted or threadlike growth. This finally develops to give the characteristic dark or coloured colonies covering small or large areas which may even lead to the gradual decomposition of the films (Fig. 7.50) by secretion or undermining or under particularly unfavourable circumstances, of the objects which the paint films are supposed to protect. On walls growth is often accentuated over the mortar [2].

Paint films containing zinc oxide in high concentrations (up to 40% on the pigment content) are less given to fungus growth. Zinc oxide has proved effective both in emulsion paints and alkyd enamels and in the oil based house paints so popular in the United States. Care must be taken when using reactive media to see that they do not react with zinc oxide and lead to thickening, seeding or gelation of the paint. Indirect process and acicular types of zinc oxide appear to give the best results without detrimental effect on other film properties. Zinc oxide has also been shown to act as a synergist and improve the activity of fungicides present in the film.

Lead paints are not poisonous enough to avoid fungoid growth; they are even rather susceptible owing to their slow hardening. Lithopone and (zinc free) titanium white are worse than white lead, and antimony oxide is hardly any better. It is stated that blanc fixe [4], carbon black, bone blacks and other pigments containing adsorbed salts of potash or phosphorus assist mould growths, and that most other pigments except those containing copper, mercury or organotin compounds and combinations thereof have practically no effect on the growths. Fatty oils such as linseed oil and soya bean oil, especially those containing 'foots' or traces of nitrogeneous matter, appear to further mould growth; stand oils, however, less so than raw or boiled oils [4].

Wood-oil paints are claimed to be superior to linseed stand-oil paints. This is, no doubt, due to their faster and harder drying. Water paints and distempers with organic constituents such as glue or casein, which are nutrient media, are, if damp and exposed to sufficiently high temperature, open to attack, and this may even begin in the container if no efficient fungicides are present.

Chlorinated media such as chlorinated rubber, polyvinyl chloride (p.v.c.) and chlorinated waxes are fairly resistant to mould growth and bacteria. However, mould growth has been observed on a chlorinated rubber paint in a sugar refinery.

Additions of chlorinated paraffins as plasticizers will increase resistance to fungi. Good results are also obtainable with some other synthetic products, particularly those which polymerize by heat or by catalytic action to very hard films, as do urea formaldehyde coatings.

Llewellyn and Eldridge [7] point out that, according to their experience at the British Building Research Station, mould growth is frequent in reconstruction work in old buildings because dormant spores from old building materials or infected new building materials are spread about. They also suggest that lining paper may be responsible for carrying and spreading of mould growth on to new plaster-work, because size, paste and paper are favourable media for growth.

It has to be borne in mind that moulds may remain dormant for long periods under rather dry conditions, but that above a certain degree of humidity they are activated again.

Thus the base, say plaster or damp wood, may already be in an infected condition before painting, and this has to be met with effective measures.

W.P.K. Findlay stresses that various plasters have varying degrees of effect upon subsequent growth [4].

Sources of dampness should, if possible, be eliminated before painting. The following fungicidal washes are recommended:

(i) Solution of sodium hypochlorite containing 1–2% active chlorine [8]. This destroys the spores on the surface and also bleaches the stains caused by moulds. It is sometimes possible to clean the painted surface sufficiently to avoid redecoration altogether.
(ii) 2% solution of sodium pentachlorphenate in water. This has limited permanency.
(iii) 2% aqueous solution of sodium salicylate. Limited effect but often used on interior surfaces.
(iv) 10% solution of magnesium silicofluoride in water is sometimes used on exterior walls.
(v) 2% formaldehyde solution. Effective, but used on exterior walls only because of toxicity.
(vi) 1% solution of *p*-chloro-*m*-xylenol in an alcoholic solvent is sometimes suggested.

Where algae are the principal infestation on masonry and concrete or on paintwork, solutions of copper salts are very effective, as is pentachlorophenol and sodium pentachlorophenate, and can be used for washing down. Amine salts of pentachlorophenol will kill algae and mosses. Dichlorophen, an algicide with a wide application spectrum and low phytotoxicity is strongly recommended by some authorities [9]. Cement based products like concrete and asbestos cement seem to be more prone than other substrates to support algal growth [6].

Badly infected films have to be completely removed, possibly burned off, e.g. to the bare wood. For less severe cases a good sterilizing wash with soft soap, soda, etc., fungicidal solutions or, better, the above antiseptics, is recommended before repainting, otherwise fungus growth may redevelop underneath a new paint film and lead to trouble, e.g. peeling. Where decorations are to follow, no moisture must be left. Any excess of fungicides has to be removed and the dried surface wiped carefully, otherwise the drying and colour of the subsequent paint may be upset. Much can be done in this respect by the selection of the most suitable fungicide. Paints which already contain poisons or fungicides for preservation after application are available.

The necessity for regular good ventilation of infected rooms is obviously of the utmost importance, and it will sometimes pay to create some new sources of daylight in otherwise dark places.

Plaster, stone and brickwork are also treated by washes with petrifying agents which kill the remains of old growths and thus prevent to a certain extent new growth.

The risk of mould growth is less on alkaline materials such as lime plaster or lime-gauged plaster, than on materials which are neutral or slightly acid, such as many types of calcium sulphate plaster.

Organo-mercurials have probably been used for the protection of paint films against mould growth longer than any other types and on a cost/effectiveness basis they still compare favourably with many more recently developed products, although their protective effect declines fairly fast when exposed to the weather. The compounds most frequently incorporated in paint formulations were phenyl-mercuric-acetate, phenyl-mercuric-propionate and phenyl-mercuric-oleate. All these are very toxic and for this reason their use is banned in some countries and looked upon with disfavour in many others. Similar considerations apply to pentachlorophenol, but in any case its effectiveness has been shown to be limited [10].

Phenyl mercury compounds can react with sulphide pigments or hydrogen sulphide in the atmosphere, forming mercuric sulphide which causes dark stains and inactivates the fungicide [11].

A very large number of other fungicides have been examined and numerous papers have been published, but no universal claims can be substantiated. The reason is that fungus growth is much dependent on the micro-climate in the vicinity of any location and the performance of a fungicidal paint can vary from one part of a building to another. Hoffmann has investigated the inhibition of mould growth under different environmental conditions and tested many fungicides [8]. The continued search for non-mercurial fungicides is well exemplified by a series of papers published by the C–D–I–C Society of Paint Technology in 1967, 1968 and 1974, which reached the conclusion that 'most of the non-mercurial fungicide protected paints had better mildew resistance than the unprotected control, but none were as good as the control containing zinc oxide plus phenyl-mercury-propionate, based on exposure in Miami, Florida and Columbus, Ohio' [12]. However, there is now ample evidence that good protection at relatively high cost can be obtained, albeit for limited periods of two to three years. Even compounds of very low solubility in water decline in their fungicidal effectiveness very often after short exposure and those with higher solubility are unsuitable.

Some of the newer fungicides used with success in paints [13] are described below. It should be noted that some of the compounds owe their evaluation in paints to their notable performance as agricultural fungicides and in some cases the attempt to take advantage of their active nature by incorporating them in paints has been justified by results.

(i) *Barium metaborate* [14]: mainly useful in exterior alkyd paints at concentrations of 15–20% on the solids content of the paint.

(ii) *Copper-8-hydroxyquinolate* [15]: because of its yellow colour not suitable for white or pale colours. A concentration of 5% on paint solids has been particularly useful in breweries, but also in other locations.

(iii) *Tributyltinoxide*: useful in indoor emulsion paints at 0·3–0·5% on paint solids. Although the organic tin compounds are toxic, unlike mercurials they are not cumulative poisons.

(iv) *Tetramethyl-thiuramdisulphide* (*TMTD* or *Thiuram*) [16] : a good broad-spectrum fungicide, useful in emulsion paints at concentration of 3–8% on paint solids. It is a strong anti-oxidant and destroys the drying ability of oil-based paints. It can also prevent drying of paints used for overcoating emulsion paints containing TMTD.

(v) *N(trichloromethylthio) phthalimide* [17] : a plant fungicide, is effective in solvent-based coatings at 2–3% on paint solids.

(vi) *N(fluorodichloromethylthio)phthalimide, Preventol A3* (Bayer, Leverkusen) [18]: effective in solvent-based and emulsion paint, but mainly in the former. Use concentration 1·5–2·5% on paint solids.

(vii) *N-dimethyl-N phenyl N′ (fluorodichloromethylthio) sulfamide* [18] : known as dichlofluanide, Preventol A4 (Bayer, Leverkusen). Normally not used in emulsion paints. In lacquers and coatings has been used with success in concentrations ranging from 0·5 to 2% on paint solids.

(viii) *Tetrachloropyridine-4-methylsulfone* [13], *Dowicide S 13* (Dow, U.S.A.): good long term efficiency in solvent based formulations at 2–4% on paint solids.

Complex *guanidine* derivatives have been patented by the Paint Research Association as fungicides and algicides and are claimed to be effective at between 0·25 and 5% in paints (British Patent No. 1,447,541, U.S. Patent No. 3, 915, 918.)

(b) Algae thriving on the paint films of buildings when they are frequently subjected to long exposure to moisture can be an unsightly nuisance (Figs. 7.51 and 7.52).

Mosses which have grown on old plaster or damp organic coatings such as casein paints may develop again after repainting if they have not been thoroughly removed. If enough moisture is present it even may grow beneath some finishes, such as oil paint and varnish films. At first greyish patches appear in the film and within a few months the moss penetrates and thus completely destroys the film in places. The majority of fungicides are also algicides to varying degrees.

The removal of moss can be done either mechanically or by chemical means (see sub-section (a)).

(c) The activity of bacteria may lead to black discoloration of films and to corrosion of metals.

Bacteria, too, need food for their development and multiply only in the water phase. But more moisture and a higher optimum growth temperature (for many bacteria about 37°C (99°F)) is required than by fungi. Little damage from bacteria (and moulds) should be experienced at a relative humidity below 70 to 75% [2]. The commonest types of bacteria found in water paints, including oil-bound water paints, are ‘short non-sporing rods, although sporing types do sometimes occur’.

The species, *Vibrio desulphuricans*, which flourishes only in the absence of oxygen, has the ‘ability to reduce sulphates to sulphides and so, e.g. in the presence of lead or iron, cause the formation of black sulphide' [2]. A slight smell of hydrogen sulphide may be noticeable.

Such organisms occur in sands, clays and earthy materials generally and can be introduced into building material [2] which has not been subjected to heat as is the case with bricks, cement and some plasters.

Bacterium fluorescens, both the liquefying and the non-liquefying strains, are common in water paints. In some cases there are gas-producing spore-formers which, besides producing carbon dioxide, produce acetic and butyric acids [2].

The use of hard-drying paints and synthetic materials or of antiseptics as described in connection with mould growth in sub-section (a) above may be advisable. The antiseptics are, however, not necessarily the same. Acriflavine has been mentioned as a good bactericide.

(d) Ants and termites are a serious menace to paint film in the tropics. Long-oil films are unsuitable. It is considered that hard-drying films containing cross linking resins, e.g. epoxy resins or phenolic resin/wood oil vehicles are less endangered. Incorporation of pentachlorophenol in coatings may be helpful.

Insecticidal lacquers containing DDT, Aldrin, Dieldrine or benzene-hexachloride are available. These compounds can also be used, either dissolved in fuel oil or emulsified in water, to poison the soil in the vicinity of timber structures [19].

The best protection is achieved by impregnating timber by pressure or diffusion before painting. Creosote is very effective, but can only be used where colour and tendency to stain are acceptable. Pentachlorophenol, copper naphthenate or a blend of zinc chloride and potassium chromate are other suitable impregnating chemicals. Arsenicals are very reliable, but toxicity restricts their use severely. Timbers reputed to be resistant to termite attack are listed in Forest Research Bulletin No. 24 [20]. (See Section 7.19.)

Notes

[1] Carter, G. (1973), *JOCCA,* **56**, 302.
[2] Bunker, H.J. (1946), *JOCCA,* **29**, 202.
[3] Turner, M.B. (1942), *Paint Manuf.,* **12**, 130, 137.
[4] Findlay, W.P.K. (1940), *JOCCA,* **23**, 217.
[5] British Patent No. 709, 721.
[6] Whiteley, P. (1973), *JOCCA,* **56**, 383.
[7] *The Decoration of New Plaster and Cement.* 1939, p. 179.
[8] Hoffmann, E. (1971), *J. Paint Tech.,* **43**, No. 558, 54.
[9] Rechenberg, W. (1972), *Betontechnische Berichte,* **22** (6), 249.
[10] Hoffmann, E., Saracz, A. and Barned, J.R. (1966), *JOCCA,* **49**, 551.
[11] Hoffmann, E. and Georgoussis, O. (1960), *JOCCA,* **43**, 779.
[12] *J. Paint Tech.,* **46**, No. 588, 54, 1974.
[13] Pauli, O. (1973), **56**, *JOCCA,* 285.
[14] Hoffmann, E. and Saracz, A. (1970), *JOCCA,* **53**, 680.

[15] Hoffmann, E. and Saracz, A. (1970), *JOCCA*, **53**, 792.
[16] Hoffmann, E. and Saracz, A. (1969), *JOCCA*, **52**, 193.
[17] Hoffmann, E. and Saracz, A. (1969), *JOCCA*, **52**, 841.
[18] Pauli, O. (1967), *Deutsche Farben Z.*, **21**, 64.
[19] Forest Products Research Laboratory Leaflet No. 38, 1965.
[20] H.M.S.O., 1971.

7.19 Decay or rotting of wood beneath paint films

Fungal decay of timber can only take place if the water content of the wood is high [1]. But if the wood is dry when painted, and the coating is sufficiently impermeable to moisture, decay is impossible. If repainting is delayed, however, until the old paint has cracked or peeled, moisture will penetrate to the wood, particularly on horizontal surfaces such as window sills. The need for painting those parts in contact with brick or masonry is also obvious. Once fungal attack is advanced, little can be done.

Water can also enter the unpainted endgrain and through cracked joints and cause rapid decay. The most prevalent fungi are wet rot (*Coniophora cerebella*) and dry rot (*Merulius lachrymans*) and 'blue stain' fungi (see Fig. 7.49), *Ceratocystis* and *Leptographium* [2]. The latter enter the timber between felling and drying of the sawn timber unless suitable precautions are taken. Dry rot is unique as it can spread rapidly beyond its origin in wet areas across brick, steel and concrete and infest areas of low moisture content. Preservative treatment of timber before painting is essential to avoid decay unless one is absolutely certain that there is no access of moisture to the timber. Many preservative treatments are available and they can be applied by vacuum or pressure impregnation, by dipping, spraying or brushing, the last two methods being less desirable. A number of effective chemicals are in use:

Pentachlorophenol in a light oil solvent to aid thorough penetration, and *tributyl tinoxide* are pre-eminent, but *copper naphthenate* (2% copper content) in kerosene or white spirit and *zinc naphthenate* in kerosene or white spirit enjoy some popularity. A mixture of potassium or sodium salts of chromic, hydrofluoric and arsenic acid in aqueous solution is known as Wolman salts (BS 3453; 1962); in spite of their toxicity they are still in use, as are copper/chrome/arsenic preservatives to BS 4072: 1974. Creosote is very effective, but colour, smell and problems with overpainting make it undesirable if painting is contemplated. Water repellent wood preservatives have some advantages. The best known types are derived from the 'Madison' formula which contains paraffin wax, zinc stearate and pentachlorophenol and boiled linseed oil, dissolved in white spirit [3]. These preparations are often pigmented with earth colours such as red oxide, umber and sienna to match the colour of the timber.

In an effort to overcome the disadvantages of mere physical impregnation, an interesting attempt has been made to produce reactive fungicides which are designed to bind the fungicide (organotin compounds) chemically direct to the wood in the manner of some dyestuffs bound to a textile fibre [4]. Some success has been reported by using a diepoxide and a di-isocyanate as coupling agents. Paints containing organotin compounds (tributyl-tinoxide) as preservatives are recommended by the Tin Research Institute.

Notes

[1] Edwards, W.A. (1954), *JOCCA*, **37**, 655.
[2] Savory, J.G. (1973), *JOCCA*, **56**, 247.
[3] Hill, R.R. (1973), *JOCCA*, **56**, 251.
[4] Heertjes, P.M. and de Jong, J. (1971), *JOCCA*, **54**, 642.

7.20 Coloured stains on paintwork (as distinct from mould growth)

Continuous exposure to strong sun (sunny side of buildings, etc.) or to moisture (weather side) may induce or accelerate the development of the above-mentioned faults.

(a) Resinous exudations, knots, etc., in wood surfaces. The following types of wood exhibit many knots: Douglas fir, yellow pine and Californian pine. Wood cells comprising a knot provide channels leading radially through the annual growth rings and make it a simple matter for both resin and water to pass to the surface of the painted board. There are:

(i) Extremely small (pin) knots.
(ii) Knots which only partly penetrate through the wood, so that the actual surface appears knot-free.
(iii) Loose and resinous knots, which give the most trouble [1].

The paint-holding qualities of resin-bearing wood is poor.

Apart from stains, blistering (see Fig. 7.28), peeling and alligatoring can be expected. Interaction of resin acids with pigments of a paint film may reduce its mechanical properties [2]. The brownish discoloration of paint films covering knots is increased by the dark self-colour and the small absorption of paints by the knots.

Seasoning of resinous wood at 70°C (158°F) is advisable. The resin will thus become hard and exudations be removed during the machining. Wood treated by this method does not exude much when exposed to the sun.

Resinous areas on the wood or black streaks of pitch should either be rubbed off and then washed with white spirit or turps or removed with some neutral organic paint remover (alkaline agents may cause undesirable roughening of the wood surface). The most satisfactory solution of this problem is to avoid the use of an oil-base primer, but to insulate one inch around the areas in question with one, or better two, layers of thin shellac solution (knotting) for dark, or of nitrocellulose lacquer for light-coloured paints, or with an aluminium paint or metal foil. Excessively thick layers of knotting may cause the paint to peel off on weathering.

In bad cases both washing with solvent and insulation should be carried out.

The use of a blow-lamp may even worsen the defect.

A thorough job can also be made by mechanically rubbing off any rosin with steel wool or an oil-resistant abrasive paper wetted with a 1:1 mixture of white spirit and raw linseed oil. Any surplus of oil in the latter case has, of course, to be removed with the help of white spirit. Big occlusions are sometimes also cut out with a knife. The surface has then to be levelled either with a filler or with a piece of new wood. Knots on joinery timber placed in contact with wet bricks should receive a coat of knotting also on the reverse side to prevent moisture entering from there. It should become a general procedure to treat knots on both ends.

(b) The discoloration can also be induced by sap stain, especially on the end grain of coniferous wood, in particular pine. Sap-stain fungi cause no decay of wood. They need a moisture content of over 25% for their spores to develop, and a temperature of 15–25°C (60–77°F) [1]. This may show through transparent or translucent films. If the wood is and remains dry, no further trouble will be caused by the blue (or black) stain; if it becomes damp, moulds penetrate the paint layer and discolour it, and in grave cases the staining will be followed by peeling.

Whilst the defect is mostly found on coatings of old and bad wood, it may also develop on moist brickwork, etc.

Paint only dry wood and during dry weather. Allow rooms of new buildings to ventilate thoroughly and as long as possible previous to and after application—as long as the walls exude moisture. Paints intended for outside application or for humid rooms must form a fairly hard film. To avoid sap-stain discoloration ('blue stain') of wood, the varnish etc. film must be water resistant and thick enough for all wood fibres to be fully coated [1]. It is claimed that certain proprietary bleaching agents will remove sap-stain discoloration for good; any paint film present will be removed in the process. Zinc oxide paints are preferable. It may be necessary to sterilize the wooden base before painting.

(c) Yellow discoloration of paint resulting from migration of anti-oxidant from wall-board adhesive has been reported by D.A. St. John, and K.R. Markham [2].

(d) Dark brown or black stains can appear on white or light coloured paints over steel nail heads. These may be the result of electrochemical corrosion or reaction of the iron with tannins or other acidic components of the wood. Steel nail heads should be spot primed with a zinc phosphate primer. The trouble can be avoided by the use of galvanized nails.

(e) Traces of copper in rain water can produce dark spots on bright red paints based on cadmium reds [3]. The copper can originate in overhead wires etc. The staining can be prevented by incorporating a chelating agent for copper such as benzotriazole.

(f) Dark stains can appear on emulsion and flat wall paints when phenyl mercury compounds are used as fungicides [4]. They originate from reaction between the mercury and H_2S which is present in many dwelling rooms.

(g) Lead-containing paints darken in the presence of H_2S due to the formation of black lead sulphide.

(h) In areas subject to heavy condensation, paints containing lithopone (zinc sulphide) can darken by reaction between zinc sulphide and lead driers leading to formation of lead sulphide. Lead sulphide stains on white paint films can often be bleached by acidified hydrogen peroxide.

(i) Rust stains and dark copper stains have been observed when iron and/ or copper particles, arising from plumbing or other construction work during or shortly after painting have settled in paint films. The development of the stains may take several months.

Notes

[1] Zycha, H. (1961), *Farbe Lack,* **10**, 628.
[2] St. John, D.A. and Markham, K.R. (1976), *JOCCA,* **59**, 331.
[3] Bullett, T.R. (1971), *JOCCA,* **54**, 841.
[4] Hoffman, E. (1959), *Aust. Paint. J.,* **4**, 15.

7.21 Staining through all kinds of paint films

(a) Contamination of cement, plaster, stone or wood surfaces due to water, soot, smoke and tobacco, etc., will stain through paint films if no precautions are taken (Fig. 7.53).

(i) Stains caused by water usually have a well-defined borderline. In cases which are not too bad, insulate when dry, after having scraped off all paint for some distance around the damp patch. Use a thin, hard-drying varnish, chlorinated rubber or shellac solutions with or without pigment, or with good proprietary brands of insulating media. (Test for dampness with cobalt paper or with a moisture meter).

In the long run, insulation can only be successful if the dampness which caused the patchiness had a chance to evaporate, either before repainting or at least through the other side, e.g. of a wall, and if the cause responsible for the humidity had been removed. Otherwise the staining through will reappear gradually, perhaps within weeks or months. Moisture is retained in brickwork and in ceilings in moderate climate up to at least two years, whether the source of it has been removed or not, but even with the aid of warmth it may be there up to about one year. It is advisable to remove badly contaminated plaster and to apply a new layer. For minor repairs, an area of at least 30·5 cm (1 ft) diameter in excess of the patches should be covered. This is the minimum up to which the old paint and decomposed plaster should be scraped off.
A repaired patch of plaster may again become visible so long as it is comparatively new.

Ceilings should never be filled with material which on becoming wet, as happens occasionally by leakage, tends to stain.

It is a widespread and wise practice to paint new walls with distempers or emulsion paints which are porous enough to allow the evaporation of building moisture. Oil and other sealing paints or wallpaper can, under normal conditions, safely be applied after about two years. With modern, rapid setting, low density plasters some weeks only may be necessary.

(ii) Patches caused by soot and smoke have first to be removed mechanically, if possible. Staining through caused by soot and smoke shows up yellowish-brown. Insulation with pigmented spirit varnishes, or one of the other media mentioned under (i), is effective.

Unsatisfactory burning off of paint on wood may by carbonization produce or worsen the appearance of stains in the finishing coats. If the discoloration has been caused by a fire and the above treatments are not regarded as satisfactory, apply a steam jet and wire brush.

(iii) Nicotine stains from tobacco smoke on plaster have also to be removed carefully before repainting. They can be removed by an alkaline cleaner.

(b) Bleeding into new paintwork may be caused by the presence of certain organic pigments [1], [2] or dyestuffs in old paint coatings or stained wood, such materials being soluble in the solvents of the new paint. The old surface should be insulated with a coat of a spirit varnish. If the bleeding is intense, a spirit lacquer pigmented with aluminium is usually effective. On wood which has been treated with water stains, little trouble is usually encountered with conventional paint coatings. However, application of a thin 'wash' coat of spirit varnish is an additional safeguard.

(c) Asphalt, bitumen, coal-tar and creosote oil stains on the base surfaces may give a blackish or brownish bleed-through. Tarry bases or bituminous paint films are able even to cause delay in drying and cracking of superimposed film (Fig. 7.54).

In the case of staining through of bituminous bases use two coats of knotting, aluminium bronze paints, or of another suitable resin solution in alcohol, to insulate. Bituminous stains can often be sealed off by emulsion paint.

(d) Brown tobacco-juice-like coloured stains are easily formed on light-coloured paints applied on imitation red cedar, red beech and other woods containing natural dyes, especially if the wood becomes moist.

Discoloration of paints over timber which is normally suited for painting is possible if it is unseasoned. This is due to the penetration of moisture and sap. Stains from cedar can be effectively sealed by a primer pigmented with a zinc phospho-oxide (Nalzin Sc–1). The acidic colouring matter forms an insoluble zinc compound [3].

Before painting, suspicious wood surfaces should be rubbed off with a mixture of equal parts of water and methylated spirit. This method frequently eliminates the trouble.

Where wood is in contact with, e.g. wet plaster, insulation is necessary. Well-primed wood, such as cedar, rarely shows such defects.

Some woods, e.g. cherry, larch, pear, tend to assume more or less intense discoloration, depending on age, under the influence of lacquer containing acids, such as, for instance, the cold-hardening synthetic resin lacquers which contain acid catalysts.

These woods should be pretreated with a pore filler and sealer, e.g. of the nitro-cellulose or polyester type.

(e) Oil patches in plaster may cause staining of superimposed films of emulsion paint or distempers. Patches which are caused by non-drying oils, such as machine oils, edible fats, milk products, are often fairly dark and tend to cause great trouble by penetrating again and again.

Insulate the surfaces with thin layers of a spirit varnish.

(f) Damp wood, which possesses an unusually high content of tannic acid, may develop awkward black spots if it comes in contact with iron nails. Discolorations due to the content of tannic acid alone are possible on green, damp oak. Both types of stains spread from the wood to the paint films and impair their adhesion.

Prior to painting, such inky stains can sometimes be bleached with hydrogen peroxide, although with difficulty. A sealer based on calcium plumbate is often effective.

(g) Brown stains forming on newly plastered brick walls may be due to the presence of large proportions of iron compounds in the brick. These stains are usually accentuated by the use of water-based paints on the surface of which they are noticeable. They may be washed off with warm water and soap or detergent. The stains usually reappear due to leaching to the surface.

Since the stains are water soluble, satisfactory sealing down by the use of a plaster primer of the tung-oil varnish type may be achieved. Where the plaster has become heavily contaminated it may be necessary to remove it, seal the brickwork and then replaster.

Notes

[1] Oil and Colour Chemists' Association (1976), *Introduction to Paint Technology*, 4th Edition, p. 89, London.
[2] Geissler, G. (1973), *Defazet*, **27**, No. 10, 464.
[3] Davidson, S.L. (1975), *JOCCA*, **58**, 435.
Proprietary material: Nalzin SC–1 from National Lead Industries Inc., USA.

7.22 Tarnishing and other blemishes of metals coated with transparent lacquers

The defects treated in this section may occur with air drying, e.g. cellulose or with stoving synthetic lacquers.

(a) The stains may have been caused by insufficient washing of the metallic surfaces after pickling and by applying the lacquers on the unclean surfaces. Incomplete washing of, e.g. silver-plated steel for removal of residues left over from the silver salts used in the plating baths can produce tarnishing somewhat similar to that caused by remnants of fats, etc. (see also below in sub-section (c).

(b) 'Oxidized' metals (such as, for instance, silver) containing arsenic may cause staining as well as peeling of the lacquer films.

Particular care must also be exercised in lacquering sulphur-treated or brass-plated iron or steel parts.

In the first place surfaces must be metal-clean before lacquering. Beware of metals containing arsenic. Metals which have been coloured by chemical or galvanic processes should be treated with special lacquers. The faults in question do not always appear at once, but may become visible some time after the application of the lacquer.

(c) Insufficient pretreatment of metals contaminated with residues (sometimes only slight traces of them) of fat, grease, oil, polishing agents, sweat from the finger-tips or similar substances which may in the course of time react beneath the film with certain metals, such as polished brass.

Iridescence as a consequence of incomplete degreasing of metallic surfaces. This fault also appears only after a considerable time in many instances.

(d) Staining may have been caused by the addition of unsuitable thinners, such as those containing acids, or of insufficiently refined benzene homologues (toluol, xylol, solvent naphtha) and probably also cruder petroleum distillates, containing sulphur compounds which tend to react with some metals. The metals mostly endangered by such thinners are copper, brasses, tombac, nickel and silver, on which black stains (sulphides) develop.

The possibility also exists that stains can have been caused by acidity of the medium, or by alkalinity, e.g. of a water–thinnable coating. According to the metal such stains may be coloured, e.g. green on copper. The surface of zinc under an acid coating can take on a crystalline structure.

It is always advisable to procure the thinners from the firms supplying the lacquers in order to avoid difficulties due to unsuitable mixtures.

A more neutral surface coating should be selected, if necessary.

(e) Staining of cellulose ester films may be caused by 'rust-spotting'. Ethyl-cellulose is stabilized against discoloration by iron and water by the addition of small quantities (0·5%) of a nitrobenzoate [1].

Note

[1] Dow Chemical Co. British Patent 1,449,854: 1976.

7.23 Susceptibility to heat and pressure. Film hardness generally

There is considerable variation between different types of organic binder in respect of resistance to heat. Darkening and embrittlement are the common outcome and the number of binders which will withstand prolonged heating to temperatures above 120–130°C without appreciable change is relatively small. A study of the heat resistance of a number of finishes was carried out by W. Berger [1] who exposed coated metal panels at 90°C and at 120°C for 1600 hours. Silicone resin only remained unchanged and except for polyurethanes based on aliphatic isocyanate, other coatings developed yellow or brown discoloration. With alkyds and with polyurethanes based on aromatic isocyanate considerable discoloration developed, but the protective properties were fully retained. Chlorinated and isomerized rubbers became very brittle at 90°C but 75°C is normally accepted as the safe upper limit for chlorinated rubber paints.

The majority of organic binders of the oxidizable or thermosetting types decompose when heated to temperatures in excess of 250°C. A notable exception is silicone resin which requires heating to about 200°C to harden but will then withstand temperatures up to about 320°C at which point decomposition commences.

Silicone resins are used as binders for high-temperature resisting aluminium paints which are used on steel chimney stacks, exhaust manifolds, etc. where temperatures as high as 450–500°C are reached. It is likely that, although the silicone resin decomposes at about 320°C, the success of these paints is due to a sintering or partial fusion of the lamellar aluminium to the steel surface, thereby providing good protection.

The use of chlorinated rubber, epoxy resins etc. in primers for steel has stimulated interest in the products of combustion of these resins as the primed metal often has to undergo welding or flame cutting. With chlorinated rubber, hydrochloric acid is the main product but small amounts of chlorine and phosgene also are produced if the paint contains an oxidizing pigment. Before welding or flame cutting is carried out a strip of primer, of any type, should be removed by solvent or abrasive. Manufacturers of metal primers can provide Safety Data Sheets which state the products of combustion of their products.

The methods in common use for the measurement of film hardness can be classified into:

(i) Scratch methods [2].
(ii) Rocker methods including the various pendulum methods [3].
(iii) Impact methods using a falling weight [4] or ball [5].
(iv) Indentation methods. Among the indentation methods the use of the micro-indentation apparatus [6] deserves special mention in view of the amount of information it provides. The instrument is capable of very precise measurements. A ball-ended needle is applied to the paint film and the depth of penetration under constant load is recorded as a function of time. The apparatus can be used to follow changes in films during weathering or ageing and permits measurements to be made over a range of temperatures.

Low film hardness can be due to one or more of the following causes. The film has not oxidized sufficiently, due to:

(a) An excessively thick coating resulting in slow diffusion of oxygen through the film. This is important when using the 'high build' types where the specified film thicknesses should not be exceeded.

(b) Insufficient drier. This may have arisen through adsorption of drier by the pigment during storage. The manufacturer should be consulted.

(c) Solvent retention in the film (particularly in 'high build' systems) preventing access of oxygen to the unsaturated centres. This happens when a proportion of high boiling solvent has been used and the fault is accentuated by low temperature. At 4°C (the lowest temperature for application of most paints) the drying time is roughly double that at 14°C. The drying and hardening of small objects can often be accelerated by application of mild heat ('forced drying') up to 80°C.

(d) High polymer resins, as applied in the surface-coating field, are prone to solvent retention. This can be very persistent, and the presence of residual solvents even of a fast evaporation rate has been found in films months and longer after their application. The rate of evaporation is retarded and film flow ceased after 50–70% of the volatiles has evaporated; the remaining volatile matter reaches the film surface only by diffusion. For this reason care must be taken with 'high build' chlorinated rubber coatings. Slow evaporating solvents should be avoided as far as possible.

(e) Solvent retention may produce the wrong impression of a tough and elastic film. This illusion lasts, however, only as long as the solvent, mostly of the slow type, is present, which takes on somewhat the role of a plasticizer. It can, of course, also deceive by keeping a film soft too long. Also, it can prevent a film over a long period from becoming marproof. High polymer lacquer resins coming under this heading are p.v.a., p.v.c. and their copolymers, p.v. butyral, polystyrene, polymethacrylates and acrylates. For the latter two the combination with nitrocellulose has been claimed to result in films which release solvents quickly [4]. Thick films release solvents more slowly than thin ones. Bituminous coatings, too, do not release easily.

Force drying or stoving helps greatly, though not always. Solvents and thinners must be carefully selected. Amongst media which possess a rapid solvent release are styrenated and vinyl toluenated alkyds.

(f) Air-drying long-oil alkyd finishes and finishes containing thermoplastic resins rarely become completely resistant to impact and continuous rubbing even after complete drying. Such films are sensitive to mechanical cleaning and chemical detergents.

Newly painted articles such as kitchen chairs and tables require to be brought into use as soon as possible after painting. The surfaces, particularly the legs, are likely to suffer frequent impact and it is wise to use a quicker hardening finish. Among those available for this type of work are finishes based on polyurethane alkyds and polyurethane oils, both of which have given satisfactory results.

(g) Poor abrasion resistance of stoved finishes is due to insufficient hardening through, particularly of too heavily applied films, or due to insufficient stoving (case hardening). Films of this type may have at first the appearance of satisfactory adhesion and scratch resistance, but will become brittle and sensitive to shock after complete drying. If films containing a thermo-setting type of medium have not been sufficiently hardened, corrosion of the base may develop surprisingly quickly.

It is preferable to apply a number of thin films instead of one heavy layer. A short stoving treatment of up to about 130°C (265°F) can be applied for the acceleration and improvement of the hardening process even of nitrocellulose films. If the fault is exhibited by stoving enamels, it is necessary to raise the stoving temperatures or to prolong the stoving period, depending on conditions. The use of, if necessary, a portable unit for radiation heating in connection with suitable paints should be considered.

'*Mar resistance*' is a term used to express the resistance of a paint film to marking or scratching. It is usually applied to stoved industrial finishes. Mar resistance can often be improved by the addition of a small quantity of specific silicone oil to the liquid paint.

Notes

[1] Berger, W. (1967), *JOCCA,* **50**, 244.
[2] B.S. 3900, Part E2. Scratch test.
[3] B.S. 3900, Part E5. Pendulum damping test.
[4] B.S. 3900, Part E3. Impact (falling weight) resistance.
[5] B.S. 3900, Part E7. Resistance to impact (falling ball test).
[6] Monk, C.J.H. and Wright, T.A. (1965), *JOCCA,* **48**, 520.

7.24 Faults developing in paints exposed to heat, e.g. on radiators or fires

The life of paint films exposed to heat varies with the nature of the paint and the character of the exposure, particularly the temperature. Most coatings will yellow, and if exposure to heat is severe and prolonged, hardening will occur followed eventually by flaking.

On hot water radiators most [1] alkyd gloss and flat enamels will give satisfactory service. If the temperature exceeds 120°C coatings based on styrenated alkyds or semi-drying oil alkyds will retain their colour better than other alkyds and the hardening process will be slow. If a greater degree of yellowing is acceptable or darker colours are chosen, alkyds can give good service at up to 150°C. From the point of view of efficient radiation darker colours are preferable on radiators. Phenolic enamels, although prone to severe yellowing, will tolerate temperatures of 180–200°C. Yellowing in itself is no measure of the deterioration of the coating [1].

Acrylic enamels usually show good colour retention and the best results are achieved with silicone resin coatings at up 250°C. Polyurethanes cured with aliphatic isocyanates show good resistance to yellowing at moderate temperatures.

It must be emphasized that heat-resisting coatings should be specially formulated for the purpose; careful selection of pigments and additives plays an important part in achieving good results. Most binders referred to are organic materials and therefore subject to thermal decomposition when the temperature exceeds 250°C. If exposure to higher temperatures is envisaged inorganic binders or organic binders pigmented with aluminium flake must be chosen. In the case of the latter the organic constituents will decompose, but the aluminium powder is capable of alloying to the steel surface. Aluminium paints based on silicones or butyl-titanate [2] will give reasonable protection up to 600°C. Good protection is also obtained with zinc dust pigmented paints, but the temperature limit should be set at 400°C because metallic zinc melts at 420°C.

If an aluminium finish is not acceptable, coatings resistant to very high temperatures (400°C and above) pigmented with suitable white or coloured inorganic pigments can be formulated with alkali silicate or quaternary ammonium silicate binders.

It is important to clean scrupulously before painting radiator surfaces or other equipment which is to be exposed to high temperatures.

(a) If the radiators, etc., are taken into operation too soon, i.e. before the film has completely dried and while it still contains some thinners, blisters may be formed.

(b) If paints contain too short oil varnishes or too much pigment, peeling can result. Excessively thinned primers behave similarly to the short oil coatings.

(c) An originally satisfactory priming coat on, e.g. radiators may, if the latter have been stored in the open for too long a time before fitting, lose its protective effect for iron (rusting), especially if the coat was applied very thinly, and may peel off during service. Premature brittleness may also be caused by excessive stoving.

(d) Excessive thickness of shop primer and prolonged heating of the primer before overcoating leads to flaking [3].

Notes

[1] Berger, W. (1967), *JOCCA*, **50**, 244.
[2] Sidlow, J. (1956), *JOCCA*, **39**, 415.
[3] Dostler, K. (1965), *Trav. Peint.* **20**, No. 1, 10.

7.25 Lack of solvent and lubricating oil resistance

A condition for solvent resistance of a film is that for all practical purposes either the main ingredients of the binder become insoluble after air or force drying or stoving, or that they are insoluble in the solvent to which they are exposed (see Fig. 7.55). If the drying-oil-based vehicle is after oxidation to withstand aliphatic solvents and oils, attention must be paid to the phenomenon of drier absorption by pigments either shortly after manufacture, or on storage. A white spirit test is usually carried out at intervals as a check.

The best solvent-resistant coatings all round are those which have been cross-linked by thermosetting, or with the help of catalysts by cold or force-dried curing.

If the contact with the solvent is of short duration and/or it vaporizes fast, the problem is in most cases not as difficult as on extended exposure. Films are porous and, even if insoluble, can be swollen and thus lifted. Consequently several coats may be required; in extreme cases instead of the customary three or so more than twice this number may be needed.

(a) Motor fuels and lubricating oils have become more and more aggressive to paint films. Petrols may not only contain a proportion of aromatic hydrocarbons and ethyl alcohol, apart from the aliphatic gasoline, but other additives, too.

(i) Air-drying coatings must be thoroughly dry, and oil-base paints, etc., fully oxidized before being brought in contact. These and lacquers based on shellac, chlorinated rubber, certain cellulose derivatives, etc., withstand paraffin-type petrols well. Resistance to benzol-type fuel mixtures, some even in the presence of alcohol, is shown, e.g. by acetyl cellulose, vinyl polymer and copolymer and some acrylate lacquers, and a number of water-thinnable coatings, such as casein paints.

Two-pack cold curing lacquers, e.g. polyester/isocyanate or epoxy amine cured systems give exceptional resistance to modern fuels and lubricants.

(ii) Stoved enamels must be fully cured. Partial curing due to understoving can inhibit the completion of curing when the film is eventually stoved correctly. Certain high temperature blacks, alkyd amino blends and solution p.v.c. coatings will give adequate resistance to hydrocarbons. The best results are given by thermosetting phenolics, and epoxy phenolics.

(b) Solvent attack in general may either only soften or remove a film. In the former case the damage may be negligible, if recovery takes place by re-hardening fully, or almost completely, after evaporation of the solvent, perhaps with a little loss in gloss.

In a few cases air-drying lacquers will perform satisfactorily. In the majority of cases stoving or at least catalysed cold curing or force drying coatings will be required. It must be understood that the latter usually need at least one week at room temperature to cross-link, but that even then they are not as resistant as corresponding stoved films. If films can stand it, the stoving should be as long and as high as possible for maximum results.

To the most resistant coatings generally belong thermosetting phenolic, phenolic epoxy and acrylic lacquers, also activated, e.g. isocyanate/polyester resin (resistant even to ester solvents), polyamide or amine hardening epoxy resin lacquers. Vinyl plastisols and organosols resist some solvents, including certain ketones, quite well after fusing.

Amongst coatings which resist esters and ketones are nitrocellulose-urea formaldehyde resin lacquers and solutions of isomerized or cyclized rubber.

Polyvinyl acetate (and its copolymers) lacquers are insoluble in ethyl ether.

(c) Films of shellac varnishes or polishes and lacquers which are soluble in alcohols are most unsuitable for the furnishing of public houses, liquor sets, decorations on bottles and such-like.

Storage vessels for wines and beers, too, belong to this group, though they have to stand up against a number of additional actions. They are now mostly of metallic construction and are lined with pigmented epoxy phenolic coatings or even a disposable plastic liner. For use on furnishings the preferred coatings are acid catalysed amine lacquers or two-pack polyurethane varnishes.

(d) The resistance of paint films to household detergents can be important especially with domestic appliances such as washing machines. Detergents consist of aqueous solutions of surface-active agents, but frequently contain also salts, some of them aggressive if in long contact with a paint film, especially in undiluted or very hot state.

P.v.c. based solution coatings, organosols and plastisols give good resistance qualities over the range of modern detergents. The external finish for washing machines to withstand hot detergents may be based upon thermoset acrylics, epoxy amino, polyurethanes or epoxy powder coatings.

(e) Resistance to carbon tetrachloride is sometimes required for coatings on fire extinguishers, or carriers of articles which have to undergo chemical processes such as coated aluminium bobbins in the rayon industry. The latter must usually also resist strongly acid solutions.

Thin film paint systems are not really adequate to withstand the combination of hot chemical and mechanical abrasion. Thermoplastic high molecular weight polymers applied by fluidized bed methods are the preferred coatings.

7.26 Faulty resistance to acids, alkalis, etc.

When a finish is required to resist specialized chemical attack, such as that caused by acids or alkalis, it is necessary to obtain special coatings designed to resist such attack. Full details of the requirements should then be given to the paint manufacturer as follows:

(a) The chemical involved, with the concentration of the solution, or vapour in the case of volatile materials. The temperature should be stated, and complete immersion, intermittent immersion, splashing or exposure to vapour only, must be clearly differentiated. Exaggeration of the hazards should be avoided, e.g. exposure to vapour above a concentrated alkali bath (in the absence of spray from the liquid) is generally only equivalent to exposure to steam, alkali resistance being unnecessary. On the other hand, vapours above tanks containing hot acids may be strongly acidic.

(b) The nature of the substrate to be painted should be clearly stated, and its condition, e.g. 'severely corroded', 'new and as yet uncorroded' or 'already painted' (with the type of finish if known).

(c) Facilities available to accelerate drying should be stated. This may assist the paint manufacturer in that materials of improved performance could be used if stoving or force drying is available. The maximum time allowable for painting and drying is also an essential factor in paint selection.

(d) Any other special requirements, such as resistance to abrasion, etc.

A wide range of specialized coatings is now available, and most situations can be dealt with satisfactorily if the precise conditions are known by the paint manufacturer. Good resistance to acids, alkalis and other chemicals are given (but in varying degrees) by acrylic resin, bituminous coatings, chlorinated rubber, vinyl acetate-chloride copolymers, polyvinyl ether resins, post-chlorinated polyvinyl chloride, polystyrene, cyclized rubber, cellulose ethers, pure phenolic resin-tung oil varnishes, polyurethanes, epoxide coatings (epoxy ester or the amine or otherwise cured resins), etc. Additional requirements, however, will usually narrow the field for selection, e.g. necessity for a pale shade will eliminate bituminous coatings, and resistance to solvents may prevent the use of the non-convertible materials (i.e. all but the last three types in the list above).

In non-convertible coatings the choice of plasticizers is of paramount importance and ester types must be avoided because of their saponifiable nature.

Pigments vary in their ability to resist attack by acids and alkalis and it may not be possible to meet specific colour requirements for some environments. Among the pigments and extenders recommended for chemically resistant finishes and primers are:

(i) Titanium dioxide, the most suitable white pigment.
(ii) Natural iron oxides (red oxide, black oxide, ochre, micaceous iron oxide) and some synthetic iron oxides.
(iii) Carbon blacks.
(iv) Phthalocyanine blues and greens.
(v) Some cadmium yellows.
(vi) Cobalt blue is resistant to weak acids and alkalis.
(vii) Barium sulphate (barytes and blanc fixe).
(viii) Silica.

Unsuitable pigments and extenders are:

(i) All chromate and lead pigments with the exception of metallic lead. Red lead may exhibit premature breakdown in contact with sodium chloride solutions (sea water), owing to saponification of the medium by caustic soda formed at cathodic areas.
(ii) Zinc oxide, zinc dust, aluminium – and other bronze powders.
(iii) Whiting (calcium carbonate).
(iv) Iron blue (Prussian blue) is very acid resistant, but readily attacked by alkalis.
(v) Ultramarine blue is resistant to alkali, but is destroyed in contact with acid.

It is of particular importance in these specialized circumstances to obtain adequate film thickness, preferably by application of a number of coats, although ‘filled’ bitumen solutions, p.v.c. plastisols and organosols, epoxy polyamide, etc., resin and thixotropic epoxy/coal-tar pitch and chlorinated rubber compositions are available which will produce very thick films in one application.

Some of the agents e.g. treated clays, used to achieve thixotropy can diminish chemical resistance and tests should be carried out to establish whether acid or alkali resistance is sufficient for the purpose. Single coats of traditional paints are permeable to water and water vapour, and also may contain pores or other imperfections. Multiple coats will ensure that these are reduced to a minimum. Vinylidene chloride-acrylonitrile copolymers are said to give films exceptionally free from pores and thus more impermeable to water and water vapour than most coatings. These films also possess 'high resistance to solvents, fats, acids and alkalis; they are tasteless, non-inflammable and non-objectionable physiologically' [1].

Stoving or force drying of films will usually enhance their resistance, and may also prolong their life.

Note

[1] Riese, W. (1956), *Deutsche Farben Z.*, **11**, 411.

7.27 Paint failures on photographic equipment

(a) Longish black stains and dots on photographic films can be caused by small chips from brittle air-drying or overbaked enamel films which settle on the photographic emulsions. Watch the drying process carefully or employ a more elastic enamel. Sandblasting is useful, but not always required.

(b) Photographic emulsions can be influenced to some extent by a number of finishes. For instance, finishes based on oil-modified alkyd resins appear to be particularly obnoxious. Film containers, to be used for prolonged storage under unfavourable conditions, e.g. in the tropics, should be coated with specially tested enamels. Fully cured cross-linking media are the best choice.

(c) Reflection of light from enamelled parts of a camera. The film is not thoroughly matt, perhaps due to insufficient stoving or faulty formulation. Flock spraying can give good results.

(d) The matt black finish on a projector lamp housing changed to red after a short period of use.

This fault was due to an unsuitable grade of black iron oxide which was converted by heat to the more stable red iron oxide. Pigments must be carefully checked before incorporation in formulations of heat-resistant coatings.

7.28 Influence on the taste of food of stoving canning lacquers

(a) Under-cured films of stoved oleoresinous, epoxy and other synthetic finishes may be partially dissolved by the food pack. An unpleasant taste is imparted by certain vegetable oils, cellulose acetobutyrate and phenolic residues.

Solvents retained in the stoved film can affect taste and odour. The presence of even small traces of diacetone alcohol can give rise to a mouse-like odour and the high boiling components of certain aromatic hydrocarbons have a pronounced effect upon taste. In general most alcohols, simple aromatic hydrocarbons and glycol ethers are acceptable solvents for can lining lacquers.

It is obvious, of course, that all varnishes used in connection with food or fruit canning must be absolutely free of any toxic substances, including lead salts, because they may be either dissolved out, or by the stripping of part of the film, become mixed with the food. Gum-oil and phenol resin varnish types are most commonly used, p.v.c., and epoxide resin based lacquers are other possibilities.

They are many toxicity regulations relating to products in contact with food-stuffs but the universally accepted standards are contained in the U.S. Food and Drug Administration acts as defined by the 1958 amendments. Prior to this date the old law had totally prohibited the use of any poisonous or deleterious additive and had placed the burden of proof for such unacceptability on the government and their own overworked testing laboratories. The new amendment involved two major changes. It provided for the use of additives *at safe levels* and it made industry responsible for pre-testing food additives to provide evidence for their safety under *conditions of intended use.*

(b) Canned foods contain many reactive ingredients which are particularly severe upon the can lining during the processing operation prior to can sealing. Fruit acids i.e. citric, tartaric and foods containing acetic acid are best packed in cans lined with phenolic lacquers or copal varnishes. Certain fruits are packed in uncoated tin cans to preserve the product's appearance. Meat and fish packs contain sulphurous compounds which give rise to traces of hydrogen sulphide. The internal wall of meat cans is protected by phenolic or epoxy phenolic lacquers while the ends are coated with a permeable coating of oleo-resinous varnish containing zinc oxide.

(c) It sometimes happens that lacquers sufficiently resistant to water and acids, as well as to the common tests for coatings employed in the canning industries, prove to lack resistance toward edible fats and oils (such as used for sausage and fish preserves) and to the alternating influence of fats and water (for instance, acid waters), e.g. in the canning of fish. Chlorinated rubber is softened by animal fats and must be avoided.

Specially formulated epoxy phenolic coatings pigmented with aluminium and containing meat release agents are used with solid meat packs.

(d) Beer and similar beverages need specially lacquered cans which have a roller applied base coat of epoxy phenolic and a two coat spray applied topcoat of a vinyl chloride/vinylidene chloride copolymer.

7.29 Defects caused by faulty application of electrical insulating varnishes

The requirements of an impregnating varnish vary considerably and suitability for a particular job is determined by:

(i) Type of equipment–stationary in air or in oil; rotating.
(ii) Conditions of service.
(iii) Electrical and thermal characteristics of other materials used in the construction of the equipment.
(iv) Construction of the winding to be varnished.
(v) Temperature classification of the equipment [1].

In addition to adequate dielectric strength and low loss angle, armature-impregnating varnish must possess the following properties:

(i) Elasticity and, consequently, resistance to ageing by as slow a loss of elasticity as possible under working conditions. In order to attain these properties to the fullest extent it is advisable not to demand excessively short drying time.
(ii) Resistance to hot lubricating oil.
(iii) Small development of acidity during drying to avoid corrosion of copper and tendering of fibrous covering of winding wires.
(iv) High impermeability to moisture of the dried films and a high proportion of solids to ensure complete filling of pores and capillaries and so delay the reabsorption of moisture from the atmosphere.
(v) Non-softening, non-tackiness at the working temperature of the electric machine for which the varnish is selected.

Many of these properties can be evaluated by laboratory tests [2].

Unsatisfactory insulation often arises from faulty application techniques.

(a) Conducting solvents may remain behind in the insulating layer (produced by varnish or compound) if:

(i) The varnish is too thick and could not dry through sufficiently in the usual time. Air bubbles may also be created in the insulation.
    The consistency should be adjusted by the addition of thinners recommended by the maker. High-boiling solvents such as heavy naphtha should be avoided, because they may interfere with drying. The use of light petrol fractions may create an excessive fire hazard. Do not overthin, for excessive draining may cause insufficient filling of the windings. The temperature of the varnish in the dipping tank or impregnator should not be allowed to rise above about 50°C or the varnish may become viscous by partial polymerization, in which case it cannot be thinned to its original consistency.
(ii) Insufficient drying time is allowed. Apart from increasing the time, check that the actual temperature over the whole of the oven is that registered by the oven thermometer.
(iii) The initial heating is too sudden. Only surface drying occurred and the solvents

of the lower layers could not escape, nor could the whole layer be thoroughly oxidized or polymerized.

(iv) Coils are constructed with impervious solid insulation between layers, so that subsequent removal of solvent is extremely difficult. It is usually necessary to impregnate such coils by vacuum impregnation and to remove solvent by partial vacuum at a temperature below that required for baking. The use of a solventless varnish is often helpful in these circumstances.

(v) Ventilation of the oven is inadequate. It is essential that ovens have an adequate ventilating system capable of removing solvent vapours and volatile products of drying of the varnish. Apart from the safety aspect, solvent cannot be removed from a winding if the air in the oven is saturated with solvent vapour. If the varnish used dries by oxidation, retardation of the drying process can also be due to lack of oxygen.

(b) Poor penetration of varnish.

May be due to too rapid dipping creating air inclusions; the unsuitability of the varnish for the particular type of winding or the varnish being too stout.

The rate of dipping should be about 12 mm (½ in) a minute to allow air to be driven before the advancing varnish front. Cause of unsuitability of the varnish would be wrong surface relationship between varnish and winding preventing penetration, whereas stoutness of a suitable varnish slows down the rate of penetration so that in a given time of dipping, appropriate to a varnish of correct consistency, penetration will not have been achieved.

The remedy for poor penetration is therefore to dip slowly, check consistency of varnish; if these fail, change to another varnish.

(c) Sweaty fingers of the winding operator. Some operators exude sweat more damaging to electrical insulation than others, and females are particularly suspect, particularly during menstruation.

Notes

[1] B.S. Spec. No. 2757, 1956.
[2] B.S. Spec. No. 2778, 1956.

7.30 Defects on windings and electrical apparatus due to softening of wire enamel

(a) Softening of wire enamel by varnish solvent.

Before impregnation with varnish, it had been the practice of many years standing to preheat the windings for up to 2 hours at 120–130°C to ensure that the enamel was fully cured and had reached a condition of maximum resistance to solvent. The cost of this operation has now made it more attractive to move to a better grade of enamelled wire.

Vinyl enamels have much superior resistance to solvents, and usually give no trouble provided they are fully cured.

The adhesion to copper of some types of polyester wire enamels is poor and prolonged contact with solvent, even white spirit, may cause swelling of the enamel to produce a soft sleeve completely detached from the wire. The condition is usually found in heavy field coils; the wire may cut through the swollen enamel and cause shorted turns.

When impregnating enamelled wires it is important to use only thinners recommended by the makers of the varnish.

(b) Softening of wire enamel by the non-volatile part of varnish.

The insulation provided by wire enamel may be destroyed by products of polymerization of some types of phenolic resin based varnishes. The defect is usually confined to totally enclosed apparatus which attains a high working temperature; lifting magnets are a typical example. The varnish must be changed to one designed to avoid the defect.

7.31 Defects of impregnations with varnish which develop during service

(a) Premature ageing of varnish due to abnormally high service temperature.

Prolonged exposure of a varnish to a temperature in excess of that for which it is designed leads to rapid embrittlement and cracking. On cooling moisture is absorbed into the cracks.

More flexible varnish must be used if the equipment normally works at a high temperature, but if not intended to do so electrical tests should be made to determine the cause of the abnormal temperature. For the protection of motors running permanently at temperatures of up to 180°C windings wrapped in glasscloth can be impregnated with insulating varnish based on certain silicones.

(b) Chemical action, such as by acid and ammoniacal vapours and salt deposits, may decompose films of insulating varnish.

Apply sufficient varnish designed for such circumstances to give a smooth finish.

Shellac-based varnishes must be avoided when alkaline vapours or solutions are present.

In extreme cases enveloping or even potting techniques may have to be employed.

(c) Damage to insulation by fungus growth.

Temperatures of 25°C and upwards combined with relative humidity in excess of about 80% create atmospheric conditions conducive to fungus growth. Fungus only makes its appearance on apparatus in store or during long idle periods in tropical climates, because the heat of working kills moulds.

Primary damage to the insulation usually arises from partial hydrolysis, but the presence of fungus further reduces the insulation resistance by providing leakage paths along the moisture-laden hyphal strands.

According to the severity of the attack the remedy may vary from multiple coatings of conventional insulating varnish to produce a smooth finish, followed by an air drying finishing varnish containing fungicide [1], to potting in polyester.

To aid the production of a smooth external surface of the winding, fibrous wire covering and solid insulation should be avoided.

(d) Breakdown of insulation by 'tracking'.

The mechanism of tracking is obscure and may well depend on a number of quite different causes, but in all cases it ends in the formation of a carbon path across the insulating material between conductors at different potentials; it is facilitated by the deposition of salts or other conducting deposits.

In general, impregnating varnishes have quite good track resistance characteristics, but care should be taken to ensure a smooth finish to the varnished surface to discourage lodgement. For extra protection the use of a non-spirit finishing varnish is recommended.

A method of determining tracking resistance is described in International Electro-technical Commission Publication No. 122.

Note

[1] B.S. 634: 1961. British Standards Institution.

7.32 Defects in oil immersed windings

(a) Development of sludge and/or acidity in the cooling oil.

Sludging and development of acidity of transformer oil is due to the failure of the impregnating varnish to form a perfect barrier between the cooling oil and the copper conductor.

An oleoresinous varnish should not have been used, because, since it requires oxygen for its conversion to a solid, it does not dry in the interior of the winding.

Most thermohardening varnishes are resistant to both transformer and lubricating oils, but it is advisable to ask especially for varnishes in which this characteristic is highly developed.

Given that a suitable varnish has been used, but sludge and high acidity develop:

(i) The varnish has been insufficiently dried due either to too low a temperature or to shortening of the curing time.
(ii) The varnish has become too stout and has not been fully cured.

The insulating oil should be of high quality.

7.33 Open circuits in high speed armatures

These defects which become evident during testing usually have the following causes:

(a) The stresses set up between varnish and wire, as the winding cools after baking, have been sufficient to break the wire.

The varnish dries too hard for this purpose; a more flexible varnish should be used.

(b) Some of the wires at the 'back-end' have been broken by centrifugal forces.

Presuming that the construction is not at fault, the varnish used has insufficient bond strength to hold the wires in place. Varnish for this use must cure to an extremely tough gel of good bond strength. As apparatus of this type frequently works at a high temperature the bond strength must not decrease at the working temperature.

Alternatively the wire may not break, but be loose enough to chafe the varnish and enamel insulation and so cause short circuits from one coil to another.

7.34 Various defects in connection with coating of aircraft

Anodized aluminium or titanium metals for high speed aircraft, where temperatures up to 600°C could be attained, do not require painting but they are much more costly than paint systems. It is still considered necessary to protect the structure of military and civil aircraft and the severe environmental variations can cause stresses in the paint films. Environmental effects include radiation at high altitudes of up to 12 miles, the kinetic heating of supersonic aircraft, rapid temperature changes, erosion and cracking caused by driving rain and ice, often in a salt-laden atmosphere. Lacquers are also prone to failure on flexing at low temperatures.

(a) Failures due to hydraulic fluids, fuel, lubricants, brake fluids, de-icing substances.

Many of these materials are extremely aggressive to paint films, in particular burning fuels discharged through the exhaust, skydrol, engine lubricants containing hot diester dispersants and alcohols in de-icing fluids. Oils and grease are rarely aggressive to paint films.

Finishes are chosen for their special performance properties and include acrylics, epoxies, nitrocellulose and polyurethane materials [1].

Pretreatment is essential with chromating or anodizing as the most favoured methods.

(b) Failures due to primer system.

Although aircraft are basically made of aluminium, they are, in fact, complex assemblies of dissimilar metals, including steel, stainless steel, magnesium etc. Corrosion can occur at dissimilar joints e.g. aluminium alloys and steel. These sections require the use of a special jointing compound containing chromate pigments and extenders.

Etch primers may be used without pretreatment, but this is not recommended. These materials often contain phenolic resin to improve water resistance qualities.

Epoxy primers crosslinked by polyamides to give good flexing properties are generally pigmented with strontium chromate and applied over anodized surfaces.

Corrosion of aluminium alloys at cut edges is prevented by the use of non-permeable coatings which protect by the inhibiting effect of selected chromates leaching out of the primer film. If the leaching rate is too high blistering of the film can result. Controlled leaching in keel areas, seams and rivets may be partially achieved by the use of a top coat containing barium chromate.

Magnesium alloys require multicoat applications of epoxy primers followed by chromate pigmented top coats.

Primers based upon epoxy resins tend to yellow at elevated temperatures and exhibit pronounced chalking.

(c) Failures due to finishing coats.

At speeds above 500 m.p.h. the forces encountered by the strike surfaces are such that they cannot be withstood by normal paint coatings. Stripping and erosion will take place. Thick coatings of elastomeric polyurethanes show the best resistance to these forces.

The problem of poor flexibility at low temperatures with half-second nitrocellulose has been investigated by the Hercules Powder Company and improved films obtained by use of 5–6 seconds nitrocellulose [2].

Multiple coats of methylmethacrylate lacquers perform well in contact with aircraft fluids. They are not suitable against hot oils.

For high temperature resistance, paint based on polyimides will withstand up to 400°C.

(d) Maintenance repainting might have to take place anywhere in the world and under bad conditions. It is particularly important that etch primers are not applied above 75% relative humidity while lacquers drying by evaporation are best applied below 70% relative humidity. Poor results have been obtained with etch primers when the relative humidity was less than 30%.

No defects at high humidity have been observed with polyurethane lacquers.

Recommended application temperatures are: for etch primers 15–32°C (60–90°F); nitrocellulose lacquers 10–32°C (50–90°F); acrylic lacquers 4–32°C (40–90°F).

(e) Joining of dissimilar metals, e.g. stainless steel and Duralumin on seaplane hulls and floats. Contact between stainless steel and Duralumin without careful insulation leads to corrosion.

Use special jointing compounds.

(f) Aggressive hydraulic fluids, lubricants, brake fluids, alcohol-type de-icing solutions and such-like compounds can soften and dissolve surface coatings. Lacquers based, for instance, on alkyds in combination with nitrocellulose or vinyl resins succumb fairly easily.

Catalysed lacquer systems, e.g. based on polyurethane, epoxy resins or nitrocellulose–acrylic ester lacquers over suitable primers can be completely resistant to a number of agressive fluids in use. Anti-freeze de-icing lacquers which contain free glycerol, etc., and/or silicone oils and a wax, to which ice has little adhesion, and from which speedy removal should be possible, are available.

Defects which may occur on wooden aircraft should be identifiable with the help of the various Sections dealing with plywood or wood coatings generally.

(g) Stoving finishes must have schedules in accordance with the time/temperature limits required by various light alloys. The maximum for magnesium alloys lies at one hour at 220°C per coat [3]. (See Table 4.2.)

Notes

[1] Hoey, C.E., (1968), *JOCCA*, **53**, 1026.
[2] Hercules Powder Company. Technical data sheet CSL 109.
[3] *Magnesium Data book* p. 40.

7.35 (A) Premature decay of coatings, especially on vehicles

Many coatings possess poor resistance to weathering, although quite adequate for other purposes. Examples are those based on spirit varnishes, rosin varnishes (unless the rosin content is very low compared with that of durable oils or resin), short-oil copal varnishes, tall oil (as distinct from tall-oil fatty acids of low rosin content), insufficiently plasticized nitrocellulose or chlorinated rubber and other lacquers, and any short-oil varnishes made from hard synthetic resins (e.g. rosin-modified phenolics.) Short-oil alkyd varnishes can have excellent durability.

Polyurethane two pack finishes employing an aliphatic di-isocyanate curing agent give finishes with considerably greater durability and toughness than oleo-resinous and alkyd varnishes. Styrenated alkyds have moderate durability but acrylated alkyds can show excellent improvements over standard alkyds in resistance to yellowing.

The overall performance of finishing paints applied to metallic vehicle surfaces is improved by electrophoretic priming after manufacture. Excellent chip resistance to sills and wheels is given by epoxy based powder coatings.

Many coatings will give lives of five to six years in conditions such as prevail in central Europe and Great Britain. Generally, the greater the sunshine falling upon a coating, the shorter its life. Thus alkyd finishes giving six years' life in Great Britain may fail in less than two years in tropical areas. There are also significant differences between urban, rural and coastal sites, particularly with finishes which fail by chalking. These will give longest life in urban districts where the intensity of u.v. radiation is reduced by smoke or haze, and where the paint film is rapidly coated with a protective film of dirt. Alkyd finishes, particularly the long-oil types, have been found to give excellent life, but some other types of coating approach closely to these in performance.

The life of a system may, of course, be greatly prolonged by regular attention, thus avoiding the need for expensive stripping and complete repainting for as long as possible. Elasticity, hardness, adhesion, light resistance and swelling capacity are the most important factors controlling the life of a paint film. Durability is usually greater with stoved finishes, since these can contain little or no drier metal to contribute to the degradation of the film on exposure. Stoved finishes used by the Pennsylvania Railroad have given lives at least double those of air-drying finishes available at that period. (See also Section 7.36).

Normal weathering takes place as follows:

Loss of matter in a paint film first induces hazing and dulling, followed by chalking. The pigments are thus gradually freed from the surrounding and protecting vehicle and increasingly exposed to the influence of the atmosphere. Some of them are washed away by rain, and the repetition of this process renders the film more and more permeable to the humidity of the atmosphere, inducing corrosion and destruction of the coated objects unless the protective films are renewed in time.

(a) Care has not been taken to ensure that the paint materials used were suitable for exterior application. The requirements demanded of the weather resistance also vary considerably depending on the purpose and climate, and whether the coats will be permanently exposed outdoors, or only from time to time. The after-care of the films (for instance, of automobile lacquers) also plays an important part, of course.

In buying paint other than decorators' paint it should be made a practice to inform the paint manufacturer of the intended application outdoors.

(b) One-sided exposure of the objects, i.e. a railway wagon always travelling so that one side is subjected to less favourable conditions in a severe climate (in mountains, near a seaboard, weather side, etc.). The structural design also may exert an influence, especially where rain or snow water runs down the films in some places in larger quantities than in others.

In order to obtain maximum service of the coatings of vehicles, it is necessary to ensure uniform exposure on all sides. Railway wagons, for instance, should be turned around from time to time in order to alternate the sides most exposed to weathering. The portions of the films which come most in contact with running rain water are often the first to weather.

(c) Transparent varnishes and lacquers, especially those based on wood oil and nitrocellulose, are less resistant to the action of the open air and of ultra-violet rays than pigmented coatings. Resin-modified nitrocellulose lacquers usually yellow strongly first, and this is followed by cracking. Nitrocellulose films which have been polished with certain polishing oils can develop a yellow stain which otherwise would not have occurred [1].

The use in polishing liquids of white oils and special rubbing waxes has been suggested.

(d) If paints containing oil-base media are applied during the moist or cold season (autumn or winter time), the drying outdoors naturally requires a specially long period. Under such conditions the oil can absorb water, aided by the presence of smoke, causing the film to remain soft. It is then more easily washed off by rain (Gardner) [2]. The same applies for coal and damp rooms, such as in cold storage warehouses.

In conditions such as obtain in buses, white paints including those based on long-oil alkyds slowly tend to shrivel and then crack on the folds due to condensation effects, and peeling may follow.

The paint material should be applied during suitable weather, while the conditions within closed rooms should be as dry and warm as possible during the application and drying of the paint.

The use of drying oil modified alkyd enamels is recommended for such conditions.

Only white alkyd enamels will resist the condensation and nicotine fumes without discoloration. This is not true, however, of all alkyd enamels, but only of those types formulated specially for such conditions, usually modified with soya-bean oil (or other oil of low linolenic acid content), and containing some zinc oxide.

If damp conditions cannot be avoided, quick-drying finishes, such as those based on chlorinated rubber, may be helpful.

(e) Too early exposure of new applications to weather. Particularly sensitive are synthetic resin varnishes before a minimum of polymerization has been attained. But the same is true to a greater or lesser extent with regard to all other varnishes and lacquers. The life of films exposed too soon is considerably reduced. New films of nitrocellulose lacquers and such oil-base varnishes which contain comparatively large proportions of wood oil are easily spoiled by premature exposure to frost.

Newly finished vehicles should not be taken out of the shops until completely dry. It is advisable to leave them under good cover at least one week after finishing before being taken into use, and the surprisingly long life of the varnish coats applied up to the beginning of this century is probably due not only to the selection of high-grade materials and well-matured varnishes and to good workmanship, but also to the fact that the vehicles were not taken into operation too quickly after varnishing. The time interval is frequently reduced considerably by forced-drying methods.

Pure wood-oil varnishes should not be used for exterior application. Special, e.g. nitrocellulose lacquers, resistant to rather low temperatures have been developed.

(f) Grave defects as to reduced durability have been caused by incomplete cleaning of the base.

The explanations of premature destruction of paint films on building construction, in many instances, can also be applied to those on vehicles. (See Section 7.3.)

Notes

[1] Hercules, Data Sheet CSL-119.
[2] Gardner, H.A. Circ. 317, of the Am. P. & V. Man. Ass.

7.35 (B) Motor car finishing and re-finishing

It is now fairly common practice to prime car bodies by electro-painting, the advantages being the low fire risk associated with waterthinned paints, the uniform coating thickness and covering of internal surfaces, provided the body is designed correctly. Weld joints will only be penetrated if the panel fit is imperfect.

(a) Adhesion is poor on oily surfaces (in spite of claims to the contrary), and over zinc coated metal when alkalinity may cause total loss of adhesion.

(b) Filiform corrosion is a feature of electropaint systems. Its development depends on the acid value of the binder resin and possibly residues of soluble salts in the coating. The problem can be minimized by improving control of the phosphate coating weight applied as pretreatment and by preventing stone chipping down to the bare metal.

Dip primers are less frequently used nowadays. Solvent thinned materials present a fire risk and do not normally permit accurate control of film build.

The most serious corrosion problems in vehicle bodies occur in enclosed box sections. Dipping paints suffer from solvent reflux effects in box sections during stoving, but electrocoating has generally led to better protection in these areas. The primer is usually followed by a surfacer which promotes adhesion and prevents the texture of the steel surface showing through the topcoat.

Finishing coats for vehicle bodies must not only possess outstanding light fastness and weather resistance, but also resistance to petrol, oil, hydraulic fluids and to chipping [1].

Modern finishing paints are based on:

(i) Stoving alkyd amino resins.
(ii) Thermoplastic acrylics.
(iii) Thermosetting acrylics.
(iv) NADs (Non-aqueous dispersions, usually of thermosetting acrylic resins in aliphatic solvents).
(v) Nitrocellulose (declining in use).

Very generally speaking alkyds are less prone to chipping than acrylics. NADs reduce atmospheric pollution during application. This has become a serious issue, particularly in the United States, where various regulations modelled on the original Rule 66 have led to the virtual banning of aromatic solvents, whilst permitting the use of some esters and ketones. NADs also have the advantage of high build with good solvent release.

Other high solids two-pack systems based on isocyanate cured hydroxylated acrylic polymers have shown promise, especially in the refinishing of vehicles. However, the use of isocyanates demands very careful consideration of health and safety aspects during application [2]. It is believed that in future water-base enamels and powder coatings will make notable impact on car finishing.

Underbody protection is largely restricted to bituminous type coatings. They are not only cheap, but their performance in most respects, especially their water resistance, is good. Some manufacturers use the more expensive p.v.c. plastisols which are far superior in resistance to oil and petrol but not so resistant to water. Their durability is also better and a distinct feature is their ability to accept further coats of paint.

To assist in assessing the ultimate durability of vehicle coating systems there are several testing stations available in the United States, like Subtropical Testing Services Inc. in Florida and Desert Sunshine Exposure Test Inc. in Arizona, where after a relatively short duration of exposure, results can be obtained from which, by careful evaluation, valuable conclusions can be drawn on the likely durability of finishes [3].

Specific defects on refinished vehicles

(a) Loss of gloss. This is often due to hair cracking or the formation of water containing 'microblisters' which cannot readily be detected by the naked eye. If tackled in time, polishing with a mild abrasive can restore the gloss.

(b) Discoloration. There are many substances, like road tar, lubricating greases and oils, battery acid and hydraulic fluids, which can cause this defect. If prolonged contact with such contaminants is allowed, it may be necessary to remove all traces of discoloration by wet sanding the affected areas and respraying. If tackled immediately, thorough washing with an aqueous detergent or, if necessary, white spirit will prevent deeper penetration and avoid such drastic action. Mild polishing will then restore the original colour.

(c) Water spotting. White stains caused by evaporation of water droplets on the surface. The cause can be premature exposure of the paint film to rain or abnormal weather conditions, when a rain storm is immediately followed by strong sunshine. Excessive application of wax is another cause. Removal of all wax with white spirit and polishing should remedy the fault, but in very severe cases repeated polishing may not be effective; one has then to resort to wet sanding and respraying.

(d) Industrial air pollution has been held responsible for discoloration of metallic finishes due to chemical attack on the aluminium particles in the metallic paint.

Notes

[1] Dabbs, W.F., (1974), Inst. Metal Fin. Symposium *Cars made to last,* London.
[2] Stevens, M. (1978), *PPCJ,* **168**, 134.
[3] Scott, J.L. and Anderson, T.E. (1976), *JOCCA,* **59**, 383.

7.36 Paints for railway rolling stock

The majority of rail coaches are fabricated from cold rolled steel which is phosphated by a dip process before painting. Paint application is then by brush, roller or airless spray according to circumstances.

The slow drying oil/varnish paints which were used as multi-coat systems for many years have been replaced by paints with faster drying and hardening properties and possessing greater chemical resistance. Fewer coats are also being applied.

The paints in use to-day are air-drying oxidizing types, the commonly used system consisting of epoxy ester/zinc phosphate primer, urethane oil/alkyd/modified phenolic undercoat and urethane oil/alkyd finish [1]. Such a system is claimed to give at least six years durability.

Faults which develop in such a system on rail coaches are:

(a) Chalking, which is not considered a serious defect for rolling stock in constant use.

(b) Staining, caused by brake dust and cleaning solution residues. This fault is under investigation in the British Rail laboratories [1].

In spite of these drawbacks the system described possesses advantages in drying and chemical resistance and is considered to be the best compromise at present.

Goods wagons are protected with conventional air-drying systems but rail tankers for solvents or corrosive chemicals require more resistant finishes. Two-pack epoxies and chlorinated rubber systems are widely used, the type depending on the material being transported.

Note

[1] Timmins, F.D. (1977), *JOCCA*, **60**, 291.

7.37 (A) Ships' hulls (steel) and underwater structures: preparation and protection against corrosion

The results of a great deal of work have been published during the last 30 years concerning the protection of ships' hulls against corrosion and fouling, and there is now a much better understanding of the methods of achieving long term protection. In particular there is a greater realization that the preparation of the steel and proper application of the appropriate paint systems must be given just as much attention as the formulation of the paints themselves.

Preparation

The removal of millscale by blast cleaning steel used in the construction of ships is now almost universal [1]. Automatic airless machines are generally used, in which the abrasive is directed against the steel surface by impeller wheels. It is important to ensure that the steel surface meets the required standards of cleanliness and roughness, in accordance with Second Quality of B.S. 4232, 1967, Surface Finish of Blast-cleaned Steel for Painting, or the corresponding grade Sa 2½ of Swedish Standard S.I.S. 05 59 00: 1967. This standard is the minimum required for underwater steel, particularly when a high performance paint system is to be applied. The cleanliness of the surface may be checked in the following ways:

(i) By sticking on and then detaching a strip of adhesive tape, which shows any loose dust.
(ii) Visually, using a hand lens by which residues of millscale may be seen.
(iii) Chemically by a copper sulphate test [2] which reveals residual millscale as dark spots contrasting with the bright copper deposit on clean steel.
(iv) By a reflectance measurement [3] which provides a comparison with standard surfaces.

When previously rusted steel is blast cleaned the presence of residual iron salts in pits may be detected by a spot test [4] using potassium ferricyanide solution. The surface profile [5] of the cleaned steel must not exceed 100 μm (0·004 in) and this may be checked in the following ways:

(i) By visual and tactile comparison with specially prepared samples [6] of varying roughness.
(ii) By a simple probe type instrument [7].
(iii) By taking profiles of the surface (or a replica) using a Talysurf or other stylus type surface profile instrument [8].
(iv) By a roughness gauge [3] which depends on the rate of leakage of gas from a rubber-rimmed cup placed in contact with the surface.

The reflectance method for surface cleanliness and the gas leakage method for surface roughness were developed at the U.K. Paint Research Association, and the instruments are commercially available.

Having obtained a clean, slightly rough surface, the next essential is to ensure that it does not rust during fabrication and erection of the ship or structure. This is ensured by applying a blast primer [9, 10] – also known as a shop primer or prefabrication primer – whose function is to prevent rusting during fabrication and provide a suitable basis for application of the paint system proper. The blast primer must also:

(i) Dry rapidly to permit the steel plates to be handled by normal shipyard methods after 2 or 3 minutes.
(ii) Be mechanically strong to withstand damage during normal fabrication processes such as bending.
(iii) Resist the high temperatures associated with welding or flame cutting.
(iv) Not interfere with nor affect the speed of welding or flame cutting.
(v) Not affect the quality of welds – approval must be obtained from Lloyds Register of Shipping in this respect.
(vi) Not produce obnoxious or hazardous fumes when coated steel is welded or flame cut – approval must be obtained from an independent health authority that concentrations of any fumes evolved are below the published Threshold Limit Values.

Blast primers are selected for use on the basis of performance in practical tests in shipyards, hence defects that may arise in commercial use are almost invariably the result of application faults. An exception to this general statement is that blockage of spray guns may be due to oversize particles in the paint which could be due to contamination in the paint factory or in the shipyard – the paint must be sieved at the filling point and care taken to ensure clean conditions during use. Other defects that may arise include:

(a) Slow drying – this becomes apparent when the film is marked by the rollers on which the steel plates are conveyed, causing 'tramlines' on the surface. The most usual reason for this is application of too great a thickness with the result that the film retains too much solvent when the plate meets the first set of rollers after spray painting. The remedy is to ensure that the spray guns are properly adjusted to apply the correct film thickness. The dry film thickness may be measured directly with a magnetic thickness gauge, which will reveal variations over a period although the actual readings are probably not very accurate because the film thickness of 15–25 μm is less than the surface profile of 50 μm or more. More accurate readings may be obtained by passing a glass or smooth steel panel through the automatic spray painting machine and measuring the film thickness on this panel. Often the most practical method, however, is to measure the volume of primer used over a period and from this calculate the spreading rate on the area of the steel plates coated in that time – this will reveal periodic variations. Slow drying may also be the result of insufficient ventilation leading to slow evaporation of solvents or, if warm air ventilation is used to assist rapid evaporation of solvents, insufficient heating. Excessive film thickness also becomes apparent by the degree of burning and fume evolution during welding and flame cutting (Fig. 7.57).

(b) The opposite defect, insufficient film thickness, is revealed by early rusting when blast primed steel is exposed to the weather (Fig. 7.56), or even to the atmosphere in shipyard fabrication shops. If the thickness is low over the entire surface then early rusting occurs fairly uniformly, but it is much more common for the initial rust staining to show the pattern of spray guns, indicating that these are not correctly adjusted. The fault may be either that although the correct film thickness is being applied there is insufficient overlap between successive passes, or because a low thickness is being applied and an adequete thickness is only achieved where passes overlap. These local variations in film thickness should be detected by film thickness measurements immediately after applying the blast primer, but if these are not made as a regular routine it may not be realized that the spray guns require adjustment until rusting appears as bands across the plate.

(c) Another application defect that may occur with blast primers is dry spray, particularly with airless spray application. This is liable to happen in warm weather when the low boiling solvents evaporate during spraying more rapidly than usual so that a wet film is not deposited on the steel surface – the semi-dry droplets do not coalesce to a continuous dry film but form a dusty non-adherent deposit. The remedy is to add a small amount of higher boiling thinner to the paint and so ensure application of a wet coat that dries to a continuous adherent film.

Blast cleaning in service

The outer hulls of ships in service may also be prepared by blast cleaning. Open or nozzle blasting is most often used, with expendable non-metallic abrasive such as copper slag. (Note that sand is not permitted because of the risk of the lung disease silicosis.) The surface is likely to be pitted and it is important that rust and chloride residues are not left in pits – a fine grade of abrasive [11] will ensure removal. Inspection of the surface should follow the methods outlined above. Wet blasting is also used in which the abrasive is carried in a jet of water instead of air. This removes soluble chloride residues better than dry blasting, but of course leaves a wet surface that easily becomes rust stained as it dries; addition of inhibitive salts (potassium dichromate plus trisodium phosphate) to the final rinse water prevents this. Alternatively a 2% phosphoric acid wash may be applied after wet blasting, or a polyvinyl butyral/phenolic type of blast primer applied to the surface whilst it is still slightly damp – the alcohol solvents tolerating some moisture.

The foregoing paragraphs refer particularly to ships' hulls but similar considerations apply to other steel structures immersed in the sea. These include jetties and piers which are treated in much the same way as ships and to which access is possible for recoating. In the case of offshore structures [12] such as oil drilling rigs and production platforms, first class surface preparation is essential because high performance coatings must be used – the difficulty of access and the extremely high cost of recoating *in situ* make it economical to use the very best protective measures initially, almost regardless of cost. An important point when building these structures is that the preparation and painting must be carefully supervised and inspected, preferably by an independent body.

Protection against corrosion

This requires a suitable paint system, correctly applied. Consider first the essential features of a paint system for protecting steel immersed in sea water.

Sea water is normally slightly alkaline (pH about 8.0–8.2), the surface layers have an ample supply of dissolved oxygen from the atmosphere, and the salt content provides an electrolyte of high conductivity. Steel exposed to this environment corrodes by an electrochemical mechanism [13, 14, 15] whereby excess hydroxyl ions are produced at cathodic areas, raising the pH locally to about 11. Consequently a primary requirement of protective paint systems for steel in sea water is that they must resist alkaline conditions, and it follows that alkyds and other oil based media are not suitable because they are too readily saponified. In addition the film must have a high electrical resistance to impede the flow of corrosion currents between the metal and the sea water. It is clear that many paints used successfully on steelwork exposed to the atmosphere are not suitable; in particular it may be noted that the traditional primer for structural steel – red lead in linseed oil – is not recommended for use on steel immersed in sea water.

Suitable protective paint systems may be divided into conventional and high performance types [16]. The former are based on phenolic media, pitches and bitumens, the latter on the newer types of non-saponifiable resins such as epoxies, coal tar epoxies, vinyls and chlorinated rubber. In either case it is necessary to apply an adequate film thickness to provide a sufficiently high electrical resistance to give long term protection against corrosion – conventional ship's bottom systems about 150–200 μm, high performance about 200–300 μm. On offshore structures [17] where maintenance costs are extremely high, high performance coatings are applied at thicknesses of 400–500 μm.

Paint film defects likely to be encountered on immersed steel include softening, adhesion failure (flaking or blistering) and inadequate film thickness.

(d) Softening is generally the result of alkaline attack saponifying an oil based medium. The alkalinity of sea water, particularly at cathodic areas, is sufficient to attack alkyd and many oleoresinous media. The phenolic media often used for conventional underwater protective paints – e.g., the Royal Navy's DEF 1441 Paint, Underwater, Protection for Steel (Anticorrosive No. 655) is based on a tung oil/linseed stand oil/modified phenolic resin, 1:1:1, medium – are more resistant to saponification, but are not recommended on the bottoms of ships fitted with cathodic protection because the rate of production of hydroxyl ions on the hull is then much higher than on ships not so fitted. The bituminous media also used in conventional underwater protectives are resistant to this type of attack, as are the non-saponifiable media used in high performance coatings.

(e) Adhesion failure manifests itself as flaking or blistering, and is most often the result of painting over a poorly cleaned surface. If the flaking is down to the metal it indicates that the steel was not properly cleaned, there being residues of water-soluble salts on the surface, e.g. sulphates or chlorides, or of iron oxides (rust). In the case when old ships are blast cleaned to bare metal it is particularly important to use an abrasive fine enough to clean any pits in the surface, otherwise residues of salts such as ferric chloride may remain in the pits; this salt is strongly deliquescent and painting over such residues leads to early failure as water is drawn through the film by osmotic action. Alternatively wet blasting will remove water-soluble residues from pits. Intercoat adhesion failure following repainting is often the result of painting over a surface on which sea water has been allowed to dry, leaving a film of sea salts – the remedy is always to hose down a ship's bottom with fresh water before recoating. The surface must of course be allowed time to dry, otherwise poor adhesion will again occur from painting over moisture. Another cause of intercoat adhesion failure that sometimes occurs with a cold-cured epoxy protective system is allowing too long an interval between coats; if priming and intermediate coats are allowed to cure fully before overcoating they become so chemically resistant that they are unaffected by the solvents in the next coat and poor adhesion results. The remedy is to overcoat epoxies before they are fully cured, thus permitting the solvents in the new coat to soften the film so that the two coats merge slightly and good intercoat adhesion is obtained. It may be noted that this defect does not arise with chlorinated rubber or vinyl paint systems because these types dry only by solvent evaporation; when a new coat is applied the solvents in it soften the surface of the previous coat irrespective of the interval, and good adhesion is obtained.

(f) Inadequate film thickness is a major cause of poor protection against corrosion. From the electrolytic mechanism of the corrosion of steel immersed in sea water it follows that paint coatings prevent corrosion mainly by providing a high electrical resistance between the metal and the water. The use of pigments and extenders with low contents of water-soluble salts is necessary to ensure a high electrical resistance in the film, but the main factor is to apply a film of sufficient thickness – at least 150–200 μm being required. Application of even the best paints at 100 μm or less does not provide a sufficiently high electrical resistance for long term protection and early failure occurs by rusting beneath the coating, forçing off the paint film. When investigating cases of such breakdown it is important to distinguish between rust left on the surface when the paint was applied and rust formed by corrosion since the paint was applied. The normal rate of corrosion of steel in sea water [18] is about 125 μm per year, which would give a thickness of about 200 μm of rust, assuming it all remained in contact with the metal. If then, for instance, after 6 months' immersion in sea water a thickness of rust appreciably more than about 100 μm is found under a paint film, the indications are that there must have been rust on the surface when the paint was applied. As it is essential to apply thick paint systems to ensure good protection it is important that the physical properties of the films are suitable; in fact their toughness, flexibility and adhesion are just as important as their chemical properties. This applies particularly to off-shore structures on which, as already mentioned, specially thick coatings are required; suitable types are epoxies, coal tar epoxies, chlorinated rubbers and vinyls, often applied over an inorganic zinc rich primer.

(g) Another factor which must be taken into account is the effect of cathodic protection [19, 20], since most large ships and practically all offshore rigs are fitted with this equipment. On ships 'sacrificial' or 'galvanic' anodes of high-purity zinc or of aluminium alloy may be used, but more often inert anodes fed with direct current from a generator are used, such anodes being smaller and more streamlined than sacrificial types. The anodes may be of platinized titanium (titanium with a thin permeable coating of platinum which has the effect of preventing variations in potential leading to local dissolution of the titanium) or of a silver-lead alloy. The d.c. voltage applied to the anodes must be controlled to ensure that the potential of the ship's hull is not made too highly cathodic, because this makes the conditions on the hull highly alkaline and salts such as calcium and magnesium carbonates ('cathodic chalk') will then be deposited in pores in the coating. If conditions are even more highly cathodic hydroxyl ions are produced at the cathodic surface at an excessive rate and gaseous hydrogen is formed in the pores. Deposition of these salts or evolution of hydrogen gas in the pores will disrupt the film and cause detachment. These effects can be prevented by controlling the cathodic potential of the hull electronically; the hull potential is monitored against a reference electrode, often of pure zinc, set in the hull at a position remote from the anodes, and the reading used to control the voltage applied to the anodes. The hull should be maintained at a safe cathodic potential of –0.80 to –0.85 volts against a silver/silver chloride electrode, or +0.25 volts against a zinc/sea water electrode. On stationary structures the anodes may be galvanic or impressed current types; coastal structures where power supplies are readily available often use impressed current systems whilst on offshore structures sacrificial anodes are frequently used since installation is simpler and the size and weight of the anodes required is not disadvantageous on static structures. In the latter case there is usually some degree of over-protection, this being accepted because it is important to ensure that bare steel remains protected when exposed at areas where the coatings become removed by mechanical damage. The coatings normally applied to these structures are resistant to the effects of cathodic protection because their thickness makes them practically pore-free and they have a high electrical resistance.

7.37 (B) Ships' hulls and underwater structures: protection against fouling

Underwater surfaces are subject to the attachment and growth of marine plants and animals [21–26] which have serious effects on the roughness of ships' bottoms, leading to a considerable increase in resistance [27, 28] to movement through the water. This increase leads to a reduction in service speed unless the vessel has sufficient reserve power to overcome it, at the expense of increased fuel consumption; in either case the ship operating costs are increased. In the case of large supertankers or bulk carriers speed losses are very expensive because of their high running costs – a loss of 0.1 knot on a service speed of 16 knots is equivalent to a loss of over 2 days a year in service, and at a total running cost of £40 000–50 000 a day this amounts to more than £80 000–100 000 a year [29]. On stationary structures these growths do not have such serious effects and as a general rule no steps are taken to prevent them.

In the case of ships various methods of preventing these growths, usually known as 'fouling', have been proposed. These include ultrasonic methods [30], the use of radioactive paints emitting β rays [31], and processes involving the release of poisonous compounds from perforated pipes fitted to ships' bottoms [32], but these have not been commercially successful. Only the special paints known as antifouling compositions have proved successful in keeping ships free from marine growths. They contain chemicals poisonous to the settling stages of marine plants and animals, the poisons being released into the sea water at a rate sufficient to maintain a thin layer of water next to the hull in which the small spores and larvae of the organisms involved cannot survive. This prevents attachment and further growth of the plants and animals. The biology of ship fouling has been studied for at least 70 or 80 years, but only in the last 30 years or so has information on the formulation of antifouling compositions been published. Animal fouling organisms, the most important being the barnacle (Fig. 7.57), reproduce by releasing larvae which go through a free-swimming stage and eventually cement themselves to a firm substrate and change to the adult form. Plant fouling organisms – seaweeds, e.g. *Enteromopha* (Fig. 7.58), the green weed often described as 'grass'– reproduce by spores, some of them being capable of swimming, which attach themselves to a firm surface. The function of antifouling compositions is to prevent attachment of larvae or spores at their settling stages (Fig. 7.59). Copper compounds have for many years been the principal poisons used – chiefly cuprous oxide, coloured copper salts being used on yachts. These compounds are effective against most marine fouling organisms, but their action, particularly against some plant growths, may be reinforced by other poisons, e.g. compounds of mercury, arsenic, tin, lead or zinc. In the case of arsenic, tin and lead the inorganic compounds have no useful antifouling properties, but some organometallic compounds of these elements, especially tin, have proved very effective. The use of mercury, arsenic and lead compounds is now severely restricted on account of health hazards involved in the manufacture, application and subsequent removal of antifouling compositions containing these compounds, and restrictions apply in certain countries to some organotin compounds. Hundreds of purely organic compounds have been tested for possible antifouling properties and although some have proved effective against specific organisms, e.g. the insecticide DDT is selectively effective against barnacles, none has yet been found to be poisonous to so wide a range of organisms as is cuprous oxide. The possibility exists of organic compounds being discovered that are effective against all the marine plants and animals, but they would only be used in antifouling compositions if they (1) could be incorporated into paints suitable for use under commercial drydock conditions, (2) could be handled without impracticable precautions against toxic hazards during manufacture and application of the paint, and also during removal from a ship's bottom after a period in service, and (3) showed some technical or economic advantage over known antifouling poisons.

There are several different types of antifouling compositions, differing in the mechanism [16, 29] by which they release poisons into the seawater.

(a) In the traditional rosin-based type the slightly acidic binder dissolves slowly in the slightly alkaline sea water, exposing particles of poison to the action of the water. This type is often described as the soluble-matrix type. Cuprous oxide is the poison most widely used in these compositions, the amount required depending on the rate of dissolution of the binder. The cuprous oxide content, the amount and type of other pigments or extenders, and the rosin/plasticizer ratio must be adjusted to give a minimum poison release rate, or leaching rate as it is usually called, of 10 μg copper/cm^2/day. This is the critical leaching rate [22, 33] for copper compounds, less than this being insufficient to prevent settlement of all the common fouling organisms whilst values much in excess of 10 $\mu g/cm^2$/day are wasteful as the poison in the composition becomes depleted in an unnecessarily short period.

(b) In contact type antifouling compositions the poison content is high enough to ensure that the particles of poison are in contact throughout the dry film. This ensures that as the poison particles near the outside of the film are dissolved so fresh particles become exposed to the action of the sea water. The binder of a contact type of antifouling need not be soluble in sea water because the particles of poison will dissolve leaving an intact skeleton or honeycomb of insoluble binder – in fact these compositions have sometimes been described as insoluble matrix types. In practice, however, all commercial antifouling compositions of this type contain some soluble binder, usually rosin. The 'insoluble' portion of the binder is essentially non-saponifiable, e.g. vinyl resin, chlorinated rubber or tung oil/phenolic resin, with plasticizers to adjust the physical properties of the film. The poison leaching rate depends upon the poison content, the amount and type of other pigments and extenders (if any) and the insoluble resin/plasticizer/rosin ratio. The principal poison used in this type of composition is again cuprous oxide, the content being higher than in soluble matrix types because the particles must be in contact throughout the film. As a general rule the high cuprous oxide content and the 'insoluble' medium produce a harder film than the soluble matrix types, and these compositions are used on high speed vessels where it is specially important to maintain a smooth surface. The U.S. Navy uses a composition of this type containing vinyl resin and the Royal Navy one containing phenolic resin medium.

(c) Antifouling compositions of the foregoing types do not give very good service on the very large supertankers and bulk carriers that have come into service during recent years. On these vessels animal fouling growths such as barnacles are not a problem because their larvae require about 48 hours without appreciable water movement to attach themselves firmly to a surface, and any that settled on the bottom of a supertanker during a normal 20–30 hours stay in port would be washed off when the ship sailed. In fact most supertankers do not use an antifouling composition on their flat bottoms where, because of the very low light intensity, plant growths cannot survive. On the vertical sides, however, plant growths, in particular the green weed *Enteromorpha,* have proved to be a major problem. The spores of this plant can attach themselves very firmly to a surface in only a few hours, and the conditions on the sides of these vessels in service, with exposure to strong sunlight in tropical waters and to damp atmospheric exposure on ballast voyages, encourage vigorous growth. If soluble matrix or contact types are used on these areas they do not perform well, probably because:

(i) Their rosin content does not resist atmospheric exposure well, particularly under tropical conditions of strong sunlight, and they tend to craze and crack.
(ii) Cuprous oxide is not fully effective against *Enteromorpha* possibly because the spores at the time of settling are only about 0.1 μm long and may become attached to areas of reduced toxicity between particles of cuprous oxide which are normally larger than 1 μm.

A new type of antifouling composition has been developed for use on the sides of these large vessels, based on acrylic, vinyl or synthetic rubber binders, and the poisons are organometallic compounds that form solid solutions with the binders. Vinyl resin paints containing organotin compounds such as tributyltin oxide, tributyltin fluoride or triphenyltin chloride have been introduced in recent years [29, 34, 35]. Being in solid solution the poisons diffuse through the binder, maintaining a uniform concentration throughout the film. Then as poison is leached from the surface of the film more diffuses from deeper down to replace that lost to the sea water. Since the poisons are in solid solution, i.e. molecularly dispersed, there are no areas of reduced toxicity on which *Enteromorpha* spores may settle. These 'diffusion' type antifouling compositions, although introduced initially for large supertankers are also finding applications on other vessels.

(d) Another type of antifouling composition has been developed during the last few years [36, 37, 38]. In this the poison, an organometallic type (generally organotin) is copolymerized with a film forming resin such as an acrylic. The copolymers thus produced may be pigmented and the compositions so produced have proved effective. The poison is released by dissolution/erosion of the surface and this action is claimed to maintain a smooth finish on ships' bottoms. Experience of the performance of this type of compostion (self-polishing) in service is still scarce, but there are doubts about fouling settlement that may occur when a ship is stationary for more than a few days, e.g. if the ship is delayed in port.

Since all types of antifouling composition function by releasing poison into the sea water they can only remain effective for a limited period, until the poison in the film becomes reduced to a level where the critical leaching rate cannot be maintained – fouling can then settle. This is the most usual reason for an antifouling composition ceasing to function – it has reached the end of its useful life and must be renewed. The life is affected by formulation; the poison content must be selected in relation to the type of binder, in particular the rosin content, the amount and type of other pigments or extenders. The life can be extended by increasing the film thickness and hence the poison per unit area. Leaching rates, and hence effective lives, are affected by the speed of the vessel and by the temperature, salinity and pH of the sea water – factors dependent upon the service conditions of the vessel, not on the formulation or application of the antifouling composition.

(e) Problems likely to occur with antifouling compositions, apart from those common to any paint coating, e.g. application over a dirty or damp surface, include the following:

(i) Application of an insufficiently thick coat, so that the poison per unit area is too low.
(ii) Use of an antifouling composition unsuitable for the service conditions of the vessel, e.g. an antifouling intended for the cold waters of the North Atlantic on a ship trading in the Tropics where the warmer waters lead to an excessively high leaching rate and hence early exhaustion of poison.
(iii) Use of a soluble matrix antifouling composition with a comparatively soft film on a high speed vessel – the increased turbulence leads to early erosion of the film.
(iv) Use of a soluble matrix or contact antifouling composition on the sides of a supertanker where the frequent to tropical atmospheric conditions leads to crazing and cracking of paints containing rosin.

All these causes lead to a shortened effective antifouling life, so that the ship must be drydocked sooner than normal to apply a fresh coat of antifouling composition.

Notes

[1] *Recommended Practice for the Protection and Painting of Ships* (1973). The British Ship Research Association and the Chamber of Shipping of the United Kingdom, Wallsend, Northumberland, p. 30.
[2] Singleton, D.W. (1968), *Iron and Steel,* **41**, 17.
[3] Bullett, T.R. (1968), *Br. Corros. J.,* Supplementary Issue, 5.
[4] McKelvie, A.N. (1977), *JOCCA,* **60**, 236.
[5] British Standard B.S. 4232: 1967.
[6] (U.S.) National Assoc. of Corrosion Engrs. (1970), *Mats. Performance* **9**, (8), 49.
[7] Chandler, K.A. and Shak, B.J. (1966), *Br. Corros. J.,* **1**, 307.

[8] Wilson, R.W. and Zonsveld, J.J. (1962), *Trans. N.E. Cst. Instn. Engrs. and Shipb.*, **78**, 277.
[9] Banfield, T.A. (1970), Br. Joint Corro. Grp., Corrosion Conference.
[10] Reference [1] above, p. 188.
[11] Singleton, D.W. and Wilson, R.W. (1968), *Br. Corros. J.*, Supplementary Issue, 12.
[12] Blackburn, P.G. and Dunkley, F.G. (1974), *Trans. I. Mar. E.*, Series B, Conference Report, p. 16.
[13] Reference [1] above, p. 159.
[14] Mayne, J.E.O. (1957), *JOCCA*, **40**, 183.
[15] Mayne, J.E.O. (1970), *Br. Corros. J.*, **5**, 106.
[16] Banfield, T.A. (1974), *Paint Manufacture*, **44**, (5), 18; **44**, (6), 22.
[17] Fowles-Smith, J.E. (1976), *Bull. I. Corros. S.T.*, No. 58, p. 2. (Also reference 12 above).
[18] Hudson, J.C. (1950), *J. Iron Steel Inst.*, **166**, 123.
[19] Reference [1] above, p. 265.
[20] British Standard Code of Practice, C.P. 1021:1973.
[21] Woods Hole Oceanographic Institution (1952), *Marine Fouling and its Prevention.* United States Naval Institute, Annapolis, Maryland.
[22] Harris, J.E. (1946), *J. Iron Steel Inst.*, **154**, 297P.
[23] Ragg, M. (1955), *Schiffsbodenfarben und Schiffs-Anstrichmittel, 2 Auflage.* Pansegrau-Verlag, Berlin.
[24] Rathsack, H.A. (1967), *Schiffsanstriche.* Akademie-Verlag, Berlin.
[25] van Londen, A.M. (1963), Verfinstituut T.N.O., Delft, Holland: Report 54C.
van Londen, A.M. (1964), Verfinstituut T.N.O., Delft, Holland: Report 62C.
de Wolf, (1964), Verfinstituut T.N.O., Delft, Holland, Report 64C.
[26] De La Court, F.H. and De Vries, H.J. (1973), *Progress in Organic Coatings*, **1**, 375.
[27] Lackenby, H. (1962), *Proc. Inst. Mech. Eng.* (London), **176**, 981.
[28] Aertssen, G. (1957), *Trans. Inst. Nav. Arch.*, **99**, 502.
[29] Banfield, T.A. (1972), Oceanology 72 Conference, Brighton; also *Pig. and Resin Tech.*, **1**, (4), 33.
[30] Arnold, M.H.M. *et al.*, (1954), Br. Patent 719 650.
[31] Morley, C.O. *et al.* (1958), *JOCCA*, **41**, 445.
[32] Anon (1959), *Shipbuilding and Shipping Record*, **94**, 543.
Anon (1960), *Shipbuilding and Shipping Record*, **96**, 45.
Anon (1960), *Shipbuilding and Shipping Record*, **95**, 315.
[33] Ketchum, B.H. *et al.* (1945), *Ind. Eng. Chem.*, **37**, 456.
[34] Bennett, R.F. and Zedler, R.J. (1966), *JOCCA*, **49**, 928.
[35] Mearns, R.D. (1973), *JOCCA*, **56**, 353.
[36] James, D.M. (1968), Br. Patent 1 124 297.
[37] Montemarano, J.A. and Dyckman, E.J. (1975), *J. Paint Tech.*, **47**, (600), 59.
[38] Christie, A.O. (1977), *JOCCA*, **60**, 348.

7.38 Ships' holds and superstructures

In Section 7.37 the blast cleaning and blast priming of the steel used for building ships was described and possible paint film defects discussed. Modern practice is to prepare all steel used in the construction of ships in this way, hence in this section it is assumed that paint systems are applied to a clean, dry, blast primed surface in good condition, any defects having been rectified.

The paint systems used on the topsides and superstructures do not differ in any fundamental way from those used on structural steelwork ashore. Thus conventional paint systems commencing with oleoresinous primers containing inhibitive pigments such as red lead, zinc chromate or zinc phosphate are used, although the use of lead and chromate pigments is declining because of health hazards. The traditional structural steel primer, red lead in linseed oil to B.S. 2523, 1966, Type B, is still used although it is slow drying; quick drying alkyd or oleoresinous media are now more usual. Zinc chromate primers [1] based on alkyd or oleoresinous media are suitable for use on light alloys as well as steel. Zinc phosphate primers have been developed in more recent years; they are non-toxic and can be made in white or pale colours, an advantage when the finishing paints are to be white or a pale tint. The conventional white or tinted finishing paints are generally medium/long alkyd types based on 'non-yellowing' oils with a high linoleic content, e.g. soya bean oil or tall oil. The grey paint specified [2] for British Naval vessels is of this type, containing rutile titanium dioxide in a 'non-yellowing' alkyd. White is used extensively on ships and must be made 'lead-free' to avoid sulphide staining when subjected to H_2S fumes – evolved from the polluted waters of some harbours and estuaries or from some grades of crude petroleum. Black is also widely used – it may be alkyd based although phenolic media are also suitable.

Possible defects include the yellowing of whites due to the use of linseed alkyds instead of non-yellowing types, sulphide staining of whites due to the inclusion of lead driers, chalking due to the use of anatase instead of rutile titanium dioxide or wrinkling due to the use of unsuitable drier combinations. The most usual defects noted on ships' topsides and superstructures, however, are either premature rusting due to inadequate film thickness – a minimum of approx. 200 μm (0.008 in) is normally recommended – or intercoat adhesion failure in the form of blistering or flaking due to painting over a dirty or damp previous coat. The latter defect is sometimes found with maintenance painting in service because ships are frequently painted in bad weather or without washing off salt-laden deposits, the over-riding factor being the need to return the ship to service in the shortest possible time.

High performance paint systems used on ships' topsides and superstructures include chlorinated rubbers and epoxies. These are basically the same as those used on the outer hull except that the final coat is required to have a good appearance; for this reason a two-pack polyurethane enamel is generally used as a finish over an epoxy system, being much more chalk resistant. Possible defects are essentially the same as for conventional paint systems; particular attention must be paid to the overcoating intervals with epoxy systems, otherwise there is a risk of intercoat adhesion failure as discussed in Section 7.37.

When considering ships' holds it is necessary to distinguish between dry cargo holds in which general cargo, containers or dry bulk cargo (e.g. grain, coal, iron ore) are carried and tanks in which crude petroleum, refined oil products or water ballast are carried. In the former case conventional systems consist of alkyd or oleoresinous primers as used in topsides or superstructures (see above) followed often by an aluminium finishing coat – this being to improve the illumination. Possible defects are the same as on superstructures plus the chance that the aluminium finish may have 'leaded' (i.e. lost its brightness) before application, due to the use of a medium of too high an acid value or to the inclusion of lead driers. An alternative hold paint system consists of two coats of high build bituminous aluminium paint; these are normally not as bright as alkyd or oleoresinous aluminium paints and in service may tend to lose their aluminium appearance if the bitumen used is too soft – the aluminium flakes sinking into the medium. When bulk cargoes are carried, abrasion is a problem and a high performance system such as a coal tar epoxy is required; in this case again special attention must be paid to the intervals between coats if intercoat adhesion problems are to be avoided.

Coal tar epoxies are also used in crude oil/ sea water ballast tanks and the possible defects with these coatings have already been mentioned. For tanks carrying refined oil products (e.g. white spirit, gasoline, kerosene, central heating oil) and sea water ballast, coal tar epoxies are unsuitable because the cargoes may become discoloured by pitch leaching out of the coating. Pure epoxies must be used, resisting the action of the cargoes (without causing discoloration) and also the corrosive effects of sea water, provided the system is properly applied to an adequate thickness, say about 250 μm (0.010 in). These coatings may be attacked, however, if cargoes of particularly aggressive solvents are carried – examples being benzene, methanol, acetone, ethyl glycol – which may soften and detach the paint film from the substrate. Other possible defects in these epoxy tank coatings are premature rusting due to inadequate film thickness or poor intercoat adhesion due to excessively long intervals between coats.

Another type of coating used in refined oil cargo tanks consists of zinc silicate – the aqueous types based on sodium, lithium, potassium or quaternary ammonium silicates being more commonly used here than the solvent types based on ethyl silicate. The coatings are not overcoated when used in tanks. Possible defects include premature rusting due to too low film thickness and mud-cracking due to too high film thickness – the optimum thickness being between about 75 and 100 μm (0.003 and 0.004 in). Zinc silicates must be applied to freshly blast cleaned steel or to a zinc silicate blast primer in good condition – application over other types of blast primer leading to poor adhesion.

Notes

[1] U.K. Ministry of Defence D.E.F. 1115A. *Paint, Priming, Zinc Chrome, Fire Retardant.*
[2] U.K. Ministry of Defence D.G.S. 168. *Paint System for Ships' Weatherwork.*

7.39 Destruction of the hulls of wooden ships, harbour constructions and marine piling by marine borers [1]

There are two distinct types of marine borers:

(i) Woodboring molluscs which include *Teredo, Bankia* and *Martesia.*
(ii) Crustacean borers, related to shrimps, which include *Limnoria, Chelura* and *Sphaeroma.* These are much smaller than the mollusc borers and rarely bore deeper than 15 mm below the surface of the wood.

Teredo and *Bankia* are of wormlike appearance, their length varying from a few centimetres to as much as a metre. They are popularly known as 'shipworms' and they can bore through any thickness of timber. External evidence of their presence may be slight. The common name for crustacean borers is 'gribble'. Their size does not often exceed 4 mm. All borers occur only in sea or brackish waters and are widely distributed, but are particularly destructive in subtropical or tropical waters. It is said that *Teredo* can only live in water with a salinity exceeding 0.5–0.9%, whereas *Limnoria* require a minimum salinity of 1.6–2.0% (seawater averages 3.5%).

Shipworms are found in the coastal districts of the North Atlantic up to Greenland, in the North Sea from Spitzbergen down to the coast of Scandinavia, Germany and Holland, in the Baltic sea, along the French coast and the coasts of the Mediterranean and the Black sea, and along most tropical coasts. The common species in British waters are *Teredo navalis, T. norvegica* and *T. megotara* which often attacks floating timber.

It is one of the dangerous characteristics of the activities of these 'worms' that while the outer surface of a wooden part of the structure may still appear perfectly sound, the interior of the wood can be completely perforated and burrowed and so weakened. The depth of attack of the ship-worms increases with decreasing depth of penetration of the protective paint applied and with increasing wear of the film by by abrasion, e.g. by hauling in of fishing nets. (See Fig. 7.60).

Since ship-worms thrive only in clear waters containing at least a minimum of salts, it has been found advisable to keep wooden ships during the winter time in harbours lying at the mouths of rivers which are rich in fresh water. The worms develop abnormally during the seasons of smaller rainfall and die in large numbers during rainy seasons.

The mollusc does not enter smooth, planed wood bottoms as easily as rough ones.

Protective coatings containing toxic compounds are effective, but difficult to maintain in perfect condition. Once damaged, marine borers can attack the timber. Cladding with thin sheets of metal has been suggested and copper proved the best. Concrete has been used successfully for encasing marine piling.

The surest method of protection is to impregnate the timber deeply by pressure treatment; a surface treatment is inadequate. The most successful preservative is creosote although one species, *Limnoria tripunctata,* is tolerant to creosote. However, creosote reinforced with pentachlorophenol has proved effective in this case.

Tributyltin compounds are effective against *Teredo*, but not against *Limnoria.*

Impregnation with copper/chrome/arsenic to British Standard No. 4072 has on the whole given good results, but some anomalies have recently been reported. Copper naphthenate solutions in white spirit also give effective preservation.

If the timber is to be painted after preservative treatment it is still necessary to use a final antifouling coat to prevent marine organisms like barnacles, algae etc., from settling on the surface. Two-pack polyurethane coatings which give a very hard and smooth finish have been suggested as making the removal of marine growth by scouring easier without damage to the coating or the wood. Some timbers are listed by the Forest Products Research Laboratory as resistant to marine borers. None of them is considered immune from attack. Sapwood should always be removed from timbers which are not to be treated with a preservative.

Note

[1] Building Research Establishment, Princes Risborough Laboratory, Technical Note No. 59, HMSO, 1972.

7.40 Paints for steel bridges

Steel bridges are, by their nature, located in exposed situations in marine, industrial or rural atmospheres. They often cross canals, rivers or estuaries where condensation on the underside can be an additional problem. The steelwork of bridges carrying roads which are treated with de-icing salt can be exposed to highly corrosive salt solutions as a result of percolation.

Corrosion is the major problem encountered in the protection of bridges and, in view of the high labour costs involved in repainting, the surface preparation and paint systems employed must be of the highest order. The long bridges required for many motorways have posed problems which have necessitated a close study of paint performance in relation to exposure conditions [1].

All road bridge painting, although carried out by Local Authorities, is under the ultimate control of the Ministry of Transport which issues detailed notes for guidance to engineers and others concerned. All paints used must have been previously registered with the Ministry. Both British Rail and the British Iron and Steel Research Association (BISRA) have also carried out 'in depth' studies of the problems involved in the protection of steel structures generally.

In modern bridge construction as much protection as possible is given to the steel at the fabricator's works so that it is no longer in a vulnerable condition when exposed to the weather during assembly. Scale is removed by blast cleaning and the paint system depends on the kind of environment to which the structure is to be exposed. Even so, several systems have been found to give satisfactory protection in each type of environment but it is absolutely essential for every stage to be completed in accordance with the specification requirements. This requires vigilant inspection at all stages; this is often carried out by independent inspecting organizations [2].

Design is an important factor. There should be no pockets in which water can be retained and all surfaces should as far as possible be accessible for maintenance painting.

Paints for new bridges

As mentioned above, as much protection as possible is given to the steel at works before erection. Blast cleaning to the appropriate standard [3] is followed by a blast primer of which three types are in common use – 2-pack epoxy/zinc phosphate, 2-pack epoxy/metallic zinc (zinc-rich') and phenolic-modified 2-pack etch primer. The primer is followed by the required number of undercoats which can be based on epoxy ester, tung/phenolic/micaceous iron oxide, chlorinated rubber/alkyd or chlorinated rubber. Finishes are usually applied after erection. The total film thickness is of the order of 150 μm for rural conditions and 300 μm or more for severe industrial and marine exposure. Such systems should require only minor maintenance after 5 years and no major maintenance before 8 years. Areas to which access is very difficult should be given coatings to last 12 to 15 years.

Maintenance painting

The procedure adopted follows good painting practice, namely removal of rusted or otherwise affected areas to bare metal, preferably by blast cleaning. Where this is not possible, manual or power brushes are used, care being taken to avoid polishing the surface. The edges of the sound paint should be bevelled and should show no tendency to lift.

Cleaning is followed by patch primer (and, if necessary, a stripe coat) and the system built up with undercoats and finish(es) of the type used before.
Note: This may not always be possible. Many of the older wrought iron bridges were primed with red lead/linseed oil [4], the undercoats and finishes containing a high proportion of white lead. Under these circumstances repainting is best carried out with chlorinated rubber paint for the reasons given below.

Chlorinated rubber paints are being used to an increasing extent for maintenance painting. Apart from their outstanding resistance to water and chemicals, the strong solvents used tend to 'bite' into earlier coats and so give a good adhesive bond. The earlier paint must be sufficiently hard and mature to resist 'lifting' by the strong solvents and it is advisable to check this on a small area. Chlorinated rubber paints possess the advantages of rapid drying, keying well to existing paint systems (with the proviso of not lifting) and, as non-convertible coatings, showing perfect inter-coat adhesion.

Painting of galvanized steel

The number of primers which will adhere satisfactorily to a new galvanized surface is very limited and traditional practice has consisted in exposure of such surfaces to the weather for a period of 6 to 9 months. Slight corrosion of the surface provides a key for paints. An alternative treatment was etching the zinc surface by treatment with a 5% solution of copper sulphate ('mordant solution') but results were often erratic due to indifferent cleaning of the surface. In order to obtain a satisfactory result from an etching treatment it is essential for the surface to be completely free from contamination in the form of oil and grease, dirt or zinc corrosion products. A modern form of mordant solution is 'T Wash' developed by British Rail. This consists essentially of an aqueous solution of phosphoric acid containing copper phosphate. When it is applied to a zinc surface, blackening of the surface takes place, any contaminated areas showing no change. Apart from its etching properties, T Wash is therefore an excellent indicator of surface cleanness. It must be remembered that application of T Wash is an etching process only and it must be followed by primer e.g. zinc chromate or zinc phosphate before the desired undercoats and finish.

One- and two-pack etch primers are also widely used on zinc surfaces. They show excellent adhesion provided the zinc has been cleaned thoroughly but they must not be exposed to the weather or high humidity before being overpainted.

Notes

[1] Ferguson, P.J. (1976), *JOCCA,* **59**, 253.
[2] Dunkley, F.G. (1976), *JOCCA,* **59**, 39.
[3] B.S. 4232: 1967, *Surface finish of blast-cleaned steel for painting.* Swedish Standard SIS 055900, 1967.
[4] Ferguson, *loc. cit.*

General reference

B.S. 5493, 1977. *Code of practice for protective coating of iron and steel structures against corrosion.* (Replaces CP 2008.)

7.41 Flame retardant paints

A flame retardant substance is defined (B.S. 4422, 1971) as one that 'applied to a combustible material decreases its tendency to propagate flame across its surface'. Flame retardant paints are designed to protect combustible substrates as far as this is possible and to worsen as little as possible the behaviour of non-combustible substrates against the effects of fire. Such products must therefore always be judged on the appropriate substrate and the B.S. 476 provides three tests, ignitability, fire propagation and surface spread of flame, which, when taken together, permit a judgment to be formed on the suitability of a paint as a flame retardant. The evolution of toxic fumes occurs in most fires in any case and paints are not likely to increase this hazard significantly. The formation of dense black smoke ought to be taken into account, as this can seriously hamper rescue operations. The National Bureau of Standards (U.S.A.) has designed a smoke-density chamber capable of testing this important property. There are several ways in which flame retardancy can be achieved in paints:

(i) By providing a heat barrier and forming carbonaceous materials, carbon dioxide and water instead of flammable degradation products.
(ii) By forming an impervious barrier which excludes oxygen from the substrate.
(iii) By evolving non-combustible gases, usually hydrogen chloride or bromide.
(iv) By forming a thermally insulating layer of foam.

Types (i) and (ii) are often based on the sacrificial destruction of polymers yielding carbonaceous char and non-flammable gases or on silicates forming glassy barriers when heated. The former products can be based on phenol-formaldehyde and polyimide resins. Silicates are usually very water sensitive and are not now used to any large extent. The gas producing coatings referred to under (iii) are based on binders containing a high proportion of chlorinated or brominated compounds like chlorinated alkyds or chlorinated paraffin wax and a substantial quantity of antimony oxide in the pigment portion. In dark shades antimony sulphide can be used. This combination is believed to evolve chloride radicals and eventually hydrogen chloride. This type of paint can be formulated on water-thinned emulsion binders like p.v.a. or solvent thinned binders like alkyds or damar resin. One serious disadvantage is, that although such coatings retard the spread of flame effectively, hydrogen chloride can cause considerable damage through corrosion of machinery and instruments.

Another class of flame retardant coatings, currently gaining more favour, is known as *intumescent.* These products form a thick layer of foam when exposed to fire, thus insulating the substrate often for quite long periods of time. These paints can be either emulsion or solvent based, the main requirement being that the greater part of the binder is thermoplastic. Other necessary ingredients are a source of carbon, like starch or pentaerythritol, a dehydrating agent such as phosphoric acid in the form of ammonium phosphate or melamin phosphate and a blowing agent, giving a non-combustible gas to expand the foam. Dicyandiamide, urea-formaldehyde resins and similar compounds have been recommended.

The most prevalent defect of most fire retardant paints is their water sensitivity and locations where they are exposed to heavy condensation or similar conditions must be avoided.

7.42 Failures due to inadequate specifications

A painting specification can be defined as a document which sets out a customer's requirements clearly and precisely; it should cover all factors involved in the operation [1]. The factors of importance are materials, surface preparation, method of application and the properties required in the finished product i.e. film thickness, degree of gloss, freedom from contamination and so on. For site work a note should also be included setting out the ambient conditions under which the painting should not be carried out.

Painting specifications (which, in practice, are often too vague and imprecise), should be drawn up by persons with a sound knowledge of painting materials, techniques and procedures. Co-operation between the engineer and paint technologist is of inestimable value. Some useful suggestions on the design of painting specifications are contained in B.S. 5493 [2].

On large painting contracts where failure of a system can be extremely costly it is now common practice for independent painting inspectors to be employed. Every stage of the operation is inspected so that the completed work can be certified to comply with the specification [3].

Similar considerations apply to the painting of industrial products from motor vehicles to small domestic appliances. Even closer co-operation between customer and paint manufacturer is needed and particular attention must be paid to cleaning and pretreatment of metal when setting up a production line. The coating will, in many cases, be 'tailor made' to meet the end use requirements.

Weaknesses in specifications are often the following:

(a) Incorrect paint system. Here the paint technologist can offer useful advice. For example, for normal decorative/protective work in a mild environment the conventional primer/undercoat/finish is usually adequate and few problems arise. Where conditions can be severe, as in marine or heavy industrial atmospheres, then more chemically resistant systems such as chlorinated rubber, vinyls or 2-pack epoxies may be necessary. Such systems may appear costly but the cost of early repainting of large structures will dwarf that of the best possible initial system. Between the extremes of environment quoted there are moderate conditions for which the paint technologist is best qualified to suggest the most suitable paints which may not be the most costly types. Suggested systems for steel and other metals in various types of environment are included in B.S. 5493 [2].

(b) Inadequate film thickness. It is generally accepted that there is a direct correlation between film thickness and durability but no advantage accrues from an excessively thick coat.

Particular attention must be paid to film thickness when painting blast cleaned steel, where the amplitude may be as much as 100 μm. Peaks should be covered to a depth of at least 100 μm and the specification should set out the minimum dry film thickness of each coat as well as the minimum thickness of the complete system.

The checking of film thickness on site can be carried out by two methods:

(i) The wet film thickness can be determined by the calibrated wheel and the dry film thickness calculated using the solids volume figure.
(ii) The dry film thickness can be measured on the dry film using one of the several instruments available.

The Elcometer [4] operates on a magnetic principle and is widely used on steelwork. Recommended film thicknesses for various environments are given in B.S. 5493 [2].

(c) Inadequate surface preparation. Very many paint failures have been traced to this cause, and have occurred on a variety of surfaces. The methods for surface preparation of the commonly used substrates have been described (see Index) and for metals they are also set out in B.S. 5493 [2].

The painting specification should lay down the method to be used for surface cleaning as well as the degree of cleanliness required. This would apply to both new and maintenance work.

A great deal of new work is prepared and primed at works where conditions can be closely controlled but this is more difficult in maintenance work. Here it is often not easy to decide on the amount of preparation required and the best remedial measures. For example, old paint work may be affected by rust in isolated areas or it may show chalking, mould growth, cracking or flaking. The remedial measures necessary are best agreed by the parties concerned. In general, areas affected by rust should be blast cleaned to bare metal and primed immediately. It is important that all loose or flaking paint should be removed and that edges of sound paint should be bevelled.

On masonry, all powdery or friable material should be removed. If this is not possible a test should be carried out with a penetrating sealer in an effort to bind friable material to the substrate.

Old gloss paint surfaces should be rubbed down and it is considered good practice to repaint with the same type of paint as was previously used. Undercoat should be used to avoid intercoat adhesion failure since the old paint is likely to be very hard. In repainting large steel structures chlorinated rubber paints are being increasingly used in repaints as the strong solvents used 'bite' into the old paint. Nevertheless, care must be taken to see that the old paint is sufficiently hard to resist lifting.

(d) Application. The application method to be used should be stated in the specification so that the correct type of paint can be supplied. For site work the specification should also specify conditions under which painting must not be carried out, e.g. when the ambient temperature falls to 4°C or below, when the relative humidity is 90% or over or in the presence of rain, fog, snow, etc. Particular attention should be paid to two-pack systems in which the curing reaction can be slowed or stopped at low temperatures. Two-pack epoxies, for example, should not be applied when the ambient temperature falls below 7°C.

In polluted atmospheres such as marine and industrial, the time interval between coats should be specified and rigidly enforced on site. Contamination of surfaces of primer or undercoat can seriously impair the adhesion of following coats and lead to premature failure.

Notes

[1] Gay, P.J. (1975), *JOCCA*, **58**, 443.
[2] B.S. 5493: 1977. *Code of practice for protective coating of iron and steel structures against corrosion.* British Standards Institution.
[3] Dunkley, F.G. (1976), *JOCCA*, **59**, 39.
[4] Elcometer Instruments Ltd.

7.43 Cost of painting and excessive consumption of material

In assessing the economic aspects of painting the following cost factors must be taken into account:

(i) Surface preparation.
(ii) Material.
(iii) Labour.
(iv) Durability.

It is estimated that the total costs can roughly be divided under three headings when steel surfaces are to be protected against corrosion [1]:

(i) Surface preparation 35%.
(ii) Paint cost 30%.
(iii) Labour cost 35%.

On other surfaces the cost of surface preparation will be lower, but labour costs tend to be much higher.

All the above factors, with the exception of surface preparation, depend in large measure on the properties of the paint and the method of application, and according to Sadowski [2] it is necessary to consider the following factors.

(a) *Spreading rate of the paint.* Knowing the solids content of the coating in % by volume, the area covered equals solids volume divided by film thickness, e.g. if the solids content is 50% by volume and the film thickness is 50 μm (2 thou), the area covered by one litre equals 10 m^2. Normally spreading rates of between 10 and 14 m^2 per litre are recommended by manufacturers. The actual rate achieved will depend on the type of paint, method of application and nature of the surface to be coated.

(b) *Required film thickness.* A decorative coating needs only a film thick enough to hide the underlying surface, whereas a protective paint system should normally comprise three coats, yielding an overall film thickness of 140–180 μm. For protective work paints with excessively high hiding power can be dangerous as they may induce the operator to spread the paint out too thinly and cover a greater area than specified with a given amount of paint. It is always advisable to specify not only the number of coats to be applied but also the overall film thickness to be achieved.

(c) *Method of application.* Labour costs and material consumption will depend significantly on the method available for applying the coating. In some cases brushing, although very labour intensive, is desirable for the application of priming coats particularly on wood, stone and intricate structural steel work because it enables the operator to work the paint more thoroughly into crevices and pockets.

Hot spraying reduces the usage of thinners, airless spraying enables one to obtain a greater film thickness per coat, thus reducing the number of coats required. Dipping and electrocoating are also low in labour content and generally speaking the rise in the cost of labour has led to the development of automatic application techniques which are well suited to modern mass production. This can be exemplified by 'shop priming' where often automatic sprays guns are used on a production line to coat sheet steel or other objects. This avoids costly manual painting on site. Building components such as window frames and doors are often primed in the factory by dipping or flow coating. This practice also helps to prevent deterioration when storing on site.

Table 7.7

	Large areas	Small areas	Broken areas
Brushing and roller	4%	5%	6%
Air assisted spray	30%	40%	50%
Airless spray	20%	30%	40%
Electrostatic spray	10%	15%	15%
Dipping	10%	20%	30%
Electro-deposition	5%	5%	5%

A comparison of losses occurring with various application techniques is given in Table 7.7 [2]. It should also be noted that the type of paint can make an important contribution to the saving in application costs. In recent years 'high-build' coatings have been introduced which permit 100 μm or more to be deposited per coat. Those based on chlorinated rubber have been particularly successful in the field of protection against corrosion. Solvent free coatings and powder coatings have also come much to the fore as 'high-build' coatings and are furthermore very valuable in the fight against atmospheric pollution by solvent vapours.

Thixotropic paints, based mainly on modified alkyds, can make a marked contribution to the ease and speed of painting, mainly by a substantial increase in the quantity of paint transferred per brush load. Their main use is in decorative flat wall paints, undercoats and gloss finishes.

(d) *Working rhythm* [3]. A number of points coming under this heading should be considered when laying out a production line. Air-drying paints should be quick drying to reduce dust collection, thus not only reducing the rejection rate but also speeding up dispatch. Where possible force drying at 50–60°C is useful.

Stoving paints should be capable of being cured at the lowest possible temperature. The stoving time should be as short as possible. All paints used in one plant should be capable of being cured at the same temperature to avoid duplication of drying facilities. Wet on wet coating is a great advantage as no stoving between coats is needed.

(e) *Thinner requirements.* These should be kept to a minimum and confined to as few types as possible. Consideration should be given to hot spraying, 'solvent free' and powder coatings. Water-thinned coatings can play an important role and these factors should be taken into account if purification of exhaust air is necessary for environmental reasons.

(f) *Life of coating.* This will depend on:

(i) State of the surface.
(ii) Thickness of coating.
(iii) Type of coating.
(iv) Method of application.
(v) Environment at time of application.
(vi) Conditions of service.

In arriving at the overall cost of painting and considering the above parameters, it is not only the ultimate life of the coating that counts but its surface condition when repainting is intended. If the coating has eroded away after a period of service and does not crack or flake, the cost of repainting will be low as minimal preparation will suffice. On economic grounds it may therefore be advisable to repaint before a complete breakdown has occurred and still achieve a lower average annual cost of protection than by lengthening the interval between maintenance painting. It is always worth remembering that as the cost of paint materials is only a relatively small proportion of the total cost, it pays to use coatings of the highest possible quality [4].

Causes of excessive consumption of material:

(a) Excessively thick film. Thickness of a single paint film varies from 25 to 75 μm (0.001–0.003 in) [5] but can be less than 25 μm for some industrial lacquers. Too heavy a coat can give rise to blistering and early flaking, especially in prolonged contact with water, and should be avoided by the skill of the operator and by choosing the most suitable method of application. As pointed out in the introduction to this section, hot spraying and high-build coatings of various types are satisfactory if thick films without runs are wanted.

The disadvantages of too heavy coatings are striking in the case of aircraft finishes because of the extra weight.

Comparisons made with hand roller and brush application of wall paints show that where large areas are concerned the roller shows considerable time economy and good results can be obtained with semi-skilled labour. Where broken areas or considerable ornamentation occurs, more speedy results can be obtained by the use of brushes rather than rollers.

Where unskilled labour is used for the application of gloss enamels especially to large areas, economy of time and material can be obtained by the use of rollers rather than brushes. A more even coating is usually obtained by this method, but this does not achieve the high standard obtained by skilled labour using brushes.

Careful selection of the type of roller is necessary for gloss enamels due to the possible deposition of hairs, lint, etc., and due to possible aeration resulting in a loss of gloss.

The amount of paint used for the same job can vary with different operators by as much as 50%.

(b) The operator should be furnished only with the daily quantity required, since it is not advisable that unused materials be poured back into the containers, as they may skin, thicken or gel, etc.

Sensitive coatings are most advantageously bought in containers of a size suitable to the working conditions. This advice is very relevant where two pack coatings (epoxy or polyurethane, etc.) are concerned. Measuring the two components on site is fraught with difficulties and often leads to mixtures of incorrect proportions. It is also important not to mix a larger quantity than can conveniently be used within the pot-life of the mixture.

(c) On spray: excessive air pressure in the traditional hand and automatic spraying methods, particularly in proportion to the diameter of the fluid nozzles. The coating is too finely atomized, and much is lost on passage to work pieces.

Clear and not too highly pigmented lacquers, e.g. based on cellulose or on some synthetic resins, e.g. amino resins modified with alkyds, can be hot sprayed with advantage. Savings of the order of two coats for three are possible, depending on materials used and circumstances. In some set-ups cold spraying of the primer and hot spraying of the finishing coat is practised, possibly stoving the two coats in one operation.

In spite of the excessive consumption of material, high-pressure spraying is practised at times in order to accelerate the work. If the objects are big, and the operators skilled, the losses may be reduced to reasonable proportions. The use of pressure pots is advisable for such work.

Airless hot spraying utilizes high hydraulic and vapour pressure, the latter from suitable solvents contained in the lacquers, in order to atomize them on release into the atmosphere practically without overspray. This method can also be combined with hot spraying.

For electrostatic spraying by the Ransburg No. 1, Fischer, etc., processes low pressures are used; the atomized particles of the coatings must remain wet for the somewhat longer time taken before arriving at the point of precipitation. In the bell, reciprocating disc, 'Electropinsel' and similar electrostatic processes there is little overspray.

In flowcoating no atomizing at all takes place, the paint being passed through hoses directly on to the various areas of large objects, and the surplus collected and returned into a collecting tank for re-use.

(d) Excessive speed of exhaust air currents may act in a similar manner, besides inducing faulty flow. If as a result or due to other causes films suffer from orange peel, labour costs will go up because of the increased rubbing-down operations.

In flowcoating solvent losses are greatly reduced because drain and flow chambers are employed to assist in faster drainage and better finish. The condensed solvents and diluents are either re-used or subjected to controlled burnings for boosting the temperature of the drying oven.

(e) Too big a nozzle. Unnecessarily large quantities of the coating are applied or if the objects are small pass by without hitting them.

(f) Unsuitable shape of the spraying jet. For instance, if small parts are sprayed with a flat jet, excessive material is consumed by passing the objects.

(g) Awkward arrangements of the objects on, say, the spraying tray or conveyor, causing part of the spray to miss the mark. Often large voids in articles to be coated create waste. Automatic spraying can sometimes solve such problems.

Modern spray guns can be adjusted, e.g. for round and fan spray, the latter also for vertical and horizontal application. Besides some models of guns have an adjustable spray, which can be widened or narrowed in the course of operation. Trigger the gun after each stroke and hold the gun at a proper angle and distance. Certain articles are possibly better coated by either the dipping or flowcoating method.

Progress in the saving of labour, material and speed of application is claimed for automatic and especially electrostatic spraying. This process in which the paint particles are differently charged from the objects to be sprayed allow even spraying over edges with a remarkably small loss by overspray.

(h) Powder coatings have no volatile components and are theoretically capable of 100% utilization. In practice a single colour, automated powder plant with recovery system has an efficiency approaching 98% utilization of powder.

Powder coating cost per square metre

$$= \frac{\text{Film thickness } (\mu\text{m}) \times \text{specific gravity} \times \text{price per kilo}}{1000}$$

(One m^2 = 10.764 f^2.)

Powder usage in grams per square metre

$$= \text{Thickness of film (in microns)} \times \text{specific gravity.}$$

Notes

[1] Hermelin, R. (1975), *JOCCA,* **58**, 298.
[2] Sadowski, F.H.J. (1971), *Ind. Lack. Betrieb,* **39**, 133.
[3] van der Bruggen, B. (1966), *Ind. Lack. Betrieb,* **34**, 485.
[4] Straub, E. (1966), *Ind. Lack. Betrieb,* **34**, 295.
[5] Jolly, V.G. (1944), *JOCCA,* **27**, 95.

7.44 Practical rules for prevention of fires and explosions in painting shops

(1) Enforce the non-smoking rules.
(2) Ensure good ventilation in working shops and spray booths.

(3) Waste pipes, ventilators, fans and spray booths must be earthed.
(4) Heating should be by indirect means.
(5) Do not store more than a day's supply of paint in the painting shop.
(6) Dip tanks should, wherever possible, be fitted with a lid which closes automatically in the event of fire.
(7) Heated tanks should be fitted with efficient thermostatic control.
(8) The use of convection stoves with open flames in the drying chambers should be avoided.
(9) Drying stoves should not, as a rule, be located in the same room as stationary painting equipment.
(10) Electric bulbs or other sources of radiant energy in a conveyorized infra-red drying plant should be automatically switched off whenever the belt stops.
(11) Separate flues for combustion gases of stoving ovens and for vapours from paints should be provided.
(12) At least one-third of the exhaust area of vapour exhaust dampers of drying stoves or conveyor ovens should always be open.
(13) Fire-resisting materials should be used in the construction of painting shops.
(14) For the general protection of buildings an automatic sprinkler system is most desirable.
(15) All electric motors should be totally enclosed; avoid any sparking equipment and the build-up of static electricity from driving belts etc.
(16) Faulty electric cables, conduits and switches must be repaired at once.
(17) The transfer of flammable liquids from bulk store to point of use should be by pump and pipeline. Pipe lines should be colour coded and fitted as frequently as necessary with isolating valves.
(18) Remove leaking containers at once from the painting shop.
(19) Some solvents, vegetable and mineral oils, can accelerate the combustion of materials like paper, cleaning rags, spray dust etc. in the presence of air and can even cause self-ignition.
(20) Avoid dust explosions by regularly removing deposits of finely dispersed particles, e.g. carbon black, spray dust etc., especially from exhaust pipes and flues.
(21) Keep the shop tidy.
(22) Observe the provisions of the Health and Safety at Work Act, 1974. When laying out a new painting shop, consult the officials of the Health and Safety Executive at an early stage.

8
Hazards to the paint user

This chapter deals exclusively with those hazards which may confront anyone having to handle paint, particularly when applying it, and also those hazards associated with its storage. Any reference to the hazards centred on the actual manufacture of paint, has been deliberately omitted as outside the scope of this book. It aims to provide health and safety information relating to paint whether the user be a do-it-yourself enthusiast, a professional house decorator or maintenance man, or the employer or employee of a works manufacturing products such as cars and engineering components where paint is utilized not only for decorative but for protective purposes.

By 'hazard' is meant danger to health or even to life itself. Paint, in general, presents a two-fold potential hazard: chemical poisoning, and fire. The accent however is on the word 'potential'. *By exercising due care in the light of what is known of the hazards of a particular paint no one need ever be harmed by it whether he be a handyman making spasmodic use of the material or one who depends for his livelihood on painting and decorating.*

Returning now to the toxicity and fire aspects of paint, we are today witnessing a changing scene. Frequently associated in past years with the hazards of poison has been lead, more often in the form of the compound white lead. The toxic hazards of lead and its compounds are dealt with in more detail later on, together with the toxicity of other pigments, as well as of solvents and numerous other components of modern paints. It is conceivable that the benefits of a decreasing usage of white lead might have been neutralized by the introduction of numerous other substances, some with properties, hitherto unencountered, capable of harming mankind. Happily, better education and increasing information on the toxicological properties of paint raw materials have seen to it that such hazards as are associated with painting are much better controlled than they were in the 19th century.

A similar situation appertains to the fire hazard. Although the advent of emulsion finishes and other water-based paints has done much to reduce the danger of fire, numerous new solvent-borne paints, e.g. the acrylics, have been coming along to perpetuate the risk. One must not forget, too, that the highly flammable nitro-cellulose finishes are still with us, albeit principally for industrial usage. Again, one can record with satisfaction that legislation and authority, information and education, have all played their part, not only in controlling the fire hazard, but in ensuring that the paint user knows precisely what the attendant fire risks are and what he has to do to minimize them.

8.1 Toxic hazards

The word 'toxic' covers a multitude of sins. In the extreme case it is applied to anything which causes fatal poisoning but has also come to mean substances which are merely injurious to health to varying degrees.

8.1.1 Types of toxicity

(a) *Poisoning*

May be sub-divided further into:

(i) *Acute* poisoning, i.e. attended with symptoms of severity, as opposed to
(ii) *Chronic* poisoning which is recurring or of long continuance.
(iii) *Insidious* poisoning, i.e. working secretly, so that the victim is unaware his health is gradually deteriorating and, as is so often the case, is unprepared for remedial action until it is too late.

High concentrations of phenol taken internally or accidently splashed in large volume on the skin will quickly cause *acute* poisoning whereas long exposure, say by inhalation, to low concentrations of lead compounds will lead to *chronic* lead poisoning.

There are three distinct classes of insidious poison all with what be termed delayed effects. They are the carcinogens, mutagens and teratogens. Chemical carcinogens are substances which induce cancer in man or animals. Less than a century ago there were no recognized chemical carcinogens. Today there are many hundreds of known animal carcinogens; however, only a handful of these are proven human carcinogens. Oddly enough, not every human carcinogen has been demonstrated to be an animal carcinogen, a notable example being arsenic and its compounds.

Vinyl chloride is sometimes met with in the form of its co-polymer with other vinyl compounds. Until recently, vinyl chloride was not implicated as a chemical carcinogen. Now, it is recognized as a causative agent in angiosarcoma, a comparatively rare form of liver cancer, which has been found in a statistically significant number of workers exposed to vinyl chloride monomer during the production of p.v.c.

Again, whilst the inhalation of asbestos particles has been known as liable to lead to asbestosis, which like silicosis is a crippling lung disease, it is only within recent years that it has been ascertained that exposure to asbestos, particularly the crocidolite mineral, can also lead to malignant mesothelioma, or cancer of the pleura (lining of the lung cavity). Mesothelioma invariably takes a fatal course; fortunately it is a comparatively rare condition. Asbestos powder was at one time a frequent component of various types of industrial paints; now its use as a paint raw material is actively discouraged. It could of course be argued that provided the hazards associated with the handling of asbestos powder during paint manufacture are kept under control, the final product, containing asbestos particles thoroughly dispersed in the paint medium, will present only a minimal health hazard. The health hazard could however be regenerated if an asbestos-containing paint film came to be incinerated or even dry-flatted.

Not only carcinogens, but mutagens too, are of increasing concern, even at the international level. Mutagens are substances which are capable of altering the body's genetic material, i.e. the genes and chromosomes of the cells. Only within the last 15 years has the relationship between man and his environment become a demanding issue. It is environmentally-orientated research which has contributed greatly to the literature on chemical mutagenesis. For instance, by 1974, the Environmental Mutagens Information Centre, of the Oak Ridge National Laboratory (U.S.A.) had on file over 10 000 data entries containing information on approximately 4000 different compounds, many of which were environmental pollutants. Are any of such substances met with in paint? No-one can give a categorical 'no'. For example, ethylene oxide, a non-carcinogen, has been found to have mutagenic properties. This mono-epoxide goes into the manufacture of numerous non-ionic surfactants. In addition, claims have been made that ethylene oxide monomer has been detected in streams contaminated by trade effluent, presumably containing minute amounts of unreacted ethylene oxide!

A better understanding of the properties of naturally-occurring substances and, more importantly, the rapid growth of the chemical industry, have contributed to the development of the new science of *teratology*. Teratogens are substances which, taken by a mother-to-be, can adversely affect her unborn child. The classic example of an extremely teratogenic substance is the drug thalidomide. The tragic consequences of its introduction subsequently stimulated considerable research activity to the point that monthly publications now exist devoted solely to abstracting the literature on teratology. Again, what has this to do with paint? Around the early 1970s, those industries such as plastics and paints, which made use of plasticizers were disturbed to learn that gross foetal abnormalities resulted when rats were injected with various phthalate esters, notably di-2-ethyl-hexyl phthalate. Top level discussions subsequently pointed to the fact that the experiments involved excessive amounts of injected phthalate and eventually the hue and cry died down early in 1973, when authoritative opinion (including the American F.D.A.) declared that even plastic containers in which phthalate ester was incorporated did not present an indirect food hazard.

What can we learn from all this? The paint user cannot be expected to be fully conversant with carcinogens, mutagens, and teratogens. More and more, with technical information and toxicological data at their call, paint manufacturers are becoming his guardian, protecting him from the potential health hazards associated with the vast array of new chemicals being made available to the paint industry. The corollary to this is that the purchaser of paint is well advised to rely on paint produced by reputable manufacturers, many of whom strive to minimize the hazards of their products, over and above what is required by legislation.

(b) *Skin effects*

The greatest potential of paint to harm the user probably lies in the adverse effects it can have on the skin. There are three main contributive factors for this:

(i) The heterogeneous composition of paints, containing, as they so often do, skin-degreasing solvents as well as a great variety of pigments, and miscellaneous additives, some of which can be physiologically active chemicals.
(ii) The fact that, more often than not, paint involves the use of brush, roller or other hand-operated equipment and in the greater number of instances no hand protection is used.
(iii) The regrettable fact that so many painters, even those who operate spray-guns, will often, and against good advice, clean their hands with solvent.

Human skin is a complex membrane, extensive in area, and subject to many diseases. Skin consists of several anatomically-recognized layers but there are only two main divisions:

(i) The outer layer, called the epidermis, epithelium, or cuticle.
(ii) The true skin or dermis, whose thickness can range from about 3 mm–0·5 mm.

Throughout the skin are numerous sebaceous (oil- or fat-secreting) glands and sweat glands. The latter secrete watery fluid, and terminate visibly as skin pores. In addition, the skin carries hair follicles, which originate deep in the true skin, as well as nerve and muscle fibre.

Skin complaints brought on by contact with paint range from mild skin irritation to extreme forms of dermatitis (inflammation often accompanied by flaking or fissuring). There are two principal causes of dermatitis. First and foremost are the primary irritants, of which very many are known. They include not only acids and alkalis, and chemicals such as formaldehyde (a frequent component of urea, melamine and phenolic resins), but most of the common organic solvents. The latter act simply by dissolving out the natural grease of the skin, whereas alkalis dissolve the keratin in skin. Other irritants are sensitizers. Some such as maleic acid (or its anhydride), which is occasionally used in the manufacture of certain classes of synthetic resins, do not necessarily cause skin changes at first contact but the effect is observed on repeated contact a few days later. Sensitizers of a different type are the photo-sensitive agents, i.e. substances whose action is triggered-off by sunlight. They include coal tar and pitch and are also met with among synthetic resins (e.g. *p*-tertiary butyl phenol/formaldehyde resin) and natural resins (e.g. rosin).

Whithin the United Kingdom, dermatitis is a prescribed industrial disease and could mean the sufferer being entitled to disability benefit. The Factory Inspectorate issues valuable booklets on 'Industrial Dermatitis' which are periodically updated.

The skin, whether it is abraded, broken or even intact is also a route for internal poisoning – see Section 8.1.2 (c). For protective equipment see Section 8.3; first-aid measures are dealt with in Section 8.4.

(c) *Eye effects*

The delicate structure of the eye plus the irreplaceable sense of sight renders eye-care of paramount importance. The commonest eye disorder met with in the handling of paint is undoubtedly eye irritation from volatile paint solvents. Sometimes, however, the irritation is primarily due to the presence of minor amounts of volatile chemicals such as unreacted monomer (e.g. acrylates) or pungent additives (e.g. allyl compounds).

A degradation product frequently arising from thermally disintegrated drying oils is acrolein (acraldehyde). This chemical, as well as others such as the previously mentioned allyl compounds, formaldehyde, etc., have the property of being lachrymatory (i.e. they stimulate the tear ducts, making the eyes 'water'). Formaldehyde is also the alleged agent responsible for methylated spirits drinkers becoming blind. Methyl alcohol (methanol) taken internally is oxidized to formaldehyde and routed via the blood stream to the optic nerve which it ultimately destroys. Other recorded bizarre effects on the eye which result from chemicals are double vision from exposure to maleic anhydride and hallucinatory symptoms from acrylamide. If the paint (or unreacted component) is strongly acid or strongly alkaline the term corrosive is applicable, and describes their extremely hazardous nature. It is held that, generally speaking, alkalis are more damaging than acids and with a greater capacity to lead to blindness. The argument here is that strong acids tend to precipitate a protein barrier that prevents further penetration into the tissue, whereas alkalis do not do this but continue to soak into the tissue as long as they are allowed to remain in the eye. Particularly destructive of eye tissue are the class of chemicals known as peroxides which demand great care in their handling, even when they are used in small quantities.

No less harmful to the eyes are solid particles such as paint dust from flatted surfaces, as well as liquid particles from the spray-gun or aerosol can or from mere splashes. Commonly, this results in conjunctivitis or inflammation of the conjunctiva, the moist red layer or mucous membrane in the eye.

Other potential eye hazards reside in carelessly handled spray guns. High pressure jets impinging on the eye can result in the equivalent of mechanical injury, even leading to permanent blindness.

Finally, mention should be made of a common source of eye disorders, rubbing the eye thoughtlessly with hands contaminated with paint chemicals.

(d) *Sensitization*

Certain substances, drugs, chemicals and even natural products from the vegetable and animal kingdoms can all induce a special sensitiveness in individuals, meaning the capacity to produce an allergic response. The difference between a reaction caused by an allergy and one by a toxic compound is recognized by the fact that the toxic reaction is dose dependent whereas even minute quantities of allergens can provoke severe symptoms.

Sensitization is most often associated with skin disorders and has already been referred to under 'skin effects' in Section 8.1.1 (b). However, allergens can also be responsible for many other conditions such as fever, arthritis, asthma, headache and dyspepsia. Singled out here for special mention, as being potentially of significance to users of certain types of paint, are the isocyanates with their multi-hazardous properties. Not only are these chemicals irritant (skin and eye), dermatitic and toxic but their vapour, on inhalation, is capable of leading to severe attacks of asthma. Being readily able to bond to proteins, the isocyanates are potent sensitizers.

Whilst exposure to high concentrations of isocyanates may cause asthmatic symptoms in individuals who have not become allergic to them, exposure to concentrations even lower than 0.2 p.p.m. could produce respiratory distress in sensitized persons. Hence the medical profession's current practice of disqualifying from jobs involving exposure to di-isocyanates such as toluene 2.4 di-isocyanate (TDI), workers with a history of asthma, bronchitis or previous allergy to isocyanate.

Di-isocyanates are an essential component of polyurethanes. Sometimes the latter may contain 'free', i.e. unreacted isocyanate, hence introducing a potential sensitization hazard. See also 'Polyurethanes' – Section 8.1.4.

8.1.2 Routes available for poisoning

Leaving aside the artificial routes of injection which are used chiefly in biological experiments, there are three main pathways whereby poisons can enter the body. These are through the mouth (i.e. by swallowing or ingestion), through the lungs (by inhalation) and through the surface of the skin (percutaneously).

(a) *Ingestion*

Accidental poisoning, *per os,* is comparatively rare, although there is always the possibility of poisoning from food or cigarettes placed on a dirty work-bench or contaminated by unwashed hands which have been in contact with paint containing toxic substances. Would-be suicides seldom use paint although it has been known for prison inmates to imbibe the material deliberately in order to seek hospitalization. More common incidents involve distraught parents who have been responsible for leaving an open can of paint accessible to their young child and, in addition, are unsure whether the child has partaken much, some or none of the contents of the can. Equally important in such an emergency is an estimate of the toxicity of the sample of paint in question. Here, toxicology has gone a long way to help and for many, if not most, of the raw materials which go into paint manufacture, toxicity ratings are now available. On the basis of the ultimate composition (by weight) of the paint and having available toxicity ratings for the individual components, some assessment can then be made of the paint's overall potential to harm thereby giving the first-aider and the doctor valuable information.

What is this toxicity rating and how is it arrived at? The one most commonly adopted is denoted as LD_{50} and refers to the dose of a substance which can kill 50% of one class of animal. For toxicity tests selected strains of animals are held in captivity thereby eliminating as many variables as possible, and for a particular series of tests one might take dozens of animals to obtain eventually a single mean lethal dose (LD_{50}). It is expressed as so many grams (g) or milligrams (mg) per kilogram (kg) body weight of animal. Thus it is generally held, although there are many exceptions, that the larger the animal, the more poison it can withstand. Likewise, a two-year old child is much more prone than an adult to fatal poisoning by a given weight of toxic substance or quantity of paint of a poisonous nature. Paint components with LD_{50} acute toxicity values (per kilo body weight) of 1 mg (or less), 100 mg, 1 g and 15 g (or more) would be regarded respectively as highly toxic, moderately toxic, slightly toxic, and relatively innocuous.

(b) *Inhalation*

The respiratory system is the most significant route for poisoning so far as industry is concerned. Substances which can be inhaled range from gases, and the vapours of volatile liquids and solids, to liquid particles in the form of mists and droplets, and solid particles represented by dusts.

Evolution has seen to it that lungs are adapted to inhale air whose approximate composition is nitrogen 78% and oxygen 21%. Among the minor components of air is carbon dioxide which is present to the extent of 0.03%. Moist air will, of course, contain up to a few per cent of water vapour. Inhalation of any gaseous mixture with a composition differing substantially from that of air would be definitely harmful and sometimes fatal. Thus one could not even survive for long in an atmosphere of 100% pure oxygen. Nevertheless oxygen is vital to the existence of human life. Inhaled oxygen is transported from the lungs to the cellular tissue. This movement of oxygen, as well as the reverse transportation of carbon dioxide, is accomplished largely by means of haemoglobin in the red cells of the blood. The latter represents some 8% of one's body weight and the whole volume of this blood (6–7 litres) passes through the lungs in about one minute.

Poisonous gases and vapours can be irritants, asphyxiants, anaesthetics and systemic poisons. Volatile irritants are capable of preferentially affecting specific areas such as the upper respiratory tract, the lung tissue, or the terminal respiratory passages and air sacs. For example, the upper respiratory tract is sensitive to such substances as ammonia, hydrochloric and chromic acids, and alkaline dusts.

There are different types of asphyxiant gases. Firstly, an excess of an inert gas like nitrogen, hydrogen or carbon dioxide will act merely mechanically, by starving the victim of oxygen. Carbon monoxide is highly toxic because it combines with haemoglobin to form a stable compound, thereby depriving oxygen of its means of transportation. High percentages of carbon monoxide are formed when organic materials, such as wood, paper, rubber, resins, etc are burnt in an atmosphere deficient in oxygen. This will explain why the death of a significant proportion of victims of conflagrations is due primarily to carbon monoxide poisoning.

Organic nitriles, one such is acrylonitrile, a monomer used in the manufacture of polymers and copolymers on which some paints are formulated, are poisonous in the same way as hydrogen cyanide. That is, they combine with cellular catalysts (enzymes), thereby inhibiting tissue oxidation.

Virtually all nitro- and amino- aromatic compounds, many of which (such as aniline, toluidine, dimethyl aniline and nitrobenzene) have been used in paint products, are characterized by their ability to form methaemoglobin in man. Methaemoglobinaemia results from the oxidation of haemoglobin and leads to such symptoms as cyanosis (blueness about the face and extremities), increase in pulse rate, and dizziness.

Anaesthetic substances, by a depressant action on the central nervous system, affect the supply of blood to the brain. In descending order of activity they include:

(i) Olefinic hydrocarbons, i.e. the alkylenes.
(ii) Ether.
(iii) Paraffinic hydrocarbons, e.g. propane, butane and *n*-hexane.
(iv) Ketones including acetone.
(v) Aliphatic alcohols such as ethanol and propanol.

Air contaminants, which on inhalation will cause systemic poisoning, include:

(i) Most halogenated hydrocarbons, such as carbon tetrachloride (liver and kidney poisons).
(ii) Benzene and various phenols.
(iii) Carbon disulphide and methanol (nerve poisons).
(iv) Various derivatives of toxic elements such as mercury, cadmium, manganese, lead, arsenic, phosphorus, selenium and fluorine.

Whereas gases and vapours have ready access to all parts of the respiratory tract, the anatomy of the latter acquires a greater significance when solid particles are inhaled. Inhalation makes use of a passage way starting with the nose, pharynx and larynx and continuing through to the trachea and bronchus. This leads to branching into bronchioles, further branching into alveolar ducts (5 or 6 per bronchiole) and alveolar sacs (3–6 per duct) and finally to the alveoli which are about 0.1 mm in diameter. Large inhaled particles are trapped in the upper portions of the respiratory tract and are subsequently removed towards the mouth by the wave-like motion of ciliated cells. The latter are particularly numerous in the trachea and larger bronchioles. The lung retains particles which range in diameter from 0.3 to 7 μm, most of them being a little over 1 μm in size.

The development of silicosis is largely attributable to the particle size (approx. 0.6 μm) of silica and not to its sharp-edged physical characteristic. Asbestos, on the other hand causes asbestosis because of its fibrous nature. Fibres with lengths, ranging from 5–100 μm are deposited in the lungs where they are trapped and subsequently encapsulated in protein. Because of the resulting engorgement of the lungs, the heart has more work to do supplying oxygen to the blood.

For many countries the best guide of comparative values of inhalational toxicity is that provided by the American Conference of Governmental Industrial Hygienists' (ACGIH) annual lists of Threshold Limit Values (TLVs). These refer to concentrations of air-borne substances below which nearly all workers may be exposed repeatedly, day after day, without adverse effect. Time-weighted average concentrations permit excursions above this limit, provided they are compensated by equivalent excursions below the limit during the working day. Exceptions are certain substances against which are quoted 'ceiling values' which should never be exceeded, e.g. for formaldehyde a ceiling value of 2 p.p.m. is given.

Published TLV's for contaminants range from 5000 parts per million (p.p.m.), by volume, for carbon dioxide down to no exposure as detected by the most sensitive methods for certain specified human carcinogens. Sometimes the TLV is given on a weight basis as so many mg of particulate per cubic metre of air. In the case of perfect gases the conversion of TLVs from a weight basis to a volume basis is facilitated by the knowledge that 1 g mol of the gas has a volume of 24.45 litres at 25°C and 760 mm Hg pressure.

For Great Britain, H.M. Factory Inspectorate of the Health and Safety Executive reissues the ACGIH list of TLVs but also adds a short list of substances (including asbestos and vinyl chloride) for which different values appertain in the United Kingdom.

Other industrial countries have their own lists and include those of the National Swedish Board of Occupational Safety and Health (Stockholm), the Institute National de Recherche et de Securité (Paris), the Czech TLV Committee and the German Research Association/Commission for Testing Harmful Industrial Substances. Values issued by the U.S.S.R. tend to be lower than those of the U.S.A., and Western Europe because they are based on a different physiological assessment. The World Health Organization, in their report No. 415 (Geneva 1969) has listed Safe Concentration Zones for 24 substances, with a recommendation for their adoption internationally.

(c) *Poisoning subcutaneously*

Mention has already been made under 'Skin effects' (Section 8.1.1b) of the ability of certain chemicals to act on the skin or immediately below its surface, but nothing has so far been said of passage through the skin as a route to poisoning, alternative to swallowing and inhalation. Abraded skin, cuts and lesions are all naturally liable to facilitate the absorption of poisons. In varying degrees, intact skin also has this capacity. Perhaps it is not so surprising, from what has already been said of the structure of the skin, to learn that it is penetrable by a great variety of toxic substances, which thus find their way into the blood stream and become systemic poisons.

Poisoning by skin contact has been referred to over the ages. Contaminated clothing, gloves, shirts and other garments, are alluded to in Portigliotti's *The Borgias* [1]. To the naturally-occurring poisons, particularly during this century, industry has added a considerable number of synthetic chemicals capable of acute or chronic poisoning. Even the insidious action of carcinogens is capable of being initiated through the skin.

Not until as recently as 1968 was the chemical propane sultone found to be a powerful carcinogen. By this time it had become widely used in industry and was becoming recognized as of potential value in the manufacture of paint vehicles. H. Druckrey *et al.* [2] found that even single doses of propane sultone, given subcutaneously to rats, produced a high yield of sarcomas at the site of injection. A little later, Van Duuren *et al.* reported [3] tumours in 21 out of 30 mice, at the site of injection with propane sultone. This chemical is also a skin sensitizer and repeated applications to the skin can cause extreme irritation. In other words, propane sultone can be expected to lead to cancer on skin contact, at least with animals, and without the artificial aid of injection.

Again, it is known that both the simple chemicals ethylene imine and propylene imine are readily absorbed by the skin. Ulland *et al.* [4] have pointed out that propylene imine is an important chemical intermediate in the production of coating compositions etc., and that it is closely related to its parent homologue ethylene imine which is known to have carcinogenic, mutagenic and teratogenic activity!

How significant skin absorption is as a route for lead poisoning, is the sort of question many who handle paint frequently might ask. The answer is, it all depends which compound of lead is involved. Thus inorganic lead compounds, which in general have poor solubility characteristics, are not readily absorbed by the skin. Certain lead salts however, such as the acetate and oleate are known to be absorbed, possibly by acid (radical) interchange with the natural oils and fats in the skin. Organo-lead compounds, notably lead tetra ethyl (TEL), the ubiquitous petrol anti-knock, are readily absorbed by the skin, often in toxic amounts. Marchenko and Beilikhis [5] report that this property of TEL applies both to the liquid and the vapour.

Another metallic element of interest in the context of paint, and which has skin absorptive properties, is mercury. Even the metal itself, in a fine state of subdivision, is absorbed by the skin. It was, for instance, formerly employed as an inunction (rubbed into the skin after being pre-mixed with oil or fat) in the treatment of skin diseases. On the other hand, through skin absorption, mercury has given rise to chronic poisoning among industrial workers. Experimenting with rabbits, Schamberg *et al.* [6] came to the conclusion that the respiratory absorption of mercury was of far less importance than its cutaneous absorption.

Silberberg [7, 8] and others have found evidence for the percutaneous absorption of the inorganic compound mercuric chloride and J. Barnes [9] successfully treated a case of acute mercurial poisoning which followed inunction with an ointment containing 1 part in 3 of mercuric chloride.

Direct contact with organic mercurials can cause severe (second degree) burns and, states Goldwater [10], can result in absorption of measurable amounts of mercury into the body from underclothing and bed linen. An organic mercury compound sometimes used as a paint fungicide is phenyl mercury acetate. Brana Jovicic [11] describes cases of mercury intoxication, believed due to absorption through the skin, among workers handling a fungicide based on a mixture of phenyl mercury acetate and ethyl mercury chloride. More potent are the highly toxic mercury alkyls but these have no particular relevance to paint.

Whereas inorganic tin compounds are of negligible percutaneous toxicity, a totally different picture emerges with the organo-tins. Zeman *et al.* [12] observed that four cases of poisoning resembled those from lead tetra-ethyl, the tin compounds being readily absorbed by the skin. Percutaneous poisoning by tetrabutyl tin and related compounds in experiments on rabbits was found by Yanagibashi [13] to be reversible within certain limits upon discontinuation of their administration. Local skin washing with 5% potassium permanganate directly after the application of the organo-tin was effective in cases of subacute poisoning. Barnes and Stoner [14] confirmed that many alkyl tins are toxic to animals but drew attention to the fact that di-octyl and di-nonyl derivatives are relatively harmless when applied to the skin in large doses. In experiments with tributyl tin oxide (TBTO), the organo-tin compound which has been used as stabilizer for polyvinyl chloride, as a paint fungicide, and as an active component of anti-fouling paints, Elsea and Paynter [15] found that repeated dermal application to rabbits resulted in no evidence of sensitization. They concluded, however, that because TBTO is a primary skin irritant and can be absorbed through the skin, contact should be avoided. In the event of exposure, contaminated skin areas and clothing should be thoroughly cleaned with soap and water.

Understandably, because of their generally greater lipid solubility, organic compounds have a more enhanced capacity for skin absorption than inorganic compounds, and many of them have been responsible for serious percutaneous poisoning. Particularly vicious is phenol, as well as certain substituted phenols. A report of H.M. Inspector of Factories [16] quotes the case of a 23-year old technical assistant who was splashed by phenol in attempting to unblock a pipe-line containing the material and his trousers on both thighs became soaked. Within a few yards of a medical centre, he collapsed, and despite first aid and medical treatment did not recover. This is but one of a series of similar incidents recorded over the years, involving substantial contamination of the skin with phenol many of which have terminated fatally.

Nitro- and halogenated phenols are probably even more hazardous and much smaller quantities can lead to systemic poisoning. Thus, low concentrations of the well-known fungicide pentachlorophenol (or its sodium salt) have been responsible for numbers of fatalities in many parts of the world. Truhaut *et al.* [17] described cases of poisoning which occurred in men engaged in dipping planks of wood in a 3% solution of 'penta'; two cases ended fatally.

Amines, too, are capable of being absorbed through the skin; this applies both to aromatic compounds such as aniline, dimethyl aniline, *o*-toluidine, diamino diphenyl methane, anisidine, benzidine and *p*-nitraniline as well as to aliphatic amines including butylamine, diethylene triamine, dipropylamine and triethylene tetramine.

A great variety of solvents, particularly those with good solvent power for fats, are known to penetrate the skin; among these are benzene, nitrobenzene, methanol, furfural, dioxan, dimethyl sulphoxide, dimethyl formamide, and the highly poisonous carbon di-sulphide. Not surprisingly, chlorinated solvents like carbon tetrachloride and tetrachlorethane are readily absorbed by the skin. Of even greater significance in paint formulations are the mono-ethers of ethylene glycol, such as the 'cellosolves' and 'oxitols'. As a class they can all penetrate intact skin; their degree of toxicity varies according to the etherifying alcohol content. Least toxic is the ethyl ether. More toxic, on the basis of published animal experiments, are the butyl, methyl and isoproyl ethers of ethylene glycol.

A number of toxic polymerizable monomers such as acrylonitrile, methyl and ethyl acrylates and acrylamide can act percutaneously but are only of any consequence if unreacted monomer should remain in the final paint. The last named monomer has been described as 'dangerous'. Acrylamide can penetrate the skin in toxic amounts; toxicity is cumulative. It has unusual effects on the central nervous system; the neurologic syndrome can embrace lack of muscle control and hallucinations (eye and ear).

Sometimes pesticides are incoporated into paint. Although many of these fall into twò classes, halogenated hydrocarbons and organo-phosphorus compounds, the latter are not met with in paint, partly because of their instability (ease of hydrolysis) and partly because many are extremely toxic (through cholinesterase blocking). Most of the halogenated hydrocarbon pesticides are, from their chemical structure, poisonous and furthermore most are capable of being readily absorbed by the skin. They include Aldrin, Dieldrin, Endrin and Lindane (the isomer of benzene hexachloride). Despite its good lipid solubility, however, DDT is not appreciably absorbed through the skin of man and mammals. Most organo-phosphorus pesticides are poisonous because they are cholinesterase blockers. Typical preparations are Demeton (Systox), Parathion, Malathion, EPN, Dichlorvos (DDVP), Diazinon and Phosdrin (Mevinphos).

Under 'Ingestion' (Section (a)) reference has been made to the toxicity rating LD_{50}. Comparable values for percutaneous poisoning are available for many compounds but are not so numerous as those for ingestion. The rabbit is more commonly used for LD_{50} (skin) determinations.

8.1.3 Toxic effects of paint materials

In general terms, paint is inevitably of heterogeneous composition, necessitating for the requirements of colour and opacity, a great variety of insoluble solids (pigments and extenders), a liquid medium to provide continuity of film, and a volatile liquid (solvent) to assist in the application (flow, etc.), the solvent then evaporating to enable the film to dry. Thus paint is mostly two-phase (solids dispersed in a liquid), and in the case of emulsion products, there is the additional complexity of the liquid itself being two-phase. Each type of raw material can bring its own particular hazard and the end-product can be poisonous or externally (eye and skin) harmful. Furthermore, the possibility exists of two or more components combining to introduce a third hazard, whether synergistically or by chemical interaction. Added to all this, most of the organic components of paint are combustible (liquid media and plasticizers) or flammable (solvents).

To deal briefly but systematically with the hazardous nature of the more important substances which are used in paint, this section has been subdivided into:

(a) Inorganic solids (mainly pigments and extenders).
(b) Organic solids (e.g. resins, gums, organic pigments, and polymers).
(c) Liquid media (mostly drying oils).
(d) Solvents.
(e) Plasticizers.
(f) Miscellaneous additives.

(a) *Inorganic solids*

In the case of many inorganic compounds, such as those based on the elements arsenic, mercury or lead, the nature of this element will have a profound influence on the toxicity of the derived compound. Hence this sub-section is classified according to the parent element, and for the sake of those readers who are not too familiar with the periodic classification of the elements, the latter are listed alphabetically.

Aluminium
Aluminium compounds are generally non-toxic. Aluminium's principal usage in paint is in the form of the powdered metal for 'silver' paint and 'metallic' finishes. A potential hazard, however, does exist if such finishes are improperly formulated with an acidic or alkaline medium. The aluminium reacts with the acid or alkali to cause gassing (evolution of hydrogen) and pressure build-up in closed containers. Such cans must therefore be opened with caution.

Antimony
Although certain compounds of antimony are toxic and have even led to fatalities, the oxides and sulphides were removed from the Poisons List in 1970. Nevertheless,

soluble antimony compounds, i.e. soluble in dilute hydrochloric acid, are prescribed a limit to their use in paints for toys in the 1974 Toys (Safety) Regulations.

The oxide (Sb_2O_3), known as antimony white or its proprietary name of Timonox [18] can be a skin irritant, particularly to sensitive skins. What is not so widely known is the fact that industrial grades of antimony oxide contain small percentages of both arsenic and lead.

TLV (Notice of intended changes for 1975):	
Antimony trioxide, handling and use, as Sb	0.5 mg/m^3

Arsenic

Because of the high oral toxicity of arsenic compounds their use in paint has been discouraged, although a few are still used in wood preservatives. Arsenites such as Paris Green, Schweinfurt Green and Scheele's Green are virtually pigments of the past, as is the sulphide Orpiment. The presence of arsenic in antimony oxide has already been referred to (previous paragraphs). Sometimes measurable amounts are also found in phosphoric acid. Like chromium compounds (p. 347), volatilized arsenic has been held responsible for cases of perforation of the nasal septum. Particularly poisonous is arsine (hydrogen arsenide, AsH_3) which has been responsible for some fatalities. A typical situation is the evolution of this gas from reducing solutions (e.g. acid + zinc or galvanized metal $\rightarrow$ hydrogen) acting on even low concentrations of arsenic trioxide (As_2O_3). Organo-arsenic compounds have had a specialized outlet in anti-fouling paints for ships.

TLVs:	
Arsenic trioxide, handling and use	0.5 mg/m^3
Arsine	0.05 p.p.m. (0.2 mg/m^3)

Barium

Barium is much used in paint, and pigments and extenders include lithopone (a co-precipitate of barium sulphate and zinc sulphide), barium chromate and barytes (barium sulphate). The chromate is very insoluble – see also under 'Lead'. The Toys (Safety) Regulations, in 1967, specified a limit for 'soluble' barium in paint for toys but this limit was subsequently raised to allow the continued use of pure barium sulphate. After all, this compound is used in medicine as barium meal! Attitudes differ between the U.K. and U.S.A. in so far as that in the U.K., solubility in dilute hydrochloric acid (comparable with the acidity of stomach juice) is considered as of more significance than water solubility. Thus the use of barium metaborate, a fungicide which is appreciably soluble in dilute hydrochloric acid, is discouraged in paints which are liable to be in contact with foodstuffs. This is contrary to practice in the U.S.A.

TLV:	
Barium (soluble compounds)	0.5 mg/m^3

Cadmium

For some years now, the cadmium compounds, notably the sulphide (yellow), selenide (red) and sulpho-selenides (red) have, for their valuable technical properties, become established as paint pigments. In 1950, Gabby [19] fed rats with such compounds and noted that apart from some loss of appetite over the selenium-containing pigments, at no time during the tests were any adverse symptoms noted. He concluded that if a baby consumed all the paint from the inside of its cot once every six months, over a period of several years, the effect of such cadmium pigments might be considered negligible.

In more recent times there has been growing concern, based on more toxicological data, over the hazards of cadmium, which have in consequence been the subject of several reviews, among which are those of Flick *et al.* [20], Turner [21], Kendry and Roe [22] and Webb [23]. Cadmium is yet another element on which a limit (Cd soluble in dilute HCl) has been placed in the Toys (Safety) Regulations of 1974.

The Factory Inspectorate point out [24] that cadmium (Cd) poisoning occurs primarily through inhalation of Cd dust or fume but that the ingestion route is not of significance industrially. Poisoning by cadmium compounds is, within the U.K., a notifiable industrial disease. The American Conference of Governmental Industrial Hygienists has accumulated sufficient information to list the production of cadmium oxide as an occupational carcinogenic hazard.

Thus, although many cadmium pigments have low solubility characteristics, and hence low (ingestion) toxicities, these could be a potential hazard when metal substrates coated with paints based on these compounds are subjected to burning off or to welding, as for instance in the car and ship-building industries. This would be the case particularly where thermal disintegration ultimately led to the evolution of cadmium oxide, the agent responsible for cadmium fume fewer.

TLV:	
Cadmium dusts and salts, as Cd	0.05 mg/m^3

Calcium

Calcium is ubiquitous, being present in drinking water, bone (as calcium phosphate) and numerous minerals such as chalk, and is not regarded as a toxic element *per se.* It is also met with in paint in the form of pigments (calcium plumbate), extenders (whiting and gypsum) and driers (naphthenates, etc). Water paints have often been formulated on hydrated lime, which could be made by slaking anhydrous (or dry) lime (calcium oxide). It must be stressed that calcium oxide is a powerful dehydrating agent and can be dangerous if in contact with tissue (skin and eyes) or if the powder is inhaled.

Inert calcium compounds such as the carbonate, silicate and sulphate are termed 'nuisance particulates' for which threshold limit values have been introduced.

TLV:	
Nuisance particulates	30 m.p.p.c.f.*, or 10 mg/m^3
(* Millions of particles per cubic foot of air.)	

Carbon

Because of contamination with traces of carcinogens resulting from its mode of manufacture, repeated contact with lamp-or carbon black could lead to skin cancer.

The introduction in recent years of carbon fibres has led to its use in plastics and composites and even as a paint pigment. There is no record, however, of carbon fibres causing any adverse toxicological reaction.

TLV:	
Carbon black	3.5 mg/m^3

Chromium

Chromium compounds are capable of presenting a variety of health hazards. In spite of that, however, chromates and other Cr compounds are much used as paint raw materials – mainly as pigments but also in stains, etch primers and metal pre-treatment products. The most biologically active chromium compounds are chromic acid (CrO_3) and such salts as sodium, potassium and ammonium dichromates. These are all powerful oxidizing agents which can not only cause injuries to eyes, tissue and breathing passages but are responsible for skin ulceration as well as a significant incidence of perforation of the nasal septum. Within the U.K., chrome ulceration is a notifiable industrial disease and in 1931 the Chrome Plating Regulations were promulgated to protect workers from the liquid, vapour or spray of electrolytes containing such compounds. The Toys (Safety) Regulations lay down a limit of 250 parts of 'soluble' chromium in one million parts of the dry paint film.

Chromium compounds have also caused numerous causes of dermatitis and skin sensitization, not only in the building industry (Cr is a common contaminant of cement) but because of their use as pigments and anti-rust agents in primers for metal substrates (cars, foundry operations, ship-building and aircraft). Such primers are frequently based on zinc, calcium or strontium chromates all of which are in the class of slightly soluble pigments. Both chromic acid and zinc chromate have been responsible for cases of asthma, as reported in 1969 by C.D. Williams [25].

In addition to all this, chromium is associated with a carcinogenic hazard. Baetjer [26] referred to numerous cases of cancer of the respiratory tract among chromate workers in Germany and the U.S.A. Later Payne [27] observed sarcomas in mice implanted intramuscularly with calcium chromate, and Hueper and Payne [28] produced sarcomas in rats with implantations of calcium chromate. No tumours were obtained in controls nor with the insoluble barium chromate. Not surprisingly, in view of the close chemical relationship between strontium and calcium, Hueper [29] induced sarcomas in 15 out of 33 rats, using strontium chromate. British and American specifications for aircraft primers call for the use of strontium chromate and calcium chromate respectively. In 1972, the ACGIH listed 'certain insoluble chromates' under 'Human Carcinogens' but later (1975) under the same heading quoted instead 'chromite ore processing'. However, under a separate list of 'Occupational Substances Suspect of Oncogenic Potential for Workers' the ACGIH mentioned zinc chromate specifically.

Thus care must be exercised in the handling of both soluble (e.g. chromic acid and dichromates) and less soluble (e.g. those of calcium and zinc) chromates. For example both wet scuffing (skin contact) and dry scuffing (inhalation of dust) of zinc chromate primer is to be discouraged. Alternatively, when wet flatting wear water-impervious gloves and for dry flatting wear a dust mask and (preferably) with the additional aid of dust removal by forced ventilation.

Even the highly insoluble lead chromate (solubility is 0.007 mg/100 ml) which is generally held to be a 'safe' pigment, whether ingested or inhaled, raises queries. For some years now it has been common practice within the U.K. to add barium chromate or zinc chromate to lead-containing paints, ostensibly to reduce the lead (ingestion) hazard. In 1946, Watson and Clay [30] claimed on the basis of their analytical studies that the addition of zinc chrome with the object of reducing the soluble lead content of a lead chrome should be regarded as a useful method of decreasing the toxicity of the pigment as measured by the Home Office test. Lead chrome referred to here is a mixture of lead sulphate and lead chromate.

However, one may well ask whether it is ethical, in an effort to reduce the lead hazard, to introduce the hazard of zinc chromate, or that of soluble (to whatever degree) barium? Another question centres on the biological activity of lead chromate, despite its exceptionally low solubility. Again, the U.S. Environmental Protection Agency ordered a ban on interstate shipments of paints (domestic use) containing more than 0.06% lead, thus ignoring the solubility of lead chromate.

All the preceding statements on the hazards of chromium compounds refer to hexavalent chromium (VI). It is generally accepted, however, that chromium (III), particularly such compounds as the highly insoluble green pigment chromium sesquioxide (Cr_2O_3) present only a negligible health hazard. Nevertheless Fregert and Rorsman [31] do claim from their researches that a person allergic to hexavalent chromium is regularly allergic also to trivalent chromium.

The Factory Inspectorate's Technical Data Note No. 8., entitled *Chromium – Health and Safety Precautions* covers such aspects as 'Control', 'Personal Protection' and 'First Aid'.

TLVs:	
Chromic acid and chromates (as CrO_3)	0.1 mg/m^3
Chromium, sol. chromic, chromous salts as Cr	0.5 mg/m^3
Zinc chromate, (as CrO_3)	0.1 mg/m^3 *

(* Intended change for 1975.)

Cobalt

Cobalt pigments, such as cobalt blue and cobalt violet are of decreasing interest, mainly because of their relatively high cost and the availability of alternative pigments of improved technical performance. The principal cobalt compounds met with in paint are cobalt driers, or cobalt salts of such acids as acetic, octoic, linoleic and naphthenic. As a result, many conventional paints (based on drying oil), will be found to contain measurable amounts of cobalt. Although there is an accumulated literature on the toxicity of cobalt, the indications are that many cobalt compounds are less hazardous than the metal itself.

For cobalt naphthenate, Rockhold [32] obtained an acute oral LD_{50} of 3.5–4.4 g/kg, i.e. giving it a rating of only slightly toxic. Evidence that cobalt compounds have a tumorigenic potential is equivocal.

TLV:

Cobalt, metal fume and dust	0.1 mg/m^3*

(* For 1976 an intended reduction to 0.05 mg/m^3 was suggested.)

Copper

Copper is more commonly encountered in bronze paint, being made up from bronze powder, although copper compounds such as cuprous oxide and copper thiocyanate, (both used in anti-fouling compositions), copper chromate which has fungicidal properties, and copper chromite, have all been used in paint. Copper sulphate and copper naphthenate are valuable components of some wood preservatives.

Berger [33] reports cases of constipation among operatives handling bronze powder, whilst Browning [34] asserts that copper itself is less toxic than its salts and also declares that there is little evidence that copper poisoning (acute or chronic) is a serious industrial hazard.

TLVs:

Copper fume	0.2 mg/m^3
Copper dusts and mists	1 mg/m^3

Fluorine

Gaseous, liquid and solid compounds of fluorine have all been used in paint and include catalysts (Boron trifluoride and derivatives), propellants (fluorinated hydrocarbons), metal pretreatment ingredients (fluorides), pesticides (e.g. fluoroacetates), and polymers (polyvinylidene fluoride, Teflon and Fluon). These compounds can range from substantially inert to highly dangerous. See also under 'Additives'.

TLVs:

Fluoride (as F)	2.5 mg/m^3
Hydrogen fluoride	3 p.p.m. or 2 mg/m^3
Sodium fluoroacetate	0.05 mg/m^3

Lead

For many years, lead poisoning has been rife among those who have had to handle lead compounds and lead-based products and in addition lead has been recognized as a serious pollutant of the environment. Thanks partly to the accumulation of toxicological data on lead, partly to the publicity given to its hazards and partly to the protective measures introduced by legislation to combat lead poisoning, the incidence of the latter is under much greater control and, in certain areas of lead usage, has been reduced to insignificant proportions. An additional but nevertheless important factor contributing to the decline of lead poisoning among those who handle paint was the introduction of the white pigment titanium dioxide, which, because it is virtually non-toxic and at the same time has superior technical properties, has rapidly replaced white lead.

The Annual Report of H.M. Chief Inspector of Factories, for 1900, listed 38 cases of industrial lead poisoning. In 1970 there were no deaths attributable to lead poisoning and of the 70 non-fatal cases, two were identified against 'Paint and Colour Works', and none against 'Coach and Car Painting', 'Ship-building', 'Painting of Buildings' and 'Paint used in other Industries'. Nowadays, too, the Factory Inspectorate arranges for medical examinations of thousands of workers who handle lead compounds, including operatives who make 'flaked litharge' and 'red and orange lead'.

Not surprisingly, this century has seen a spate of literature relating to the toxic hazards of lead. Among the first of the more important publications was that of Legge and Goadby [35] with their book *Lead Poisoning and Lead Absorptions*. More recent reviews include *Occupational Lead Exposure and Lead Poisoning* [36] which covers the relevant literature for the period 1850–1943 and *Lead Poisoning* by Chisholm [37] which makes an extensive examination of the biological effects and treatment of lead poisoning, particularly in children.

Because weight, rather than volume, has a closer bearing on the toxicity rating of a substance, and because lead has a high atomic weight, quantities of lead-containing paint could in consequence have significantly greater toxicity than might be apparent from their volume.

Another important point to remember in connection with lead is that it is a cumulative poison and 'sub-acute' quantities, absorbed over a prolonged period can lead to chronic lead poisoning and even fatalities.

The more characteristic symptoms of lead poisoning are anaemia (pallor), blue line on the gums, colic (painters' colic) and peripheral neuritis ('wrist drop'). One of the most unfortunate aspects of lead poisoning is that young children are particularly liable to suffer permanent brain damage (encephalopathy) leading so often to mental retardation. Added to this is the incidence of fatal lead poisoning among young children (principally 15–30 months of age) suffering from pica, which in their case takes the form of habitual chewing of old lead paint flaking from window sills etc. With this in mind, U.S.A. Congress passed in 1971 the 'Lead-Based Paint Poisoning Prevention Act' which included provision for grants to assist in the 'prompt elimination of lead-based paints from all interior surfaces, porches, and exterior surfaces to which children may be commonly exposed ...'.

Within the U.K. the Toys (Safety) Regulations placed an upper limit of 5000* parts of lead in 1000 000 parts (i.e. 0.5%) of dry paint film. Unlike the controls over toxic metals, the lead limit was for total lead and ignored the solubility factor. Also, the Paintmakers Association of Great Britain recommended to their member-companies discontinuance of the production or sale of lead primers for wood by 11th January 1975.

At this juncture a word of caution to the do-it-yourself painter planning to re-finish a wooden cot. Of what value is a lead-free undercoat and top-coat on an old cot originally coated with a lead primer containing at least 50% of lead? Teething children could easily gnaw the whole paint system down to the bare wood, not differentiating between the lead-containing and lead-free layers of paint.

See also Chronic lead poisoning, p. 333; Skin absorption of lead, p. 341; Organo-lead, p. 341; and Lead chromate and lead 'suppression', p. 348.

TLV:	
Lead, inorg., fumes and dusts, as Pb	0.15 mg/m^3

Magnesium

Magnesium is not a toxic element *per se* but is an essential element in the tremolite chrysotile and anthophyllite forms of asbestos (See under 'Silicon'). There is one magnesium compound, often used as an extender in paint, with an alleged hazard peculiar to itself. The substance is talc or hydrous magnesium silicate and the disease talcosis.

Nam and Gracey [38] reported the findings of an autopsy on someone who had died from diffuse peritonitis. Although the cause of the latter was not established, the autopsy did reveal evidence of pulmonary talcosis with gross contamination of the lungs with talc. Apparently the patient had been given to excessive use of cosmetic talcum powder and had even liberally dusted his bed sheets nightly with the powder.

In 1967 Kleinfeld *et al.* [39] reported on an increased incidence of certain forms of cancer among talc miners and millers in New York State, but ascribed this as possibly due to asbestos contamination of the talc (or talc diluted asbestos). Blejer and Arlon refer to Kleinfeld's paper in their review [40] of talc as a possible occupational and environmental carcinogen.

Manganese

Because the main use for manganese in paint is in the form of driers it means that in general the manganese compound, such as manganese naphthenate, would be a solution and in low percentage.

Symptoms of manganese poisoning are mainly neurological and most cases result from inhalation of the dioxide which is more often handled in other industries such as the manufacture of dry batteries.

(* Reduced in 1975 to 2500 parts, or 0.25%).

TLV:
Manganese and compounds, as Mn 5 mg/m^3

(This TLV is a Ceiling Value (C) which should not be exceeded even for limited periods of time.)

Mercury
Mercury compounds have been utilized in paint chiefly for their biocidal properties; the majority of such substances being the phenyl mercurials. Although organo-mercurials are still used as in-can preservatives because of their relatively low cost, the tendency has been to move away from these compounds partly because of increasing concern in recent years over pollution problems and the fact that waste mercury from paint manufacture or paint usage has a strong potential as a pollutant. Still fresh in the minds of many are the disasters which took place in Japan, notably at Minamata and Niigata in the 1950s and 1960s, where numerous cases of severe neurological poisoning and even fatalities occurred among the local population. The incidence of poisoning was traced to the eating of contaminated fish; mercury waste being converted to the highly toxic methyl mercury which has the ability to penetrate the blood-brain barrier.

Reference has already been made (pp. 341–342) to poisoning by skin absorption of mercury metal, as well as by inorganic and organic mercury compounds. Battigelli [41] has reviewed mercury toxicity from industrial exposure and Whitehead [42] makes particular reference to phenyl mercury acetate when dealing with certain organic mercurials. More recently, Mann [43] has referred to mercurial biocides as 'paint's problem material'.

TLVs:

Mercury (all forms except alkyl) as Hg	0.05 mg/m^3
Mercury (alkyl compounds*) as Hg	0.001 p.p.m., 0.01 mg/m^3

(* The contribution of percutaneous poisoning is referred to by the word 'skin'.)

Molybdenum
Industrial poisoning from molybdenum compounds is almost negligible, the main evidence for molybdenum being physiologically active results from animal tests. Here the overall indications are that molybdenum has relatively low toxicity (by ingestion) with decreasing solubility being associated with a lowering of the toxic hazard, as for instance in the case of molybdenum di-sulphide. Nor are molybdenum compounds absorbed through the skin. As a result, molybdenum pigments such as the molybdenum chromes which include Scarlet Chrome are more likely to present a toxic hazard by virtue of other elements in their composition, such as chromium or lead.

TLVs:

Molybdenum, as Mo	5 mg/m^3
Soluble compounds	10 mg/m^3

Nickel

Nickel compounds are not a common feature of paint compositions, although nickel enters into a few pigments such as Nickel Yellow (calcined nickel phosphate) and the light-fast pigment nickel dimethyl glyoxime.

Systemic poisoning by nickel is not a significant industrial hazard. A much greater problem is the frequency of nickel dermatitis arising from contact with the metal or its salts. Furthermore, sensitization to nickel is well authenticated, such sensitization tending to become permanent.

Apart from the potent carcinogen nickel carbonyl, the evidence for a carcinogenic potential among nickel derivatives is equivocal.

TLV:	
Nickel, metal and insoluble compounds (as Ni)	1 mg/m^3

Potassium

Mention need only be made here of caustic potash (potassium hydroxide) and potash (potassium carbonate), which are sometimes used as catalysts and sometimes as cleaning agents. Care is required in handling these materials or their solutions to avoid skin contact. Eye protection is strongly recommended.

Selenium

Selenium enters the composition of certain pigments, notably the insoluble cadmium selenide and sulpho-selenides. Mention of these has already been made under 'Cadmium'. Because of their poor solubility characteristics their toxicities are very low, particularly when compared with the more soluble selenium compounds such as selenium dioxide and the selenites. Industrial selenosis and dermatitis are mainly associated with the soluble compounds of selenium. Additionally it is such compounds which, in the presence of acid, and in contact with zinc, galvanized metal or tin are capable of reduction to hydrogen selenide. This gas is not only more poisonous than hydrogen sulphide but is also highly irritant.

TLVs:	
Selenium compounds (as Se)	0.2 mg/m^3
Hydrogen selenide	0.05 p.p.m. or 0.2 mg/m^3

Silicon

Since it is one of the most abundant elements in the earth's crust, many of the compounds of silicon are low cost and much utilized in the paint industry. Although the majority of them are chemically inert they are nevertheless able to present hazards by virtue of their physical nature as in the case of fibrous asbestos and crystalline silica. The more important compounds of silicon associated with paint are discussed below.

(1) *Silica*. There are many types of silica ranging from the crystalline quartz to amorphous material including the diatomaceous earths Kieselguhr, Fullers Earth, and Tripoli powder. Partly because of the variety of these forms of silica, partly because of their generally low specific gravity and absorbent qualities and partly because of their inertness, they have such diverse uses as extenders and anti-settling agents for paints, essential components of polishing and cutting compounds, and filter aids for varnishes.

The two factors of toxicological significance with silica are its respirable size and physical shape. The crystalline variety (notably quartz) is more hazardous than the amorphous variety and can, over a long period of exposure (inhalation) lead to silicosis. The latter is one of the group of pneumoconiosis group of lung diseases which embraces those substances which react with the lung and destroy its natural elasticity as the result of formation of fibrous tissue. The blood cannot absorb its normal requirements of oxygen and severe shortage of breath (emphysema) ensues.

The health of workers in a plant making highly dispersed silica has been investigated by Volk [44]. In sandblasting operations there is not only the inhalation hazard associated with silica particles of respirable size leading possibly to silicosis, but the mechanical hazards of high-pressure jets of abrasive particles as well as the possibility, in the case of sandblasting of lead-paint films, of lead poisoning. It is interesting to note that Part II of The Blasting (Castings and other Articles) Special Regulations, 1949, stipulates that 'No sand or other abrasive substance containing free silica shall be introduced as an abrasive into any *blasting* apparatus'. Under some conditions however, the Chief Inspector may, by certificate in writing, grant an exemption to this ban on sandblasting. The ACGIH, in their list of TLVs, devote much attention to differentiating between the hazards of crystalline (various forms) and amorphous silica, to silicates (various) and to silicon-based 'nuisance particulates'.

(2) *Asbestos.* Asbestos has already been referred to (p. 334), particularly in connection with mesothelioma (cancer of the pleura). The inhalation of asbestos particles can also lead to lung cancer and to asbestosis, the latter being, like silicosis, a lung-disabling disease. It is with all these hazards of asbestos in mind that within the U.K. the Asbestos Regulations were introduced in 1969. See also H.M. Factory Inspectorate's Technical Data Note 13 [45] and publication [46] on *Problems Arising from the Use of Asbestos.*

(3) *Mica.* This aluminium silicate is not regarded as carcinogenic, nor does it present an inhalational hazard on the same scale as asbestos or quartz (crystalline silica).

(4) *Talc* (hydrated magnesium silicate). If free from contamination with fibrous material (asbestos) it is regarded as no more harmful than mica. See also under 'Magnesium'.

(5) *Glass fibre.* Sometimes used as a texturing material and sometimes in composites, glass fibre would appear to many on first throught, to be particularly hazardous. Nevertheless Hill *et al.* [47] found no evidence for any respiratory hazard due to glass fibre. In tests on adults, Heisel and Hunt [48] showed that only the coarse fibres result in any significant cutaneous reactions and these were mild, transitory and interpreted as being mechanical in nature.

(6) Silicates. In addition to the naturally occurring silicates such as talc, mica and asbestos a number of synthetic silicates are used in paint products.

Sodium and potassium silicates (water glass), which find outlets in water paints for cement and plaster, are strongly alkaline and require the same care (eye and skin protection) in handling as do the caustic alkalis (sodium and potassium hydroxides). The fluorsilicates (or silicofluorides) also are dangerously alkaline.

The pigment lead silicochromate is very insoluble and what hazards it possesses are more likely to be associated with the lead and chromium in the molecule.

For Silicones see sub-section 8.1.3 (p. 373) under 'Additives'.

TLVs (ACGIH):
Crystalline silica. Quartz, Cristobalite, Tridymite, fused Silica and Tripoli.

The ACGIH list different formulae for determining the TLV of the above substances, such formulae taking into account the percentage respirable material, and these are set out in Table 8.1.

Table 8.1

Amorphous silica	20 m.p./ft^3 *
Silicates containing less than 1% of quartz	
Asbestos	5 fibres/cc $>$ 5 μm in length†
Mica	20 m.p./ft^3 *
Perlite/Volcanic glass rock	30 m.p./ft^3 *
Portland Cement (clay and chalk)	30 m.p./ft^3 *
Soapstone	20 m.p./ft^3 *
Talc (non asbestiform)	20 m.p./ft^3 *
Nuisance particulates, including calcium silicate, glass fibre ($<$ 7 μm in diam.), glass dust, silicon, and silicon carbide	30 m.p./ft^3 *

* Millions of particles per ft^3 of air

† The ACGIH also list asbestos as an occupational carcinogen. This ACGIH value for asbestos is not applicable to Great Britain and reference should be made to Technical Data Note 13 [45].

Sodium

Sodium hydroxide, otherwise known as caustic soda (an essential ingredient of some paint strippers) and to a lesser extent sodium carbonate (soda), are strong alkalis, comparable with caustic potash and potash respectively (see under 'Potassium'). Accordingly, the need is again stressed for eye protection when handling these compounds and care is called for to prevent skin contact with caustic soda and its solutions.

Strontium

Strontium chromate is the only strontium pigment of any consequence. Its association with a potential carcinogenic hazard has already been mentioned. See under 'Chromium'.

Tin

Apart from evidence that tin fume and stannic oxide can cause stannosis, a benign sort of pneumoconiosis, inorganic tin compounds are not regarded as significantly hazardous. Many of the organo-tin derivatives, on the other hand, are toxic, some highly so. The chemical structure of the organo-tin compound profoundly influences its toxicity. Thus, the trialkyl tins mainly act on the central nervous system whereas the dialkyl compounds cause general illness with damage to the biliary tract. The percutaneous toxicity of organo-tin compounds has already been discussed (p. 342); their toxicity by ingestion can be expected to be at least as great. Piver [49] has reviewed the industrial applications and toxicity of organo-tin compounds.

TLVs:

Tin (inorganic compounds, except SnH_4 and SnO_2) as Sn	2 mg/m^3
Tin (organic compounds) as Sn	0.1 mg/m^3*
Tin oxide	30 m.p.p. c.f. or 10 mg/m^3

(* Also taking account of the contribution of percutaneous poisoning.)

Titanium

The introduction this century of titanium dioxide as a white pigment has without doubt contributed greatly to the reduction in the incidence of lead poisoning. Despite this and the fact it is chemically inert, Elo *et al.* [50] were able to state, as the result of experiments on rats, that prolonged industrial exposure to titanium dioxide dust alone, or in conjunction with silica or other dusts, may provoke mild pulmonary irritation. The ACGIH regard it as a 'nuisance particulate'.

TLV:

Titanium dioxide (nuisance particulate)	30 m.p.p.c.f. or 10 mg/m^3

Vanadium

Vanadium compounds are toxic with such symptoms as respiratory, cardiac and neurological disorders, as well as skin and eye irritation. Nevertheless most of the reported cases of vanadium poisoning have been industrial and due chiefly to inhalation of vanadium pentoxide or vanadium dust. Vanadium is not too well absorbed by mouth, nor, according to Bertrand [51] is it a cumulative poison.

The principal association of paint with vanadium has been with vanadium driers such as the naphthenate. On the available evidence, the burning-off of substantial quantities of vanadium-driered paint might prove a potential inhalation hazard but still not so harmful as fumes from a comparable lead-driered paint.

TLVs:

Vanadium (V_2O_5), as V dust	0.5 mg/m^3
Vanadium (V_2O_5), as V fume	0.05 mg/m^3 (ceiling value)

Zinc

The presence of zinc, either as metal powder, or combined in the pigment molecule, is not regarded as conferring a significant toxic hazard on derived paints and a large tonnage of zinc and its compounds is absorbed by the paint industry. Zinc dust is the major component of zinc-rich primers and commonly used zinc pigments are zinc oxide (sometimes known as Chinese White), lithopone (co-precipitate of zinc sulphide and barium sulphate) and zinc chromate. The latter is potentially harmful, but see under 'Chromium'.

Zinc fume fever, from the inhalation of freshly generated zinc oxide is a common enough illness in industry but its effects are usually transitory and recovery rapid.

Noro and Uotila [52] have reported one fatal case of pneumoconiosis from the inhalation of zinc stearate powder.

TLVs:	
Zinc oxide fume	5 mg/m^3
Zinc stearate (nuisance particulate)	30 m.p.p. c.f. or 10 mg/m^3

Zirconium

Zirconium is yet another element which has found an outlet in paint driers. Although contact with zirconium salts could in some case lead to skin effects and possible sensitization, zirconium is one of the safer elements mainly due to the biological inertness of its compounds.

TLV:	
Zirconium compounds (as Zr)	5 mg/m^3

(b) *Organic solids*

This section is devoted to reviewing the potential hazards of solid organic compounds including polymers and synthetic resins. Solutions of such substances will, of course, introduce the hazard of the solvent(s) used. Solvents are dealt with separately (Section 8.1.3 (c)). See also Epoxides, Polyurethanes etc. under 'Special Paints' in Section 8.1.4. It should be remembered that in the case of polymers there is the possibility of residual monomer in the polymer and with the synthetic resins there can also be present harmful quantities of unreacted material, which in the case of some phenolic resins could be substantial.

Common also to polymers and synthetic resins is the possibility of evolution on pyrolysis, not only of volatile unreacted material but of gaseous and other volatile products of thermal degradation. Some of these products, such as hydrochloric acid, hydrogen fluoride, hydrogen cyanide, formaldehyde and carbon monoxide are of course very toxic. Ball and Boettner [53] have determined the combustion products of a wide range of plastics.

Olefinic polymers such as polyethylene and polypropylene can be regarded as chemically and biologically inert. Vinyl polymers, in general, are fairly innocuous.

However, polystyrene (and vinyl toluene polymer), as well as styrenated (and vinyl toluenated) oils and alkyds will in all probably contain some unreacted monomer. For further information on styrene and vinyl toluene see under 'Aromatic Hydrocarbons' in Section 8.1.3 (c). Vinyl chloride polymers and copolymers can also contain measurable amounts of monomeric vinyl chloride whose particular hazards have already been referred to (p. 333). Vinylidene chloride polymers and copolymers could conceivably raise similar problems. Also previously noted are the hazards of some fluorine containing polymers (see under 'Fluorine', p. 349). Not until recent years was vinyl acetate monomer regarded as meriting an inhalational TLV (30 p.p.m.) on the basis of toxicological data which had become available. The ACGIH make special mention of polytetrafluorethylene (Fluon or Teflon) whose thermal decomposition (in air) could lead to fluorine-containing products.

The acrylic polymers and copolymers are in general more hazardous than the vinyls and include polymers of acrylamide (see p. 339), acrylonitrile (p. 343) and various acrylates and methacrylates. Again, speaking generally, acrylate monomers (acid and ester) are stronger irritants (skin and eye) than the corresponding methacrylates. A particularly interesting feature of ethyl acrylate is its characteristic odour which can be detected at exceedingly low dilution.

Of the synthetic resins, the 'straight' or unmodified phenolics, i.e. those based on phenol and its lower homologues the cresols and xylenols, are significantly more hazardous than the formaldehyde condensation products with urea or melamine. This is largely because of the presence of unreacted phenols (for 'phenol' see also p. 333). Para tertiary butylphenol/formaldehyde resins have already been mentioned (p. 336) because of their potential photo-sensitivity properties The alkyd resins are, generally speaking, among the more innocuous of synthetic resins. Notable exceptions are those based on maleic anhydride. Even minor amounts of unreacted maleic acid (or its anhydride) are capable of inducing both skin irritation and sensitization. Silicone resins are fairly harmless.

Adverse skin effects are possible following contact with chlorinated paraffin wax, particularly in persons with sensitive skins. Another hazard associated with this and other types of chlorinated hydrocarbons such as chlorinated rubber is that these materials can contain measurable amounts of the harmful solvent carbon tetrachloride (see under 'Solvents', Section 8.1.3 (c), p. 366).

Natural resins such as shellac do not pose any significant toxic hazard. The property of rosin to induce skin photo-senstivity has already been commented upon (p. 336).

For toxicological data on the organic pigments one is very much in the hands of the manufacturer who would know their precise chemical structure. However, the vast majority of organic pigments are accepted as reasonably safe paint raw materials with the possible exception of the chromium modified dyestuffs which may cause skin irritation.

(c) *Solvents*

In addition to inflammability, the principal hazards associated with paint solvents centre on their capacity to remove the natural oils from the skin, thus leading to dermatitis and other skin defects and on their volatility which introduces the inhalation hazard.

Whilst care must therefore be exercised to avoid undue skin contact with these solvents, nevertheless there are occasions when soap and water or even strong detergent is unable to remove some stubborn skin contamination, say resulting from resinous material. In such cases it may prove expedient to make use of minimal amounts of a suitable solvent such as alcohol or acetone. However, these solvents, particularly acetone, must not be used in situations where there is a fire risk. After using the solvent, further washing with soap and water is recommended, followed by the use of a hand cream if such is available.

Alcohols

The most toxic of the simpler alcohols is methanol and the fact that taken internally it can lead to blindness, has already been mentioned (p. 336). Although some paint manufacturers now deliberately refrain from using methanol as a paint thinner, it should be remembered that within the U.K., industrial methylated spirits (IMS) contains 5% of wood naphtha. Since this denaturant is required to have a minimum content of 72% of methanol, IMS will as a result contain 3.5% of methanol! Furthermore such paint resins as phenol/formaldehyde, urea/formaldehyde and melamine/formaldehyde are often manufactured from formalin which invariably contains small proportions of methanol. In a review [54] with 284 references, Koivusalo has covered the various effects of methanol poisoning.

On occasions, paint sprayers and others who handle industrial paints, have, when charged with driving whilst having more than the statutory limit of alcohol in their blood (80 mg per 100 ml), pleaded (but in vain), that some if not all of this alcohol was occupational in origin. It is well known that the inhalation of alcohol can lead to effects similar to those resulting from ingestion of the liquid but it has been calculated that breathing air containing alcohol at concentrations in excess of the TLV of the latter would only lead to about 7–8 mg of blood alcohol per 100 ml of blood.

TLVs:	p.p.m.	mg/m^3
Methanol (s)	200	260
Ethyl alcohol (ethanol)	1000	1900
Propyl alcohol(s)	200	500
Isopropyl alcohol(s)	400	980
**n*-Butyl alcohol(s)	(100)	(300)
sec-Butyl alcohol	150	450
tert-Butyl alcohol	100	300
*Isobutyl alcohol	(100)	(300)
Isoamyl alcohol	100	360
Methyl isobutyl carbinol	25	100

(s) Denotes skin absorption effects taken into account in assessing TLV.

* For 1975, the TLV was intended to be changed to a ceiling value of 50 p.p.m. in the case of *n*-butyl alcohol and to a TLV of 50 p.p.m. in the case of isobutyl alcohol.

Ethers

Apart from the fact it is very highly inflammable, ether (diethyl ether) is also a hazardous solvent because of the ease with which large volumes can quickly evaporate to lead to high concentrations of a powerful anaesthetic.

The cyclic ether dioxan, which can also be regarded as a dimer of ethylene oxide, is an eye, nose and throat irritant and a systemic poison. Other cyclic ethers which have had some limited use as solvents are furan and tetrahydrofuran. The former is a strong anaesthetic and is readily absorbed by the skin; the latter solvent is narcotic. Against neither, according to E. Browning [55], is there any record of injury from industrial use.

Dichloroethyl ether which has sometimes been used as a special paint solvent, is a strong respiratory irritant and capable of being absorbed by the skin in toxic amounts.

In recent years evidence has accumulated that bis (chloromethyl ether) is carcinogenic and the ACGIH now list it as an 'Occupational Carcinogen'. Since this ether can, under some conditions, be formed from the interaction of the simple chemicals formaldehyde and hydrochloric acid there is the possibility, however slight, of adventitious formation of bis (chloromethyl ether) in paint compositions which may contain both free formaldehyde and, say as a catalyst, hydrochloric acid.

TLVs:

	p.p.m.	mg/m^3
Ethyl ether	400	1200
Isopropyl ether	250	1050
Dioxan (s)	50	180
Tetrahydrofuran	200	590
Dichloroethyl ether (s)	5	30
Bis (Chloromethyl) ether	0.001	(Occupational carcinogen)

(s) Denotes skin absorption effects taken into account in assessing TLV.

Glycol ethers

As already indicated (p. 343) the mono-ethers of ethylene glycol (i.e. the 2-alkoxy ethanols), are systemic poisons, even by percutaneous absorption, when the etherifying alcohol is methyl, isopropyl, or butyl. Because of this, many of the leading paint manufacturers in the U.K. label thinners containing any quantity of any one or more of these three solvents with the injunction: *Warning. Avoid contact with skin; use with adequate ventilation.*

The mono ethyl ether of glycol, known also by such proprietary names as Cellosolve (Union Carbide), Oxitol (Shell Chemicals), or Ethoxol (ICI) is much safer than the three previously mentioned mono-ethers of glycol. The acetates of all these mono-ethers are comparable in toxicity to the parent ether. Mono-ethers of diethylene glycol, sometimes known as the Carbitols (Union Carbide), are, in general, less hazardous than the corresponding ethers of ethylene glycol but have restricted usage as paint solvents because of their lower volatility.

TLVs:

	p.p.m.	mg/m³
Ethylene glycol mono-methyl ether	25	80
Ethylene glycol mono-methyl ether acetate	25	120
Ethylene glycol mono-ethyl ether	100	370
Ethylene glycol mono-ethyl ether acetate	100	540
Ethylene glycol mono-butyl ether	50	240

Esters

The toxicity of the esters is in the main related to the nature of their constituent alcohols and acids to which they can hydrolyse or metabolize. Care should be exercised in using old samples of methyl acetate since it readily hydrolyses in moist air to methanol and acetic acid. Ethyl acetate, on the other hand, is one of the safer lacquer solvents; limited amounts of this ester are even permitted in foodstuffs. In general, the acetates, butyrates and benzoates are not regarded as particularly hazardous. The formates are more irritant than the corresponding acetates.

Ethyl oxalate is moderately toxic and can be absorbed through the skin, whereas ethyl lactate is relatively non-toxic but is narcotic in high concentrations. Butyl lactate is somewhat more poisonous than the ethyl ester.

TLVs:

	p.p.m.	mg/m³
Methyl acetate	200	610
Ethyl acetate	400	1400
n-Propyl acetate	200	840
Isopropyl acetate	250	950
n-Butyl acetate	150	710
sec-Butyl acetate	200	950
tert-Butyl acetate	200	950
Isobutyl acetate	150	700
n-Amyl acetate	100	525
sec-Amyl acetate	125	650
Isoamyl acetate	100	525
sec-Hexyl acetate	50	300
Methyl formate	100	250
Ethyl formate	100	300
Butyl lactate	5*	25*

* TLVs as proposed for 1975

Ketones

The widely used acetone is one of the least toxic of lacquer solvents. Three of its derivatives are the keto-alcohol diacetone alcohol, the unsaturated compound mesityl oxide and the cyclic ketone isophorone. All three are useful lacquer solvents but all are more toxic than the parent ketone.

An outbreak in 1973 of neurological disorders (peripheral neuropathy, etc.) among workers in a coated-fabrics plant (U.S.A.) was alleged to be due to exposure to methyl *n*-butyl ketone (MBK). These incidents probably contributed to the ACGIH's intention to reduce for 1975 the TLV for MBK from 100 p.p.m. to 25 p.p.m.

TLVs:

Ketone	Alternative name(s)	p.p.m.	mg/m^3
Acetone	Dimethyl ketone	1000	2400
MEK	2-Butanone	200	590
Methyl propyl ketone	2-Pentanone	200	700
MBK	2-Hexanone (s)	100*	410
MIBK	Methyl isobutyl ketone, Hexone (s)	100	410
Methyl iso-amyl ketone		100	475
Methyl *n*-amyl ketone	2-Heptanone	100	465
Cyclohexanone	Sextone	50	200
Diacetone alcohol	4 Hydroxy-4 methyl-2-pentanone	50	240
Isophorone	3.5.5 Trimethyl-2-cyclohexene-1-one	10†	50
Mesityl oxide	Methyl isobutenyl ketone	25	100

(s) Denotes capacity for poisoning by skin absorption.
* For 1975, reduction to 25 p.p.m. intended.
† For 1975, reduction to 5 p.p.m. intended.

Hydrocarbons

A great variety of hydrocarbon solvents are met with in paint. They include chemical entities as well as petroleum and coal tar distillates. Being low cost and ranging widely in volatility and solvent power, they are used to good effect in numerous paint and lacquer formulations.

A common property of all hydrocarbons is their readiness to dissolve oils and fats. Because of this, careless and excessive skin contact with these solvents will lead to skin troubles such as dermatitis. In addition, the higher boiling hydrocarbon distillates such as the cutting oils and other mineral oils are associated with an increased incidence of skin cancer. Generally speaking, however, the hydrocarbons are not considered a toxic class of solvent. One notable exception is the simplest aromatic hydrocarbon, benzene; it is not to be confused with the much less hazardous petroleum product benzine.

(1) *Benzene* (*benzol, benzole*). This substance is unique among hydrocarbon solvents in its capacity to cause bone-marrow disease and hence lead to aplastic anaemia (reduction in red blood corpuscles). So toxic is it – brief inhalation of an atmosphere containing 20 000 parts per million could prove fatal – that it is the subject of legislation, promulgated and pending, at the national and international level, controlling its usage as an industrial solvent. One problem legislation has to face is the fact

that benzene is a useful starting material for the manufacture of many important organic compounds and it would be impractical to ban its use as an industrial raw material.

In 1967, the International Labour Office convened a meeting, in Geneva, of experts to discuss the safe use of benzene, and of solvents containing benzene. The meeting was unanimous in recommending that solvents which may contain benzene should not have a content of the latter exceeding 1%. Within the U.K. chronic benzene poisoning is a notifiable industrial disease.

TLV for Benzene (with skin absorption contributing to the total toxicity):

1974 (Ceiling value):	25 p.p.m., 80 mg/m^3
1975 (Intended Change):	'Occupational substance suspect of Oncogenic Potential for Workers' 10 p.p.m.; 30 mg/m^3

(2) (*Higher*) *Aromatic hydrocarbons.* In addition to benzene, the commoner aromatic hydrocarbon solvents embrace toluol (toluene), xylol (xylene) with its three isomers (ortho, meta and para), ethyl benzene, cumene (isopropyl benzene) and the trimethyl benzenes pseudo cumene (1.2.4), mesitylene (1.3.5) and hemimellitene (1.2.3). The vapours of all these alkyl benzenes may produce irritation of the respiratory tract and also cause dizziness and nausea. The frequently used toluene and xylene are significantly less toxic than benzene but like the latter are capable of cutaneous absorption. Care should be taken to see that solvents such as toluene and certain petroleum distillates do not contain benzene in excess of 1%.

Indene, which belongs to the cyclopentadiene ring system and is a component of solvent naphtha, is rated fairly toxic. Cyclopentadiene is apparently less hazardous than indene.

The polymerizable aromatic hydrocarbons styrene (phenyl ethylene, vinyl benzene), alpha methyl styrene, and vinyl toluene are comparable in toxicity to xylene. Aromasol H, which contains about 70% of trimethyl benzenes, is also, according to its supplier ICI, similar in toxicity to xylene.

TLVs:

	p.p.m.	mg/m^3
Toluene (s)	100	375
Xylene (*o, m, p* isomers) (s)	100	435
Ethyl benzene	100	435
Cumene (s)	50	245
Trimethyl benzene	25	120
p-tert-Butyl toluene	10	60
Styrene	100	420
Alpha methyl styrene	100*	480
Vinyl toluene	100	480
Indene	10	45
Cyclopentadiene	75	200

(s) Denotes skin absorption toxicity included
* Ceiling value

In order to arrive at an approximate TLV for petroleum distillates for which no specific TLVs are listed, the ACGIH provide a formula which takes into account the percentage of aromatic and aliphatic components present as well as the mean boiling point of the distillate. The formula is not applicable however to distillates whose mean boiling point is above 200°C, nor if its benzene content exceeds 1%.

(3) *Non-benzenoid hydrocarbons.* These include the saturated and unsaturated aliphatic hydrocarbons and the saturated alicyclic compounds. As with the aromatic hydrocarbons, particularly the more volatile, they are strong degreasing agents, leading to dry skin and the possibility of dermatitic effects. Generally speaking, however, they are noticeably less toxic systemically and the high boiling and highly purified 'white oil' is known also as 'medicinal paraffin'.

The alkynes (acetylene derivatives) and alkenes (olefines) comprise the unsaturated aliphatic hydrocarbons, but are not much used as paint solvents. The alkanes include the paraffins (methane derivatives) such as pentane, hexane, heptane and octane and are notable components of the petroleum ethers. They can irritate the respiratory system and are potentially narcotic. This too applies to the alicyclic hydrocarbons such as cyclohexane and methyl cyclohexane which have comparable toxicities.

The high boiling hydrocarbon solvents white spirit (U.S.A. mineral spirits or Stoddard solvent) and kerosene (paraffin, mineral oil) contain relatively low percentages of armoatic hydrocarbons but substantial quantities of the naphthenes which are high boiling alicyclics. Because of their lower volatility solvents such as white spirit and kerosene present a reduced vapour inhalation hazard but can be very harmful if aspirated into the lungs. Kerosene poisoning among children is said to be more common in the developing countries [56]. It is irritant to the gastrointestinal tract and if absorbed depresses the central nervous system.

The partially unsaturated cyclohexene is similar in toxicity to its saturated analogue cyclohexane. Another partially hydrogenated hydrocarbon, namely Tetralin (tetrahydronaphthalene) is said [57] to produce narcotic effects, headache and fatigue, as well as irritation of the upper respiratory tract and a characteristic blue or green discoloration of the urine. The fully saturated compound Decalin (decahydronaphthalene) also produces a similar discoloration of the urine [57] but has no TLV listed by the ACGIH. In Bulgaria, Poland and the U.S.S.R. the official TLV is 17 p.p.m. whilst in Rumania it is 26 p.p.m. Patty [58] says 'a value of 25 p.p.m. appears to be a reasonable figure' on the basis of published toxicity studies.

TLVs (ACGIH):

	1974		Intended changes, 1975	
	p.p.m.	mg/m^3	p.p.m.	mg/m^3
Pentane	500	1500	600	1800
Hexane (*n*-Hexane)	500	1800	100	360
Heptane (*n*-Heptane)	500	2000	400	1600
Octane	400	1900	300	1450
Cyclohexane	300	1050		
Methyl cyclohexane	500	2000	400	1600
Cyclohexene	300	1015		
Stoddard solvent (white spirit)	200	1150	100	175

(4) *Terpene hydrocarbons.* Turpentine and its three main components α and β pinene and dipentene are skin irritants, and if ingested will also cause irritation of the gastro-enteric tract. According to McCord [59] wood turpentine is more irritating to the skin than gum turpentine. Kidney injury has been reported following exposure to turpentine vapour [60]. Large doses taken internally will also result in kidney damage.

TLV:
Turpentine 100 p.p.m., 560 mg/m^3

Chlorinated hydrocarbons
Chlorination of hydrocarbon solvents results in increasing solvent power, and virtual elimination of inflammability, but introduces a variety of toxic hazards including cutaneous absorption, narcotic properties, and the possibility of liver and kidney damage. Some of these chlorinated solvents, however, are not significantly more toxic than aromatic hydrocarbons such as xylene and have therefore become widely used as industrial paint solvents and in paint remover and degreasing compositions.

When chlorinated hydrocarbon solvents are in contact with surfaces which are at a temperature of 400°C or higher, the highly poisonous gas phosgene is formed in measurable amounts. Because of this, it was at one time widely assumed that smoking cigarettes in atmospheres contaminated with these solvents would result in the inhalation of phosgene. However, J. Little [61] exploded this idea when he demonstrated, in the case of cigarettes smoked in solvent-contaminated atmospheres, that phosgene could not be detected in the exit gas from the inhaled cigarette, nor in the atmosphere in the vicinity of the glowing tip.

Another hazard associated with chlorinated hydrocarbon solvents, this time well-substantiated, is that of 'potentiation', i.e. an enhanced health hazard when certain of these solvents are inhaled by individuals who have partaken of alcohol or of some drugs.

Some of the more important chlorinated hydrocarbon solvents are itemized below:

(1) *Trichloroethylene.* Widely used because of its relatively low cost and excellent solvent power, this solvent is a skin and eye irritant and well-known too for its anaesthetic properties. It potentiates with alcohol [62] and adrenaline [63] but not apparently with phenobarbitone [64]. G.F. Smith [65] has reviewed and discussed the physical, chemical and toxicological properties of trichloroethylene.

1.1.1 Trichloroethane (methyl chloroform), commonly known in the U.K. as Genklene (ICI) and as Chlorothene (Dow), is more costly than trichloroethylene, but with a similar boiling point and solvent power and a significantly reduced inhalation hazard. It is not to be confused with its much more toxic isomer 1.1.2 trichloroethane. Unlike trichloroethylene, 1.1.1 trichloroethane does not potentiate with alcohol [62].

(2) *Methylene chloride (dichloromethane).* This important solvent, and component of many paint strippers, can produce symptoms of central nervous depression but is believed to be less of an inhalation hazard than trichloroethylene. However, because it is much lower boiling (40°C) than trichloroethylene (87°C), dangerous concentrations can be more quickly reached on evaporation of bulk quantities. Steward *et al.* [66] suggest from their studies on human exposure to methylene chloride that carbon monoxide may be a metabolite of this solvent.

(3) *Perchloroethylene (tetrachloroethylene; 'Perklone').* This solvent has an inhalation hazard comparable with that of trichloroethylene but is somewhat higher boiling (121°C). Nor does it potentiate with alcohol [62].

(4) *Carbon tetrachloride (tetrachloromethane).* Because of its high toxicity its usage as a solvent has diminished whilst equally efficacious but less hazardous chlorinated hydrocarbons have been made available. However, chlorinated rubber is usually found to contain residual amounts of carbon tetrachloride (see also p. 358). This solvent also potentiates with alcohol and with phenobarbitone.

In animal tests, carbon tetrachloride has been shown to have a carcinogenic potential but there is no evidence of it being a human carcinogen.

TLVs:

	p.p.m.	mg/m^3
Methyl chloride	100	210
Methylene chloride (dichloromethane)	100*	360*
Ethylene dichloride (1.2 dichloroethane)	50	200
Carbon tetrachloride (tetrachloromethane)(s)	10	65
Trichloroethylene	100	535
Tetrachloroethylene (perchloroethylene; Perklone)	100	670

TLVs:

	p.p.m.	mg/m^3
1.1.1 Trichloroethane (Genklene; methyl chloroform)	350	1900
1.1.2 Trichloroethane (s)	10	45
1.1.2.2 Tetrachloroethane (s)	5	35

(s) Signifies skin absorption effect taken into account
* Intended to be changed in 1975 to 200 p.p.m. and 720 mg/m^3

Miscellaneous solvents
Most of the paint and lacquer solvents which are physiologically active have already been dealt with. The more important of the unclassified solvents are the nitro-paraffins as well as nitrobenzene, dimethylformamide, dimethyl sulphoxide, and carbon disulphide.

(1) *Nitroparaffins.* These powerful solvents are respiratory tract irritants but with little evidence of skin absorption. They could, on ingestion, cause liver damage.

TLVs:

	Boiling points	p.p.m.	mg/m^3
Nitromethane	101°C	100	250
Nitroethane	115°C	100	310
1-Nitro-propane	131.5°C	25	90
2-Nitro-propane	120°C	25	90

(2) *Nitrobenzene.* This liquid, with a pronounced bitter-almond odour, is employed more often as a solvent for polishes rather than paints but has also been used as a preservative for spray paints. It is a central nervous poison and is readily absorbed by the skin. A notable toxic effect is methaemoglobinaemia – see also p. 339.

TLV (skin):
Nitrobenzene 1 p.p.m., 5 mg/m^3

(3) *Dimethylformamide.* Widely used in industrial paints, this solvent is moderately irritating to the skin and eyes, and whilst it is capable of percutaneous absorption there is no evidence of skin sensitizing properties. There is the possibility of cumulative systemic (liver) injury following repeated inhalation or repeated skin absorption.

TLV (skin):
Dimethylformamide 10 p.p.m., 30 mg/m^3

(4) *Dimethyl sulphoxide.* This solvent and paint remover is a colourless hygroscopic liquid which is capable of penetrating intact skin. It also acts as a skin irritant and furthermore is known to lead to cases of nettle rash (urticaria). Moderately toxic by ingestion it can cause vomiting and nausea.

(5) *Carbon disulphide.* Not only is this strong solvent, with a disagreeable odour, highly volatile (b. p. 46°C) and highly inflammable but it is acutely toxic. It is a nerve poison and also capable of causing liver and kidney damage. Poisoning can take place percutaneously. Not surprisingly, the usage of carbon disulphide as a lacquer solvent has declined to negligible proportions.

TLV (skin):	
Carbon disulphide	20 p.p.m., 60 mg/m^3

(d) *Plasticizers*

Of the various classes of external plasticizers which are used in paints and lacquers, three are worthy of special mention, viz. the phosphates, phthalates and chlorinated diphenyls.

Phosphates

Of the phosphate esters it is the aryl compounds, as opposed to the alkyl esters which have proved troublesome toxicologically. In particular, misuse of tri-ortho cresyl phosphate (tri-*o*-tolyl phosphate) has resulted in a number of incidents of mass poisoning. For instance, some 10 000 Moroccans suffered tricresyl phosphate intoxication as a result of eating food cooked in culinary oil to which had been added commercial cooling oil. Fehr [67] deals with the symptoms and mechanism of intoxication associated with this particular catastrophe. Again poisoning in eleven African patients in Durban was traced to the industrial use of tri-*o*-cresyl phosphate as a plasticizer [63]. Although as a class the aryl phosphates are skin and eye irritants the specific toxic effects, viz. paralysis (polyneuritis), of tri-ortho cresyl phosphate appear to be limited to this particular compound and are not exhibited by the meta or para isomers. According to Henschler [69] the toxicity of trixylenyl phosphate is even less than that of tricresyl phosphate poor in ortho isomer content.

Advisedly, tricresyl phosphate should only be used as a plasticizer for paints and lacquers if it is free of the ortho isomer or if the content of the latter is minimal and specified.

TLVs:	mg/m^3
Tributyl phosphate	5
Triphenyl phosphate	3
Triorthocresyl phosphate	0.1

Phthalates

Allegations of teratogenicity against phthalate plasticizers, particularly, di-2-ethylhexyl phthalate, have already been described (p. 334). Otherwise the phthalates are generally of a low order of toxicity. Colley [70] quotes LD_{50} values (intraperitoneal in mice) from which it will be seen that increasing molecular weight appears to bring about a reduction in toxicity:

	LD_{50} (gm/kg)
Dimethyl phthalate	1.58
Diethyl phthalate	2.83
Dibutyl phthalate	4.00
Diisobutyl phthalate	4.50
Di (2-ethylhexyl) phthalate	14.19

TLV:	
Dibutyl phthalate	5 mg/m^3

Chlorinated diphenyls

The chlorinated diphenyls (Aroclors), which range from mobile oils to brittle resins, according to their degree of chlorination, are chemically related to the environmental pollutants the PCBs. All are capable of accumulation in the liver to lead to systemic poisoning. The chlorinated diphenyls are also skin irritants; a characteristic effect is chloracne (skin eruptions). The toxic hazards of the chlorinated diphenyls appear to be related in proportion to their chlorine content.

TLVs:	
Chlorodiphenyl (42% chlorine) – skin	1 mg/m^3
Chlorodiphenyl (54% chlorine) – skin	0.5 mg/m^3

(e) *Liquid media*

Vegetable oils

Edible oils like soyabean and cottonseed are of course non-toxic. Fatalities have been recorded against the ingestion of a few castor beans due to the presence of the highly toxic protein ricin. Castor oil, however, is free of ricin and its purgative action is due merely to the component triglycerides. Linseed oil is also laxative but its disagreeable taste limits its medicinal use to veterinary science.

Chlorinated rubber

The hazards of finishes based on chlorinated rubber are more likely to be associated with the solvent(s) employed in preparing the paint vehicle. In addition, and as already pointed out (see under 'Chlorinated hydrocarbons'), minor quantities of the toxic solvent carbon tetrachloride might be introduced because residual amounts could derive from the manufacture of the chlorinated rubber.

Cashew nutshell liquid
In the process of removing edible cashew kernels by roasting cashew nuts, a liquid by-product is obtained from the shell. This oil contains anacardic acid and cardol, a phenolic substance. Anacardic acid on heat treatment decarboxylates to another phenol, namely cardanol. Cashew nutshell liquid as well as cardol and cardanol yield a variety of coating compositions. Because of their phenolic character, however, all these substances, as well as uncured products derived from them, are toxic and irritant and can produce blisters (vesicant action). Skin and eye contact must therefore be avoided.

(f) *Additives*

Substances, excluding pigment, oil, resin and solvent, which have been added to paints and lacquers to improve the quality and performance of the product, are too numerous to list here. Instead, some of the more notable additives, both with respect to usage and potential health hazard, are briefly mentioned below.

Acids
(See also under 'Eye Effects' on p. 336).

Sulphuric acid as well as the simpler (organic) sulphonic acids, are all corrosive, and skin and eye contact must be avoided. Never add water to concentrated sulphuric acid in a container; the heat evolved can cause the acid to erupt violently. Instead, reverse the procedure, and dilute the acid by adding it gradually to the water.

Hydrochloric acid is another strong mineral acid; avoid skin and eye contact with concentrated solutions or with the gas itself. Also avoid breathing the gas; it has a low TLV (ceiling value) of 5 p.p.m. or 7 mg/m^3. With formaldehyde, hydrochloric acid can under some conditions form the carcinogen bis (chloromethyl) ether (see under 'Ethers', p. 360).

Boron trifluoride (gas) and many of its liquid derivatives such as the hydrate and complexes including borfluoracetic acid and boron trifluoride/butanol are not only powerful acid-type catalysts, but in the presence of moisture can partly break down (hydrolyse) to the highly corrosive hydrofluoric acid.

Orthophosphoric acid, which is widely used in paint products as a catalyst or component of etch primers and de-rusting agents, is also a strong mineral acid which can burn skin and eyes or cause severe irritation and damage if taken internally.

Formic acid, often used as a component of paint removers, is poisonous, irritant and corrosive and causes painful eye burns and skin blisters.

Alkalis
The severe effects of splashes in the eye with strong alkalis have already been described (p. 336). Also, for caustic soda and caustic potash see respectively under 'Sodium' (p. 355) and 'Potassium' (p. 353).

Not only are ammonia gas and concentrated ammonia solutions corrosive to skin and eye, but they are highly irritant to all parts of the respiratory system. Small quantities of ammonia are best handled in a fume cupboard, or under forced ventilation; for handling large quantities the use of respiratory equipment is advised.

Both aliphatic and aromatic amines are capable of absorption through the skin (see p. 343). However, whereas the simpler aliphatic amines are strongly alkaline and can cause severe eye damage, most of the aromatic amines do not behave as caustic alkalis but instead can lead to systemic poisoning (methaemoglobinaemia) which has been described previously (p. 339).

In addition, a number of aromatic amines are now recognized as human carcinogens, and include benzidine, para amino diphenyl, and β-naphthylamine. Aniline, although highly toxic, is not held to be a carcinogen. The aromatic diamine 'MOCA' (methylene-bisorthochloroaniline), which has been used as a curing agent for iso-cyanate polymers, is listed by the ACGIH as an 'occupational substance suspect of oncogenic (tumour-forming) potential for workers'.

Anti-skinning agents

Frequently used as anti-skinning agents have been the oximes, compounds obtained by the action of hydroxylamine on ketones and aldehydes.

Following the claim that workers handling cyclohexanone oxime became drowsy, and that this compound had a narcotic but otherwise harmless effect on rats, Hauschild [71] studied a number of oximes and determined their LD_{50} as well as their media narcosis-producing doses.

Kurita [72] later (1967) found that 'methyl ethyl ketoxime has many toxic effects on the living body' but its toxicity 'is not so severe as far as the lethal dose is concerned'. He concluded that 'hygienic precautions are essential in the industrial handling of methyl ethyl ketoxime'. In case of the possibility of skin absorption or of eye irritation, skin and eye contact should be guarded against. Contaminated areas can be washed with soap and water. Inhalation of the vapour should be kept minimal.

Butyraldoxime, another oxime used as a paint antioxidant, has skin absorption and skin irritation properties no less severe than those of the ketoximes; aldoximes are generally held to be more toxic than the ketoximes.

Driers

See under the appropriate metal in Section 8.1.3 (a), namely Cobalt (p. 349), Lead (p. 350) and Manganese (p. 351).

Fungicides

In addition to such chemical entities as pentachlorophenol (see p. 343), a host of newer fungicides have been introduced in the past few decades, some of these (such as Captan, Thiram, Folpet, Shirlan, and Mertectik), having valuable properties in paint formulations (see Section 7.18). For the great majority of these fungicides which are sold under proprietary names, one is very much in the hands of their manufacturer or, in turn, the paint manufacturer who uses these substances. Reputable manufacturers see to it that toxicological data is available before marketing the fungicide; often such data is readily offered, usually in brochure form, to *bona fide* enquirers. Additionally useful data, both on composition and toxicological, concerning the above-mentioned agricultural fungicides and many others, are to be found in the British Crop Council's Pesticide Manual, which is frequently updated.

Peroxides

In recent years a considerable number of organic peroxides have been made available for use in paint and other industries. The U.K. Petroleum (Organic Peroxides) Order (1973) lists between 60 and 70 such peroxides. As a class, peroxides not only present potential fire and explosion hazards but in contact with skin tissue can cause severe burns (oxidative destruction). Not surprisingly, therefore, even very small quantities of peroxide in the eyes could lead to permanent blindness if not attended to promptly. Hence the exhortation (see also p. 336) for eye protection when handling such compounds.

A fatality from the ingestion of a solution of organic peroxides has been recorded by Burger and Chandor [73].

Laporte [74] have drawn up a Code of Practice to assist users of organic peroxides. Particular emphasis is placed on the importance of eye protection. In the U.S.A., the American Society for Testing and Materials [75] has published a guide to the safe handling and storage of peroxy compounds.

Pesticides

Pesticides such as DDT have sometimes been incorporated into paint. Some are so toxic that their use has been the subject of governmental control, or even, at times a complete ban.

Two major classes of insecticide are the chlorinated hydrocarbons (which include DDT) and the organo-phosphorus compounds. Characteristic of most of the substances within these two groups is their capacity to be absorbed percutaneously in toxic amounts (for further information see p. 343).

Gaines [76] has determined LD_{50} values for 98 pesticides including Aldrin, Chlordane, Heptachlor and Mirex, and Deichmann [77] has reviewed the toxicology of DDT and related chlorinated hydrocarbon pesticides. A text-book by Fest and Schmidt [78] covers the chemistry of organo-phosphorus pesticides as well as their reactivity, synthesis, mode of action and toxicology.

Propellants

For use as propellants in aerosol cans, the fluorocarbons have come very much to the fore in recent years. In a 200-page publication, Clayton [79] has reviewed the health effects of fluorocarbons.

Still more recently, high-flying american planes have found evidence, thirteen miles up, that the gaseous fluorocarbons 11 and 12 were creating a new health hazard. They drift into the upper atmosphere and, it is claimed, gradually destroy the layer of ozone which blankets the sun's ultraviolet radiation. If this radiation reaches the earth it could cause skin cancer. As from 1 March 1977, the sale of aerosol spray cans with fluorocarbons as propellant was banned in the state of Oregon [80]; this State accounts for one quarter of U.S.A. production.

Silicones

The word silicone embraces a great variety of compounds based on the element silicon. Although most silicone resins and fluids are virtually innocuous, having been used for prosthetic reasons and as barrier creams, a few silicones, because of their chemical structure, can present skin, eye and inhalation hazards. Reference to the particular hazards of such compounds should be made back to their suppliers. Some silane-type adhesion promoters are also in this category.

Viscosity-control agents

Among the more important, physiologically, of the additives which are used to control paint viscosity are the various grades of finely-divided silica. Their toxic hazards, mainly inhalational, are related to the percentage of crystalline material present, as well as to their particle size. See also p. 340 (silicosis) and p. 354 (silica).

8.1.4 Special paints

A selection of the more notable is given below.

(1) *Antifouling compositions*

Antifouling paints are by intent toxic; they contain various biocidal compounds, organic as well as inorganic, which function by being gradually leached out by the sea water. Such biocides are to be found among the inorganic and organic arsenicals and mercurials (see pp. 345 and 341–342 respectively), inorganic copper (p. 349), organo-tin (p. 342) and organo-lead (see 'Lead', p. 350). In addition, purely organic pesticides have sometimes been used (see pp. 343 and 372).

The great majority of such antifouling paints can be expected to be toxic by ingestion. Sometimes there will be a skin absorption or eye hazard, and an inhalation hazard can arise either from droplets (spray application) or from fume. The latter could result from the burning-off of paint, welding repairs or ship-breaking. It has been claimed for at least one organo-lead compound that it is virtually non-toxic by ingestion. However, an inhalation hazard could still arise from inorganic lead fume evolved during the thermal disintegration of ship's coatings containing such organo-lead compounds. Because of the exceptional toxic hazards of anti-fouling paints and the large volumes which normally need to be applied, it is important that the user carefully heeds all warning labels displayed on the can or conveyed in the supplier's literature.

(2) *Bituminous paints*

Because of the possibility of skin cancer following protracted usage, avoid skin contact. If areas of skin should become contaminated, clean off immediately.

(3) *Epoxide paints*

Some of the early epoxide resins were not only dermatitic but among them were some with a carcinogenic potential which had been ascertained from animal tests. Carcinogenicity appeared to be related both to chemical structure and to molecular size. Tumorigenic properties were encountered more among di- and polyepoxides than with the mono-epoxides and were also more prominent among smaller molecules (low molecular weight) than larger molecules, derived say from the reaction between epichlorhydrin and the phenolic compound diphenylopropane. Many of these latter types of epoxide resin are currently components of epoxide paints.

Nowadays, the hazards of epoxide paints are more associated with the agents used to cure (polymerize or cross-link) them. These curing agents range from strong acids, such as acid anhydrides and boron fluoride complexes (see p. 370), to strongly alkaline substances such as the aliphatic amines (see pp. 343 and 371). Also frequently employed are aromatic amines (avoiding those which are known to be carcinogenic) and the neutral polyamide resins. These polyamides are amc the 'safer' of epoxide curing agents.

Also regarded as 'safe' epoxides are epoxidized drying oils, such as epoxidized soyabean oil, which find outlets as plasticizers.

In addition to a publication on the hazards of handling epoxy paints [81], the Paintmakers Association of Great Britain has also issued a document [82] on the toxicity of the curing agents which are used for epoxy resins.

(4) *Japanese Lacquer*

Rhus vernicuflua, the lac or varnish tree, is the source of Japanese Lacquer. This tree which is common to China, provides a thick emulsion which, after purification, dries in thin films to dark but very glossy coatings. However, this emulsion is comparable to Cashew Nutshell Liquid (see p. 370) in being highly irritant to the skin due to the presence of a phenolic substance, in this case Urushiol (3-*n*-pentadecylcatechol). Although this irritant property has, since the turn of the century, militated against the introduction and development of Japanese Lacquer in Great Britain, it is claimed it can be safely used if handled with care.

(5) *Polyurethane paints*

The toxic hazards of isocyanates, the essential components of polyurethanes, have already been referred to in the paragraph on 'Sensitization' on p. 337. In 1965, the hazards of di-isocyanates were reviewed in some detail by Munn [83]. In the face of growing evidence of the serious respiratory effects of di-isocyanates an opportunity to re-appraise the situation was afforded in 1969 by the convening of a meeting in London by the Royal Society of Medicine. At this meeting Peters [84] reported on the continued impairment of lung function on exposures to toluene-di-isocyanate (TDI), even at concentrations below the TLV of 0.02 p.p.m. (see also Peters *et al.* [85]. McKerrow *et al.* [86] at the same meeting gave evidence of acute respiratory symptoms and chronic lung disorders among 36 firemen exposed to a high concentration of TDI at a factory fire and in a Fire Research Station report on a study of the thermal decomposition of flexible polyurethane foams, Woolley *et al.* [87] found nitrogen-containing compounds such as hydrogen cyanide, acetonitrile and pyridine among the high temperature decomposition products of the foams.

Isocyanates enter into the composition of a variety of urethane-type products. Some of these such as urethane alkyds, urethane oils and 'blocked' isocyanates contain no unreacted (original) isocyanate whereas polyisocyanates can be expected to contain free isocyanate in order to react with atmospheric moisture or with added curing agents. The Paintmakers Association of Great Britain, after discussion, with representatives from government departments and organizations, such as the British Medical Association issued a document [88] on the hazards arising from the use of polyurethane (PU) paints and varnishes and precautions to be adopted. This publication differentiates between the hazards of these various isocyanate compositions, between the hazards of different polyisocyanates and to a certain extent, even between the different hazards of the di-isocyanates (toluene di-isocyanate, hexamethylene di-isocyanate and diphenyl methane di-isocyanate) on which most polyurethanes are based. The PMA document also makes recommendations for the safe usage and application of PU paints as well as for the labelling of such products.

It is thus extremely important that the user of polyurethane paints and varnishes, particularly if they are moisture curing or two-pack, pays special heed to recommendations given on the can label and/or the suppliers' literature, in order to avoid inhalation of spray mist and, where advocated, to ensure that skin, eyes and clothing do not come in contact with the polyurethane.

Still more recently, the Manufacturing Chemists Association (U.S.A.) has issued a revised data sheet [89] for toluene di-isocyanate and Smith of Chrysler (U.K.) Ltd. [90] in a paper in 'Occupational Health' outlines procedures to be adopted for the safe handling of isocyanates.

(6) *Powder coatings*

The absence of solvent from powder coatings results in a reduced fire risk but an increased explosion hazard; the latter is dealt with in Section 8.2.7. The principal toxic hazard is that of inhalation of powdered material; the skin (and dermatitic) hazard is considered to be quite low. Even non-toxic powders are of course irritant to eyes and mucous membranes, and so-called 'nuisance particulates' are given a TLV of 10 mg/m^3.

Because powder coatings have been introduced long after epoxide resins first became available, the intervening years have allowed for the accumulation of much data on the chemistry and toxicology of epoxides. As a result, epoxides (and their curing agents) selected for powder coatings are usually of a low order of toxicity (refer also to the supplier for confirmation).

Other compositions can of course bring other hazards; for example, the liberation of caprolactam during the curing of polyester polyurethane powder. Caprolactam is rated as possessing a significant inhalation hazard (low TLV) by the ACGIH.

TLVs:	
Caprolactam dust	1 mg/m^3
Caprolactam vapour	5 p.p.m., 2 mg/m^3

Two useful documents on powder coatings are the cod of safe practice for the application of powder coatings by the Paintmakers Association [91] and a report on powder coating by Korf [92]. The PMA also publish a checklist of Working Precautions suitable for display at work places.

(7) *Welding primers*

Understandably, the welding of painted metal surfaces introduces an inhalation hazard which is dependent on the composition of the paint. Thus, the welding of surfaces coated with lead paint could lead to plumbism, whilst cadmium-based paint has yielded cadmium oxide fumes, which in the absence of adequate exhaust ventilation have resulted in fatal cases of respiratory disease.

The development of shop (or welding) primers designed to protect freshly shot-blasted steel up to the time it was welded, without interfering with the welding process, meant increased opportunities for the evolution of fumes of metal oxides resulting from the high-temperature disintegration of the paint film. It was perhaps fortuitous that among the more successful and popular types of welding primers were those containing high proportions of zinc dust. Although these could lead to zinc fume fever, the ill-effects of the latter, as already pointed out (p. 357), although unpleasant, are not of a permanent nature.

(8) *Wood preservatives*

Depending on their mode of impregnation and whether they are solvent- or water-borne, wood preservatives contain pesticides ranging from inorganic substances such as copper salts, arsenicals, fluorine derivatives, zinc compounds and boric acid to organic compounds including chlorinated hydrocarbons, naphthenates, tar oils, organo-tin compounds and phenol derivatives such as pentachlorophenol (and its sodium salt), dinitrophenol and ortho-phenyl phenate.

The great majority of these wood preservatives are well-known for their toxic hazards some of which are more detailed under such specific headings as Arsenic (p. 345), Copper (p. 349), Tin (pp. 342 and 356), Pesticides (p. 372) and Pentachlorophenol (pp. 343 and 371).

8.1.5 Legislation and labelling for toxic hazards

Over the past few decades and mainly throughout the industrial nations, there has been an upsurge in legislation aimed to protect man from the various hazards, toxic and otherwise, which beset him, both at his work-place and in the environment.

(1) *Great Britain*

Health and Safety at Work Act, 1974

For many years the health and safety of people carrying out work activities in Britain has been the subject of control by not one but several government departments. Such a situation had its obvious shortcomings which were to be rectified in the first instance by the establishment in 1970 of a Committee on Safety and Health at Work, under the chairmanship of Lord Robens. The activities of this committee, reported on in 1972, led to the Health and Safety at Work Act, 1974.

The far-sighted provisions of this Act endeavour to secure the health and safety, not only of persons at work, but also to protect other persons against health and safety risks stemming from the activities of those at work. The Act is also fair in distributing the responsibilities governing health and safety at work, between employer, supplier and employee. Thus:

Section 2: (1) 'It shall be the duty of every employer to ensure, so far as is reasonably practicable, the health, safety and welfare at work of all his employees'.

Section 6: (4) 'It shall be the duty of any person who manufactures, imports, or supplies any substance for use at work – (a) to ensure, so far as is reasonably practicable, that the substance is safe and without risks to health when properly used'.

Section 7: 'It shall be the duty of every employee while at work – (a) to take reasonable care for the health and safety of himself and of other persons who may be affected by his acts or missions at work'.

Useful, independent publications which comment on the 1974 Act are those of Jackson [93] the Royal Institute of Chemistry [94] which deals with the professional responsibilities of the chemist, and Mitchell [95] who sets out to provide an employer's guide to the Act.

Many aspects of the new Act are already covered by existing legislation and call for no revision for years to come. Some of the more notable features of existing Acts, Regulations and Orders, which have some bearing on the toxic hazards of paint, or its raw materials are as follows.

The Factories Acts, 1961

Section 63: Removal of dust or fumes. 'In every factory in which ... there is given off any dust or fume ... to such an extent as to be likely to be injurious or offensive to the persons employed ... all practicable measures shall be taken to protect the persons employed against the inhalation of the dust or fume ... .

(For an interpretation of injurious dust or fume reference is made to the annual American Threshold Limit Values, which are re-issued by H.M. Factory Inspectorate, Department of Employment).

Section 65: Protection of eyes in certain processes. 'In the case of any such process ... which involves a special risk of injury to the eyes from particles or fragments thrown off in the course of the process, suitable goggles ... shall be provided to protect the eyes of the persons employed in the process'.

Section 74: Prohibition of employment of women and young persons in certain processes connected with lead manufacture. Specific to the manufacture, rather than usage of certain named lead compounds (e.g. oxide, carbonate and sulphate).

Section 82: Notification of industrial diseases. Requires the medical practitioner to inform the Chief Inspector of Factories of patients whom he believes to be suffering from poisoning by certain substances such as lead, phosphorus, mercury and arsenic compounds.

Section 128: Employment of women and young persons in places other than factories in processes connected with lead manufacture or involving the use of lead compounds.

Section 129: Use of lead paint in connection with buildings. Includes reference to the Lead Processes (Medical Examinations) Regulations, 1964.

Section 130: Power to take samples of paint, etc.

Section 131: Prohibition of employment of women and young persons in painting buildings with lead paint.

Section 132: Provisions supplementary to Sections 129–131. Defines as 'lead paint' any paint, paste etc. which 'yields to an aqueous solution of hydrochloric acid a quantity of soluble lead compound exceeding, when calculated as lead monoxide, 5% of the dry weight of the portion taken for analysis'.

Section 151: Appointed factory doctors. Outlines the duties of such doctors.

The Asbestos Regulations, 1969
Provide for strict controls to prevent the inhalation of asbestos dust in factories where asbestos is handled.

The Carcinogenic Substances Regulations, 1967
Prohibits the use of the bladder carcinogens β-naphthylamine, benzidine, 4-aminodiphenyl and 4-nitrodiphenyl and their salts and controls the use of certain other aromatic amino-compounds.

Other legislation

The Chromium Plating Regulations, 1931 (and Amendment 1973).
The Manufacture of Chromate ... Regulations, 1913 (revoked by the Chemical Works Regulations, 1922).
The Notification of Chrome Ulceration (part of order No. 1775, 1919).
The Dyeing (Use of Bichromate of Potassium or Sodium) Welfare Order 1918.
The Tanning Welfare Order, 1930.

The existence of the above items, although not specific to the usage of chromium in paint, are symptomatic of the concern of authority over the hazards of chromium compounds.

The Lead Compounds Manufacture Regulations, 1921.
The Lead Smelting, and Manufacture Regulations, 1911.
The Paints and Colours Manufacture Regulations, 1907.

These three Regulations are more relevant to the *manufacture* of lead pigments and derived paints than to the usage of such paints.

For the definition of 'lead paint' referred to in the 1927, 1926 and 1960 Regulations quoted below, see Section 132 of the Factories Acts, 1962 (p. 378 of this book).

Painting of Buildings with Lead Paint, Regulations, 1927

Section 2: 'Lead paint shall not be applied in the form of spray in the interior of buildings'.

The Vehicle Painting Regulations, 1926

Section 3: 'Lead paint shall not be used in any spraying operation, except in a special compartment provided with an efficient exhaust draught ... '.

Section 4: 'A surface painted with lead paint shall not be rubbed down or scraped by a dry process'.

The Shipbuilding and Ship Repairing Regulations, 1960

Section 78: 'Lead paint shall not be applied in the form of a spray in the interior painting of any part of a ship or vessel'.

The Consumer Protection Act, 1961
The Toys (Safety) Regulations 1974 (revoking the 1967 Regulations) place a limit (p.p.m. on dry paint film) on toxic elements in paints for toys, thus: Lead, 2500; soluble* barium, 500; soluble antimony or chromium, 250; soluble arsenic, cadmium or mercury, 100.

* Extractable from the paint film by shaking with 50 times its weight of 0.25% hydrochloric acid.

Petroleum (Consolidation) Act, 1928
The Corrosive Substances (Conveyance by Road) Regulations, 1971 relate to safe procedures (container size, marking of containers etc.) for the road transport of such substances as chromic acid, hydrofluoric acid and sulphuric acid.

The Organic Peroxides (Conveyance by Road) Regulations, 1973 provide similarly for the safe transport of organic peroxides, all of which are named in an extensive list.

Pharmacy and Poisons Act, 1933
The Poisons Rules (updated frequently). Although these Rules relate to restrictions on the sale of poisons among which are listed salts of barium, compounds of lead, and phenols, general exemption is provided for paints, varnishes and antifouling compositions.

Labelling [96]
The Health and Safety Commission has issued draft proposals for British legislation for the labelling of dangerous substances, based on EEC directives. There is also agreed labelling as between various authorities in Britain for the transport of dangerous substances and products. For example there is one label with the wording 'CORROSIVE' and acceptable, where appropriate, under any of the following regulations:

The Corrosive Substances (Conveyance by Road) Regulations, 1971.
British Rail Regulations (Dangerous Goods by Freight Train and by Passenger Train or similar service).
Regulations for the Carriage of Dangerous Goods in Ships ('Blue Book' of Department of Trade and Industry).
International Air Transport Association Restricted Articles Regulations.

(2) *EEC and other European countries*

As the result of recommendations made by the Council of Europe (Strasbourg) with the first issue in 1962 of their multilingual, so-called 'Yellow Book', the EEC decided on harmonizing legislation relating to the classification and labelling of various groups of materials, such as dangerous substances, paints and solvents. For Britain, the Health and Safety at Work Commission, in 1976, followed up the EEC directive on dangerous substances with proposals for legislation (1977). Also in 1976 the EEC published their directive on solvents [97]. EEC directives on Paints and Varnishes, as well as on Toys are under active consideration.

Many of the individual countries of Europe have long had legislation relevant to the hazards of paints and varnishes. Austria, Belgium, Finland, Greece and Switzerland, to name but a few, exert controls on lead in paint. Many European countries also legislate for protection against the hazards of such paint solvents as benzene, methanol and carbon tetrachloride. Holland, West Germany and

Switzerland are among European countries insisting on labelling for isocyanate hazards. In addition to Great Britain, Denmark, West Germany and the Netherlands have 'Toys' regulations. Legislation in Northern Ireland follows on that of Great Britain.

(3) *U.S.A. and other non-European countries*

Playing a dominant role in the U.S.A., comparable with the Health and Safety at Work Act in Britain, the Occupational Safety and Health Act (OSHA) was passed in December 1970. For controls it relies on Standards, based on toxicological data prepared by such U.S. government organizations as the U.S. Environmental Protection Agency (Washington), the National Library of Medicine (Bethesda, Maryland), and the Oak Ridge National Laboratory (Tennessee). The National Institute of Occupational Safety and Health's list of toxic substances, published annually, is the largest in the world; computerized data is stored at Rockville (Maryland). Some sections of OSHA are of course of particular relevance to paint [98].

The American 'Lead-Based Paint Poisoning Prevention Act', 1971, has already been referred to (see p. 350). Five years later, clarification is still awaited on the future of lead limits in paint; is it to be 0.5% or as low as 0.06%? Other American legislation with some relevance to paint of a hazardous nature is the Fair Packaging and Labelling Act, 1966, and the Federal Hazardous Substances Act (as amended January 1971).

Independent of U.S. Federal Law are the various State Laws, some of which relate to the hazards of paint, or of paint ingredients such as solvents, propellants (see p. 372) and lead and other toxic compounds. Whilst Federal legislation on the hazardous components of paint for toys is still awaited, some individual States already have such laws. Outside the U.S.A. and Europe, countries such as Canada and Japan have also legislated for controls on paints formulated for use on toys.

Of particular value to manufacturers, suppliers and users of coating compositions for food cans and wrapping materials for food are (American) Food and Drug Administration decisions on petitions relating to 'indirect' food additives. Such decisions are widely respected not only because they are authoritative but because they are of a high standard. Occasionally, however, criticism is voiced of their stringency.

8.2 Hazards of fire and explosion

Conventional solvent-borne paints as well as cellulose lacquers are notorious for fire risks which are associated with substances described as 'inflammable', 'spontaneously inflammable' or merely 'combustible'.

(1) '*Inflammable*' (*flammable*). Gases which can be ignited by a flame, and those liquids and solids whose vapours can be so ignited. Except for halogenated hydrocarbons, and high-boiling liquids, nearly all the organic solvents fall into this class.

(2) '*Spontaneously inflammable*'. Substances which can ignite without the application of a flame. Ignition can result from elevated temperatures, from highly oxidative conditions, or as the result of certain chemical reactions.

(3) '*Combustible*'. Liquids and solids which will burn on the application of a flame. Most organic liquids (drying oils, plasticizers and high-boiling solvents), organic solids (resins and organic pigments) and a few inorganic solids (e.g. aluminium powder) are in this category.

8.2.1 Flash points

For the practical control of inflammable substances, legislation has to classify them into degrees of fire risk and has resorted to flash points which are mostly determined in equipment described as 'closed cup' or 'open cup'. The latter gives readings which are usually, but not invariably, higher than those obtained by the closed cup method.

Details of the Abel test apparatus for determination of 'closed cup' flash points are to be found in the 2nd Schedule to the (Home Office) Petroleum (Consolidation) Act, 1928. British Standards also exist for both closed [99] and open cup [100] methods. Flash points of the numerous solvents used in paints are to found in various text books. A useful publication is that produced by British Drug Houses [101] wherein are listed flash points (closed and open cup methods) for some hundreds of substances. From their publication, a selection has been made of flash points (closed cup) of the more commonly used paint solvents; these are listed in Table 8.2.

8.2.2 Regulations for the control of fire hazards

(1) (*Home Office*) *Petroleum* (*Consolidated*) *Act 1928*

Consolidated all earlier (1871 onwards) enactments relating to petroleum spirit and defined the latter as petroleum giving off an inflammable vapour at a temperature less than 73°F when determined by the Abel closed cup method.

Petroleum (Mixtures) Order, 1929
Relates to inflammable mixtures of petroleum with other substances (any percentage).

Petroleum (Inflammable Liquids and other Dangerous Substances) Order, 1947
Application of the Petroleum (Consolidated) Act to a list of non-hydrocarbon substances (named).

Table 8.2 Flash points (closed cup) of commonly used paint solvents

	°C	°F
Acetone	−17	1
n-Amyl acetate	25	77
n-Amyl alcohol	43	110
Benzene	−11	12
n-Butyl acetate	23	73
n-Butyl alcohol	29	84
iso-Butyl alcohol	28*	82*
Cyclohexane	−17	1
Cyclohexanol	68	154
Cyclohexanone	47	117
Decalin	57*	135*
Diacetone alcohol (tech.)	9	48
Diacetone alcohol (purfd.)	59*	138*
Dioxan	12	54
Ethyl alcohol 95%	14*	57*
Ethyl alcohol (absolute)	12*	54*
2-Ethoxy-ethanol (Cellosolve etc.)	40	104
n-Heptane	−4	25
Isophorone	96	205
Mesityl oxide	31	87
Methanol	10	50
Methyl acetate	−9	15
Methylated spirit	15	59
Methyl cyclohexanone (tech.)	48	118
Methyl ethyl ketone	−7	19
Methyl methacrylate (monomer)	11	52
α-Methyl styrene	43*	110*
Solvent naphtha	32 (approx)	90 (approx)
Styrene (monomer)	31	88
Tetrahydrofuran	−17	1
Toluene	4	40
Xylene (mixed isomers)	24	75

* Figures marked with an asterisk were determined by BDH in the closed cup (Pensky-Martens apparatus).

The Petroleum (Inflammable Liquids) Order, 1971
Revoked the 1968 Order, which in turn followed on the previously-mentioned 1947 Order, with a much extended list of non-hydrocarbon inflammable liquids.

The Inflammable Liquids (Conveyance by Road) Regulations, and The Inflammable Substances (Conveyance by Road) Labelling Regulations.
The scope of these two Regulations, both of which are related to the Petroleum (Inflammable Liquids) Order, is indicated in their respective titles.

(2) (*Factories Acts*) *Highly Flammable Liquids and Liquefied Petroleum Gases Regulations, 1972*

Controls the usage, storage and labelling of highly flammable liquids which are defined as substances which give off a flammable vapour at a temperature of less than 32°C, when tested by the Abel closed cup method *and* in addition support combustion when tested by the method described in Schedule 2 of the Regulations.

Although these Regulations revoke the Cellulose Regulations, 1934, they nevertheless still cater for the hazards of nitrocellulose, thus:

Section 9 (3): 'Where any cellulose nitrate is present in any highly flammable liquid, all practicable steps shall be taken to prevent the deposit of any solid residue resulting therefrom on any surface which is liable to attain a temperature of 120° Celsius'.

Section 13 (2): 'No removal of any such residue containing cellulose nitrate shall be effected by the use of an iron or steel implement'.

(3) *Miscellaneous transport regulations*

As in the case of hazards other than fire, the transport of inflammable substances by rail, sea, air and postal services, are controlled respectively by British Rail, The Department of Trade and Industry, I.A.T.A. and the Post Office. For example, quoting the Post Office Guide [102], 'Paints and so on which flash below 32°C (90°F) – closed cup – are totally prohibited'.

8.2.3 Labelling for fire hazards (U.K. requirements)

Because of the proliferation of regulations relating to inflammable (flammable) substances, paint cans are liable to bear a number of warning labels, although sometimes the warning label need only appear on the outer package. These are set out in Table 8.3

Table 8.3

Wording on label	When required
'Petroleum mixture giving off an inflammable heavy vapour'.	Products which flash below 73°F and contain any quantity of petroleum hydrocarbon.
'Inflammable'	For the road transport of: (i) All products bearing the petroleum mixture label. (ii) Products which flash below 73°F and contain at least one of the liquids named in the Petroleum (Inflammable Liquids) Order, 1971.
'Highly Flammable' *or* 'Flash Point below 32°C' *or* 'Flash Point in the range of 22°C–32°C' *or* otherwise with an appropriate indication of flammability.	Storage of products subject to the 'Highly Flammable Liquids and LPG Regulations, 1972.
'Inflammable Liquid, Flash Point below 0°F' 'Inflammable Liquid, Flash Point 0–72°F' 'Inflammable Liquid, Flash Point 73–141°F' 'Spontaneously Combustible'	Products subject to Regulations for the Carriage of Dangerous Goods in Ships (Department of Trade and Industry) or Products subject to Regulations governing 'Dangerous Goods by Freight Train and by Passenger Train or Similar Service' (British Rail)

8.2.4 Spontaneous Combustion

Rags, paper etc., contaminated with oils, whether of vegetable or animal origin, or with certain solvents such as turpentine, can sometimes ignite spontaneously in air. Collections of oily rags should be stored air-tight in metal containers; single rags are best left spread out when not in use.

Some forms of carbons, such as lamp black of vegetable origin, have been known to ignite spontaneously. Regulations for the stowage on board ship of substances liable to spontaneous combustion are given under Class 4b in the Department of Trade and Industry's 'Blue Book'. See also Section 8.2.3 for the labelling of such materials.

Many substances, even so-called non-inflammable solvents, are capable of auto-ignition, if heated in air or oxygen to a sufficiently high temperature. For instance, benzene which has a flash point of –11°C (12°F) will ignite spontaneously if heated in air to 580°C (1076°F). Methylene chloride, which has not a recognized flash-point will ignite in air at 641°C (1188°F) and in oxygen at 606°C (1123°F).

8.2.5 Explosions of solvents/air mixtures

All combustible gases and vapours are capable of exploding on ignition in air provided there is a minimum concentration of solvent vapour and a minimum concentration of air (or oxygen) in the mixture. In other words, an explosion in air of the vapour of an inflammable solvent is only possible between what are known as the lower explosive limit (LEL) and upper explosive limit for that solvent. The lower limit is more important from a practical point of view.

Explosive limits (or limits of inflammability), on a percentage volume basis, for some paint solvents, are given in Table 8.4

Table 8.4

Solvent	%
Acetone	3–13
Benzene	1.35– 8
Butyl alcohol	1.4–11
Butyl acetate	1.4– 7.6
Cyclohexane	1.3– 8.4
Diacetone alcohol (pure)	1.8– 6.9
Dioxan	2–22
Dimethyl formamide	2.2–15.2
Ethyl alcohol	3.3–19
Ethyl acetate	2.2–11.4
Heptane	1– 6
Methyl ethyl ketone	1.8– 9.5
Methyl methacrylate (monomer)	2.1–12.5
Methyl isobutyl ketone	1.3– 8
Isopropanol	2.3–12.7
Styrene (monomer)	1.1– 6.1
Tetrahydrofuran	2.3–11.8
Toluene	1.3– 6.7
Xylene	1– 7 ca.

8.2.6 Dust explosions

As with inflammable gases and vapours, mixtures of inflammable dusts and air can also burn with explosive violence. Such cloud explosions have been the cause of numerous incidents in recent years, demolishing factories and resulting in many fatalities.

The Factory Inspectorate has issued a booklet [103] on dust explosions wherein are listed such data as 'Ignition temperature of dust cloud', 'Minimum explosive concentration', and 'Maximum explosion pressure' (p.s.i.) for numerous substances. The latter include such paint ingredients as aluminium powder, various synthetic resins, phthalic anhydride and adipic acid.

'Dust Explosions and Fires' is the title of a book by Palmer [104] which discusses the theoretical and practical aspects of dust explosions and fires as well as the test methods used in various countries to assess these hazards. Explosion properties of many dusts are tabulated.

8.2.7 Electrostatic spraying of powder coatings

The two more important hazards associated with the electrostatic spraying of powder coatings are those of explosion and of toxicity. The latter has already been dealt with p. 375) whilst the explosion hazard has in part been covered by the previous section (8.2.6) which is devoted to 'Dust explosions'.

The possibility of explosions arising in the course of powder spraying depends mainly on:

(i) The concentration of the powder in air.
(ii) The presence of sources of energy which could ignite the powder if its concentration lay between its explosive limits, and it is to these two factors that regulations and recommendations for the safe spraying of powder coatings direct most of their attention.

German regulations governing the application of powder coatings are reproduced in a Shell Research publication [105] and are also listed by Korf [92]. There is also a publication, issued by the European Committee of Paint Manufacturers Associations [106], which outlines recommendations for safety in the handling of powder coatings.

A valuable document on the application of powder coatings by electrostatic spraying is that of the Paintmakers Association of Great Britain [91]. This publication in the preparation of which H.M. Factory Inspectorate and other organizations assisted, is issued as a 'Code of Practice' and includes a check list of principal precautions.

8.2.8 Miscellaneous explosion hazards

In addition to the explosion hazards of solvent/air and dust/air mixtures, referred to in the preceding paragraphs, three other areas of explosion risks deserve mention, namely nitrocellulose, the trichloroethylene/caustic alkali reaction, and peroxides.

(a) *Nitrocellulose (N/C)*

Whether wet or dry, N/C is regarded in the U.K. as an explosive material if its nitrogen content is in excess of 12.3%. The N/C used in lacquers has a nitrogen content ranging from 10.5 to 12.3% and is only considered as an explosive when in the dry state. Thus the restrictions of the Explosives Act, which date back to 1875, are not applicable to industrial N/C provided the latter is damped to a minimum of 25% (on the wet weight). Butanol, isopropanol or ethanol are usually used as damping media for N/C required for lacquer manufacture.

The lacquer film itself, although virtually free of solvent, nevertheless still contains enough plasticizer, resin, etc. to render it non-explosive.

See also p. 384 for references to N/C under the 1974 Highly Flammable Liquids Regulations.

Although more of a fire risk than an explosion hazard, it is not inappropriate to mention here that dry N/C can be ignited by contact with solid caustic soda and by a variety of amines. Gørtz [107] found N/C to ignite when in contact with dry morpholine, dibutylamine or diethylene triamine. With these and other amines he also noted temperature rises when N/C paste (33% solvent) was added.

(b) *Trichloroethylene explosive mixtures with caustic soda and with epoxides*

Trichloroethylene is capable of dehydrochlorination with caustic alkali to yield the highly unstable dichloracetylene. The latter compound, when pure, is a colourless oil which boils at 32–33°C, and is capable of ignition on exposure to air and exploding even when stirred.

Furthermore, Dobinson and Green [108] reported, apparently for the first time on record, that certain epoxides can cause the formation of dichloracetylene from trichloroethylene, even in the absence of alkali.

Hence trichloroethylene must not be allowed to come in contact with caustic soda or caustic potash, whether hot or cold, and whether solid or in solution. Weaker alkalis such as sodium carbonate or ammonia do not present a notable hazard in this respect. Additionally, and for similar reasons, trichloroethylene must not be used in conjunction with certain epoxides such as epichlorhydrin and butane diol mono- and di-glycidyl ethers.

(c) *Peroxides*

Peroxides are important additives in paint compositions. Some of their hazards have already been referred to (see p. 372) and special mention given under (Eye effects' (p. 336).

As a class, peroxides are powerful oxidizing agents and will decompose with the evolution of oxygen at rates which range from very slow decomposition to explosive violence, depending on such factors as their chemical constitution, temperature, concentration and whether any catalytic agents are present.

It is important to store organic peroxides away from heat, naked flames or sparks and at temperatures recommended by the supplier. They must also be kept from many other substances which include strong mineral acids, vanadium and cobalt compounds, organic amines, and, in general, metal oxides and reducing compounds.

The transport of organic peroxides is also subject to special regulations such as the 'Organic Peroxides (Conveyance by Road) Regulations', 1973, and in the 'Blue Book' (Carriage of Dangerous Goods in Ships) peroxides are given a special classification (Class 5b).

Another hazard associated with organic peroxides resides in their inadvertent formation when certain organic compounds, many of which are common solvents or even polymerizable monomers such as styrene, are in prolonged contact with oxygen or air. Notably prone to autoxidation are ethyl ether, dioxan and tetrahydrofuran. When autoxidized solvent is distilled the residue can be a dangerously explosive concentrate. Ferrous sulphate is a popular reagent for the removal of peroxide.

8.3 Protective equipment

To combat the effects of noxious substances on the human body protection is required for the three areas, eyes, skin and lungs. Skin will include clothing since contaminated clothing can lead to dermatitis and other skin effects; with lungs is associated the whole of the respiratory tract.

In the interests of impartiality, specific reference to manufacturers of protective gear are avoided. Instead, the names and addresses of such manufacturers and suppliers are readily obtainable from advertisements appearing regularly in appropriate trade and technical journals, and advertisers usually respond promptly to enquiries with illustrated literature. Within the U.K. advice on the selection of protective equipment can be sought from a number of organizations and institutions. At the Industrial Health and Safety Centre, London [109] may be seen a comprehensive display of respirators and comparable equipment. In the U.S.A. such organizations as the National Institute for Occupational Safety and Health [110] and the National Fire Association [111] have the know-how to advise on personal protection against industrial hazards.

Some idea of the needs for protection and the types of equipment available may be gathered from what is outlined below.

8.3.1 Skin protection

(a) *Barrier creams*

In the first instance and whether or not the skin has already been contaminated, barrier creams should be applied after washing and drying the hands, thereby establishing better contact between skin and barrier cream. Various types of barrier cream are available, some with water-repellent properties and others which are designated as oil-repellent (or water miscible). The latter are obviously preferred when handling solvents and solvent-borne paints. In addition, barrier creams have been formulated (e.g. see British Pharmacopoeia Codex) to protect against skin-sensitizing dusts and powders such as are to be met with among certain pigments and other solid ingredients of paint, although opinion is divided on the value of barrier creams for protection against sensitizers.

To protect against the effects of 'chrome liquid or dust' on uncovered skin, H.M. Factory Inspectorate has recommended an ointment made up from three parts of soft paraffin (B.P.) and one part of lanolin.

(b) *Gloves*

These can range from heavy-duty gauntlet-type to light-weight disposable polythene, and surgical (rubber) gloves. Some gloves, too are reinforced with fabric for added strength and comfort and to absorb perspiration.

The chemical resistance of the various polymers and elastomers used in the manufacture of protective gloves depends very much on the concentration and temperature of the chemical being handled. Some generalizations on the resistance of these polymers are given in Table 8.5. However, by suitable modification of the polymer its resistance can be upgraded to withstand the attack of certain chemicals. Thus manufacturers are able to offer gloves of improved resistance to specific substances, solvents etc.

(c) *Protective clothing*

Protective clothing can in the extreme case imply head to foot coverage. For instance, in the spraying of some anti-fouling compositions protection would be of the same order as stipulated for the large-scale spraying of (agricultural) pesticides. Sometimes the recommendations here would include, among other items, the use of an overall and rubber apron or a mackintosh, or a hood, or a rubber coat and a sou' wester. Rubber boots and apron too are desirable when handling bulk of strong acids or alkalis. Protection of the operator's clothing, as well as of exposed skin, is called for in other situations as for instance in the spraying of polyurethanes containing appreciable amounts of free iso-cyanate. A review by Vallaud [112] classifies protective clothing by material (of fabrication) and by hazard.

8.3.2 Eye protection

On many occasions even the wearing of ordinary spectacles has provided physical protection for someone's eyesight; the use of safety spectacles (with side shields) should be mandatory where there is a risk of flying material (e.g. resulting from accidental pressure release).

Chemical goggles or eye shields are recommended when handling peroxides and other corrosive substances as well as solvents, and gas-tight goggles where the material is lachrymatory. Other means of eye protection are afforded by face masks, face shields and protective helmets, as well as, of course, by 'full-face' respiratory equipment (see below).

See also the International Labour Office Encyclopedia [113] for examples of protective equipment for the eyes and face.

Table 8.5

Polymer (elastomer)	Resistance properties
Natural rubber	Resistant to acids, alkalis and aqueous solutions. Unsuitable for use against most organic solvents.
Synthetic rubber (styrene/butadiene type)	Similar to natural rubber (above).
Synthetic rubber (nitrile type)	Good resistance against petrol, diesel oil and lubricating oils. Limited resistance, unless thick, against degreasing solvents such as toluene and chlorinated hydrocarbons.
Chlorinated rubber (Neoprene)	Good resistance against vegetable oils and fats, acids and alkalis and some organic solvents such as esters and alcohols.
Polythene	Resistant to acids and alkalis, aqueous solutions and glycerol but poor against degreasing solvents.
Polypropylene	Similar resistance properties to polythene (above) but more suitable for the manufacture of packaging materials than for gloves.
Polyvinyl chloride	Although unplasticized p.v.c. is resistant to most chemicals and solvents, it is the plasticized polymer which is used in the manufacture of flexible p.v.c. gloves. Nearly all organic solvents are capable of extracting the plasticizer, often with the result of leaving the gloves more rigid and reduced in toughness. Plasticized p.v.c. gloves are however excellent for use in handling many inorganic compounds (e.g. acids and alkalis) and such organic compounds as isopropanol, glycerol and acetic acid (dilute).

8.3.3 Respiratory protection

To avoid the inhalation of harmful gases, vapours and particulate (spray mist, liquid drops and dusts) two procedures can be adopted. In the first instance recourse can be made to fume-cupboards, hoods or spray-booths, all fitted with exhaust ventilation. It is important that the extracted fumes or vapours are not re-admitted to work places where they could be inhaled, nor ventilated to the environment in harmful concentration (see also 'Emissions into the Atmosphere', Section 5 of the Health and Safety at Work Act, 1974.

However, although fume-cupboards and spray-booths are ideal for handling hazardous paint ingredients or for spray-painting on a small-scale they are out of the question for the spraying of large vehicles, ships or aircraft. For such operations and for dealing with gross spillages of (harmful) volatile liquids or accidental emissions of poisonous gases or vapours, breathing equipment is needed so that the operator is not inhaling atmosphere contaminated with noxious substances which are at a concentration in excess of their threshold limit value or, more important, their ceiling value (see also p. 340). For example, Section 8 (1) (a) of the Asbestos Regulations (1969), calls for the provision of 'approved respiratory equipment' where it is impracticable to comply with the requirements of exhaust ventilation (Section 7).

Many varieties* of breathing apparatus are available, some more suitable than others for particular situations. They can be resolved into two main types:

(i) Those which purify inhaled air by filtering out contaminant.
(ii) Those which ensure the user is supplied with only fresh air (or oxygen).

(a) *Filter-type equipment*

Dust mask (half mask)
Useful against solid particles (dusts). Even the more elaborate rubber face mask, where the wearer has to inhale through filter cartridges fixed to the mask, and which have an efficiency of 90% (against 0·5 μ particulate), are unsuitable protection against asbestos. The dust mask is also useless against gases and vapours and is not recommended as protection against toxic mist or droplets. The principal advantage of dust masks, more so the simple gauze filter rather than the cartridge type, is their minimum interference with breathing.

Canister respirator (full face piece)
Must be fitted with the correct canister to absorb the specific vapour or gas. Convenient for dealing with relatively low concentrations or for short periods of time. Where the canister is replaced by a disposable cartridge (to absorb gases and vapours) the working life of the equipment is even shorter.

Combined filter/canister
More versatile in its ability to protect from solid, liquid and gaseous material.

(b) *Air (or oxygen) – supply types*

Air line (lungs as pump)
Suitable where no more than about 30 minutes of protection is required. The

* There is also a British Standard (BS4275:1974) which gives recommendations for the selection, use and maintenance of respiratory protective equipment.

face-piece must fit well. The long hose-line needs to be kept free of entanglement.

Air line (pressurized)
Air is blown in by means of a hand pump or blower at 5–25 p.s.i. at the rate of 6 ft^3/min. This has to be controlled with a warning device. The equipment includes either a helmet or a face-piece. Again, good fitting is essential when using the helmet for protection against toxic material. A half-mask face-piece is more convenient to use than full-face equipment but the latter provides eye protection in addition.

Self-contained
Comparatively simple piece of equipment if air supplied, when it can be used for up to 30 minutes. Becomes more complex if oxygen supplied, but with a longer service life (up to about 2 hours).

8.4 First-aid

First-aid procedures associated with paints and painters will differ, depending on the situation. Painting of domestic premises could involve very young children innocently sampling the paint pot. Industrial painting will involve sophisticated and often very hazardous paint compositions but at the same time could mean a medical centre or even a first-aid post is at hand, and frequently a number of individuals trained in first-aid could be within calling distance.

8.4.1 Procedure

(a) *If swallowed*

For the great majority of harmful substances which have been swallowed it is advisable first to rinse the mouth with water and then to give quantities of water to drink (dilution effect), particularly if the ingested substance is corrosive.

If an organic peroxide has been swallowed, dilute (2%) sodium bicarbonate solution, if available, may be given in place of water.

Emesis (vomiting) is not resorted to as frequently as in the past. Unconscious patients might inhale their vomit with serious complications. Also, emetics *must not* be used if the substance swallowed is caustic, corrosive, or contains substantial amounts of hydrocarbon solvents (as do nearly all solvent-borne paints). The use of an emetic, if considered necessary (e.g. against isocyanates) should be left to the discretion of a qualified first-aider or medical officer.

In any case, where the nature and extent of the poisoning necessitates medical attention, details of the description of the paint (or paint ingredient) as well as the manufacturer's name should be noted and given to the doctor. He may, if he thinks fit, call for further information from the manufacturer or enlist the aid of a national poisons information service. British and many overseas poisons centres will already have information stored on the (generalized) composition of leading (branded) paints and lacquers. Within Britain there are National Poisons References Centres at Belfast, Cardiff, Dublin, and Edinburgh. London (Guy's Hospital) is the focal point for this Service. Unlike the majority of comparable European and American centres (in the U.S.A. well over 100 are functional), the British Poisons Information Service is normally only available to hospitals and doctors.

(b) *Eyes*

For almost every case of eye contamination with paints or paint ingredients (including solvents), flushing with water is the accepted procedure, followed by continuing irrigation with water (for a minimum of 10 minutes). Where the offending material is an organic peroxide, oils or oily liquids *must not* be applied to the eye.

If the contaminating product or substance is exceptionally dangerous, such as a strong acid, caustic alkali, phenolic compound, pesticide or peroxide, medical attention must in addition be sought.

(c) *Skin*

If the contaminant is water-soluble or water-miscible (e.g. an emulsion), flush or drench with quantities of water. If water-insoluble (e.g. hydrocarbon solvent or solvent-borne paint) wipe off and wash area of contaminated skin with soap and water.

Phenol contamination is a special case following the findings (animal experiments) of Conning and Hayes [114] that the percutaneous toxicity of phenol was increased by the diluents water, methylated spirit or olive oil. They recommend swabbing all phenol-contaminated skin with cotton wool soaked in glycerol, polyethylene glycol or a polyethylene glycol/methylated spirit mixture for at least 10 minutes (see also [115]. Medical attention is imperative.

(d) *Clothing*

For significant contamination with noxious substances, e.g. most solvents, remove clothing immediately. Urgent for phenol and isocyanate splashes. Do not re-use until thoroughly cleaned and dried.

(e) *Gassing*

In the case of most such incidents, e.g. following excessive exposure (inhalation) to solvents or isocyanates, remove patient to fresh air and keep him rested and warm. Send immediately for first-aider and doctor.

Note: Patients who have been gassed with solvents such as trichloroethylene and perchlorethylene must avoid over-excitement or strong physical exertion for the next 24 hours.

8.4.2 First aid procedure – quick reference

If swallowed

Rinse mouth with water and give quantities of **water to drink** (or dilute (2%) sodium carbonate, if available, if organic peroxide swallowed).

Do not give emetic if **corrosive, caustic alkali** or **hydrocarbon solvent** or **solvent-borne paint** has been swallowed.

In any case leave use of emetic to discretion of doctor, or qualified first-aider. If a doctor needs to be called note nature or description of paint ingredient or product (and manufacturer).

Eyes

Flush with water and **continue irrigation** with water for **minimum** of **10 minutes.**

Eyes contaminated with **organic peroxide** – **do not use oils** or **oily liquids.**

Eyes contaminated with **caustic alkali, strong acid, phenolic substance, pesticide** or **peroxide** – **seek medical aid immediately.**

Skin

Contaminant	*Action*
Water-soluble or **water miscible**	**Flush with quantities of water. If inflamed, seek medical attention**
Oils, hydrocarbon solvents and other **water-insoluble** material	**Wipe off** and **wash contaminated area** with **soap and water. If inflamed, seek medical attention.**
Phenol, cresols, cresylic acid	**Swab** with **cotton wool soaked** in **glycerol, polyethylene glycol (P.E.G)** or a **P.E.G./methylated spirits** mixture for **at least 10 minutes. Send for doctor immediately.**

Clothing

Significant contamination with **noxious substances (solvents** and **solvent-borne paints)** – **remove clothing immediately. Urgent** for **phenol** and **isocyanate splashes. Thoroughly clean** and **dry before re-use.**

Gassing

Remove patient to **fresh air** and **keep him rested** and **warm. Send immediately** for **first-aider** or **doctor. Trichloroethylene** exposure – patient must **avoid over-excitement** or **strong physical exertion** for **next 24 hours.**

8.4.3 Further information

It is impossible here to give detailed first-aid procedures for the hundreds of potentially dangerous substances which nowadays enter into the composition of paints and lacquers. Additional information, however, can be obtained from numerous publications, including first-aid charts. A selection of these is given below:

(i) *Hazards in the Laboratory*, ed. Muir, G.D., The Chemical Society, London, (2nd Ed. 1977). Chap. 7, pp. 89–110, 'Medical Services and First Aid'. Chap. 8, pp. 111–435, 'Hazardous Chemicals', listing numerous chemicals with data including first-aid recommendations.

(ii) *Marking Containers of Hazardous Chemicals,* Chemical Industries Association, London (1970) [96].
For numerous chemicals provides warning labels on many of which are given first-aid instructions.

The CIA also publish a booklet *Exposure to Gases and Vapours* containing prepared labels to accompany patients sent to hospital suffering from the effects of poisons or corrosives, as useful information for the doctor attending the patient.

(iii) *Toxicology of Drugs and Chemicals,* Deichman W.B. and Gerarde, H.W., Academic Press, New York and London (1969). Chap. 1, pp. 1–34, 'First Aid and General Suggestions for Treatment'. Chap. IV pp. 61–663, 'Drugs and Chemicals, Signs, Symptoms and Treatment of Intoxications'.
Treatment given for poisoning, for very many substances.

(iv) *Laboratory First Aid,* British Drug Houses, Poole, England.
Wall chart with numerous details.

(v) *Emergency Procedures for Dangerous Materials,* Science Related Materials Inc., Evanston, Ill. 60704, U.S.A.
Emergency procedures given against many hazardous liquids, solids and gases.

8.5 Spillages and pollution

8.5.1 Spillages

It is important to deal with spillages as soon as possible, before they cause further trouble.

If they are liquid – someone could slip.
If inflammable – a fire or explosion hazard is created.
If corrosive or emitting toxic vapours – a serious health hazard is introduced.

Spillages are in effect 'one-off' incidents of limited pollution. Their disposal must not result in wider pollution (i.e. of the environment) nor must local or national regulations governing pollution be infringed. The disposal of substantial spillages could bring the problem into the same difficult area experienced by paint manufacturers over paint waste. See also Section 8.5.2, following, on 'Pollution'.

Before dealing with spillages which are:

(i) Inflammable – shut off all sources of ignition.
(ii) Otherwise harmful (toxic vapour, corrosive, irritant) – use appropriate (personal) protection.

(a) *Paints*

According to quantity and consistency, shovel up or mop up and transfer to a bin or other suitable container. If the paint is inflammable or otherwise hazardous the nature of the hazard must be clearly marked on the container. Residual traces can be flushed away with water, if water-soluble or water-miscible; or with water plus detergent or emulsifying liquid, if insoluble or immiscible with water.

(b) *Solvents*

If water soluble or water miscible – drench and wash away with water.

If water-insoluble, limited amounts may be treated with emulsifying agent or detergent and washed away with large quantities of water.

(c) *Pigments* (*and many other solid chemicals*)

The majority may be safely dealt with by damping with water, sweeping up and transferring to bins (or other suitable containers). Spillages of *cellulose nitrate* (not to be confused with N/C lacquer) *must be* kept *moist or damp.*

(d) *Chemicals*

Covering most liquids including strong acids*, and water-soluble solids including caustic alkalis.

Drench and flush away, using added detergent where water insolubility is creating a problem.

Some chemicals like allyl chloride, methacrylic acid and ethyl hexyl acrylate are more safely dealt with by adding inert absorbent (sand, earth, etc.) and destroying by incineration.

* For *hydrofluoric* and *fluoroboric* acids, add, if available, chalk or lime, before drenching. For *chromic* acid, add, if available, sodium carbonate or soda ash, before drenching.

(e) *Special cases*

For the disposal of spillages of particularly hazardous substances additional reference should be made to one or other of the many text-books or other sources of information which are available on the disposal of spillages. These include:

(i) *Laboratory Waste Disposal Manual,* Manufacturing Chemists' Association, U.S.A.

(ii) *Marking Containers of Hazardous Chemicals,* Chemical Industries Association, London (1970).
Advice on handling spillages given on many of the labels for the chemicals which are listed.

(iii) *Hazards in the Laboratory,* The Chemical Society, London (2nd ed., 1977). Spillage disposal instructions are given for the many hazardous chemicals, which are listed alphabetically.

(iv) *How to Deal with Spillages of Hazardous Chemicals,* Chart, British Drug Houses Ltd., Poole, England.

Two groups of particularly dangerous substances, which are ingredients of certain industrial paint compositions and whose spillage demands special handling precautions are:

(1) Isocyanates
In the first instance it is important that spillages of unreacted toxic isocyanate are not allowed to run to sewers or drains and that personnel dealing with the spillage are wearing self-contained or air-supplied breathing apparatus (see also p. 392).

Although one should heed the supplier's or manufacturer's advice over spillages, substantial differences are to be noted in the recommendations given by various authorities for the disposal, for instance, of the toxic material toluene diisocyanate (TDI or diisocyanatotoluene). Some references to the disposal of spillages of TDI include:

(i) *Marking Containers of Hazardous Chemicals* p. 63 Chemical Industries Association, London (1970).

(ii) *Hazards in the Laboratory,* p. 232, The Chemical Society, London (1977).

(iii) *Toluene Diisocyanate,* Chem. Safety Data Sheet SD-73, Manufacturing Chemists Association, (1971).

(iv) *Isocyanates: Hazards and Safe Handling Procedures,* Tech. Inform. U93, ICI (3rd Ed. 1974).

(v) *Handling Isocyanates Safely,* Smith, D.C. [90].

Simplified procedure: Cover spillages with a thin layer of isocyanate neutralizing powder made up from 1 part of sawdust and 2 parts of inert powder such as china clay or a diatomaceous earth (see p. 354). Leave for 15 minutes, sweep up and discard to bins or other suitable container. Wash contaminated area with 5% ammonia solution containing a little detergent.

(2) Peroxides [74, 75]
Do not mop up spillages with sawdust, rags or any other organic material. Cover spillages with wet sand or earth, or drench with water.

8.5.2 Pollution

(a) *United Kingdom*

The year 1974 saw two Acts introduced, one in part and one as a whole, concerned with pollution in the U.K. The first, the Health and Safety at Work Act, has already been referred to (p. 377). Section 5 (1) of Part I of this enabling Act states quite clearly that:

'It shall be the duty of the person having control of any premises of a class prescribed for the purposes of Section 1 (i) (d) to use the best practicable means for preventing the emission into the atmosphere from the premises of noxious or offensive substances and for rendering harmless and inoffensive such substances as may be so emitted'.

In one comprehensive piece of legislation the Control of Pollution Act brought together the numerous earlier Acts which had had some bearing on pollution. It could be said that those concerned with pollution in the U.K. were therefore fortunate in being presented with rationalized legislation which although it was voluminous (running to 136 pages), nevertheless repealed the whole of the three Acts: The Rivers (Prevention of Pollution) Act, 1951; The Clean Rivers (Esturaries and Tidal Waters) Act, 1960, and The Deposit of Poisonous Waste Act, 1972.

The 1974 Act also repealed the larger part of several other Acts and amended numerous other Acts, e.g. the Clean Air Act, 1956. The four parts of the Control of Pollution Act, 1974, cover respectively waste on land, pollution of water, noise, and pollution of the atmosphere. Three of the parts, i.e. excluding the one on noise, are particularly relevant to industrial painting, e.g. the disposal of accumulated paint waste from spray booths and the emission of volatile material from stoving ovens (and their fume stacks).

(b) *U.S.A.*

The paint industry, not only in the U.S.A., but throughout the industrial world, was perturbed with the stringency of the now notorious Rule 66 which was introduced in 1966 in Los Angeles. It related to atmospheric pollution in that locality of the U.S.A. where for various reasons conditions were frequently conducive to smog formation. The smog was allegedly due to photo-chemical agents among which were many which are paint solvents, e.g. xylol, turpentine and dipentene. Rule 66, revised in August 1971, was designed to exercise control on the daily emissions into the atmosphere of such photo-chemical agents.

The Toxic Substances Control Act, first passed by the Senate in 1972 and finally passed, with amendments, in 1976, included a controversial measure, namely the pre-market screeing of all new commercial chemicals. Data to be supplied to the Environmental Protection Agency would include the testing required for the chemical's impact on the envionment.

Among the controls introduced by the Occupational Safety and Health Act, which was passed at the end of 1970 (see also p. 381), are some governing atmospheric pollution.

(c) *Europe*

Most European countries have for long been concerned with pollution. As an example it is interesting to record the existence for many years of the International Commission for the Protection of the Rhine against Pollution whose member countries are Switzerland, Germany, France and Holland. A useful survey is that of J.B. Davis [116] entitled *Pollution – the European Scene.* The EEC too has been active with a proposal in 1976 [117] for a Directive which would cover the disposal (among other aspects) of toxic waste.

(d) *Global*

It is not inappropriate to conclude here with a mention of the existence of the United Nations Environment Programme (UNEP) which commenced functioning, in a preliminary way, in Nairobi, being subsequently (end 1975) transferred to Geneva. Among its objectives is the formation of an International Register of Potentially Toxic Chemicals (IRPTC), the establishment of which, was recommended by the United Nations Conference on Human Environment, which was held in Stockholm in June 1972.

The ultimate aim of the Register (IRPTC) is to serve mainly the following users [118]:

(i) The Member States of the United Nations.
(ii) Authorities responsible for human health, e.g. governing the transportation of harmful chemicals.
(iii) Scientists engaged in research.
(iv) Industries and other organizations involved in activities that could release harmful chemicals into the environment.

This fourth category of user would be expected to embrace many who make use of the more hazardous types of industrial coating compositions.

8.6 References

[1] Portigliotti (1925), *The Borgias,* p. 95 (Notes to Chapter V) Milan, Treves. First published in English (translation by Bernard Miall) in 1928.

[2] Druckrey, H. *et al.* (1970), *Z. Krebsforsch.* **75**, 69–84.

[3] Van Duuren, B.L. *et al.* (1971), *J. Natn. Cancer Inst.* **46** (1) 143–149.

[4] Ulland, B. *et al.* (1971), *Nature* **230**, 460.

[5] Marchenko, E.N. and Beilikhis, G.A. (1946), Non-occupational poisoning by lead tetraethyl, *Gig. i Sanit.* **11**, 17.

[6] Schamberg, J.F. (1918), *J. Amer. Med. Assoc.*, **70**, 142.

[7] Silberberg, I. (1968), *J. invest. Derm.*, **50**, 323.

[8] Silberberg, I. *et al.* (1969), *Arch. envir. Health,* **19**, 7.

[9] Barnes, J. (1939), *Lancet,* **89**.

[10] Goldwater, L.J. (1971), *Sci. Am.* **224**, (5) 15–21.

[11] Brana Jovičic, (1968), *Ann. Occup. Hyg.*, **11**, 305–307.

[12] Zeman, W. *et al.* (1951), *Dent. Arch. klin. Med.*, **198**, 713–721.

[13] Yanagibashi, H. (1962), *Tokyo Ika Daigaku Zasshi,* **20** (1), 1–50.

[14] Barnes, J.M. and Stoner, H.B. (1958), *Brit. J. industr. Med.*, **15**, 15.

[15] Elsea, J.R. and Paynter, O.C. (1958), *Arch. indust. Health,* **18**, 214–217.

[16] H.M. Chief Inspector of Factories, Annual Report 1970, pp. 69–70.

[17] Truhaut, K., L'Epee, P. and Boussemart, E. (1952), *Arch Mal. prof.*, **13**, 567.

[18] Associated Lead Manufacturers Ltd.

[19] Gabby, J.L. (1950), *Arch. Ind. Hyg. Occ. Med.*, **1**, 677–684.

[20] Flick, D.F. *et al.* (1971), *Environmental Research,* **4**, 71–85.

[21] Turner, F.M. (1957), U.K. Atomic Energy Authority, Harwell, Established MED/M-17.

[22] Kendrey, G. and Roe, F.J.C. (1969), *Lancet,* I, 1206–1207.

[23] Webb, M. (1975), *Br. Med. Bull.*, (3), 246–250.

[24] Department of Employment and Productivity, H.M. Factory Inspectorate, Technical Data Note 11 (1970), *Cadmium. Health and Safety Precautions.*

[25] Williams, C.D. (1969), *N. Carolina Med. J.*, **30**, 482–490.

[26] Baetjer, A.M. (1960), *Arch. Ind. Hyg. and Occ. Med.*, **2**, 487.

[27] Payne, W.W. (1960), *Arch. indust. Health,* **21**, 530.

[28] Hueper, W.C. and Payne, W.W. (1959), *Amer. industr. Hyg. Ass. J.*, **20**, 274.

[29] Hueper, W.C. (1961), *Cancer Res.*, **21**, 842.

[30] Watson, V. and Clay, H.F. (1946), *JOCCA,* **29** (313), 151–154.

[31] Fregert, S. and Rorsman, H. (1964), *Arch. Derm.*, **90**, 4–6.

[32] Rockhold, W.T. (1955), *Arch. Ind. Health,* **12**, 477–482.

[33] Berger, H. (1938), *Chem. Ind.*, **10**, 262.

[34] Browning, E. (1969), *Toxicity of Industrial Metals,* 2nd edn, Butterworth and Co Ltd, London.

[35] Legge, T.M. and Goadby, K.W. (1912), *Lead Poisoning and Lead Absorption,* Longmans and Green, London and New York.

[36] *Occupational Lead Exposure and Lead Poisoning,* Ind. Hyg. Sect. of the American Public Health Assocn., 1970 Broadway, N.Y. (1943).

[37] Chisholm, J.J. Jr. (1971), Lead poisoning, *Sci. Am.* **224** (2), 15–23.

[38] Nam, K. and Gracey, D.R. (1972), *J. Am. med. Ass.*, **221**, 492.

[39] Kleinfeld, M. *et al.* (1967), *Arch. Environ. Health,* **14**, 663–667.

[40] Blejer, H.P. and Arlon, R. (1973), *J. Occ. Med.,* **15** (2), 92.
[41] Battigelli, M.C. (1960), *J. Occ. Med.* **2**, 337–344 and 344–399.
[42] Whitehead, K.P. (1965), *Ann. Occ. Hyg.,* **8**, 85–89.
[43] Mann, A. (1971), *Paint Varnish Prod.* **61** (3), 26–37.
[44] Volk, H. (1960), *Achiv. Environ. Health,* **1**, 125–128.
[45] H.M. Factory Inspectorate, (1970), Department of Employment and Productivity, Tech. Data Note 13 (3/70), *Standards for Asbestos dust concentration for use with the Asbestos Regulations 1969.*
[46] H.M. Factory Inspectorate (1967), *Problems Arising from the Use of Asbestos.*
[47] Hill, J.W. *et al.*, (1973), *Brit. J. Ind. Med.,* **30** (2), 174–179.
[48] Heisel, E.B. and Hunt, F.E. (1968), *Arch. Environ. Health* **17**, 705.
[49] Piver, W.T. (1973), *Environ. Health Perspect.* (4), 61–79.
[50] Elo, R. (1972), *Archs. Path.,* **94**, 417.
[51] Bertrand, D. (1942), *C.R. Acad. Sci. (Paris),* **215**, 150, 205.
[52] Noro, L. and Uotila, U. (1954), *Inter Cong. industr. Med., Naples,* p. 116; Uotila, U. and Noro, L. (1957), *Folia med. Napoli,* **40**, 245.
[53] Ball, G.L. and Boettner, E.A. (1973), *Am. Chem. Soc. Div. Polymer Chem., Pre-prints,* **14** (2), 986–989.
[54] Koivusalo, M. (1970), *Int. Encycl. Pharmacol. Ther. 20 Alc. Deriv.,* V2, pp. 465–505, Pergamon, Oxford.
[55] Browning, E. (1965), *Toxicity and Metabolism of Industrial Solvents,* Elsevier, Amsterdam, London and New York.
[56] Matthew, H. and Lawson, A.A.H. (1971), *Treatment of Common Acute Poisonings,* E. and S. Livingstone, Edinburgh and London.
[57] *Handbook of Organic Industrial Solvents,* National Association of Mutual Casualty Companies, 1958.
[58] Patty, F.A. (1962), *Industrial Hygiene and Toxicology,* Vol. II, 2nd ed. Wiley Interscience, New York.
[59] McCord, C.P. (1926), *J. Am. Med. Assoc.,* **86**, 1978.
[60] Chapman, E.M. (1941), *J. Ind. Hyg. Toxicol.,* **23**, 277.
[61] Little, J. (1955), *Brit. J. industr. Med.,* **12**, 304.
[62] Cornish, H.H. and Adefuin, J. (1966), *Am. Ind. Hyg. Ass. J.,* **27**, 57–61.
[63] Hoschek, R. (1962), *Arch. Gewerbepath and Gewerbehyg.,* **19**, 318.
[64] Cornish, H.H. and Ling, B.P. (1972), *Am. Ind. Hyg. Assoc. J.,* **33** (2), 81.
[65] Smith, G.F. (1966), *Brit. J. Ind. Med.,* **23**, (4), 249–262.
[66] Steward, R.D. *et al.* (1972), *Arch. Environ. Health,* **25** (5), 342–348.
[67] Fehr, I.K. (1963), *Schweiz. Med. Wochschr.* **93** (11), 436–444; (12) 465–471.
[68] Susser, M. and Stein, Z. (1957), *Brit. J. Ind. Med.,* **14**, 111–120.
[69] Henschler, D. (1960), *Zent. für Arbeitsmed. u. Arbeitsschutz,* **10** (4), 86–88.
[70] Colley, D. (1966), *J. Pharm. Sci.,* **55**, 58.
[71] Hauschild, F. (1962), *Conf. Hung. Therap. Invest. Pharmacol.* **2**, Budapest, 37–39.
[72] Kurita, H. (1967), *Bull. Hyg.,* **42**, (10), 1078.

[73] Burger, L.M. and Chandor, S.B. (1971), *Arch. Environ. Health,* **23**, (5), 402–404.
[74] *Code of Practice for the Storage of Organic Peroxides,* Laporte Industries, Widnes, 1970.
[75] Castrantas, H.M. and Banerjee, D.K. (1970), *Laboratory Handling and Storage of Peroxy Compounds,* A.S.T.M., Philadelphia.
[76] Gaines, T.B. (1969), *Toxicol. Appl. Pharmacol.,* **14**, (3), 515–534.
[77] Deichmann, W.B. (1972), *J. Occup. Med.,* **14**, (4), 285–292.
[78] Fest, C. and Schmidt, K.-J. (1973), *The Chemistry of Organo-Phosphorus Pesticides,* Springer, New York.
[79] Clayton, J.W. Jr. (1974), *Health Effects of Fluorocarbons,* Chem. Rubber Co., Cleveland, Ohio.
[80] *New Scientist,* p. 575, 10th March (1977).
[81] Paintmakers Association of Great Britain Ltd. (1973), *Handling Hazards in the Use and Manufacture of Epoxy Paints and Precautions to be Adopted.*
[82] Paintmakers Association of Great Britain Ltd. (1971), *Toxicity of Various Curing Agents for Epoxy Resins.*
[83] Munn, A. (1965), *Ann. Occup. Hyg.,* **8**, 163–169.
[84] Peters, J.M. (1970), *Proc. Roy. Soc. Med.,* **63**, 373–375.
[85] Peters, J.M., *et al.* (1970), *Arch. Environ. Health,* **20**, (3), 364–367.
[86] McKerrow, C.B., Davies, H.J. and Parry–Jones, A. (1970), *Proc. Roy. Soc. Med.,* **63**, 376–378.
[87] Woolley, W.D., Wadley, A.I. and Field, P., Fire Research Memorandum No. 881, Borehamwood, England.
[88] Paintmakers Association of Great Britain Ltd. (1970), *The Hazards Arising from the Manufacture of Polyurethane Paints and Varnishes.*
[89] Manufacturing Chemists Association, Washington, D.C. *Toluene Di-isocyanate* Chemical Safety Data Sheet SD-73 (revised 1971).
[90] Smith, D.C., Chrysler (U.K.) Ltd. (1973), *Occup. Health,* **25**, 92–96.
[91] Paintmakers Association of Great Britain Ltd. (1975), *Code of Practice – Application of Powder Coatings by Electrostatic Spraying,* 3rd Impression.
[92] *Powder Coating – A report on Materials, Testing, Plant and Application,* (1975), Chr. Korf, Dutch Consulting and Testing Labs. for the Paint and Finishing Industries, Haarlem, Holland; Pubs. R.H. Chandler Ltd., Braintree, Essex.
[93] Jackson, J. (1974), *Health and Safety – The New Law,* The Commercial Publishing Co. London.
[94] *Health and Safety – The Social and Professional Responsibilities of the Chemist,* The Royal Institute of Chemistry (1975).
[95] Mitchell, E. (1975), *The Employer's Guide to the Law on Health, Safety and Welfare at Work,* Business Books Ltd.
[96] *Marking Containers of Hazardous Chemicals,* Chemical Industries Association Ltd. London.
[97] *Official Journal of the European Communities,* **16**, (L189), 1973.

[98] *Modern Paint and Coatings,* **66**, (8), Supplement (1976).
[99] British Standards Institution, BS3442 Pt. 2 (1974).
[100] British Standards Institution, BS4688-9, (1971).
[101] British Drug Houses Ltd. Poole, England, (1962), *Flash-Points.*
[102] Post Office Guide, Nov. 1976, p. 76 (Inland Post).
[103] Department of Employment, Booklet No. 22, *Dust Explosions in Factories.*
[104] Palmer, K.N. (1973), *Dust Explosions and Fires,* Chapman and Hall, London.
[105] Shell Research Ltd. (1972), *Powder Coatings: Explosive Hazards.*
[106] *Safe Powder Coatings,* European Committee of Paint, Printing Ink and Artists' Colours Manufacturers' Association, Brussels, (2nd edn) 1975.
[107] Gørtz, J.V. (1966), *Farg och Lack,* **12**, (1), 18.
[108] Dobinson, B. and Green, G.E. (1972), *Chem. and Ind.* 4th March, p. 214.
[109] The Industrial Health and Safety Centre (Dept. of Employment and Productivity) 97 Horseferry Road, London S.W.1.
[110] NIOSH Certified Personal Protective Equipment, Publications DTS, NIOSH, Cincinnati, Ohio, 42226.
[111] National Fire Protection Association, 60 Batterymarch Street, Boston, Mass. 02110.
[112] Vallaud, A. (1972), Protective clothing, *Occupational Health and Safety,* p. 1127, International Labour Office, Geneva.
[113] Miura, T. *ibid.* pp. 501–503.
[114] Conning, D.M. and Hayes, M.J. (1970), *Brit. J. Ind. Med.,* **27**, (2), 155–159.
[115] Anon, (1974), *Quarterly Safety Summary,* **45**, (177), 5.
[116] Davis, J.B. (1972), *Poly. Paint and Col. J.,* **161**, 82.
[117] Official Journal of the European Communities, **19**, (C194).
[118] The International Register of Potentially Toxic Chemicals, United Nations Environment Programme, September 1975.

Bibliography

This list gives details of books mentioned in the text, or recommended for further reading.

Abraham, H., *Asphalt and Allied Substances,* (4 vols.), D. van Nostrand, New York, 1960–1962.

Barton, K., *Protection against Atmospheric Corrosion,* Wiley, Chichester, 1976.

Bernendsen, A.M., *Shipbuilding Painting Manual,* De Boer Maritien/Verfinstituut TNO, Delft, 1975.

Blom, A.V., *Organic Coatings in Theory and Practice,* Cleaver–Hume Press, London, 1947.

Booth, G.H., *Microbiological Corrosion,* Mills and Boon, London, 1961.

Boxall, J. and von Fraunhofer, J.A., *Concise Paint Technology,* Elek Science, London, 1977.

Browning, E., *Toxicity of Industrial Metals,* 2nd Edn, Butterworth, London, 1969.

Browning, E., *Toxicity and Metabolism of Industrial Solvents,* Elsevier, Amsterdam, 1965.

Burns, R.M. and Bradley, W.W., *Protective Coatings for Metals,* 3rd Edn, Reinhold, New York, 1967.

Chatfield, H.W., *Varnish Constituents,* Leonard Hill, London, 1953.

Chatfield, H.W., *Science of Surface Coatings,* Ernest Benn, London, 1962.

Chemical Industries Association, *Marking Containers of Hazardous Chemicals,* London.

Clayton, J.W., *Health Effects of Fluorocarbons,* Chemical Rubber Co., Cleveland, Ohio, 1974.

D.S.I.R. Forest Products Research, *A Handbook of Hardwoods,* London, 1956.

Durrans, T.H., *Solvents,* 8th Edn, revised by E.H. Davies, Chapman and Hall, London, 1971.

Evans, U.R., *The Corrosion and Oxidation of Metals,* Arnold, London, 1960. 1st Supplementary, volume 1968; 2nd Supplementary volume, 1976.

Fancutt, F., Hudson, J.C., Rudram, A.T.S. and Stanners, J.F., *Protective Painting of Structural Steel,* Chapman and Hall, London, 1968.
Fest, C. and Schmidt, K.J., *The Chemistry of Organophosphorus Pesticides,* Springer, New York, 1973.
Gardner, H.A. and Sward, G.G., *Physical and Chemical Examination of Paints, Varnishes, Lacquers and Colours,* 12th Edn, Gardner Lab., Bethesda, 1962.
Gynes, N.I. and Danziger, G.N., *Formulation of Organic Coatings,* D. van Nostrand, New York, 1967.
Harris, S.T., *Technology of Powder Coatings,* Portcullis Press, Redhill, 1976.
Holland L., *Vacuum Deposition of Thin Films,* Chapman and Hall, London, 1956.
Hudson, J.C., *The Corrosion of Iron and Steel,* Chapman and Hall, London, 1940.
International Technical Information Institute, Japan, *Toxic and Hazardous Industrial Chemicals Safety Manual,* Japan, 1975.
Jackson, J., *Health and Safety – The New Law,* Commercial Publicity Co., London, 1974.
Jordan, O., *The Technology of Solvents,* Leonard Hill, 1940.
Judd, D.B., *Color in Business, Science and Industry,* Wiley, New York, 1963.
Kraus, A., *Handbuch der Nitrozelluloselacke,* Parts 1–3, Berlin, 1955–1961.
Llewellyn, H.M. and Eldridge, H.J. (1939), *The Decoration of New Plaster and Cement,* 29th Bulletin of the Research Association of British Paint, Colour and Varnish Manufacturers, Teddington, Middlesex.
Marsden, C. and Mann, S., *Solvents Guide,* Cleaver-Hume Press, London, 1963.
Matthew, H. and Lawson, A.A.H., *Treatment of Common Acute Poisonings,* Livingstone, London, 1971.
Mills, M.R., *An Introduction to Drying Oil Technology,* Pergamon Press, Oxford, 1952.
Mitchell, E., *The Employer's Guide to the Law on Health, Safety and Welfare at Work,* Business Books Ltd, London, 1975.
Moore, N.L. (ed.), *Radiation Drying of Paints, Inks and Related Systems,* Hertis, Hatfield, 1973.
Morgans, W.M., *Outlines of Paint Technology,* Charles Griffin, London, 1969.
Morgans, W.M., *Pigments for Paints and Inks,* S.I.T.A. Ltd, London, 1977.
Morrell, R.S. (ed.), *Synthetic Resins and Allied Plastics,* Oxford University Press, Oxford, 1943.
Myers, R.R. and Long, J.S. (Eds.), *Treatise on Coatings,* 5 vols. Marcel Dekker, New York, 1968 onwards.
Narcus, H., *Metallizing of Plastics,* Reinhold, New York, 1960.
National Association of Mutual Casualty Companies, *Handbook of Organic Industrial Solvents,* 1958.
Neufeld, P. (ed.), *Elementary Aspects of Corrosion,* Portcullis Press, Redhill, 1975.
Nylen, P. and Sunderland, E., *Modern Surface Coatings,* J. Wiley and Sons Inc., New York, 1965.
Oil and Colour Chemists' Association, *Introduction to Paint Technology,* 4th Edn, 1976.
Oil and Colour Chemists' Association, *Paint Technology Manuals,* Parts 1–6, 1961–1969, Chapman and Hall, London.

Paintmakers Association of Great Britain Ltd., *The Hazards Arising from the Manufacture of Polyurethane Paints and Varnishes,* London, 1970.
Idem, *Handling Hazards in the Use and Manufacture of Epoxy Paints and Precautions to be Adopted,* 1973.
Idem, *Toxicity of Various Curing Agents for Epoxy Resins,* 1971.
Idem, *Code of Practice – Application of Powder Coatings by Electrostatic Spraying,* 3rd Impression, 1975.
Palmer, K.N., *Dust Explosions and Fires,* Chapman and Hall, London, 1973.
Parker, D.H., *Principles of Surface Coating Technology,* John Wiley, New York, 1965.
Patton, T.C., *Paint Flow and Pigment Dispersion,* Wiley, Interscience, New York, 1965.
Patty, F.A., *Industrial Hygiene and Toxicology,* Wiley, Interscience, New York, 1962.
Payne, H.F., *Organic Coatings Technology,* 2 vols., John Wiley, New York, 1961.
Plaster, H.J., *Blast Cleaning and Allied Processes,* 2 vols., Portcullis Press, Redhill, 1973.
Ragg, M., *Schiffsboden and Rostschutzfarben,* Berlin, 1954.
Ramney, M.W., *Powder Coatings Technology,* Noyes Data Corpn, 1975.
Rathsack, H.A., *Schiffsanstriche,* Akademie-Verlag, Berlin.
Royal Institute of Chemistry, *Health and Safety – The Social and Professional Responsibilities of the Chemist,* London, 1975.
Sax, N.I., *Dangerous Properties of Industrial Materials,* 4th Edn, van Nostrand and Reinhold, New York, 1975.
Schreir, L.L. (Ed.), *Corrosion,* 2nd Edn, 2 vols., Newnes-Butterworth, London, 1976.
Tatton, W.H. and Drew, E.W., *Industrial Paint Application,* Newnes, London, 1964.
Turner, G.P.A., *Introduction to Paint Chemistry,* Chapman and Hall, London, 1967.
Tysall, L.A., *Industrial Paints,* Pergamon Press, Oxford, 1964.
Wampler, R.H., *Modern Organic Finishes,* Chemical Publishing Company, New York, 1946.
Wehlte, K., *The Materials and Techniques of Painting,* van Nostrand, New York, 1975.
Woods Hole Oceanographic Institute, *Marine Fouling and its Prevention,* U.S. Naval Institute, Annapolis, Maryland, 1952.
Wright, W.D. *The Measurement of Colour,* 4th Edn, Hilger and Watts, 1969.

Standards Institutions and Research Associations

British Standards Institution (BSI), 2, Park Street, London, W.1.
American Society for Testing and Materials (ASTM), (Standards), Philadelphia, Pa. U.S.A.
National Bureau of Standards, Circulars, News Bulletins, Journals. Washington, D.C. U.S.A.
Shipbuilding Research Association of Japan, (Standards).
Standardiseringskommissionen i Sverige (SIS), Stockholm.
Paint Research Association (PRA), Teddington, Middlesex.

Abbreviations of titles of journals frequently cited in the text

Journals	*Abbreviations*
American Chemical Society	*Am. Chem. Soc.*
American Industrial Hygiene Association Journal	*Am. indust. Hyg. Ass. J.*
American Paint Journal, St Louis	*Am. Paint J.*
American Paint and Varnish Manufacturers Association Circular	*Am. P. and V. Man. Ass.*
Angewandte Chemie, Weinheim	*Angew. Chem.*
Anti-Corrosion, Methods and Materials, London	*Anti-Corrosion*
Annals of Occupational Hygiene, London	*Ann. Occup. Hygiene*
Archives of Dermatology (U.S.A.)	*Arch. Derm.*
Archives of Environmental Health (U.S.A.)	*Arch. Envir. Health*
Archiv der Gewerbepathologie und Gewerbehygiene, Berlin	*Arch. Gewerbepath. und Gewerbehyg.*
Archives of Industrial Health (U.S.A.) – now Archives of Environmental Health	*Arch. Ind. Health*
Archives of Industrial Hygiene and Occupational Medicine– now Archives of Environmental Health (U.S.A.)	*Arch. Ind. Hyg. and Occ. Med.*
Archives des Maladies professionelles de medecine du travail et de Securité Sociale (France)	*Arch. Mal. Prof.*
Archives of Pathology (U.S.A.)	*Arch. Path.*
Australasian Corrosion Engineering	
Australian Paint Journal, Sydney	*Austral. Paint. J.*
British Corrosion Journal, London	*Brit. Corr. J.*
British Journal of Industrial Medicine	*Brit. J. Ind. Med.*
British Medical Bulletin	*Brit. Med. Bull.*
British Polymer Journal (Society of Chemical Industry), London	*Brit. Pol. J.*
Building, London	*Building*
Building Research Establishment Digest/News, Garston	*Building Res. Estab. Dig/News*
Bulletin of Hygiene (London)	*Bull. Hyg.*
Bulletin of the Institution of Corrosion Scientists and Technologists	*Bull. I. Corros. S.T.*
Comptes rendus de l'académie des sciences (Paris)	*C.R. Acad. Sci. (Paris)*
Canadian Paint and Finishing, Toronto	*Can. Paint and Fin.*
Cancer Research (U.S.A.)	*Cancer Res.*
Chemical Age International, London	*Chem. Age Intl*
Chemistry in Britain (incorporating Journal of the Royal Institute of Chemistry and Proceedings of the Chemical Society, London)	*Chem. Britain*

Journals	*Abbreviations*
Chemistry and Industry, London	*Chem. Ind.*
Continental Paint and Resin News, Braintree, Essex, U.K.	
Corrosion Science, Oxford	*Corr. Sci.*
Decorating Contractor, London	
Derwent Patents Reports, London	*Derwent*
Deutsche Farben-Zeitschrift, Stuttgart	*Defazet (formerly D.F.Z)*
Deutsches Archiv für Klinische Medizin	*Deut. Arch. Klin. Med.*
Double Liason, Paris	
Environmental Health Perspectives, (U.S.A).	*Environ. Health Perspect.*
Environmental Research (U.S.A.)	
Farbe und Lack, Hanover	*Farbe Lack*
FATIPEC Congress Book	
Folia Medica Napoli, Naples	*Folia med. Napoli*
Gigiena i Sanitarya (U.S.S.R.)	*Gig. i Sanit.*
Gummi, Asbest und Kunststoffe, Stuttgart	*Gummi Asbest. Kunst.*
Industrie-Lackier-Betrieb, Hanover	*Ind. Lack. Betrieb.*
Industrial and Engineering Chemistry, Philadelphia	*Ind. Eng. Chem.*
Industrial Finishing and Surface Coatings, Watford	*Ind. Fin. and Surf. Coat.*
Industrial Finishing, Indianapolis	*Ind. Fin. (U.S.)*
Iron and Steel, London	
Journal of the American Chemical Society	*J. Am. Chem. Soc.*
Journal of the American Medical Association	*J. Am. Med. Ass.*
Journal of the American Oil Chemists Society, Chicago	*J. Am. Oil. Chem. Soc.*
Journal of Applied Chemistry and Biotechnology, London	*J. Appl. Chem. Biotech.*
Journal of Applied Polymer Science, New York	*J. Appl. Pol. Sci.*
Journal of The Chemical Society, London	*J. Chem. Soc.*
Journal of Coatings Technology (formerly Journal of Paint Technology), Philadelphia	*J. Coat. Tech.*
Journal of Industrial Hygiene and Toxicology	*J. Ind. Hyg. Toxicol.*
Journal of Investigative Dermatology (U.S.A.)	*J. Invest. Derm.*
Journal of the Iron and Steel Institute, London	*J. Iron and Steel Inst.*
Journal of Institute of Metals, London	*J. Inst. Metals*
Journal of the National Cancer Institute (U.S.A.)	*J. Nat. Cancer Inst.*
Journal of the Oil and Colour Chemists' Association, Wembley, Middlesex, U.K.	*JOCCA*
Journal of Occupational Medicine (Chicago)	*J. Occ. Med.*
Journal of Paint Technology, (formerly Official Digest, and now Journal of Coatings Technology)	*J. Paint Tech.*
Journal of Pharmaceutical Sciences (U.S.A.)	*J. Pharm. Sci.*

Journals	*Abbreviations*
Journal of The Plastics Institute, London	*J. Plastics Inst.*
Journal of Polymer Science, New York	*J. Pol. Sci.*
Journal of Scientific Instruments, London	*J. Sci. Instr.*
Journal of The Society of Chemical Industry, London	*J. Soc. Chem. Ind.*
Journal of The Society of Dyers and Colourists, Bradford	*J.S.D.C.*
Kolloid-Zeitschrift und Zeitschrift für Polymere, Frankfurt	*Kolloid-Z.*
Lancet, London	
Materials Protection and Performance, Houston	*Mat. Prot.*
Metal Finishing, New Jersey	*Met. Fin. (U.S.)*
Metal Finishing Journal, London	*Met. Fin. J.*
Metalloberfläche	
Modern Paint and Coatings, (U.S.A.), (formerly Paint and Varnish Production	*Mod. Paint and Coatings*
Nature, London	
New Scientist, London	
North Carolina Medical Journal	*N. Carolina Med. J.*
Occupational Health	*Occup. Health*
Official Digest (now Journal of Coatings Technology)	*Off. Dig.*
Official Journal of the European Communities	
Paintindia, Bombay	
Paint Industry Magazine	*Paint Industry Mag.*
Paint Manufacture, London	*Paint Manuf.*
Paint Technology, London (now Pigment and Resin Technology)	*Paint Tech.*
Paint and Varnish Production, Washington, (now Modern Paint and Coatings)	*Paint Var. Prod.*
Painting and Decorating Journal, Southport	*Paint. Decor. J.*
Peintures, Pigments, Vernis, Paris	*Peint. Pig. Vernis*
Plaste und Kautschuk, Leipzig	*Plaste u. Kautschuk.*
Plastics and Polymers, London	
Polymers Paint Colour Journal, Redhill, Surrey	*PPCJ*
Proceedings of the Institute of Mechanical Engineers	*Proc. Inst. Mech. Eng.*
Proceedings of the Royal Society of Medicine	*Proc. Roy. Soc. Med.*
Product Finishing, London	*Prod. Fin.*
Progress in Organic Coatings, Lausanne	*Progr. Org. Coat.*
Przemsyl Chemiczny, Warsaw	*Przem. Chem.*
Resin notes, Cray Valley Products Ltd, Orpington	
Review of Current Literature, Paint Research Association (now World Surface Coating Abstracts)	
Schweizerische Medizinische Wochenschrift, Basel	*Schweiz. Med. Wochsehr.*
Scientific American, New York	*Sci. Am.*
Skandinavisk Tidskrift för Färg och Lack, Copenhagen	*Färg och Lack*

Journals	*Abbreviations*
Tokyo Ika Daigaku Zasshi	
Toxicology and Applied Pharmacology (U.S.A.)	*Toxicol. Appl. Pharmacol.*
Transactions of the Institute of Marine Engineers, London	*Trans. I. Mar. E.*
Transactions of the Institute of Metal Finishing, London	*Trans. I. Met. Fin.*
Transactions of the Institute of Naval Architects, London	*Trans. Inst. Naval Arch.*
Transactions of the NE Coast Institute of Engineers and Shipbuilders	*Trans. NE Cst. Inst. Eng. and Shipbrs.*
Travaux des peintures, Paris	*Trav. Peint.*
Verfkroniek, Amsterdam	
World Surface Coating Abstracts, Paint Research Association	*WSCA*
Zeitschrift für Angewandte Chemie, Berlin	*Z. Angew. Chem.*
Zentralblatt für Arbeitsmedizin und Arbeits-schutz, Darmstadt	*Z. für Arbeitsmed u. Arbeitschutz*
Zeitschrift für Krebsforschung und klinische Onkologie, W. Germany	*Z. Krebsforsch.*
Zinc Abstracts, Zinc Development Association, London	*Zinc Abs.*

Author index

Subject index

A number in bold type indicates the page where the main section on the topic begins.